Lecture Notes in Computer Science 8504

Commenced Publication in 1973
Founding and Former Series Editors:
Gerhard Goos, Juris Hartmanis, and Jan van Leeuwen

Joachim Gudmundsson   Jyrki Katajainen (Eds.)

Experimental Algorithms

13th International Symposium, SEA 2014
Copenhagen, Denmark, June 29 – July 1, 2014
Proceedings

 Springer

Volume Editors

Joachim Gudmundsson
University of Sydney, School of IT
Building J12, Sydney, NSW 2006, Australia
E-mail: joachim.gudmundsson@sydney.edu.au

Jyrki Katajainen
University of Copenhagen, Department of Computer Science
Universitetsparken 5, 2100 Copenhagen East, Denmark
E-mail: jyrki@di.ku.dk

ISSN 0302-9743   e-ISSN 1611-3349
ISBN 978-3-319-07958-5   e-ISBN 978-3-319-07959-2
DOI 10.1007/978-3-319-07959-2
Springer Cham Heidelberg New York Dordrecht London

Library of Congress Control Number: 2014940382

LNCS Sublibrary: SL 1 – Theoretical Computer Science and General Issues

Typesetting: Camera-ready by author, data conversion by Scientific Publishing Services, Chennai, India

Printed on acid-free paper

Springer is part of Springer Science+Business Media (www.springer.com)

Preface

The papers in this volume were selected for presentation at the 13th International Symposium on Experimental Algorithms (SEA 2014) held from June 29 to July 1, 2014, in Copenhagen, Denmark. This year the conference was co-located with the 14th Scandinavian Symposium and Workshops on Algorithm Theory (SWAT) and the 41st International Colloquium on Automata, Languages, and Programming (ICALP).

The Annual International Symposium on Experimental Algorithms is a forum for researchers in the area of design, analysis, and experimental evaluation and engineering of algorithms, as well as in various aspects of computational optimization and its applications. Typical topics covered by this conference include but are not limited to: algorithm engineering, algorithmic libraries, approximation techniques, combinatorial structures and graphs, computational optimization, data structures, distributed and parallel algorithms, experimental techniques and statistics, information retrieval, and experimental aspects of implementation, testing, evaluation and fine-tuning algorithms.

This year there were 81 papers submitted to the conference. Each paper received a minimum of three independent expert reviews by the Program Committee members or sub-reviewers. The reviews were thoroughly discussed by the Program Committee, and 36 papers were selected for presentation at the conference. In addition to these presentations, the program also included three invited presentations, by Robert Bixby, Jon Bentley, and Rasmus Pagh.

We would like to express our gratitude for the authors of all papers submitted to SEA 2014, the invited speakers, the Program Committee members, the external reviewers, and the members of the Organizing Committee for their contribution to making this conference possible.

April 2014

Joachim Gudmundsson
Jyrki Katajainen

Organization

Program Committee

David G. Andersen	Carnegie Mellon University, USA
Maike Buchin	Ruhr University of Bochum, Germany
Camil Demetrescu	Sapienza University of Rome, Italy
Benjamin Doerr	Max-Planck-Institut für Informatik, Germany
Joachim Gudmundsson	University of Sydney and NICTA, Australia
Jyrki Katajainen	University of Copenhagen, Denmark
David Mount	University of Maryland, USA
Petra Mutzel	Technical University of Dortmund, Germany
Rolf Niedermeier	Technical University of Berlin, Germany
Erwin Pesch	University of Siegen, Germany
Simon Puglisi	University of Helsinki, Finland
Martin Skutella	Technical University of Berlin, Germany
Matthias Stallmann	North Carolina State University, USA
Philippas Tsigas	Chalmers University of Technology, Sweden
Suresh Venkatasubramanian	University of Utah, USA
Toby Walsh	NICTA and University of New South Wales, Australia
Yusu Wang	Ohio State University, USA
Gerhard Woeginger	Technical University of Eindhoven, The Netherlands

Steering Committee

Edoard Amaldi	Kurt Mehlhorn
David A. Bader	Ian Munro
Josep Diaz	Sotiris Nikoletseas
Giuseppe F. Italiano	Jose Rolim (**Chair**)
David Johnson	Pavlos Spirakis
Klaus Jansen	

Organizing Committee

Susan Nasirumbi Ipsen	University of Copenhagen
Inge Hviid Jensen	University of Copenhagen
Jyrki Katajainen	University of Copenhagen (**Conference Chair**)

Additional Reviewers

Anagnostopoulos, Aris
Arulselvan, Ashwin
Baertschi, Andreas
Bast, Hannah
Belazzougui, Djamal
Berkman, Omer
Berthold, Jost
Bevern, René Van
Bingmann, Timo
Bober, Jonathan
Bodlaender, Hans L.
Buchin, Kevin
Bulteau, Laurent
Bökler, Fritz
Cela, Eranda
Cseh, Ágnes
D'Elia, Daniele Cono
Dahlgaard, Søren
Dasler, Philip
Delling, Daniel
Ehrgott, Matthias
Eisenbrand, Friedrich
Elmasry, Amr
Erickson, Lawrence
Fafianie, Stefan
Fischer, Johannes
Fisikopoulos, Vissarion
Fredriksson, Kimmo
Froese, Vincent
Gao, Jie
Gerrits, Dirk H.P.
Giaquinta, Emanuele
Gog, Simon
Gross, Martin
Gutwenger, Carsten
Hartung, Sepp
Hiary, Ghaith
Höhn, Wiebke
Hüffner, Falk
Jepsen, Jacob
Kempa, Dominik
Kobitzsch, Moritz
Kobourov, Stephen

Komusiewicz, Christian
Koster, Arie
Kress, Dominik
Kriege, Nils
Kroller, Alexander
Kurz, Denis
Kärkkäinen, Juha
Kötzing, Timo
Labourel, Arnaud
Letchford, Adam
Liptak, Zsuzsanna
Manduchi, Roberto
Mannino, Carlo
Mattei, Nicholas
Matulef, Kevin
Meiswinkel, Sebastian
Meyerhenke, Henning
Moradi, Farnaz
Mulzer, Wolfgang
Müller-Hannemann, Matthias
Na, Joong Chae
Narodytska, Nina
Neumann, Adrian
Nichterlein, André
Nossack, Jenny
Parsa, Salman
Peng, Pan
Pettie, Seth
Polishchuk, Valentin
Querzoni, Leonardo
Rademacher, Luis
Raghvendra, Sharath Kumar
Reinelt, Gerhard
Ribichini, Andrea
Safro, Ilya
Salazar González, Juan José
Salmela, Leena
Satti, Srinivasa Rao
Schieferdecker, Dennis
Schulz, André
Schäfer, Till
Sidiropoulos, Anastasios
Silveira, Rodrigo

Sorge, Manuel
Stiller, Sebastian
Storandt, Sabine
Strasser, Ben
Tamir, Tami
Telle, Jan Arne
Toma, Laura
Turpin, Andrew
Ventura, Paolo
Verschae, José

Vitale, Fabio
Wagler, Annegret
Wassenberg, Jan
Welz, Wolfgang A.
Wenk, Carola
Werneck, Renato
Wetzel, Susanne
Wiese, Andreas
Zey, Bernd
Zheng, Yan

Abstract of Invited Papers

Computational Progress
in Mixed-Integer Programming

Robert E. Bixby

Gurobi Optimization, Inc., USA

Abstract. A mixed-integer program (MIP) is a linear programming (LP) problem (an optimization problem with linear equality and inequality constraints and a linear objective function) plus the condition that some or all of the variables must take on integral values. This added integrality condition takes a problem that is convex and very well understood, and transforms it into a non-convex, NP-hard problem. At the same time, it is precisely this integrality condition that allows am enormous variety of real-world business decision problems to be modeling using MIP, situations in which, for example, setting a variable to 1 or 0 may mean making a decision to do something or not do it, and setting it to a fractional value has no meaning. The remarkable fact, and the main topic of this talk, is that modern MIP codes can now robustly solve a large fraction of MIPs that arise in practice, and do so "out of the box".

Selecting Data for Experiments: Past, Present and Future

Jon Bentley

Avaya Labs Research
211 Mount Airy Road
Basking Ridge, NJ 07974 USA
JonBentley@gmail.com

Abstract. This essay describes three different contexts in which algorithm designers select input data on which to test their implementations. Selecting input data for *past* problems typically involves scholarship to assemble existing data and ingenuity to model it efficiently. Selecting data for *present* problems should be based on active discussions with users and careful study of existing data. Selecting data to model problems that may arise in the *future* is the most interesting and delicate of the tasks that we will consider.

High-Performance Pseudorandomness

Rasmus Pagh*

IT University of Copenhagen
Rued Langgaards Vej 7, 2300 København S, Denmark
pagh@itu.dk

Abstract. For many randomized algorithms there is a gap between the
properties that can be shown if the algorithm has access to idealized random objects — such as hash functions whose values are fully independent — and the properties that can be shown with efficiently computable
pseudorandomness such as universal hash functions.

This talk starts with a survey of recent methods for constructing
hash functions and pseudorandom sequences that are highly efficient,
yet preserve the properties of the corresponding fully random objects
with respect to many randomized algorithms.

Based on this overview some challenges for algorithm engineering
of high-performance pseudorandomness are presented. Finally, it is discussed how experimental algorithms might aid in our understanding of
the properties of pseudorandom objects.

* Supported by the Danish National Research Foundation under the Sapere Aude
program.

Table of Contents

Graph Drawing

Shortest Path

Strings

Graph Algorithms

Suffix Structures

Selecting Data for Experiments: Past, Present and Future

Jon Bentley

Avaya Labs Research
211 Mount Airy Road
Basking Ridge, NJ 07974, USA
JonBentley@gmail.com

Abstract. This essay describes three different contexts in which algorithm designers select input data on which to test their implementations. Selecting input data for *past* problems typically involves scholarship to assemble existing data and ingenuity to model it efficiently. Selecting data for *present* problems should be based on active discussions with users and careful study of existing data. Selecting data to model problems that may arise in the *future* is the most interesting and delicate of the tasks that we will consider.

1   Introduction

So you've designed a beautiful new algorithm for a classic problem and implemented it in elegant code. Now you want to run it to see how it really performs. What input data do you use? Or maybe you're consulting on a big system, and you've incorporated a well-known algorithm deep in its innards. You want to ensure that the implementation is as efficient as your analysis predicts. What input data do you use? Or maybe you're working on a new problem that you're sure will become important in years to come. What input data do you use?

Researchers have long considered the question of selecting inputs for experiments on algorithms. McGeoch [8] devotes her Section 2.2.1 to the topic, and presents an extensive list of different types of inputs: stress-test, worst-case, bad-case, random and real are a few of the base types that she considers. She then extends those with generators of various forms, including hybrid, algorithm-centered and reality-centered. She finally wraps up those various components into public test beds. All of those are very useful ways to think about data for algorithmic experiments.

This essay takes an alternative view of looking at data for experiments: *past*, *present* and *future*. Section 2 considers how we can encapsulate our *past* experiences and insights into input data. Section 3 addresses how we can learn from our *present* data and users to build inputs that accurately model the essential problem. Section 4 proposes ways that we can mold our predictions about the *future* into appropriate inputs. Conclusions are offered in Section 5.

J. Gudmundsson and J. Katajainen (Eds.): SEA 2014, LNCS 8504, pp. 1–9, 2014.

2   Encapsulating the Past

In algorithms, as in most fields, much work is devoted to performing an existing task more efficiently. When we test such an algorithm experimentally, it is important to bring to bear on that problem as much of the past work as possible.

2.1   Sorting I

In 1991, Doug McIlroy and I began to work on a new qsort function for our C library. That general-purpose sort function is usually implemented as a variant of Hoare's [7] Quicksort algorithm. Our attention was first brought to the problem by a disastrous failure of a 20-year old qsort: a plausible set of inputs drove it to quadratic behavior. As we studied that function, we found that we could beat its speed on most inputs, in addition to producing a sort that was never driven to excessive time. We eventually published our work in Bentley and McIlroy [2].

We had learned long before that because sorting is so common, our primary goal for the program should be its speed on typical inputs. Thus our first "Qsort Requirement":

QR1: The code should be fast on large arrays of (mostly) distinct values.

As we developed the code, our baseline test was to time it on uniform data that was large enough to consume a few seconds of CPU time, yet small enough so that we could get accurate estimates of run times in a few minutes. We left to others the job of building a fast sort for very small inputs. Every now and then we would time it on very small and very large inputs to ensure that our extrapolations were valid, and we were never disappointed.

We were optimistic about an early version of our program. We observed that sorting n elements took time proportional to $n \lg n$ on all the inputs that we had tested. We further observed that our code was 30 or 40 percent faster than the existing function. We released our code for testing to Tom Duff, an extremely friendly user. He quickly reported that it was so unbearably slow that he could not consider using it in his application.

When we asked about the nature of the offending input, Duff stated that it contained many duplicated elements. We timed our code as it sorted an array of identical elements, and observed only a slight slowdown. Duff agreed with our time estimates. He then pointed out that while our code was slightly slower in the presence of many duplicates, the existing code was lightning fast on such inputs. Like many users before him, he was sorting to bring identical elements together, and there were often many of them. This reasonable explanation led to our second requirement:

QR2: The code should be fast in the presence of many duplicated elements.

In this case, our very communicative present reminded us of an important lesson from the past.

We had been motivated to study qsort due to a catastrophic failure of the previous implementation. Allan Wilks and Rick Becker had found that an "organ-pipe" array of $2n$ integers of the form $123...nn...321$ took quadratic time. They expected the sort to take a few minutes, and stopped their program after several hours. They eventually learned that the sort would have gone on for weeks before terminating.

As we developed our code, we started to collect inputs that might drive an implementation of Quicksort to excessive time. We started with the motivating organ-pipe input. We added many cases that were identified in the past to slow down sloppy implementations of Quicksort, such as an array of identical or ascending elements. We found that an array of random zeros and ones slowed a competing qsort, so that went into the mix. With a dozen or more examples in our collection, we added our third performance requirement:

QR3: The code should not slow down substantially on particular test input.

We felt that we could never collect enough examples to give us confidence that this requirement had been met, so we took an alternative approach. We constructed a certification program that systematically generates a large number (thousands) of inputs with bizarre structure across a range of input size n, including all of the offenders that we had identified so far.

As we tuned our functions to meet requirements QR1 and QR2, we were concerned about tiny (say, one percent) differences in run times. Our "torture test" of requirement Q3 ignored slight differences in time, and tested only for extreme aberrations. We ran that test on our new qsort function, and on three of its predecessor qsort functions. We were delighted to find that our stress test identified all known bugs with the predecessors, and also identified many new issues with them. Each of the existing algorithms was easily driven to quadratic performance by many different inputs. Our new program was more robust: it exceeded $1.2n \lg n$ comparisons in less than two percent of these nasty cases, and never exceeded $1.5n \lg n$ comparisons. It has been widely used for over two decades, and we have yet to hear the first report of disastrous performance.

We were careful not to distill these three orthogonal issues into a single performance index. Whenever we timed our program, we made three different measurements:

QR1. How does it perform on mostly distinct elements?
QR2. How does it perform in the presence of many duplicates?
QR3. Does it ever fail on torture-test inputs?

The third requirement was absolute: the program had to be robust. The first two requirements required tasteful tradeoffs: we could have easily increased QR1 performance by sacrificing QR2 performance, or vice versa. The right decision called for engineering judgment.

I returned to the problem of engineering an efficient library sort in 2009 when I worked with Vladimir Yaroslavskiy and Josh Bloch to develop a Java library sort. These principles from a distant past helped to guide our development of the resulting "dual-pivot" Quicksort. We used a similar but greatly expanded torture test.

3   Learning from the Present

Sometimes a user comes to an algorithm designer with a well-formed problem and data that can serve as test inputs. Other times, an algorithm designer has to go out and find the data.

3.1   Sorting II

The previous section described how lessons from the past led to a framework for testing and timing a library sorting algorithm. Along the way, we were taught by our friend Tom Duff, a user very much in our present. He reminded us that our requirements were incomplete, and helped us to formulate requirement QR2, which strongly influenced our development. Algorithm designers always learn by listening to users.

3.2   Geometric Point Sets I

Algorithm designers can also learn by listening to the data. Bentley, Weide and Yao [5] describe optimal expected-time algorithms for a variety of closest-point problems, including nearest neighbor searching and constructing Voronoi diagrams. We started by assuming that the input consisted of n points uniformly distributed over the unit square, and placed the points in a grid of roughly n cells. We proved that we could build such a structure in linear time, and that we could then find the nearest neighbor to an input query in constant expected time. We extended the algorithms and analyses to more complex structures, such as Voronoi diagrams and thereby minimum spanning trees. We later extended some of the results from the planar case to Euclidean k-space.

The work was originally described in a technical report (Bentley, Weide and Yao [4]) that contained problems, theoretical algorithms and proofs, but not a single description of an implementation or experiment. Before the paper appeared in print, I came across some planar point sets as I was consulting on a geopolitical database, and we used that data to conduct experiments that were described in our final paper.

The algorithm was simple to code: a dozen lines to build the structure, and a few dozen lines to search it. Its run time overhead was not high: on uniform data, our constant-expected-time search broke even with the competing linear search at $n = 53$, and was much faster as n increased. We studied two data sets that represented the centroids of US political census tracts. The first set represented the $n = 318$ precincts in San Diego County, California, and the second set represented the $n = 1122$ precincts in the (roughly square) State of New Mexico.

Our analysis and experiments had been for uniform data. The real data was highly non-uniform. The points in San Diego were distributed about uniformly over just half of a roughly square region, while the majority of the points in New Mexico were clustered in a few cities. Even so, our algorithms were only about 75% and 150% slower than the times predicted on uniform data, respectively. A late addition allowed us to report these results (at slightly greater length) in our journal paper. Even though the data was from an application that we had not considered when we proved our theorems, the geopolitical data showed that our algorithms might indeed have practical importance.

We listened to the geopolitical data, and it told us useful facts about our algorithms. The data also described to us the shape of things to come; we'll return to that topic shortly.

4   Predicting the Future

A naval visionary is often praised with the phrase "He steers by the stars ahead, and not by the wake behind." It is rewarding for an algorithm designer to study the past and to listen to the small voices of the present, but it is perhaps most exciting to predict the future and then aim for the stars.

4.1   A Simple and Safe Bet

Only a few things are certain in life: death, taxes, computers getting faster, and algorithmic problems getting larger. Straightforward experiments can usually determine whether a system will grow large gracefully.

I was once a member of a research team that consulted on the performance of a communication system that would notify users of an emergency in a variety of ways: text messages, e-mail, telephone calls and more in a programmable order until the user acknowledged receipt. As the system moved towards release, the testers concentrated on increasingly subtle sequences of messages (in business hours, first try my pager, then my office phone, then my e-mail, then my home phone, but at nights ..., and then on weekends, ...).

As algorithmic consultants, we encouraged the testers to ensure that the system would remain efficient as it grew large. While we knew that some organizations might want to contact $n = 100,000$ people at some point in the not-so-distant future, most of the tests were on complex message interactions with n less than 100. Our group therefore encouraged the testers to run just one sequence of tests that ignored interactions, and measured how each communication mode grew in itself.

Most of the tests were passed with flying colors, but the experiments on e-mail performance were surprising. The system could send $n = 1000$ pieces of e-mail in 11.4 seconds. That was promising: we hoped that that the time would scale linearly to about 20 minutes for $n = 100,000$, which was right on the edge of "fast enough". Our next experiment was at $n = 2000$, where the time was 39.6 seconds, and then at $n = 4000$, where the time was 167.1 seconds. Because the time increased by four when the problem size doubled, we concluded that the e-mail algorithm that should have been linear was in fact quadratic. We estimated the time at $n = 100,000$ to be well over a day, at which time most emergencies had long since passed!

How could a system possibly use quadratic time to send n pieces of e-mail? Simple. An eager programmer decided that it would be wasteful and inefficient to send the same message twice to a user, so the system was augmented with a list of all the people that had received a given message. Before the piece of mail was sent to the next user, that list was searched by a linear scan to ensure that the new address was not redundant. This "speedup" made a linear algorithm quadratic.

Our simple tests projected the performance of a huge and complex system onto a several independent dimensions, and then measured their asymptotic growth rate of each. Most grew linearly, as we expected, but one surprised us with quadratic growth. Such simple tests of many systems have given a peek at asymptopia and thereby uncovered insidious performance bugs before they have a chance to bite.

4.2   Geometric Point Sets II

We saw earlier that a section that was a late addition to Bentley, Weide and Yao [5] described the implementation of our algorithm. The section briefly described the ease of writing the program (a few dozen lines of code), its low overhead (a break even point at around $n = 50$), and its robustness in the presence of non-uniform data. We did not, however, describe in that paper a number of the patterns that we saw in the data that were to shape our thinking about geometric inputs for years to come.

After that experience, I studied several kinds of geometric patterns that I observed in population data sets. The geopolitical data was highly non-uniform, but many of its components displayed a great deal of structure along the following lines:

> *Uniform.* This model was plausible where suburbs melted into rural areas.
> *Normal.* Relatively many people live in the centers of small towns, and the density rapidly decreases with distance.
> *Linear.* Populations were originally spread along railroad lines, which are typically now near highways. People sometimes cluster along linear political borders.
> *Grids.* In suburbs, the distribution is smoother than uniform due to zoning laws that encourage fixed lot sizes.

Our cell-based algorithm was originally designed for uniform data, and it performed quite well on such inputs. Normal and linear sets both led to a few cells stuffed with many points and many empty cells, which could slow down cell-based algorithms. The grid distribution represents a bad case for algorithms that are flummoxed by many equal coordinate values or inter-point distances.

For some time, whenever I developed a geometric algorithm that did not have good worst-case performance, I thought carefully about how it might perform on these individual inputs, and on combinations of them. A decade later, I extended that approach from thought experiments to input generators.

Bentley [1] describes a number of algorithms that employ multidimensional binary search trees (or "k-d trees", in k dimensions) to implement a variety of heuristics for approximating geometric instances of the Travelling Salesman Problem. I was not able to prove the worst-case behavior of the various algorithms, but I gave heuristic arguments about why they might be expected to perform well on uniform data. I offered extensive experiments to show that their running time on uniform inputs was in fact quite efficient.

I also conducted experiments on the highly non-uniform data sketched above. I started with generators for the individual distributions (and several more), and then used generators that mixed and matched the various distributions. The majority of the algorithms adapted gracefully to highly non-uniform data. Although far from a proof of worst-case running time, these experiments greatly increased my confidence in the robustness of the algorithms, and my optimism about their application in real systems.

Neither in 1979 nor in 1992 did I personally have access to a lot of digital geometric data. Some of the data that I did have was proprietary, and I was not able to describe it in public. But I could view a number of types of geometric structures, and construct simple stochastic models that could be embedded in code. Those guesses helped to guide me in the development of algorithms that have proven to be robust.

4.3   Data Compression Using Long Common Strings

Doug McIlroy and I were in a similar position as we considered the state of data compression in 1998. Every compression method that we knew worked locally: as it processed a string, it would use a sliding window that only took recent characters into consideration, in a size that ranged from hundreds to a few tens of thousands. Many of the methods were remarkably effective in squeezing out redundancy.

But we had the intuition that the World Wide Web and other large data sets would have a fundamentally different kind of redundancy: large common strings that were widely separated in the input data. Bentley and McIlroy [3] therefore describe a compression method that processes its input string, and looks for common strings in the entire length of the input, which can be huge. We proposed mechanisms to make the search efficient, and simple encodings to remove the redundant text. Our method was expressly developed to be used in conjunction with existing compression tools: it would find long duplicated strings that were far apart in the input, and existing methods would efficiently represent short duplicated strings that were near one another.

We could prove that our technique would remove any redundancy that was present, but how could we measure its effectiveness? We started with a simple experiment: we collected the text of the King James Bible into one large string, and compressed it. Existing methods were very efficient, while our method found little additional redundancy. We then concatenated that string with itself, to provide two copies of the Bible, back to back. Existing methods used almost exactly twice the space to represent it, while our method used just a few additional bytes. This experiment clearly illustrated the obvious.

But how would the method do on the data that we foresaw coming? Our first model of the future was a string built by concatenating all the files on the 1994 Project Gutenberg CD ROM, for a total of about 66 megabytes of data. We found that our method interacted nicely with existing methods, and further reduced the size of the data by about 14%. When we studied the data we found that common duplicates were stern legal boilerplates and multiple editions of various works. This matched our intuition. In the mature field of data compression, researchers were happy to see improvements of a fraction of a percent, and we had "real" data with an improvement of over ten percent!

We tried our method on all of the data that we could find that modeled attributes that we thought the Web would soon display: mathematical software packages, collection of newswire stories, software distributions, and the like. On some of the inputs our method found essentially no duplication, while on other collections our method worked gracefully with existing techniques to halve the file size. With this warning of wide variability, we released our paper to the world.

Chang et al. [6] describe how their Google team incorporated our compression scheme into that company's "Bigtable" distributed storage system. They credit our algorithm with replacing "typical Gzip reductions of 3-to-1 or 4-to-1" with "a 10-to-1 reduction in space". They attribute this to finding large amounts of shared boilerplate from the same host. Wikipedia – BigTable [9] describes how BigTable is used in a number of Google applications including Web indexing, MapReduce, Google Maps, Google Earth, YouTube and Gmail. We were fortunate in that our guess about the future actually came to pass.

5   Conclusions

So you want to experiment on an algorithm. What input data do you use? Here are some concrete hints.

For problems that arose in the past, study the data that was previously used and incorporate as much of it as possible, in the most efficient way you can. A public test bed built on input generators is the gold standard. Because of Moore's Law and its many corollaries, the odds are good that you can study larger inputs than your predecessors. You might also try searching for new classes of inputs that were not previously examined.

For problems in your present, listen carefully to your users and to your data. Users often enunciate subtle requirements that have serious algorithmic implications. Study particular data sets to help you choose distributions that accurately model the inputs that your code will process. Collect any nefarious inputs to incorporate into a torture test, which will then become part of your test bed.

Predicting future data is undoubtedly the most fun and the most dangerous task. A sure bet is that problem size will increase in the future; always ensure that your program displays appropriate asymptotic growth rates. The data processed by existing algorithms might substantially alter in character over time; predict and model possible changes as you develop algorithms. And always keep your eyes open for new problems, which will process remarkable new data that only you have dreamed of.

Acknowledgements. I am grateful for the insightful comments of Dan Bentley, P. Krishnan, Catherine McGeoch and Audris Mockus.

References

1. Bentley, J.L.: Fast algorithms for geometric traveling salesman problems. ORSA J. Computing 4, 387–411 (1992)
2. Bentley, J.L., McIlroy, M.D.: Engineering a sort function. Software Pract. Exper. 23, 1249–1265 (1993)
3. Bentley, J.L., McIlroy, M.D.: Data compression with long repeated strings. Inform. Sci. 135, 1–11 (2001)
4. Bentley, J.L., Weide, B.W., Yao, A.C.: Optimal expected-time algorithms for closest point problems, Technical report CMU-CS-79-111. Carnegie-Mellon University Computer Science Department (1979)
5. Bentley, J.L., Weide, B.W., Yao, A.C.: Optimal expected-time algorithms for closest point problems. ACM Trans. Math. Software 6, 4 (1980)
6. Chang, F., Dean, J., Ghemawat, S., Hsieh, W.C., Wallach, D.A., Burrows, M., Chandra, T., Fikes, A., Gruber, R.E.,, B.: A distributed storage system for structured data. ACM Trans. Computer Systems 26, 2 (2008)
7. Hoare, C.A.R.: Quicksort. Computer J. 5, 10–15 (1962)
8. McGeoch, C.C.: A Guide to Experimental Algorithmics. Cambridge University Press, Cambridge (2012)
9. Wikipedia: BigTable (April 2014)

The Hospitals / Residents Problem with Couples: Complexity and Integer Programming Models*

Péter Biró[1],**, David F. Manlove[2],***, and Iain McBride[2],†

[1] Institute of Economics, Research Centre for Economic and Regional Studies,
Hungarian Academy of Sciences, H-1112, Budaörsi út 45, Budapest, Hungary
[2] School of Computing Science, Sir Alwyn Williams Building, University of Glasgow,
Glasgow G12 8QQ, UK
i.mcbride.1@research.gla.ac.uk

Abstract. The Hospitals / Residents problem with Couples (HRC) is a generalisation of the classical Hospitals / Residents problem (HR) that is important in practical applications because it models the case where couples submit joint preference lists over pairs of (typically geographically close) hospitals. In this paper we give a new NP-completeness result for the problem of deciding whether a stable matching exists, in highly restricted instances of HRC, and also an inapproximability bound for finding a matching with the minimum number of blocking pairs in equally restricted instances of HRC. Further, we present a full description of the first Integer Programming model for finding a maximum cardinality stable matching in an instance of HRC and we describe empirical results when this model applied to randomly generated instances of HRC.

1 Introduction

The Hospitals / Residents Problem. The *Hospitals / Residents problem* (HR) is a many-to-one allocation problem. An instance of HR consists of two groups of agents – one containing *hospitals* and one containing *residents*. Every hospital expresses a linear preference over some subset of the residents, its *preference list*. The residents in a hospital's preference list are its *acceptable* partners. Further, every hospital has a *capacity*, c_j, the maximum number of posts it has available to match with residents. Every resident expresses a linear preference over some subset of the hospitals, his *acceptable* hospitals.

The preferences expressed in this fashion are reciprocal: if a resident r_i is acceptable to a hospital h_j, then h_j is also acceptable to r_i, and vice versa. A many-to-one *matching* between residents and hospitals is sought, which is a

---

* A preliminary version of this paper appeared in the Proceedings of OR 2013.
** Supported by the Hungarian Academy of Sciences under its Momentum Programme (LD-004/2010) and also by OTKA grant no. K108673.
*** Supported by Engineering and Physical Sciences Research Council grant EP/K010042/1.
† Supported by a SICSA Prize PhD Studentship.

J. Gudmundsson and J. Katajainen (Eds.): SEA 2014, LNCS 8504, pp. 10–21, 2014.

set of acceptable resident-hospital pairs such that each resident appears in at most one pair and each hospital h_j at most c_j pairs. If a resident r_i appears in some pair of M, r_i is said to be *assigned* in M and *unassigned* otherwise. Any hospital assigned fewer residents than its capacity in some matching M is *under-subscribed* in M.

A matching is *stable* if it admits no *blocking pair*. Following the definition in [10], a blocking pair consists of a mutually acceptable resident-hospital pair (r, h) such that both of the following hold: (i) either r is unassigned, or r prefers h to his assigned hospital; (ii) either h is under-subscribed in the matching, or h prefers r to at least one of its assigned residents. Were such a pair to exist, they could form a pairing outside of the matching, undermining its integrity [20].

It is known that every instance of HR admits at least one stable matching and such a matching may be found in time linear in the size of the instance [10]. Also, for an arbitrary HR instance I, any resident that is assigned in one stable matching in I is assigned in all stable matchings in I, moreover any hospital that is under-subscribed in some stable matching in I is assigned exactly the same set of residents in every stable matching in I [11, 20, 21].

HR can be viewed as an abstract model of the matching process involved in a centralised matching scheme such as the National Resident Matching Program (NRMP) [18] through which graduating medical students are assigned to hospital posts in the USA. A similar process was used until recently to match medical graduates to Foundation Programme places in Scotland, called the Scottish Foundation Allocation Scheme (SFAS) [13]. Analogous allocation schemes having a similar underlying problem model exist around the world, both in the medical sphere, e.g. in Canada [9], Japan [14], and beyond, e.g. in higher education allocation in Hungary [5].

The Hospitals / Residents Problem with Couples. Centralised matching schemes such as the NRMP and the SFAS have had to evolve to accommodate couples who wish to be allocated to (geographically) compatible hospitals. The capability to take account of the joint preferences of couples has been in place in the NRMP context since 1983 and since 2009 in the case of SFAS. In schemes where the agents may be involved in couples, the underlying allocation problem can modelled by the so-called *Hospitals / Residents problem with Couples* (HRC).

As in the case of HR, an instance of HRC consists of a set of *hospitals* H and a set of *residents* R. The residents in R are partitioned into two sets, S and S'. The set S consists of *single* residents and the set S' consists of those residents involved in *couples*. There is a set $C = \{(r_i, r_j) : r_i, r_j \in S'\}$ of *couples* such that each resident in S' belongs to exactly one pair in C.

Each single resident $r_i \in S$ expresses a linear preference order over his acceptable hospitals. Each pair of residents $(r_i, r_j) \in C$ expresses a joint linear preference order over a subset A of $H \times H$ where $(h_p, h_q) \in A$ represents the joint assignment of r_i to h_p and r_j to h_q. The hospital pairs in A represent those joint assignments that are *acceptable* to (r_i, r_j), all other joint assignments being *unacceptable* to (r_i, r_j).

Each hospital $h_j \in H$ expresses a linear preference order over those residents who find h_j acceptable, either as a single resident or as part of a couple. As in the HR case, each hospital $h_j \in H$ has a *capacity, c_j*.

A many-to-one *matching* between residents and hospitals is sought, which is defined as for HR with the additional restriction that each couple (r_i, r_j) is either jointly unassigned, meaning that both r_i and r_j are unassigned, or jointly assigned to some pair (h_k, h_l) that (r_i, r_j) find acceptable.As in HR, we seek a *stable* matching, which guarantees that no resident and hospital, and no couple and pair of hospitals, have an incentive to deviate from their assignments and become assigned to each other.

Roth [20] considered stability in the HRC context although did not define the concept explicitly. Whilst Gusfield and Irving [12] defined stability in HRC, their definition neglected to deal with the case that both members of a couple may wish to be assigned to the same hospital. Manlove and McDermid [16] extended their definition to deal with this possibility (however both definitions are equivalent in the case that no pair of the form (h_p, h_p) appears in any couple's preference list). We adopt Manlove and McDermid's stability definition in this paper, and now define it formally as follows.

Definition 1 ([16]). *A matching M is* stable *if none of the following holds:*

1. *The matching is blocked by a hospital h_j and a single resident r_i, as in the classical HR problem.*
2. *The matching is blocked by a couple (r_i, r_j) and a hospital h_k such that* either
   (a) (r_i, r_j) *prefers $(h_k, M(r_j))$ to $(M(r_i), M(r_j))$, and h_k is either under-subscribed in M or prefers r_i to some member of $M(h_k)\backslash\{r_j\}$* or
   (b) (r_i, r_j) *prefers $(M(r_i), h_k)$ to $(M(r_i), M(r_j))$, and h_k is either under-subscribed in M or prefers r_j to some member of $M(h_k)\backslash\{r_i\}$*
3. *The matching is blocked by a couple (r_i, r_j) and (not necessarily distinct) hospitals $h_k \neq M(r_i)$, $h_l \neq M(r_j)$; that is, (r_i, r_j) prefers the joint assignment (h_k, h_l) to $(M(r_i), M(r_j))$, and* either
   (a) $h_k \neq h_l$, *and h_k (respectively h_l) is either under-subscribed in M or prefers r_i (respectively r_j) to at least one of its assigned residents in M;* or
   (b) $h_k = h_l$, *and h_k has at least two free posts in M, i.e., $c_k - |M(h_k)| \geq 2$;* or
   (c) $h_k = h_l$, *and h_k has one free post in M, i.e., $c_k - |M(h_k)| = 1$, and h_k prefers at least one of r_i, r_j to some member of $M(h_k)$;* or
   (d) $h_k = h_l$, *h_k is full in M, h_k prefers r_i to some $r_s \in M(h_k)$, and h_k prefers r_j to some $r_t \in M(h_k)\backslash\{r_s\}$.*

Existing Algorithmic Results for HRC. In contrast with HR, an instance of HRC need not admit a stable matching [20]. Also an instance of HRC may admit stable matchings of differing sizes [2]. Further, the problem of deciding whether a stable matching exists in an instance of HRC is NP-complete, even in the restricted case where there are no single residents and all of the hospitals have only one available post [17,19].

In many practical applications of HRC the residents' preference lists are short. Let (α, β)-HRC denote the restriction of HRC in which each single resident's preference list contains at most α hospitals, each couple's preference list contains at most α pairs of hospitals and each hospital's preference list contains at most β residents. (α, β)-HRC is hard even for small values of α and β: Manlove and McDermid [16] showed that $(3,6)$-HRC is NP-complete.

Since the existence of an efficient algorithm for finding a stable matching, or reporting that none exists, in an instance of HRC is unlikely, in practical applications such as SFAS and NRMP, stable matchings are found by applying heuristics [3,6,22]. However, neither the SFAS heuristic, nor the NRMP heuristic guarantee to terminate and output a stable matching, even in instances where a stable matching does exist. Hence, a method which guarantees to find a maximum cardinality stable matching in an arbitrary instance of HRC, where one exists, might be of considerable interest. For further results on HRC the reader is referred to [7] and [15].

Contribution of This Work. In this paper, we present in Section 2 a new NP-completeness result for the problem of deciding whether there exists a stable matching in an instance of $(2,2)$-HRC where there are no single residents and all hospitals have capacity 1. This is the most restricted case of HRC currently known for which NP-completeness holds. A natural way to try to cope with this complexity is to approximate a matching that is 'as stable as possible', i.e., admits the minimum number of blocking pairs [1]. Let MIN-BP-HRC denote the problem of finding a matching with the minimum number of blocking pairs, given an instance of HRC, and let (α, β)-MIN-BP-HRC denote the restriction to instances of (α, β)-HRC. We prove that $(2,2)$-MIN-BP-HRC is not approximable within $n_1^{1-\varepsilon}$, where n_1 is the number of residents in a given instance, for any $\varepsilon > 0$, unless $P = NP$. Further in Section 3 we present a description of the first Integer Programming (IP) model for finding a maximum cardinality stable matching or reporting that none exists in an arbitrary instance of HRC. Then in Section 4 we present elements of an empirical study of this model as applied to randomly generated instances.

2   Complexity Results

In this section we present hardness results for finding and approximating stable matchings in instances of HRC. For space reasons all of the proofs are omitted but appear in full in [8], a technical report by the same authors. We begin by establishing NP-completeness for the problem of deciding whether a stable matching exists in a highly restricted instance of HRC. Our proof involves a reduction from $(2,2)$-E3-SAT, the problem of deciding, given a Boolean formula B in CNF over a set of variables V, whether B is satisfiable, where B has the following properties: (i) each clause contains exactly 3 literals and (ii) for each $v_i \in V$, each of literals v_i and $\bar{v}_i$ appears exactly twice in B. Berman et al. [4] have shown that $(2,2)$-E3-SAT is NP-complete.

Theorem 1. *Given an instance of* (2, 2)-HRC, *the problem of deciding whether there exists a stable matching is NP-complete. The result holds even if there are no single residents and each hospital has capacity 1.*

We now turn to MIN-BP-HRC. Clearly Theorem 1 implies that this problem is NP-hard. By chaining together instances of (2, 2)-HRC constructed in the proof of Theorem 1, we arrive at a gap-introducing reduction which establishes a strong inapproximability result for MIN-BP-HRC under the same restrictions as in Theorem 1.

Theorem 2. (2, 2)-MIN-BP-HRC *is not approximable within* $n_1^{1-\varepsilon}$, *where* n_1 *is the number of residents in a given instance, for any* $\varepsilon > 0$, *unless* $P = NP$, *even if there are no single residents and each hospital has capacity 1.*

3   An IP Formulation for HRC

In this section we describe an IP model which finds a maximum cardinality stable matching in an arbitrary instance of HRC, or reports that no stable matching exists. The variables and constraints required to construct the model are shown below; a detailed proof of the correctness of the model is omitted due to space restrictions, but is presented in full in [8].

Let I be an instance of HRC with residents $R = \{r_1, r_2, \ldots, r_{n_1}\}$ and hospitals $H = \{h_1, h_2, \ldots, h_{n_2}\}$. Without loss of generality, suppose residents $r_1, r_2 \ldots r_{2c}$ are in couples. Again, without loss of generality, suppose that the couples are (r_{2i-1}, r_{2i}) $(1 \leq i \leq c)$. Suppose that the joint preference list of a couple $c_i = (r_{2i-1}, r_{2i})$ is:

$$c_i \;:\; (h_{\alpha_1}, h_{\beta_1}), (h_{\alpha_2}, h_{\beta_2}) \ldots (h_{\alpha_l}, h_{\beta_l}).$$

From this list we create the following projected preference lists for r_{2i-1} and r_{2i}:

$$r_{2i-1} \;:\; h_{\alpha_1}, h_{\alpha_2} \ldots h_{\alpha_l} \qquad r_{2i} \;:\; h_{\beta_1}, h_{\beta_2} \ldots h_{\beta_l}.$$

Let $l(c_i)$ denote the length of the preference list of c_i, and let $l(r_{2i-1})$ and $l(r_{2i})$ denote the lengths of the projected preference lists of r_{2i-1} and r_{2i} respectively. Then $l(r_{2i-1}) = l(r_{2i}) = l(c_i)$. A given hospital h_j may appear more than once in the projected preference list of a resident in a couple $c_i = (r_{2i-1}, r_{2i})$.

Let the single residents be $r_{2c+1}, r_{2c+2} \ldots r_{n_1}$, where each single resident r_i, has a preference list of length $l(r_i)$ consisting of individual hospitals $h_j \in H$. Each hospital $h_j \in H$ has a preference list of individual residents $r_i \in R$ of length $l(h_j)$. Further, each hospital $h_j \in H$ has capacity $c_j \geq 1$, the number of residents with which it may match.

Let J be the following IP formulation of I. In J, for each i $(1 \leq i \leq n_1)$ and p $(1 \leq p \leq l(r_i))$, define a variable $x_{i,p}$ such that

$$x_{i,p} = \begin{cases} 1 \text{ if } r_i \text{ is assigned to their } p^{th} \text{ choice hospital} \\ 0 \text{ otherwise.} \end{cases}$$

For $p = l(r_i) + 1$ define a variable $x_{i,p}$ whose intuitive meaning is that resident r_i is unassigned. Therefore we also have

$$x_{i,l(r_i)+1} = \begin{cases} 1 \text{ if } r_i \text{ is unassigned} \\ 0 \text{ otherwise.} \end{cases}$$

Let $X = \{x_{i,p} : 1 \leq i \leq n_1 \wedge 1 \leq p \leq l(r_i) + 1\}$. As part of the model, for all $x_{i,p} \in X$, we enforce $x_{i,p} \in \{0,1\}$. Let $pref(r_i, p)$ denote the hospital at position p of a single resident r_i's preference list or on the projected preference list of coupled resident where $1 \leq i \leq n_1$ and $1 \leq p \leq l(r_i)$. Let $pref((r_{2i}, r_{2i-1}), p)$ denote the hospital pair at position p on the joint preference list of (r_{2i-1}, r_{2i}).

For an acceptable resident-hospital pair (r_i, h_j), let $rank(h_j, r_i) = q$ denote the rank which hospital h_j assigns resident r_i where $1 \leq j \leq n_2$, $1 \leq i \leq n_1$ and $1 \leq q \leq l(h_j)$. Thus, $rank(h_j, r_i)$ is equal to the number of residents that h_j prefers to r_i plus one.

Further, for i $(1 \leq i \leq n_1)$, j $(1 \leq j \leq n_2)$, p $(1 \leq p \leq l(r_i))$ and q $(1 \leq q \leq l(h_j))$ let the set $R(h_j, q)$ contain resident integer pairs (r_i, p) such that $rank(h_j, r_i) = q$ and $pref(r_i, p) = h_j$. Hence:

$$R(h_j, q) = \{(r_i, p) \in R \times \mathbb{Z} : rank(h_j, r_i) = q \wedge 1 \leq p \leq l(r_i) \wedge pref(r_i, p) = h_j\}.$$

Intuitively, the set $R(h_j, q)$ contains the resident-position pairs (r_i, p) such that r_i is assigned a rank of q $(1 \leq q \leq l(h_j))$ by h_j and h_j is in position p $(1 \leq p \leq l(r_i))$ on r_i's preference list.

Let $A = \{\alpha_{j,q} : 1 \leq j \leq n_2 \wedge 1 \leq q \leq l(h_j)\}$ and further, for all $\alpha_{j,q} \in A$, we enforce $\alpha_{j,q} \in \{0,1\}$. Similarly, Let $B = \{\beta_{j,q} : 1 \leq j \leq n_2 \wedge 1 \leq q \leq l(h_j)\}$ and again, for all $\beta_{j,q} \in B$, we enforce $\beta_{j,q} \in \{0,1\}$. The intuitive meaning of the variables $\alpha_{j,q}$ and $\beta_{j,q}$ will be given later.

We now introduce the constraints that belong to the model. The text in bold before a constraint definition below shows the part of Definition 1 with which the constraint corresponds. Hence, a constraint preceded by 'Stability 1' is intended to prevent blocking pairs described by part 1 of Definition 1.

As each resident $r_i \in R$ is either assigned to a single hospital or is unassigned, we introduce the following constraint for all i $(1 \leq i \leq n_1)$:

$$\sum_{p=1}^{l(r_i)+1} x_{i,p} = 1. \tag{1}$$

Since a hospital h_j may be assigned at most c_j residents, $x_{i,p} = 1$ where $pref(r_i, p) = h_j$ for at most c_j residents. We thus obtain the following constraint for all j $(1 \leq j \leq n_2)$:

$$\sum_{i=1}^{n_1} \sum_{p=1}^{l(r_i)} \{x_{i,p} \in X : pref(r_i, p) = h_j\} \leq c_j. \tag{2}$$

For each couple (r_{2i-1}, r_{2i}), if resident r_{2i-1} is assigned to the hospital in position p in their projected preference list then r_{2i} must also be assigned to the hospital in position p in their projected preference list. We thus obtain the following constraint for all $1 \leq i \leq c$ and $1 \leq p \leq l(r_{2i-1}) + 1$:

$$x_{2i-1,p} = x_{2i,p}. \tag{3}$$

Stability 1 - In a stable matching M in I, if a single resident $r_i \in R$ has a worse partner than some hospital $h_j \in H$ where $pref(r_i, p) = h_j$ and $rank(h_j, r_i) = q$ then h_j must be fully subscribed with better partners than r_i. Therefore, either $\sum_{p'=p+1}^{l(r_i)+1} x_{i,p'} = 0$ or h_j is fully subscribed with better partners than r_i and $\sum_{q'=1}^{q-1} \{x_{i',p''} \in X : (r_{i',p''}) \in R(h_j, q')\} = c_j$. Thus, for each i $(2c + 1 \leq i \leq n_1)$ and p $(1 \leq p \leq l(r_i))$ we obtain the following constraint where $pref(r_i, p) = h_j$ and $rank(h_j, r_i) = q$:

$$c_j \sum_{p'=p+1}^{l(r_i)+1} x_{i,p'} \leq \sum_{q'=1}^{q-1} \{x_{i',p''} \in X : (r_{i',p''}) \in R(h_j, q')\}. \tag{4}$$

Stability 2(a) - In a stable matching M in I, if a couple $c_i = (r_{2i-1}, r_{2i})$ prefers hospital pair (h_{j_1}, h_{j_2}) (which is at position p_1 on c_i's preference list) to $(M(r_{2i-1}), M(r_{2i}))$ (which is at position p_2) then it must not be the case that, if $h_{j_2} = M(r_{2i})$ then h_{j_1} is under-subscribed or prefers r_{2i-1} to one of its partners in M. In the special case in which $pref(r_{2i-1}, p_1) = pref(r_{2i}, p_1) = h_{j_1}$ it must not be the case that, if $h_{j_1} = h_{j_2} = M(r_{2i})$ then h_{j_1} is under-subscribed or prefers r_{2i-1} to one of its partners in M other than r_{2i}.

Thus, for the general case, we obtain the following constraint for all i $(1 \leq i \leq c)$ and p_1, p_2 $(1 \leq p_1 < p_2 \leq l(r_{2i-1}))$ such that $pref(r_{2i}, p_1) = pref(r_{2i}, p_2)$ and $rank(h_{j_1}, r_{2i-1}) = q$:

$$c_{j_1} x_{2i,p_2} \leq \sum_{q'=1}^{q-1} \{x_{i',p''} \in X : (r_{i',p''}) \in R(h_{j_1}, q')\}. \tag{5}$$

However, for the special case in which $pref(r_{2i-1}, p_1) = pref(r_{2i}, p_1) = h_{j_1}$ we obtain the following constraint for all i $(1 \leq i \leq c)$ and p_1, p_2 where $(1 \leq p_1 < p_2 \leq l(r_{2i-1}))$ such that $pref(r_{2i}, p_1) = pref(r_{2i}, p_2)$ and $rank(h_{j_1}, r_{2i-1}) = q$:

$$(c_{j_1} - 1)x_{2i,p_2} \leq \sum_{q'=1}^{q-1} \{x_{i',p''} \in X : q' \neq rank(h_{j_1}, r_{2i}) \wedge (r_{i',p''}) \in R(h_{j_1}, q')\}. \tag{6}$$

Stability 2(b) - A similar constraint is required for the odd members of each couple. Thus, for the general case, we obtain the following constraint for all i $(1 \leq i \leq c)$ and p_1, p_2 where $(1 \leq p_1 < p_2 \leq l(r_{2i}))$ such that $pref(r_{2i-1}, p_1) = pref(r_{2i-1}, p_2)$ and $rank(h_{j_2}, r_{2i}) = q$:

$$c_{j_2} x_{2i-1,p_2} \in X \leq \sum_{q'=1}^{q-1} \{x_{i',p''} : (r_{i',p''}) \in R(h_{j_2}, q')\}. \tag{7}$$

Again, for the special case in which $pref(r_{2i-1}, p_1) = pref(r_{2i}, p_1) = h_{j_2}$ we obtain the following constraint for all i $(1 \leq i \leq c)$ and p_1, p_2 where $(1 \leq p_1 < p_2 \leq l(r_{2i}))$ such that $pref(r_{2i-1}, p_1) = pref(r_{2i-1}, p_2)$ and $rank(h_{j_2}, r_{2i}) = q$:

$$(c_{j_1} - 1)x_{2i-1,p_2} \leq \sum_{q'=1}^{q-1} \{x_{i',p''} \in X : q' \neq rank(h_{j_2}, r_{2i-1}) \wedge (r_{i',p''}) \in R(h_{j_2}, q')\}.$$

$$(8)$$

For all j $(1 \leq j \leq n_2)$ and q $(1 \leq q \leq l(h_j))$ define a new constraint such that:

$$\alpha_{j,q} \geq 1 - \frac{\sum_{q'=1}^{q-1} \{x_{i',p''} \in X : (r_{i',p''}) \in R(h_j, q')\}}{c_j}.$$

$$(9)$$

Thus, if h_j is full with assignees better than rank q then $\alpha_{j,q}$ may take the value 0 or 1. However, if h_j is not full with assignees better than rank q then $\alpha_{j,q} = 1$.

For all j $(1 \leq j \leq n_2)$ and q $(1 \leq q \leq l(h_j))$ define a new constraint such that:

$$\beta_{j,q} \geq 1 - \frac{\sum_{q'=1}^{q-1} \{x_{i',p''} \in X : (r_{i',p''}) \in R(h_j, q')\}}{(c_j - 1)}.$$

$$(10)$$

Thus, if h_j has $c_j - 1$ or more assignees better than rank q then $\beta_{j,q}$ may take the value 0 or 1. However, if h_j has less than $c_j - 1$ assignees better than rank q then $\beta_{j,q} = 1$.

Stability 3(a) - In a stable matching M in I, if a couple $c_i = (r_{2i-1}, r_{2i})$ is assigned to a worse pair than hospital pair (h_{j_1}, h_{j_2}) (where $h_{j_1} \neq h_{j_2}$) it must be the case that for some $t \in \{1, 2\}$, h_{j_t} is full and prefers its worst assignee to r_{2i-2+t}.

Thus we obtain the following constraint for all i $(1 \leq i \leq c)$ and p $(1 \leq p \leq l(r_{2i-1}))$ where $h_{j_1} = pref(r_{2i-1}, p)$, $h_{j_2} = pref(r_{2i}, p)$, $h_{j_1} \neq h_{j_2}$, $rank(h_{j_1}, r_{2i-1}) = q_1$ and $rank(h_{j_2}, r_{2i}) = q_2$:

$$\sum_{p'=p+1}^{l(r_{2i-1})+1} x_{2i-1,p'} + \alpha_{j_1,q_1} + \alpha_{j_2,q_2} \leq 2.$$

$$(11)$$

Stability 3(b) - In a stable matching M in I, if a couple $c_i = (r_{2i-1}, r_{2i})$ is assigned to a worse pair than (h_j, h_j) where $M(r_{2i-1}) \neq h_j$ and $M(r_{2i}) \neq h_j$ then h_j must not have two or more free posts available.

Stability 3(c) - In a stable matching M in I, if a couple $c_i = (r_{2i-1}, r_{2i})$ is assigned to a worse pair than (h_j, h_j) where $M(r_{2i-1}) \neq h_j$ and $M(r_{2i}) \neq h_j$ then h_j must not prefer at least one of r_{2i-1} or r_{2i} to some assignee of h_j in M while having a single free post.

Both of the preceding stability definitions may be modeled by a single constraint. Thus, we obtain the following constraint for i $(1 \leq i \leq c)$ and p $(1 \leq p \leq$

$l(r_{2i-1})$) such that $pref(r_{2i-1}, p) = pref(r_{2i}, p)$ and $h_j = pref(r_{2i-1}, p)$ where $q = \min\{rank(h_j, r_{2i}), rank(h_j, r_{2i-1})\}$:

$$c_j \sum_{p'=p+1}^{l(r_{2i-1})+1} x_{2i-1,p'} - \frac{\sum_{q'=1}^{q-1} \{x_{i',p''} \in X : (r_{i',p''}) \in R(h_j, q')\}}{(c_j - 1)}$$

$$\leq \sum_{q'=1}^{l(h_j)} \{x_{i',p''} \in X : (r_{i'}, p'') \in R(h_j, q')\}. \tag{12}$$

Stability 3(d) - In a stable matching M in I, if a couple $c_i = (r_{2i-1}, r_{2i})$ is jointly assigned to a worse pair than (h_j, h_j) where $M(r_{2i-1}) \neq h_j$ and $M(r_{2i}) \neq h_j$ then h_j must not be fully subscribed and also have two assigned partners r_x and r_y (where $x \neq y$) such that h_j strictly prefers r_{2i-1} to r_x and also prefers r_{2i} to r_y.

For each (h_j, h_j) acceptable to (r_{2i-1}, r_{2i}), let r_{min} be the better of r_{2i-1} and r_{2i} according to hospital h_j with $rank(h_j, r_{min}) = q_{min}$. Analogously, let r_{max} be the worse of r_{2i} and r_{2i-1} according to hospital h_j with $rank(h_j, r_{max}) = q_{max}$. Thus we obtain the following constraint for i ($1 \leq i \leq c$) and p ($1 \leq p \leq l(r_{2i-1})$) such that $pref(r_{2i-1}, p) = pref(r_{2i}, p) = h_j$.

$$\sum_{p'=p+1}^{l(r_{2i-1})+1} x_{2i-1,p'} + \alpha_{j,q_{max}} + \beta_{j,q_{min}} \leq 2. \tag{13}$$

Objective Function - A maximum cardinalilty matching M in I is a stable matching in which the largest number of residents is assigned amongst all of the stable matchings admitted by I. To maximise the size of the stable matching found we apply the following objective function:

$$\max \sum_{i=1}^{n_1} \sum_{p=1}^{l(r_i)} x_{i,p}. \tag{14}$$

Given an instance I of HRC, the above IP model J constructed from I satisfies the property that I admits a stable matching if and only if J admits a feasible solution, the full details of the proof are shown in [8]. The model has $O(m)$ binary-valued variables and $O(m + cL^2)$ constraints where m is the total length of the single residents' preference lists and the coupled residents' projected preference lists, c is number of couples and L is the maximum length of a couple's preference list. The space complexity of the model is $O(m(m + cL^2))$ and the model can be built in $O(m(m + cL^2))$ time in the worst case for an arbitrary instance.

4   Empirical Results

We ran experiments on a Java implementation of the IP models as described in Section 3 applied to randomly-generated data. We present data showing (i) the

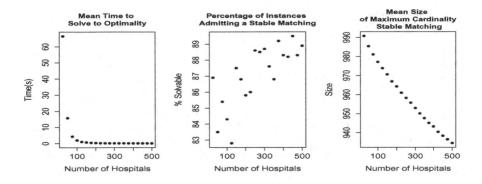

Fig. 1. Empirical Results in Experiment 1

average time taken to find a maximum cardinality stable matching or report that no stable matching exists, and (ii) the average size of a maximum cardinality stable matching where a stable matching did exist. Instances were generated with a skewed preference list distribution on both sides, taking into account that in practice some residents and hospitals are more popular than others (on both sides, the most popular agent was approximately 3 times as popular as the least popular agent).

All experiments were carried out on a desktop PC with an Intel i5-2400 3.1Ghz processor, with 8Gb of memory running Windows 7. The IP solver used in all cases was CPLEX 12.4 and the model was implemented in Java using CPLEX Concert. We have also extended the model to cope with preference lists containing ties, and we are able to find a maximum cardinality stable matching in real data derived from the SFAS application (see [8] for further details).

Experiment 1. In our first experiment, we report on data obtained as we increased the number of hospitals in the instance while maintaining the same total number of residents, couples and posts. For various values of x ($25 \leq x \leq 500$) in increments of 25, 1000 randomly generated instances of size 1000 were created consisting of 1000 residents in total, x hospitals, 100 couples (and hence 800 single residents) and 1000 available posts which were unevenly distributed amongst the hospitals. The time taken to find a maximum cardinality stable matching or report that no stable matching existed in each instance is plotted in Figure 1 for all tested values of x. Figure 1 also shows charts displaying the percentage of instances encountered which admitted a stable matching and the mean size of a maximum cardinality stable solution for all tested values of x.

Figure 1 shows that the mean time taken to find a maximum cardinality stable matching tended to decrease as we increased the number of hospitals in the instances. We believe that this is due to the hospitals' preference lists becoming shorter, thereby reducing the model's complexity. The data in Figure 1 also shows that the percentage of HRC instances admitting a stable matching appeared to increase with the number of hospitals involved in the instance.

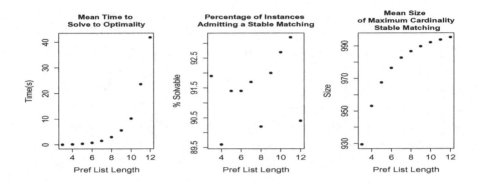

Fig. 2. Empirical Results in Experiment 2

We conjecture that this is because, as each hospital has a smaller number of posts, it is more likely to become full, and therefore less likely to be involved in a blocking pair due to being under-subscribed. Finally, the data shows that as the number of hospitals in the instances increased, the mean size of a maximum cardinality stable matching supported by the instances tended to decrease. This can be explained by the fact that, as the number of hospitals increases but the residents' preference list lengths and the total number of posts remain constant, the number of posts per hospital decreases. Hence the total number of posts among all hospitals on a resident's preference list decreases.

Experiment 2. In our second experiment, we report on data obtained as we increased the length of the individual preference lists for the residents in the instance while maintaining the same total number of residents, couples, hospitals and posts. For various values of x ($3 \leq x \leq 12$) in increments of 1, 1000 randomly generated instances of size 1000 were created consisting of 1000 residents in total, 100 hospitals, 100 couples (and hence 800 single residents) and 1000 available posts which were unevenly distributed amongst the hospitals.

The time taken to find a maximum cardinality stable matching or report that no stable matching existed in each instance is plotted in Figure 2 for all tested values of x. Figure 2 also shows charts displaying the percentage of instances encountered admitting a stable matching and the mean size of a maximum cardinality stable solution for all tested values of x. Figure 2 shows that the mean time taken to find a maximum cardinality stable matching increased as we increased the length of the individual residents' preference lists in the instances. The data in Figure 2 also shows that the percentage of HRC instances admitting a stable matching did not appear to be correlated with the length of the individual residents' preference lists in the instances and further that as the length of the individual residents' preference lists in the instances increased, the mean size of a maximum cardinality stable matching supported by the instances also tended to increase. The first and third of these phenomena would seem to be explained by the fact that the underlying graph is simply becoming more dense.

Acknowledgement. We would like to thank the anonymous reviewers for their valuable comments.

References

1. Abraham, D.J., Biró, P., Manlove, D.F.: "Almost stable" matchings in the room-mates problem. In: Erlebach, T., Persinao, G. (eds.) WAOA 2005. LNCS, vol. 3879, pp. 1–14. Springer, Heidelberg (2006)
2. Aldershof, B., Carducci, O.M.: Stable matching with couples. Discrete Appl. Math. 68, 203–207 (1996)
3. Ashlagi, I., Braverman, M., Hassidim, A.: Matching with couples revisited. Tech. Rep. arXiv: 1011.2121, Cornell University Library (2010)
4. Berman, P., Karpinski, M., Scott, A.D.: Approximation hardness of short symmetric instances of MAX-3SAT. Tech. Rep. 49, ECCC (2003)
5. Biró, P.: Student admissions in Hungary as Gale and Shapley envisaged. Tech. Rep. TR-2008-291, University of Glasgow, Dept. of Computing Science (2008)
6. Biró, P., Irving, R.W., Schlotter, I.: Stable matching with couples: an empirical study. ACM J. Exp. Algorithmics 16, section 1, article 2 (2011)
7. Biró, P., Klijn, F.: Matching with couples: a multidisciplinary survey. International Game Theory Review 15(2), article number 1340008 (2013)
8. Biró, P., Manlove, D.F., McBride, I.: The hospitals / residents problem with couples: Complexity and integer programming models. Tech. Rep. arXiv: 1308.4534, Cornell University Library (2013)
9. Canadian Resident Matching Service, http://www.carms.ca
10. Gale, D., Shapley, L.S.: College admissions and the stability of marriage. Amer. Math. Monthly 69, 9–15 (1962)
11. Gale, D., Sotomayor, M.: Some remarks on the stable matching problem. Discrete Appl. Math. 11, 223–232 (1985)
12. Gusfield, D., Irving, R.W.: The Stable Marriage Problem: Structure and Algorithms. MIT Press, Boston (1989)
13. Irving, R.W.: Matching medical students to pairs of hospitals: A new variation on a well-known theme. In: Bilardi, G., Pietracaprina, A., Italiano, G.F., Pucci, G. (eds.) ESA 1998. LNCS, vol. 1461, pp. 381–392. Springer, Heidelberg (1998)
14. Japan Resident Matching Program, http://www.jrmp.jp
15. Manlove, D.F.: Algorithmics of Matching Under Preferences. World Scientific, Singapore (2013)
16. McDermid, E.J., Manlove, D.F.: Keeping partners together: Algorithmic results for the hospitals / residents problem with couples. J. Comb. Optim. 19(3), 279–303 (2010)
17. Ng, C., Hirschberg, D.S.: Lower bounds for the stable marriage problem and its variants. SIAM J. Comput. 19, 71–77 (1990)
18. National Resident Matching Program, http://www.nrmp.org
19. Ronn, E.: NP-complete stable matching problems. J. Algorithms 11, 285–304 (1990)
20. Roth, A.E.: The evolution of the labor market for medical interns and residents: A case study in game theory. J. Political Economy 92(6), 991–1016 (1984)
21. Roth, A.E.: On the allocation of residents to rural hospitals: A general property of two-sided matching markets. Econometrica 54, 425–427 (1986)
22. Roth, A.E., Peranson, E.: The effects of the change in the NRMP matching algorithm. J. Amer. Medical Assoc. 278(9), 729–732 (1997)

Integral Simplex Using Decomposition with Primal Cuts

Samuel Rosat[1,2], Issmail Elhallaoui[1,2], François Soumis[1,2], and Andrea Lodi[3]

[1] GERAD, 2920 ch. de la Tour, H3T1J4, Montréal, Canada
`samuel.rosat@polymtl.ca`
[2] Polytechnique Montréal, 2900 bd. Édouard Montpetit, H3T1J4, Montréal, Canada
[3] DEIS, Università di Bologna, viale Risorgimento 2, 40136, Bologna, Italy

Abstract. The integral simplex using decomposition (ISUD) algorithm [22] is a dynamic constraint reduction method that aims to solve the popular set partitioning problem (SPP). It is a special case of primal algorithms, i.e. algorithms that furnish an improving sequence of feasible solutions based on the resolution, at each iteration, of an augmentation problem that either determines an improving direction, or asserts that the current solution is optimal. To show how ISUD is related to primal algorithms, we introduce a new augmentation problem, **MRA**. We show that **MRA** canonically induces a decomposition of the augmentation problem and deepens the understanding of ISUD. We characterize cuts that adapt to this decomposition and relate them to *primal cuts*. These cuts yield a major improvement over ISUD, making the mean optimality gap drop from 33.92% to 0.21% on some aircrew scheduling problems.

Keywords: Integer Programming, Primal Algorithms, Cutting Planes, Primal Cuts, Constraint Aggregation, Decomposition, Set Partitioning.

1 Introduction

1.1 The Set Partitioning Problem

We consider the set partitioning problem (SPP):

$$
\begin{aligned}
\min \ & c \bullet x \\
\text{s.t. } & Ax = e \\
& x \in \mathbb{B}^n
\end{aligned} \tag{$\mathbb{P}$}
$$

where $\mathbb{B} = \{0, 1\}$, A is an $m \times n$ binary matrix, $c \in \mathbb{R}^n$, and $e = (1, \ldots, 1) \in$. $\mathcal{M}$ and $\mathcal{N}$ respectively denote the set of indices of the rows and columns of A. $\mathbb{P}'$ is the linear program obtained from $\mathbb{P}$ by replacing the binary constraint $x \in \mathbb{B}^n$ by a nonnegativity constraint $x_j \geqslant 0, j \in \{1, \ldots, n\}$. We can assume without loss of generality that A has no zero or identical rows or columns. A_j is the jth column of A, and for any set of columns T and rows U, A_T^U is the submatrix $(A_{ij})_{i \in U, j \in T}$. Moreover, for any optimization program P, $\mathcal{F}_P$ designates its feasible set.

In this paper, we present an efficient and promising primal framework for the solution of $\mathbb{P}$. It combines and extends previous work concerning *primal algorithms* on the one hand and *constraint aggregation* on the other hand.

J. Gudmundsson and J. Katajainen (Eds.): SEA 2014, LNCS 8504, pp. 22–33, 2014.

1.2   Primal Algorithms

As noted by Letchford and Lodi [10], algorithms for integer programming can be divided into three classes: *dual fractional*, *dual integral*, and *primal*. *Dual fractional* algorithms maintain optimality and linear-constraint feasibility at every iteration, and they stop when integrality is reached. They are typically standard cutting plane procedures such as Gomory's algorithm [8]. *Dual integral* methods maintain integrality and optimality, and they terminate once the primal linear constraints are satisfied. Letchford and Lodi give a single example, another algorithm of Gomory [9]. Finally, *primal methods* maintain feasibility (and integrality) throughout the process and stop when optimality is reached. These are in fact descent algorithms for which the decreasing sequence $(x_k)_{k=1...K}$ satisfies:

(H0) $x^k \in \mathcal{F}_\mathbb{P}$ (incl. integrality constraints);

(H1) x^K is optimal;

(H2) $c \bullet x^{k+1} < c \bullet x^k$ (decreasing sequence).

In this review, we give an overview of the context of our work only (for a more extensive review, see [16]). Primal methods are sometimes classified as *augmenting algorithms* and include the so-called *integral simplex* procedures. They were first introduced in [2] and [20] and improved in [21] and [6]. In Young's method [20,21], at iteration k, a simplex pivot is considered: if it leads to an integer solution, it is performed. Otherwise, cuts are generated and added to the problem, thereby changing the underlying structure of the constraints. Young also developed the concept of a *decreasing vector* (sometimes called an *improving direction*) at x^k, i.e., a direction $z \in \mathbb{R}^n$ s.t. $x^k + z$ is integer, feasible, and of lower cost than x^k. From this notion comes the *primal augmentation problem* (**AUG**) that involves finding such a direction if it exists or asserting that x^k is optimal. Traditionally, papers on constraint aggregation and integral simplex algorithms deal with minimization problems, whereas authors usually present generic primal algorithms for maximization problems. We therefore draw the reader's attention to the following: **to retain the usual classification, we call the improving direction problem AUG, although it supplies a decreasing direction**.

Recently, there has been a renewed interest in primal integer algorithms, inspired by Robert Weismantel (a.o.). Many recent works specifically concern 0/1-programming. However, only a few papers have addressed the practical solution of **AUG**; most of them consider it an oracle.

Most of the rare computational work since 2000 on primal algorithms concerns the *primal separation problem* (**P-SEP**): given x^k a current solution of $\mathbb{P}$ and an infeasible point $x^\star$ (typically a vertex of the linear relaxation), is there a hyperplane that separates $x^\star$ from $\mathcal{F}_\mathbb{P}$ and is tight at x^k? In 2003, Eisenbrand et al. [3] proved that the primal separation problem is as difficult as the integral optimization problem for 0/1-programming. It is therefore expected to be a "complicated" problem because 0/1-programming is $\mathcal{NP}$-hard. Letchford and Lodi [10,11] and Eisenbrand et al. [3] adapt well-known algorithms for the standard separation problem to primal separation. To the best of our knowledge, very few papers present computational experiments using primal methods [13,17,10]. All these papers present results on relatively small instances. Only Letchford and Lodi presented results for an algorithm using primal cutting planes and, interestingly, they stated that degeneracy prevented them from solving larger instances.

Last but not least, *integral simplex methods* were first proposed for the SPP by Balas and Padberg [1]. They were the first to propose an augmenting method, specific to SPP, that yields a sequence of feasible solutions (x^k) satisfying (H0)–(H2) and also

(H3) x^{k+1} is a neighbor of x^k in $\mathcal{F}_{\mathbb{P}'}$.

This property does not prevent the algorithm from reaching optimality since the SPP is *quasi-integral* [19], i.e., every edge of $Conv(\mathcal{F}_{\mathbb{P}})$ is also an edge of its linear relaxation $\mathcal{F}_{\mathbb{P}'}$. Balas and Padberg were therefore able to base their algorithm on a sequence of well-chosen simplex pivots for the linear relaxation $\mathbb{P}'$. Other integral simplex methods have since been presented [22,18,14,12].

The integral simplex using decomposition (ISUD) of Zaghrouti et al. [22] adapts Elhallaoui et al.'s work on constraint reduction for linear programming [5,4] to the SPP. Zaghrouti et al. present promising numerical results for much larger instances than those in [10]. However, ISUD uses neither cutting planes nor exhaustive branching procedures and sometimes stops prematurely, often far from optimality. We will adapt primal cutting to ISUD to obtain an efficient primal algorithm for large (degenerate) SPPs. Note that despite the strong theoretical framework, implementation techniques prevent to reach optimality when solving problems that are too large to be solved with commercial solvers, which is precisely the ultimate goal of our study.

1.3   Objectives and Contributions

This paper, which focuses on the SPP, is organized as follows. In Section 2, we introduce the *maximum reduced-mean augmentation* problem (**MRA**), which is a variant of the standard *maximum mean augmentation* problem (**MMA**); we show that **MRA** corresponds to the subproblem of ISUD, and thus we demonstrate the link between ISUD and primal algorithms. We show (Theorem 1) that **MRA** yields a canonical decomposition of the search space of **AUG**, and we underline the difference between **MRA** and **MMA**. In Section 3, we show how to add a cutting-plane procedure to ISUD and demonstrate that only primal cuts adapt to this factoring. We also give a procedure to transfer the cuts to the decomposed problems. In Section 4, we present computational results for the instances in [22] and obtain considerable improvements, particularly on those for which the best solution found was far from optimality where the mean optimality gap drops from 33.92% to 0.21%. These instances are much larger than those of [10].

2   An Integral Simplex for SPP

Suppose a decreasing sequence of $\mathbb{P}$-feasible solutions ending at x^k is known. We want to determine a direction $d \in \mathbb{R}^n$ and a step $r > 0$ s.t. $x^{k+1} = x^k + rd$ is $\mathbb{P}$-feasible and of lower cost than x^k or to assert that x^k is optimal.

From now on, x^k will always denote the current (binary) solution, $\mathcal{P} = \{j \mid x_j^k = 1\}$ the set of positive-valued variables, and $\mathcal{Z} = \{j \mid x_j^k = 0\}$ the set of null variables. $A_{\mathcal{P}}$ is the *reduced basis*. For the sake of readability, $\mathcal{P}$, $\mathcal{Z}$, d, and r (and later $\mathcal{C}$ and $\mathcal{I}$) will not be indexed on k although they depend on x^k.

The next solution x^{k+1} must satisfy $Ax^{k+1} = e$ and be integer. In the first part, we will require x^{k+1} to satisfy only the linear constraints of $\mathbb{P}'$ (the linear relaxation of $\mathbb{P}$). In the second part we will give conditions under which integrality is maintained when the algorithm takes a step r along d from x^k to reach x^{k+1}.

2.1 Relaxing the Integrality Constraints

Considering only the linear constraints, **AUG** consists in determining d s.t.
   (i) d is a *feasible* direction: $\exists \rho > 0 \mid x^k + \rho d \in \mathcal{F}_{\mathbb{P}'}$;
   (ii) d is an *improving* direction: $c \bullet d < 0$.
It is easy to see that such a d exists iff the program

$$z^*_{\mathrm{MRA}} = \min c \bullet d$$
$$\text{s.t. } Ad = 0, \ e \bullet d_{\mathcal{Z}} = 1, \ d_{\mathcal{P}} \leqslant 0, \ d_{\mathcal{Z}} \geqslant 0 \qquad \textbf{(MRA)}$$

satisfies $z^*_{\mathrm{MRA}} < 0$. On the one hand, any optimal solution of **MRA** yields a solution to **AUG**; on the other hand, if z^*_{MRA} is nonnegative, x^k is optimal.

Remark 1. The specific augmentation problem **MRA** is close to the classical **MMA** (which is itself an extension of the *minimum mean cycle* [7] to more general linear programs; see [16]). It differs only in the normalization constraint: that of **MRA** concerns only the extra-basic variables, while that of **MMA** is $e \bullet d = 1$. This slight difference allows us to decompose the search space $\mathcal{F}_{\mathrm{MRA}}$ into two smaller subspaces and still guarantee **MRA**-optimality, as will be shown in Theorem 1. Theorem 1 does not hold for **MMA** (simple counterexamples are easy to find).

Given x^k and $\mathcal{P}$, the corresponding reduced basis, we refer to any vector of its linear span ($span(A_{\mathcal{P}})$) as a *compatible* vector. This is extended to the columns of A: for any nonbasic index $j \in \mathcal{Z}$, A_j is a *compatible column* if it lies in $span(A_{\mathcal{P}})$; otherwise, A_j is *incompatible*. $\mathcal{C}$ (resp. $\mathcal{I}$) denotes the set of the indices of compatible (resp. incompatible) columns at the current iteration. If we partition $\mathcal{N}$ into $(\mathcal{P}, \mathcal{C}, \mathcal{I})$ and reorder the rows of A, we can write:

$$A = \begin{bmatrix} I_p & A_{\mathcal{C}}^{\mathcal{P}} & A_{\mathcal{I}}^{\mathcal{P}} \\ A_{\mathcal{P}}^N & A_{\mathcal{C}}^N & A_{\mathcal{I}}^N \end{bmatrix}. \qquad (1)$$

From the constraints $Ad = 0$ of **MRA**, one can easily see that the aggregation of all the increasing columns (for which $d_j \geqslant 0$) $w = A_{\mathcal{Z}} d_{\mathcal{Z}}$ is compatible. As in a reduced-gradient algorithm, we are in fact looking for an aggregate column w that can enter the reduced basis with a positive value by lowering only some variables (R) of $\mathcal{P}$. We introduce the following problems (called the *Restricted-MRA* and the *Complementary-MRA*):

$$
\begin{aligned}
z^{\star}_{\text{R-MRA}} = \min_{x \in \mathbb{R}^n} \quad & c_{\mathcal{P}} \bullet d_{\mathcal{P}} + c_{\mathcal{C}} \bullet d_{\mathcal{C}} \\
\text{s.t.} \quad & A_{\mathcal{P}} d_{\mathcal{P}} + A_{\mathcal{C}} d_{\mathcal{C}} = 0 \\
& e \bullet d_{\mathcal{C}} = 1 \\
& d_{\mathcal{P}} \leqslant 0 \quad d_{\mathcal{C}} \geqslant 0
\end{aligned}
\tag{R-MRA}
$$

$$
\begin{aligned}
z^{\star}_{\text{C-MRA}} = \min_{x \in \mathbb{R}^n} \quad & c_{\mathcal{P}} \bullet d_{\mathcal{P}} + c_{\mathcal{I}} \bullet d_{\mathcal{I}} \\
\text{s.t.} \quad & A_{\mathcal{P}} d_{\mathcal{P}} + A_{\mathcal{I}} d_{\mathcal{I}} = 0 \\
& e \bullet d_{\mathcal{I}} = 1 \\
& d_{\mathcal{P}} \leqslant 0 \quad d_{\mathcal{I}} \geqslant 0
\end{aligned}
\tag{C-MRA}
$$

Lemma 1. *For all $j \in \mathcal{Z}$, A_j is compatible iff $\exists! R_j \subset \mathcal{P} \mid A_j = \sum_{r \in R_j} A_r$.*

Theorem 1. $z^{\star}_{MRA} = \min \left\{ z^{\star}_{R\text{-}MRA}, z^{\star}_{C\text{-}MRA} \right\}$.

Proof. Let $d = (d_{\mathcal{P}}, d_{\mathcal{C}}, d_{\mathcal{I}}) \in \mathbb{R}^n$ be an optimal solution of **MRA**. Suppose that the support of $(d_{\mathcal{C}}, d_{\mathcal{I}})$ is neither in $\mathcal{C}$ nor in $\mathcal{I}$. By Lemma 1, the surrogate column $A_{\mathcal{C}} d_{\mathcal{C}}$ can be written as a linear combination of columns of $\mathcal{P}$: $A_{\mathcal{C}} d_{\mathcal{C}} = -A_{\mathcal{P}} u'_{\mathcal{P}}$, $u'_{\mathcal{P}} \leqslant 0$. Let $u' = (u'_{\mathcal{P}}, d_{\mathcal{C}}, 0)$, and $u'' = d - u'$ and denote $d' = u'/\|d_{\mathcal{C}}\|_1$, $d'' = u''/\|d_{\mathcal{I}}\|_1$. Here d' and d'' are both feasible for **MRA** and resp. for **R-MRA** and **C-MRA**. $d = \|d_{\mathcal{C}}\|_1 d' + \|d_{\mathcal{I}}\|_1 d''$ is a convex combination since $\|d_{\mathcal{C}}\|_1 + \|d_{\mathcal{I}}\|_1 = \|d_{\mathcal{Z}}\|_1 = 1$. Therefore, d is an extreme point of $\mathcal{F}_{\text{MRA}}$ iff $d = d'$ or $d = d''$. It is well known that one can always find an optimal solution of a linear program that is also an extreme point of the feasible domain, which concludes the proof. $\qquad \square$

Theorem 1 extends the results of Elhallaoui et al. [4] and Zaghrouti et al. [22], and it also justifies their procedures in a trivial way. This purely primal interpretation of their less-intuitive dual approach allows us to state a precise factoring of **MRA**. This factoring naturally leads to a sequential resolution of **R-MRA** and **C-MRA** that avoids the direct solution of **MRA**. As noted in Remark 1, it also accurately describes the differences between the classical **MMA** and the algorithms of [4] and [22].

As a consequence of Theorem 1, we will consider the pair of problems **R-MRA** and **C-MRA** instead of the more complicated **MRA**, and we will solve them sequentially. In particular, we will not solve **C-MRA** if $z^{\star}_{\text{R-MRA}} < 0$ because we already know an improving direction. In the next section, we discuss how to solve **R-MRA** and **C-MRA** and in particular how to ensure that x^{k+1} is an integer solution or that x^k is optimal.

2.2   Taking Integrality into Account

Integrality Issues in R-MRA. An advantage of **R-MRA** is that it deals in a trivial way with the integrality constraints. For $j \in \mathcal{C}$, by Lemma 1,

$\exists! R_j \subset \mathcal{P} \,|\, A_j = \sum_{r \in R_j} A_r$. Therefore, any extreme point of $\mathcal{F}_{\text{R-MRA}}$ is of the form $d_k^j = 1$ if $k = j$, -1 if $k \in R_j$, 0 otherwise. Thus, taking a step in this direction is strictly equivalent to entering A_j into the reduced basis by performing a single simplex pivot. The cost of such a solution is $c \bullet d^j = c_j - \sum_{r \in R} c_r = \bar{c}_j$, i.e., the reduced cost of column j as computed in the simplex algorithm ($\bar{c} = c - \pi^T A$, with π being the dual-variable vector).

The optimal value of a linear program is always at an extreme point of its domain, so we can rewrite the problem as

$$z_{\text{R-MRA}}^{\star} = \min_{j \in \mathcal{P}} \bar{c}_j \tag{2}$$

and given $j \in \mathcal{P}$ s.t. $\bar{c}_j$ is minimal, the corresponding optimal solution is d^j. This formulation yields a search strategy equivalent to the Dantzig criterion in the simplex algorithm. Furthermore, the restriction of the search space to $span(A_{\mathcal{P}})$ prevents degenerate pivots (step $r = 1$) and, as Proposition 1 states, preserves integrality. Note that this formulation is close to that of a reduced gradient.

Proposition 1. *If j is optimal for (2) and r is the maximum feasible step along d^j at x^k, then $x^{k+1} = x^k + rd^j$ satisfies the conditions (H0)–(H3).*

Proof. x^{k+1} is reached by performing a single simplex pivot (entering A_j into the basis), thus (H3) holds. Since $r = 1$ (trivial), (H0)–(H2) are straightforward. □

Integrality Issues in C-MRA. AUG is as difficult as SPP [15]. Since SPP is $\mathcal{NP}$-complete and **R-MRA** is polynomial, **C-MRA** has to be *hard* and no simple formulation can be expected. However, we will show that **C-MRA** can be reduced to an $(m-p) \times |\mathcal{I}|$ linear program with specific additional constraints, and we will address the solution of this new formulation **RC-MRA**.

In **C-MRA**, the first p constraints are $d_{\mathcal{P}} = -A_{\mathcal{I}}^{\mathcal{P}} d_{\mathcal{I}}$. As in a reduced-gradient algorithm, the modification of the basic variables occurs via a linear transformation of the increasing nonbasic variables. With $d = D\delta$, the *transfer matrix* D summarizes the relationship between the reduced direction $d_{\mathcal{I}} = \delta \in \mathbb{R}^{|\mathcal{I}|}$ and the corresponding direction $d = D\delta$ in the original space $\mathbb{R}^n$. **C-MRA** becomes (*Reduced Complementary-MRA*):

$$z_{\text{RC-MRA}}^{\star} = \min_{\delta \in \mathbb{R}^{|\mathcal{I}|}} c \bullet (D\delta) \\ \text{s.t. } A_{\mathcal{P},\mathcal{I}}^N (D\delta) = 0, \ e \bullet \delta = 1, \ \delta \geqslant 0 \tag{RC-MRA}$$

The set of all feasible directions at x^k is a cone whose extreme directions are those of all the edges of $\mathcal{F}_{\mathbb{P}'}$ that go through x^k. Since SPP is quasi-integral, any reduced direction δ s.t. $d = D\delta$ satisfies (H0) and (H3) is an extreme point of $\mathcal{F}_{\text{RC-MRA}}$. Let $\Delta = \{\delta \in \mathcal{F}_{\text{RC-MRA}} \,|\, d = D\delta$ satisfies (H0) and (H3)$\}$. We will now give a simple characterization of Δ. A set of indices U (or columns A_U) is called *column-disjoint* if no pair of columns $A_i, A_j \in A_U$, $i \neq j$ has a common nonzero entry, i.e., $\forall i, j \in U, i \neq j \Rightarrow A_i \bullet A_j = 0$.

Proposition 2. (Propositions 6 and 7, Zaghrouti et al. [22]) *Given δ a* **vertex** *of* $\mathcal{F}_{\textbf{C-MRA}}$, $d = D\delta$ *is an integral feasible direction iff S (the support of δ) is column-disjoint. In this case, the maximal feasible step along d is $r = |S|$.*

2.3 Algorithmic Framework

From Proposition 2, $\Delta = \{\delta \mid \delta$ is a vertex of $\mathcal{F}_{\textbf{C-MRA}}$ and S is column-disjoint$\}$. Therefore, if we can solve

$$z^{\star}_{(3)} = \min_{\delta \in \Delta} c \bullet (D\delta) = \min_{\delta \in Conv(\Delta)} c \bullet (D\delta) \tag{3}$$

then we can use Algorithm 1 to solve $\mathbb{P}$.

Algorithm 1. Integral Simplex Using Decomposition for SPP

---

1. Find an initial solution of SPP x^0; $k \leftarrow 0$;
2. If $z^{\star}_{\textbf{R-MRA}} < 0$: perform a single compatible pivot to obtain x^{k+1}; $k \leftarrow k+1$; GOTO 2;
3. If $z^{\star}_{(3)} < 0$: given δ (optimal solution of (3)) and r (maximum feasible step along $D\delta$ at x^k): $x^{k+1} \leftarrow x^k + rD\delta$; $k \leftarrow k + 1$; GOTO 2;
4. Return x^k, the optimal solution of $\mathbb{P}$.

---

3 Solving RC-MRA with Cutting Planes

3.1 Theory and Generic Cutting-Plane Procedure

In this section, we characterize cutting planes for **RC-MRA**. Given $\delta^\star$ an optimal solution of **RC-MRA** that is not column-disjoint, we want to determine an inequality $\bar{\Gamma}$ that separates $\delta^\star$ from $\mathcal{F}_{(3)} = \Delta$ to tighten the relaxation. We will show that $\bar{\Gamma}$ can always be obtained from a *primal cut* (in the sense of [10]).

Consider $\bar{\Gamma} : \bar{\alpha} \bullet \delta \leqslant \bar{\beta}$, a valid inequality for Δ. Since $Conv(\Delta) \subset \{\delta \mid e \bullet \delta = 1\}$, we can assume $\bar{\alpha} \notin Span(\{e\})$. Then $\{e \bullet \delta = 1\} \cap \{\bar{\alpha} \bullet \delta = \bar{\beta}\} = F$ is of dimension $|\mathcal{I}| - 2$ (the intersection of two nonparallel hyperplanes). Thus, $span(F \cup \{0\})$ is a hyperplane of $\mathbb{R}^{|\mathcal{I}|}$ that yields the same valid inequality as $\bar{\Gamma}$ within $\mathcal{F}_{\textbf{RC-MRA}}$ and that reads $\bar{\alpha}' \bullet \delta \leqslant 0$ for some $\bar{\alpha}'$. Without loss of generality, from now on, we consider $\bar{\alpha} = \bar{\alpha}'$ and thus $\bar{\beta} = 0$.

We will now characterize $\bar{\Gamma}$ in terms of the original SPP formulation. Let $\alpha \in \mathbb{R}^{p+|\mathcal{I}|}$ be such that

$$\bar{\alpha} = D\alpha. \tag{4}$$

Such an α always exists since $(0, \bar{\alpha})$ is a trivial solution of (4). $\bar{\Gamma}$ is valid for (3) iff for all $\delta \in \Delta$ and all $r > 0$, $\alpha \bullet (x^k + rD\delta) \leqslant \alpha \bullet x^k$. Since $\{d = D\delta \mid \delta \in \Delta\}$ is the cone of all feasible directions at x^k within $Conv(\mathcal{F}_{\mathbb{P}})$, this is equivalent to the inequality

$$\Gamma : \qquad \alpha \bullet (x - x^k) \leqslant 0 \tag{Γ}$$

being a valid inequality for SPP. Since a *primal valid inequality* is a valid inequality that is tight at x^k, Γ is obviously a primal valid inequality.

Consider now the case where the optimal solution δ^* of **RC-MRA** is not column-disjoint ($\delta^* \notin \Delta$). x^* is the new fractional solution found by taking a maximal step r^* along $D\delta^*$ at x^k: $x^* = x^k + r^*D\delta^*$. We now address the problem of separating δ^* from Δ.

Proposition 3. $\bar{\Gamma}$ *is a valid inequality for* (3) *that separates* δ^* *from* Δ *iff* Γ *is a primal valid inequality for the SPP that separates* x^* *from* $\mathcal{F}_\mathbb{P}$. *In this case,* $\bar{\Gamma}$ *is a* cut *for* (3) *and* Γ *is a* primal cut *for* $\mathbb{P}$ *in the sense of [10].*

Proof. Most of the proof has been given in the previous paragraphs. We need to prove only the separation. $\bar{\Gamma}$ separates δ^* from Δ iff $\bar{\alpha} \bullet \delta^* > 0$. Equivalently, $\alpha \bullet (x^k + rD\delta^*) > \alpha \bullet x^k$ or $\alpha \bullet (x^* - x^k) > 0$, which concludes the proof.   □

What we have shown must now be seen the other way round to take advantage of previous work on primal separation. Assume that we have a primal cut Γ for the SPP that separates x^* from $\mathcal{F}_\mathbb{P}$. Then, the inequality

$$\bar{\Gamma}: \quad \alpha \bullet D\delta \leqslant 0 \qquad (\bar{\Gamma})$$

is a cut for **RC-MRA** that separates δ^* from $Conv(\Delta)$. Moreover, we have shown that any cut for **RC-MRA** can be obtained in this way. This enables us to develop a procedure based on **P-SEP**: Given x^k and x^*, is there a valid inequality for $\mathcal{F}_\mathbb{P}$, tight at x^k, that separates x^* from $\mathcal{F}_\mathbb{P}$? If it exists, it will be transferred to **RC-MRA** to tighten the relaxation of $\mathcal{F}_{(3)}$, as in Algorithm 2.

Remark 2. $\bar{\Gamma}$ is obtained from Γ by multiplying α and the transfer matrix D, as **RC-MRA** was obtained from **C-MRA** by multiplying the objective and the constraints by D. This shows the role played by D in the transformation from the directions in $\mathbb{R}^n$ to the reduced directions in $\mathbb{R}^{|\mathcal{I}|}$.

Algorithm 2. Cutting-plane procedure for (3)

---

    1. $\delta^* \leftarrow$ an optimal solution of **RC-MRA**;
    2. If stopping conditions are met, return δ^* although it may not be (3)-feasible;
    3. If δ^* is not column-disjoint, find a solution to **P-SEP** $\Gamma : \alpha \bullet (x - x^k) \leqslant 0$; transfer it to **RC-MRA** as $\bar{\Gamma} : \alpha \bullet D\delta \leqslant 0$; GOTO 1;
    4. Return δ^*, the optimal solution of (3).

Algorithm 2 requires a primal separation procedure that can solve **P-SEP**. Note that any primal algorithm, by its nature, can provide cuts only in a given family $\mathcal{O}$ and actually solves $\mathcal{O}$-**P-SEP** (a primal separation but in a given cut family).

Remark 3. There exist families of cuts for which no stopping criterion is required (e.g., Gomory–Young's cuts [20]). These families guarantee $\delta^* \in \Delta$ after a finite number of $\mathcal{O}$-**P-SEP** problems have been solved.

Remark 4. For the stopping criterion, a maximal number of iterations can be fixed prior to running the algorithm. For families for which no cut may exist even if $\delta^\star \notin \Delta$, the algorithm stops whenever $\mathcal{O}$-**P-SEP** yields no new cut. In this case, $\delta^\star$ may not be column-disjoint, and either a branching procedure is used or the primal algorithm stops prematurely at x^k.

3.2   Primal Clique Cut Separation: $\mathcal{Q}$-P-SEP

In this section, we present a well-known cutting-plane family $\mathcal{Q}$, called the clique cuts, for which there exists a relatively simple procedure that solves $\mathcal{Q}$-**P-SEP**. Unlike Gomory–Young cuts, $\mathcal{Q}$-**P-SEP** may have no solution although $\delta^\star \notin \Delta$. However, clique inequalities are usually sufficient to reach Δ-optimality and yield deep cuts. Consider $\mathcal{G} = (\mathcal{N}, E)$, the conflict graph obtained from matrix A, i.e., $\{i, j\} \in E$ iff $A_i \bullet A_j \neq 0$. Given a clique $\mathcal{W}$ in this graph, any binary solution of SPP satisfies

$$\sum_{j \in \mathcal{W}} x_j \leqslant 1. \qquad (Q_{\mathcal{W}})$$

$Q_{\mathcal{W}}$ is called the *clique inequality* associated with $\mathcal{W}$, and it is valid for $\mathcal{F}_\mathbb{P}$. Moreover, given $x^\star$, a fractional extreme point of $\mathcal{F}_{\mathbb{P}'}$, finding a clique cut is equivalent to finding a clique of total weight greater than 1 in a weighted graph $\mathcal{G}$, with weight function $w_j = x_j^\star$, $j \in \mathcal{N}$.

In our case, $x^\star$ is typically the fractional vertex of $\mathcal{F}_{\mathbb{P}'}$ that *would* be reached *if* a step were taken in direction $D\delta^\star$, where $\delta^\star$ is an extreme optimal solution of $\mathcal{F}_{\textbf{RC-MRA}}$ that is not column-disjoint ($\delta^\star \notin \Delta$).

For $Q_{\mathcal{W}}$ to be tight at x^k, it must satisfy $\sum_{l \in \mathcal{W} \cap \mathcal{P}} x_l^k = 1$, or equivalently $|\mathcal{W} \cap \mathcal{P}| = 1$. Therefore, with $\mathcal{G}_l$ being the subgraph of $\mathcal{G}$ with the vertices l and all neighbors of l, a primal separation for the clique cuts can be found using Algorithm 3.

Algorithm 3. $\mathcal{Q}$-P-SEP

---

    1. $\mathcal{K} \leftarrow \emptyset$ (set of all primal cliques found);
    2. For all $l \in R = support(D\delta^\star) \cap \mathcal{P}$: Find a clique $\mathcal{W}_l$ of maximal weight in $\mathcal{G}_l$;
        $\mathcal{K} \leftarrow \mathcal{K} \cup \{Q_{\mathcal{W}_l}\}$;
    3. Return $\mathcal{K}$.

---

4   Experimentation

4.1   Algorithm and Instances

Algorithm. To solve SPP, we use Algorithm 1 with Zaghrouti et al.'s multi-phase strategy. Problem (3) is solved using Algorithm 2. If the solution of **RC-MRA** is not column-disjoint, we generate primal clique inequalities using Algorithm 3 and transfer them to **RC-MRA** to tighten the relaxation of $\mathcal{F}_{(3)}$. For each solution $\delta^\star \notin \Delta$, at most 70 $\mathcal{Q}$-**P-SEP** are solved. However, note

that this number is seldom reached (usually, before the 70th separation procedure is launched, either δ^* is column-disjoint or Q-P-SEP has no solution).
When the cutting planes do not manage to ensure $\delta^* \in \Delta$, the nonexhaustive
branching procedure of [22] is used. After each branch, new cutting planes may
be generated.

Cut Pool. As in a standard *branch-and-bound*, all the generated cuts are kept
in a pool. Before generating any supplementary cuts, we transfer any cuts in the
pool that can be used to eliminate $\delta^* \notin \Delta$ at the current iteration to **RC-MRA**.

Instances. Tests were run on an aircrew scheduling problem from OR-Lib of
size $m = 823$, $n = 8,904$. The different instances correspond to different initial
solutions x^0. These initial solutions are created to resemble typical initial solutions for aircrew scheduling problems and are far from optimality. We chose to
focus on the hardest instances, i.e., those for which the solutions in [22] were
furthest from optimality. See [22] for more details on these instances.

4.2 Numerical Results

Table 1 shows that for the ten hardest instances, adding primal Q-inequalities
reduces the average optimality gap from 33.92% to 0.21% and decreases the
maximal gap from 200.63% to 2.06%. This improvement comes with an increase
in the number of steps and a small increase in the computational time.

On the nonoptimal instances (instances 3, 5, 8, and 10), ISUD_Clique finds a
slightly worse solution for instance 5, the same solution for instance 8, and greatly
improved solutions for instances 3 and 10. Note that the improvement comes with
a higher number of steps: +6 (resp. +11) for instance 3 (resp. 10). This means
that the cutting planes allow the algorithm to find column-disjoint combinations

Table 1. Comparison of performance of ISUD with and without clique cuts on aircrew
scheduling problems. Opt. val. 56, 137

Instance	Init. gap	ISUD_NoCut			ISUD_Clique		
		Gap (%)	Steps	Time (s)	Gap (%)	Steps	Time (s)
1	570.22	0	42	8.75	0	42	10.27
2	560.97	0	46	13.44	0	46	16.56
3	559.92	138.08	27	12.15	**0.03**	**33**	17.20
4	557.67	0	45	9.96	0	45	11.47
5	562.5	**0.5**	54	19.88	2.06	59	24.98
6	561.03	0	35	9.85	0	39	13.84
7	573.09	0	42	7.98	0	42	9.738
8	569.64	0.02	42	9.58	0.02	43	10.98
9	569.29	0	53	11.62	0	54	13.39
10	573.74	200.63	33	6.78	**0.01**	**44**	17.75
Avg.	565.81	33.92	41.9	11.00	**0.21**	44.7	14.62

Table 2. Performance of `ISUD_Clique` on the aircrew scheduling instances

Instance	Gap (%)	Steps	Cuts	Sep.	Fails	Tot. (s)	Sep. (s)
1	0	42	266	64	3	10.27	1.01
2	0	46	490	94	4	16.56	2.02
3	0.03	33	440	130	9	17.20	1.99
4	0	45	281	70	4	11.47	1.00
5	2.06	59	536	155	19	24.98	2.64
6	0	39	473	142	6	13.84	1.42
7	0	42	314	78	4	9.74	1.17
8	0.02	43	339	69	7	10.98	1.07
9	0	54	299	80	5	13.39	1.12
10	0.01	44	496	132	12	17.75	1.62
Avg.	0.21	44.7	393.4	101.4	7.3	14.62	1.50

whereas the nonexhaustive branching did not. When the cuts worsen the solution (instance 5), the algorithm performs five additional steps but reaches a more expensive solution. This indicates that the cuts changed the *path* $(x^k)_k$ earlier in the process by taking an improving direction that *locally seemed better* than that taken by `ISUD_NoCut`.

Tables 1 and 2 indicate that solving the separation problems is not very time-consuming. The cuts add iterations (K increases). The extra primal cuts, and the corresponding reoptimizations, increase the time per iteration by 24% (on average) while reducing the average gap from 33.9% to 0.21%.

Note that branching prior to cutting would have allowed `ISUD_Clique` to always find a better solution than that found by `ISUD_NoCut`. However, it seems logical to begin with cuts that do not discard any feasible solution rather than to begin with a heuristic branching. Therefore, we took the risk that the solution would deteriorate on some instances, as happened for instance 5.

5   Conclusions and Future Work

We have introduced **MRA** to show the link between the ISUD constraint reduction algorithm and primal algorithms. The factoring of **MRA** into **R-MRA** and **C-MRA** has been explained and justified, and the result is an efficient primal cutting-plane algorithm for the SPP. A proof of concept of this algorithm was provided by using the Q-cuts family.

In future work, we plan to apply our algorithm to larger instances $(1,600 \times 570,000$ as in [22]) to obtain a more complete benchmark. Other families of cuts such as primal cycle cuts will be added to improve the performance of the algorithm.

Acknowledgements. The authors thank Matthieu Delorme for his help with the implementation, Jean-Bertrand Gauthier for numerous fruitful discussions and Cem Unlubayrak who completed an internship under the supervision of Samuel Rosat.

References

1. Balas, E., Padberg, M.: On the set-covering problem: 2 - an algorithm for set partitioning. Oper. Res. 23(1), 74–90 (1975)
2. Ben-Israel, A., Charnes, A.: On some problems of diophantine programming. Cahiers du Centre d'Ét. de Rech. Opér. 4, 215–280 (1962)
3. Eisenbrand, F., Rinaldi, G., Ventura, P.: Primal separation for 0/1 polytopes. Math. Program. 95(3), 475–491 (2003)
4. Elhallaoui, I., Metrane, A., Desaulniers, G., Soumis, F.: An improved primal simplex algorithm for degenerate linear programs. INFORMS J. Comput. 23(4), 569–577 (2011)
5. Elhallaoui, I., Villeneuve, D., Soumis, F., Desaulniers, G.: Dynamic aggregation of set-partitioning constraints in column generation. Oper. Res. 53(4), 632–645 (2005)
6. Glover, F.: A new foundation for a simplified primal integer programming algorithm. Oper. Res. 16, 727–740 (1968)
7. Goldberg, A.V., Tarjan, R.E.: Finding minimum-cost circulations by canceling negative cycles. J. ACM 36(4), 873–886 (1989)
8. Gomory, R.E.: Outline of an algorithm for integer solutions to linear program. Bull. Amer. Math. Soc. 64(5), 275–278 (1958)
9. Gomory, R.E.: All-integer integer programming algorithm. Ind. Sched., 193–206 (1963)
10. Letchford, A.N., Lodi, A.: Primal cutting plane algorithms revisited. Math. Methods Oper. Res. 56(1), 67–81 (2002)
11. Letchford, A.N., Lodi, A.: Primal separation algorithms. Q. J. Belg. Fr. Ital. Oper. Res. Soc. 1(3), 209–224 (2003)
12. Rönnberg, E., Larsson, T.: Column generation in the integral simplex method. European J. Oper. Res. 192(1), 333–342 (2009)
13. Salkin, H.M., Koncal, R.D.: Set covering by an all-integer algorithm: Computational experience. J. ACM 20(2), 189–193 (1973)
14. Saxena, A.: Set-partitioning via integral simplex method. OR Group, Carnegie-Mellon University, Pittsburgh (2003) (unpublished manuscript)
15. Schulz, A.S., Weismantel, R., Ziegler, G.M.: 0/1-integer programming: Optimization and augmentation are equivalent. In: Spirakis, P. (ed.) ESA 1995. LNCS, vol. 979, pp. 473–483. Springer, Heidelberg (1995)
16. Spille, B., Weismantel, R.: Primal integer programming. In: Aardal, K., Nemhauser, G., Weismantel, R. (eds.) Discrete Optimization, Handbooks in Operations Research and Management Science, vol. 12, pp. 245–276. Elsevier, Amsterdam (2005)
17. Stallmann, M.F., Brglez, F.: High-contrast algorithm behavior: Observation, conjecture, and experimental design. In: ACM-FCRC. ACM, New York (2007), 549075
18. Thompson, G.L.: An integral simplex algorithm for solving combinatorial optimization problems. Comput. Optim. Appl. 22(3), 351–367 (2002)
19. Trubin, V.: On a method of solution of integer linear programming problems of a special kind. Soviet Math. Dokl. 10, 1544–1546 (1969)
20. Young, R.D.: A primal (all-integer) integer programming algorithm. J. Res. Nat. Bureau of Standards: B. Math. and Math. Phys., 213–250 (1965)
21. Young, R.D.: A simplified primal (all-integer) integer programming algorithm. Oper. Res. 16(4), 750–782 (1968)
22. Zaghrouti, A., Soumis, F., Elhallaoui, I.: Integral simplex using decomposition for the set partitioning problem. Oper. Res. (to appear)

A Branch-Price-and-Cut Algorithm for Packing Cuts in Undirected Graphs

Martin Bergner*, Marco E. Lübbecke, and Jonas T. Witt

Operations Research, RWTH Aachen University,
Kackertstr. 7, 52072 Aachen, Germany
{martin.bergner,marco.luebbecke,jonas.witt}@rwth-aachen.de

Abstract. The cut packing problem in an undirected graph is to find a largest cardinality collection of pairwise edge-disjoint cuts. We provide the first experimental study of this NP-hard problem that interested theorists and practitioners alike. We propose a branch-price-and-cut algorithm to optimally solve instances from various graph classes, random and from the literature, with up to several hundred vertices. In particular we investigate how complexity results match computational experience and how combinatorial properties help improving the algorithm's performance.

1 Introduction

Given an undirected graph $G = (V, E)$ with $n = |V|$ vertices and $m = |E|$ edges, every $S \subseteq V$ induces a *cut* $\delta(S) = \{ij \in E \mid i \in S, j \notin S\}$. We call S and $V \backslash S$ the *shores* of cut $\delta(S)$. We assume G to be connected and S a non-trivial subset, thus $\delta(S) \neq \emptyset$. Two cuts $\delta_1 \subseteq E$ and $\delta_2 \subseteq E$ are *disjoint*, if $\delta_1 \cap \delta_2 = \emptyset$. The *cut packing problem* is to find a largest cardinality set of pairwise disjoint cuts. The maximum is called the cut packing number $\gamma(G)$.

In combinatorial optimization, cut packing is known for its role in duality theorems [12,22]. It is NP-hard in general [6] and even in planar graphs [4], but polynomial time solvable in chordal or bipartite graphs [6] or when packing *s-t* cuts [7,22]. Interestingly, packing directed cuts in digraphs is polynomial time solvable as well [19]. Cut packing is closely related to other combinatorial optimization problems like cycle packing [4] and independent set [6]: An independent set $I = \{v_1, \ldots, v_k\} \subseteq V$ immediately translates to a (particular) cut packing $\{\delta(\{v_i\}) : v_i \in I\}$. The latter provides a strong inapproximability result [5]: for the stability number $\alpha(G)$ (the cardinality of a maximum independent set in G) it holds that $\alpha(G) \leq \gamma(G) \leq 2\alpha(G) - 1$. Again, special graph classes allow better approximation guarantees [5]. The parameterized version of the cut packing problem is W[1]-hard by a reduction from parameterized independent set [10], which is performed in the full version of this paper. Cut packing and variants have applications in bioinformatics [5] and network reliability [6]. Colbourn [7] mentions that the NP-hardness of several edge packing problems "limits their

---

* Supported by the German Research Foundation (DFG) as part of the Priority Program "Algorithm Engineering" under grants no. LU770/4-1 and LU770/4-2.

J. Gudmundsson and J. Katajainen (Eds.): SEA 2014, LNCS 8504, pp. 34–45, 2014.

applicability" for obtaining bounds for network reliability. We mitigate this argument by presenting an exact integer programming based approach to solve the cut packing problem for arbitrary graphs optimally.

Our Contribution. We are not aware of any experimental results, neither exact nor approximate, for cut packing; thus we provide the first computational study of the problem. We propose a branch-price-and-cut algorithm to optimally solve instances from various graph classes, random and from the literature, with up to several hundred vertices. This success mainly builds on an easily computable combinatorial upper bound. In particular, we investigate how theoretical complexity and approximability results match with computational experience.

2   Formulations and Properties

2.1   Compact Formulation

There are several known (integer) linear programming formulations for finding one (minimum) cut in a graph. As trivially $\gamma(G) \leq m$, we replicate the constraints of such a formulation, one set for each *potential* cut in a packing, indexed by $c = 1, \ldots, m$. A cut $\delta_c = \{ij \in E \mid i \in S_c, j \notin S_c\}$ induced by S_c is represented by binary variables u_i^c, $i \in V$, taking value 0 when $i \in S_c$ and value 1 when $i \notin S_c$. With $V = \{1, \ldots, n\}$, we normalize $1 \in S_c$, $c = 1, \ldots, m$, mildly reducing symmetry. Note that $\delta(S_c) = \delta(V \backslash S_c)$. The binary variables y_{ij}^c, $ij \in E$, take value 1 iff $ij \in \delta_c$, and the binary variables x^c are used to count the cuts in the packing. The cut packing problem can be formulated as

$$\max \quad \sum_{c=1}^{m} x^c$$

$$\begin{array}{lll}
\text{s.t.} & u_i^c - u_j^c + y_{ij}^c \geq 0 & \forall ij \in E, \forall c \in \{1, \ldots, m\} \quad (1) \\
& u_j^c - u_i^c + y_{ij}^c \geq 0 & \forall ij \in E, \forall c \in \{1, \ldots, m\} \quad (2) \\
& y_{ij}^c - u_i^c - u_j^c \leq 0 & \forall ij \in E, \forall c \in \{1, \ldots, m\} \quad (3) \\
& u_i^c + u_j^c + y_{ij}^c \leq 2 & \forall ij \in E, \forall c \in \{1, \ldots, m\} \quad (4) \\
& u_1^c = 0 & \forall c \in \{1, \ldots, m\} \quad (5) \\
& x^c \leq \sum_{ij \in E} y_{ij}^c & \forall c \in \{1, \ldots, m\} \quad (6)
\end{array}$$

$$\sum_{c=1}^{m} y_{ij}^c \leq 1 \qquad \forall ij \in E \quad (7)$$

$$x^c, u_i^c, y_{ij}^c \in \{0, 1\} \qquad \forall ij \in E, \forall c \in \{1, \ldots, m\}, \forall i \in V .$$

The metric inequalities (1)–(4) ensure compatibility between u and y variables for the c-th potential cut. Constraints (1) and (2) guarantee that if two vertices are on different shores, the edge between them has to be in the cut. On the other hand, constraints (3) and (4) ensure that the edge between vertices on the same shore will not be in the cut. Constraint (5) puts vertex 1 in S_c and constraint (6) states that a cut is counted only iff there are edges assigned to that cut. Together with constraint (3), at least one other vertex is in $V \backslash S_c$. The packing constraint (7) ensures that each edge is contained in at most one cut.

2.2 Extended Formulation

Let D denote the set of *all* non-empty cuts in G. A natural formulation is based on variables $x_\delta \in \{0, 1\}$, representing whether $\delta \in D$ is part of a packing or not.

$$\max \qquad \sum_{\delta \in D} x_\delta$$

$$\text{s.t.} \qquad \sum_{\delta \ni ij} x_\delta + \bar{x}_{ij} = 1 \qquad \forall ij \in E \qquad (8)$$

$$x_\delta, \bar{x}_{ij} \in \{0, 1\} \qquad \forall ij \in E, \forall \delta \in D \ .$$

The binary slack variable $\bar{x}_{ij}$ for edge $ij \in E$ attains value 1 iff ij is not contained in any cut in the packing. This model (in packing form) was presented in [5]. We remark that it formally results from a Dantzig-Wolfe reformulation of our compact formulation by keeping constraint (7) in the master problem and reformulating constraints (1)–(6) into m identical subproblems which are then aggregated into a single subproblem, thereby completely eliminating the symmetry from the compact formulation.

3   Algorithmic Ingredients

In this section we describe the components of a full branch-price-and-cut algorithm to solve the extended formulation to integer optimality. As D grows exponentially, the linear programming (LP) relaxation of formulation (8) needs to be solved by column generation. That is, we include a small set of variables in a restricted master problem and price positive reduced cost variables using an auxiliary optimization problem, the pricing problem. If no more positive reduced variables can be found, the relaxation is proven to be optimal. Variables are also referred to as columns in this context. See [9] for a thorough introduction to the topic.

3.1   Solving the Pricing Problem

In order to find a variable/column/cut of positive reduced cost, or to conclude that none exists we seek a maximum reduced cost cut. This pricing problem amounts to solving $\max_{\delta \in D}(1 - \pi^T y^\delta) = 1 - \min_{\delta \in D}(\pi^T y^\delta)$, where $\pi = (\pi_{ij})_{ij \in E}$ denotes the current dual solution corresponding to equation (8) and

$y^\delta = (y_{ij}^\delta)_{ij \in E}$ is a binary vector indicating whether a given edge $ij \in E$ belongs to cut $\delta \in D$.

In the resulting *minimum* cut problem, the dual variables π_{ij} are free, we thus face a min-cut problem with arbitrary edge weights. We remark however, that all dual variables at the root node are non-negative as the variables $\bar{x}_{ij} \geq 0$ can be interpreted as slack variables. This enables us to benefit from the broad range of state-of-the-art minimum cut algorithms. We use the Stoer-Wagner algorithm introduced in [25] and then generalized to hypergraphs in [17]. In case some dual variables are negative (which happens because of branching, see below), we need to solve the pricing problem as a binary program, as this results in a *maximum* cut problem, which is NP-hard. The binary program arises from the compact formulation by leaving out constraints (7), omitting the index c, and replacing the objective function by the reduced cost function for the potential cut.

Desrochers et al. remark that pricing disjoint columns helps with arriving at integral solutions [8]. In order to find such a set of disjoint columns, we use a greedy heuristic during pricing, where we compute a cut δ_1 of maximum reduced cost, fix all variables corresponding to edges $ij \in \delta_1$ to 0 and resolve the pricing problem, creating a new cut δ_2, disjoint to δ_1. We again fix the edges from δ_2 and iterate until no cut δ_{k+1} can be found. The heuristic cut packing with objective function value k is $\{\delta_1, \ldots, \delta_k\}$.

As the pricing problem in the tree resembles a max-cut problem, we employ further heuristics to price out favorable columns based on max-cut heuristics [24] inspired by the formulation of the problem. Columns that are not immediately added to the restricted master problem were collected in a column pool which was searched for positive reduced cost columns before calling any pricing algorithm.

3.2 Branching

As the restricted master problem is the linear programming relaxation of the formulation, we need to use branching schemes in order to find optimal integral solutions. We employ two different branching rules: First we try to branch on original y_{ij} variables and if that is not possible, we will use Ryan-Foster branching [23] to branch on pairs of edges.

In the first case, given an LP solution x^* of the master problem, we can calculate the value of variables y_{ij} for each edge $ij \in E$ as $y_{ij} = \sum_{\delta:ij \in \delta} x_\delta^* = 1 - \bar{x}_{ij}^*$, we can create two branches by setting the slack variables $\bar{x}_{ij} = 1$ in one and $\bar{x}_{ij} = 0$ in the other branch enforcing that either no cut contains edge ij or exactly one cut going through this edge ij. The first case can be respected directly in the pricing by setting $y_{ij} = 0$ or by solving a combinatorial algorithm on a restricted graph where ij is contracted. The remaining dual variables stay nonnegative in this case which is beneficial since we can still use a combinatorial algorithm. In the second case, the dual variables corresponding to edge ij can be negative which we cannot respect in a min-cut algorithm. This justifies solving the pricing problem using a binary program instead of using a classical combinatorial min-cut algorithm throughout the branch-and-price tree because of mixed negative and positive edge weights.

In case $y_{ij} = 1 - \bar{x}_{ij}$ is integral for all edges $ij \in E$, we can branch on pairs of edges ij and kl, analogous to Ryan-Foster [23] branching. The branching decisions are: Either $\bar{x}_{ij} = \bar{x}_{kl}$, (this is called the *same* branch) or $\bar{x}_{ij} + \bar{x}_{kl} \geq 1$ (the *diff* branch). In the *same* branch, the two edges either must appear together in a cut or neither of the edges must be part of a cut. In the *diff* branch, not both edges are allowed to be in the same cut. These conditions can easily be respected in the binary program. Theoretically, we could respect these branching decisions in a combinatorial algorithm by constructing two reduced graphs for the *same* branching decision by contracting edges ik and jl (and il and jk, resp.). Unfortunately, this is only feasible for a limited number pairs and does not scale in the case of a large number of consecutive branching decisions.

Solving the pricing problem with an arbitrary number of *diff* constraints is an NP-hard problem on its own: Given a graph G, a weight function $c : E \to \mathbb{Q}^+$ and a set of edge pairs $P \subseteq E \times E$, we want to find a minimum cut with the constraint that at most one edge from each pair $p \in P$ is active in the cut. This follows by a reduction from MONOTONE 1-in-3 SAT [11], which is performed in the full version of this paper.

After enforcing the branching decisions, the set $\emptyset \neq S \subsetneq V$ to a cut $\delta(S)$ with positive reduced cost $1 - \sum_{ij \in \delta} \pi_{ij}$ might not be connected (if S is connected but $V \setminus S$ is not, we can exchange the roles of S and $V \setminus S$). While adding this cut to the restricted master problem can potentially slow down the solution process as it will not be part of an optimal solution, this causes no further harm. In some cases, a special handling of these cuts can improve the performance which we cover in detail in the full version of this paper.

3.3 Cutting Planes

The optimality gap between the LP relaxation of the extended formulation and an optimal integral solution can be arbitrarily bad: The cardinality of an optimal cut packing for a clique $K = (V_K, E_K)$ with n nodes is 1 whereas the optimal LP solution value will be $\frac{n}{2}$, with cuts separating each vertex from the remaining of the clique. The associated master variables all take the value $\frac{1}{2}$. To improve the LP bound, we thus separate clique inequalities using the maximum weighted clique heuristic based on an algorithm by Borndörfer and Kormos [3]. This is readily available in the solver framework we use, which is SCIP.

For the column generation algorithm, we need to respect the dual variables introduced when separating the cutting planes in the master problem. Given a graph $G = (V, E)$ and a set of cliques $\mathcal{K}$, we can construct a hypergraph $H = (V, F)$ where original edges are copied and all cliques are converted to hyperedges. Formally, $F = \{e \in E : e \notin E_K \,\forall K \in \mathcal{K}\} \cup \{V_K : K \in \mathcal{K}\}$ and the dual values from the clique cuts can be transferred to the hyperedges.

In the root node, we can use the hypergraph extension of the Stoer-Wagner algorithm to solve the minimal cut problem. In the binary programming pricing case, the transformation is straightforward by adding new variables and constraints to the pricing problem. This is particularly described in the full version of this paper.

3.4   Combinatorial Dual Bounds

In order to further strengthen the dual bound, we also consider combinatorial upper bounds. An *edge clique cover* Q is a set of cliques with $\bigcup_{K \in Q} E_K = E$. Finding an edge clique cover of minimum cardinality is NP-hard [18]. In an integral cut packing, at most one cut can be active per clique. Thus, the cut packing number $\gamma(G)$ is bounded from above by the number of cliques in an edge clique cover of minimum cardinality. We solve the edge clique covering problem approximately using the algorithm in [13] in order to obtain a valid upper bound. We calculate an edge clique cover at the beginning and add all induced clique cuts to the master problem before starting the column generation algorithm.

3.5   Primal Heuristics

Besides using the generic heuristics included in SCIP, we try to find good cut packings by using a (noncrossing) maximum s-t path cut packing heuristic [6] which we will call *maxpath*. Furthermore, we use an approximation algorithm for the independent set [16] for degree bounded graphs and transfer the resulting solution $I = \{v_1, \ldots, v_k\}$ to a solution $S = \{\delta(\{v_i\}) : v_i \in I\}$ for the cut packing problem by sorting the nodes according to non-decreasing degree and sequentially adding the next available node to the independent set. In order to assess the quality of the heuristic we calculate a maximum independent set exactly by solving the textbook formulation $\max\{\sum_{i \in V} x_i : x_i + x_j \leq 1 \,\forall ij \in E, \, x_i \in \{0,1\} \,\forall i \in V\}$.

4   Computational Setup and Results

We implemented the branch-price-and-cut algorithm in SCIP 3.0.1 [1] with CPLEX 12.4.0.1 as LP-solver. All computations were performed on Intel Core i7-2600 CPUs with 16GB of RAM on openSUSE 12.1 workstations running Linux kernel 3.1.10. The default time limit is 3600 seconds unless stated otherwise.

We applied our approach to instances from the 10th DIMACS implementation challenge [2]. We expect difficult graph partitioning problems to be hard for the cut packing problem, too, as in both settings the vertex set is partitioned in some way. In addition, we collected coloring instances from [26] and investigate the performance of our algorithm on these instances, because each color class in a coloring constitutes an independent set, to which a cut packing is intimately related.

We further generated smaller random graphs using several graph generators such as Rudy by Rinaldi [21], Randgraph by Pettie and Ramachandran [20] and NetworkX 1.7 by Hagberg et al. [15]. Moreover, we used the MUSKETEER graph generator [14] to obtain a set of graphs which are similar to the graphs from the literature. We used the random seed value 1 for the generation of random graphs unless stated otherwise. In total, our test set comprises around 100 instances.

4.1 Compact vs. Extended Formulation

In Fig. 1, we compare the integrality gaps and the running times of the compact and the extended formulation on a test set generated with the `Randgraph` graph generator. A generated graph *graph-n-d* consists of a random tree on n nodes plus $\max\{0, d\% \cdot \frac{n \cdot (n-1)}{2} - n + 1\}$ additional randomly selected edges. In Fig. 1, the *all* column comprises the implementation with all presented features, in *basic*, only the bare column generation implementation is visualized.

When looking at the time needed to solve the instances to optimality, the extended formulation outperforms the compact one on almost every instance. The difference in solution time increases further if all presented features are added. In particular, the compact formulation was only able to solve 2 out of 12 instances to optimality. The basic implementation of the reformulation however solved 8 out of 12 instances and with all features turned on, all instances can be solved to optimality within the time limit.

If we look at the integrality gaps at the root node, we see that the gap obtained by the extended formulation is remarkably better than the gap obtained by the compact formulation. The bound improvement translates to a better pruning of branch-and-price nodes and the removal of the symmetry by aggregation leads to fewer branching decisions, which explains the huge difference in solution time.

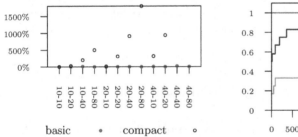 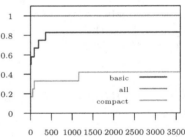

Fig. 1. Relative gap between primal and dual bounds at the root node for the compact and extended formulations for random graphs of type *graph-n-m* (left) and a performance profile for all the different formulations and settings (right) showing in how many instances (*y*-axis in percent) an algorithm is at most *x* times slower (factor on the *x*-axis).

4.2 Different Graph Classes

It is known that the cut packing problem can be solved in polynomial time on bipartite and chordal graphs. In order to investigate whether this complexity improvement translates to faster solution times also for our (exponential) algorithm, we evaluate the performance of the extended formulation on these instance types using random instances generated with `NetworkX` 1.7. We generated connected bipartite graphs and connected chordal graphs. The latter were generated from a random graph on n nodes, in which every edge occurs with probability $d\%$. To this graph, chords were iteratively added until the graph became chordal by using NetworkX to search for chordless cycles of size larger than 3. The chord was added randomly. We generated graphs with 10, 20, 40, 80, 160, and 320 nodes and edge densities of 10, 20, 40, and 80 percent. We only generated chordal graphs up to 160 nodes as using NetworkX to find all chordless cycles in larger graphs was too expensive. The results are summarized in Table 1.

Table 1. Number of solved instances and mean of solution times for the extended formulation for cut packing on random general, bipartite and chordal graphs. All reported times in s are shifted geometrical means.

Type	Nodes												mean time
	10		20		40		80		160		320		
	Solved	Time	Solved	Time	Solved	Time	Solved	Time	Solved	Time	Solved	Time	
random	4/4	0.0	4/4	0.0	4/4	25.2	0/4	3600	0/4	3600	0/4	3600	224.3
bipartite	4/4	0.0	4/4	0.0	4/4	0.1	4/4	1.0	4/4	20.8	4/4	527.8	13.9
chordal	4/4	0.0	4/4	0.0	4/4	0.1	4/4	1.3	4/4	9.3			1.7

We notice that the column generation procedure runs significantly faster on the graph classes where the problem is solvable in polynomial time. We are able to solve the cut packing problem to optimality in all generated bipartite and chordal graphs with up to 320 nodes, resp. 160 nodes, in contrast to general random graphs where we fail to solve instances larger than 40 nodes.

To evaluate the performance of our implementation on real-world data, we selected those real-world instances from the the 10th DIMACS implementation challenge with 500 nodes or less and all coloring instances from [26] with less than 5000 edges. We also tried to solve the problem on instances with hidden optimal solutions for classical graph problems [27], but we were not able to successfully solve the root node of any of the instances within the 1 hour time limit.

To compare the performance in these graphs to random graphs, we generated both a random graph and a similar graph for each real-world graph. These graphs have the same number of nodes and edges. The random graphs were generated with Randgraph, a connected graph $rand\text{-}n\text{-}m$ consists of n nodes and m edges. In order to create similar graphs, we used MUSKETEER where we edited 7.5% of the edges at level 2, 15% of the edges at level 1, and 30% of the level 0 of the coarsening phases. The new graphs, forced to be connected, are denoted by suffix $-m$. The results on these graphs are compared in Table 2.

We observe that our algorithm solved 5 out of 9 real-world instances to optimality, but only 1 out of 9 of the corresponding random graphs. In contrast, the graphs edited by MUSKETEER are more similar to the original graphs and indeed our algorithm performs better on those than on arbitrary random graphs.

Table 2. Branch-and-bound nodes and solution time (or relative gap achieved within time limit) for real-world DIMACS partitioning benchmark graphs

Name	Nodes	Time/Gap	Name	Nodes	Time/Gap	Name	Nodes	Time/Gap
karate	1	0.1	karate-m	1	0.1	rand-34-78	25	7.5
dolphins	51	317.4	dolphins-m	1	1.5	rand-62-159	>188	8.0%
lesmis	3	2.3	lesmis-m	8	21.4	rand-77-254	>77	19.2%
polbooks	4	95.9	polbooks-m	1	36.3	rand-105-441	>74	17.6%
adjnoun	>102	5.6%	adjnoun-m	>46	8.5%	rand-112-225	>304	5.2%
football	>29	9.5%	football-m	9	2327.9	rand-115-613	>57	32.3%
jazz	1	380.4	jazz-m	>1	27.25%	rand-198-2742	>1	102.9%
celegansneural	>1	4195.0%	celegansneural-m	>1	1161.3%	rand-297-2148	>1	3612.2%
celegans_meta	>1	866.7%	celegans_meta-m	>1	1646.8%	rand-453-2025	>1	4038.6%

Overall, it seems that our algorithm is effective on small real-world problems and on those graph classes where the problem is easy. On the other hand, random graphs seem to be a challenge, even when they are reasonably sparse (10% of overall edges).

4.3   Influence of Particular Implementation Parts

In addition to the basic column generation algorithm, we presented a few enhancements with the purpose of saving computation time. From our experiments, the clique cover, clique separator, and the combinatorial pricing algorithm seem to have the largest impact on the performance and we will restrict attention to those only. We plot performance profiles in Fig. 2.

It becomes evident that disabling the initial clique cover heuristic including the resulting clique inequalities causes the largest performance drop. The clique separator can partially catch this drop by separating the cliques during run time, but if both the clique separator and the clique cover heuristic are disabled, the performance decrease is significant. A similar improvement can be noticed if these features are added to a bare column generation procedure. In our case, the combinatorial pricing is a nice add-on but has, nonetheless not that much influence on the solution time as expected.

4.4   Relation to Independent Set

Motivated by the close connection to maximum independent set (MIS) we want to study the relationship between an MIS and an optimal cut packing on selected instances. For a given graph, we therefore compare the solution values of a greedy algorithm to approximate the cardinality of an MIS, the cardinality $\alpha(G)$ of an optimal MIS by using an exact integer program based algorithm to the values of the maximum s-t path heuristic and $\gamma(G)$. If an instance has not been solved to optimality, only the best known dual bound on the cut packing number $\gamma(G)$ is shown. This can be seen in Fig. 3.

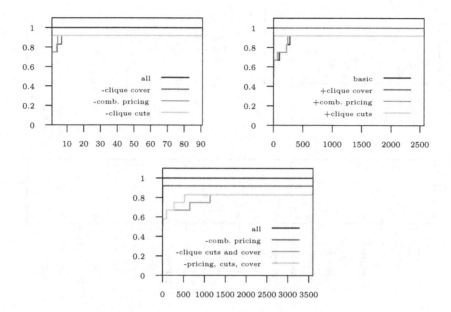

Fig. 2. Performance profiles for clique cover, clique separator, and combinatorial algorithm showing in how many instances (y-axis in percent) a given setting is at most x times slower (factor on the x-axis). The plot on the upper left states the influence of disabling each part of the algorithm, on the upper right the influence of enabling each part on a basic implementation is shown. The plot on the bottom presents the performance of a concurrent deactivation.

We notice that the value of the solution found by the s-t maximal path heuristic is often much worse than any of the other algorithms. The solution found by the MIS heuristic is better but is usually much worse than the optimal solution. In contrast, $\alpha(G)$ is mostly identical to $\gamma(G)$. This comes as a surprise as there is a theoretical gap of a factor of 2 that we do not observe in the instances in our experiments. We are aware of the fact that the gap will be closer to 2 when considering sparser graphs and in particular paths and trees.

5   Summary and Conclusions

We presented an exact algorithm for the cut packing problem and assessed its performance on about 100 random and benchmark graphs from the literature. The current size limits of our approach are up to 80 vertices on random graphs, up to 500 vertices on graph partitioning benchmark graphs, and up to 5000 edges on vertex coloring benchmark graphs. Our algorithm performs much better on instances where the cut packing problem is solvable in polynomial time. Our experiments revealed that $\gamma(G)$ is typically much closer (and often even identical) to $\alpha(G)$ than to the theoretically possible $2\alpha(G) - 1$.

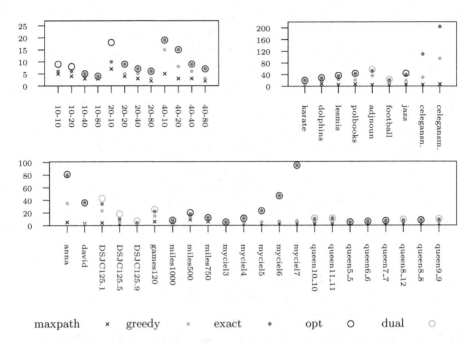

maxpath ×   greedy •   exact •   opt ○   dual ⊙

Fig. 3. Objective function values of max path heuristic, greedy and exact MIS, optimal cut packing or dual bound for random graphs (top left), DIMACS partitioning benchmark graphs (top right), and coloring graphs (bottom)

We have made algorithmic use of the combinatorial structure of the problem: we used min-cut algorithms for the pricing problem; we computed a combinatorial dual bound for tightening the relaxation; and we have exploited the problem's close relation to the independent set problem e.g., for designing primal heuristics. Future computational research on the problem should clarify whether it pays to work on (integer programming formulations for) better clique edge covers for stronger dual bounds; and whether different formulations for min-cut can speed up solving the pricing problem. In addition, one should fully exploit the known branch-and-price machinery like dual variable stabilization.

References

1. Achterberg, T.: SCIP: solving constraint integer programs. Math. Programming Comp. 1(1), 1–41 (2009)
2. Bader, D.A., Meyerhenke, H., Sanders, P., Wagner, D. (eds.): Graph Partitioning and Graph Clustering. 10th DIMACS Implementation Challenge Workshop, February 13-14, 2012. Contemp. Mathematics, vol. 588. American Mathematical Society (2013)
3. Borndörfer, R., Kormos, Z.: An algorithm for maximum cliques. unpublished working paper, Konrad-Zuse-Zentrum für Informationstechnik Berlin (1997)

4. Caprara, A., Panconesi, A., Rizzi, R.: Packing cycles in undirected graphs. J. Algorithms 48, 239–256 (2003)
5. Caprara, A., Panconesi, A., Rizzi, R.: Packing cuts in undirected graphs. Networks 44(1), 1–11 (2004)
6. Colbourn, C.J.: The Combinatorics of Network Reliability. Oxford University Press, New York (1987)
7. Colbourn, C.: Edge-packing of graphs and network reliability. Discrete Math 72(1-3), 49–61 (1988)
8. Desrochers, M., Desrosiers, J., Solomon, M.: A new optimization algorithm for the vehicle routing problem with time windows. Oper. Res. 40(2), 342–354 (1992)
9. Desrosiers, J., Lübbecke, M.: Branch-price-and-cut algorithms. In: Cochran, J. (ed.) Encyclopedia of Operations Research and Management Science. John Wiley & Sons, Chichester (2011)
10. Downey, R., Fellows, M.: Fixed-parameter tractability and completeness II: On completeness for W(1). Theoretical Computer Science 141(12), 109–131 (1995), http://www.sciencedirect.com/science/article/pii/0304397594000973
11. Fox-Epstein, E.: Forbidden Pairs Make Problems Hard. Bachelor's thesis. Wesleyan University (2011)
12. Fulkerson, D.: Blocking and anti-blocking pairs of polyhedra. Math. Programming 1, 168–194 (1971)
13. Gramm, J., Guo, J., Hüffner, F., Niedermeier, R.: Data reduction, exact, and heuristic algorithms for clique cover. In: Proc. 8th ALENEX, pp. 86–94 (2006)
14. Gutfraind, A., Meyers, L.A., Safro, I.: Multiscale network generation, arXiv:1207.4266 (2012)
15. Hagberg, A.A., Schult, D.A., Swart, P.J.: Exploring network structure, dynamics, and function using NetworkX. In: Proc. of the 7th Python in Science Conference (SciPy 2008), Pasadena, pp. 11–15 (2008)
16. Halldórsson, M.M., Radhakrishnan, J.: Greed is good: Approximating independent sets in sparse and bounded-degree graphs. Algorithmica 18(1), 145–163 (1997)
17. Klimmek, R., Wagner, F.: A simple hypergraph min cut algorithm. Tech. Rep. B 96-02. FU Berlin (1996)
18. Kou, L.T., Stockmeyer, L.J., Wong, C.K.: Covering edges by cliques with regard to keyword conflicts and intersection graphs. Commun. ACM 21(2), 135–139 (1978)
19. Lucchesi, C., Younger, D.: A minimax theorem for directed graphs. J. Lond. Math. Soc. 17, 369–374 (1978)
20. Pettie, S., Ramachandran, V.: Randgraph graph generator (2006), http://www.dis.uniroma1.it/challenge9/download.shtml
21. Rinaldi, G.: Rudy, a graph generator (1998), http://www-user.tu-chemnitz.de/~helmberg/sdp_software.html
22. Robacker, J.T.: Min-max theorems on shortest chains and disjunct cuts of a network. Tech. Rep. RM-1660-PR. Rand Corporation (1956)
23. Ryan, D.M., Foster, B.A.: An integer programming approach to scheduling. Opt. Res. Q. 27(2), 367–384 (1976)
24. Sahni, S., Gonzalez, T.: P-complete approximation problems. J. ACM 23(3), 555–565 (1976)
25. Stoer, M., Wagner, F.: A simple min-cut algorithm. J. ACM 44(4), 585–591 (1997)
26. Trick, M.: Coloring instances (1993), http://mat.gsia.cmu.edu/COLOR/instances.html
27. Xu, K.: Bhoslib: Benchmarks with hidden optimum solutions for graph problems (2010), http://www.nlsde.buaa.edu.cn/~kexu/benchmarks/graph-benchmarks.htm

Experimental Evaluation of a Branch and Bound Algorithm for Computing Pathwidth[*]

David Coudert[1,2], Dorian Mazauric[3], and Nicolas Nisse[1,2]

[1] Inria, France
[2] Univ. Nice Sophia Antipolis, CNRS, I3S, UMR 7271, 06900 Sophia Antipolis, France
[3] Aix-Marseille Université, CNRS, LIF UMR 7279, Marseille, France

Abstract. *Path-decompositions* of graphs are an important ingredient of dynamic programming algorithms for solving efficiently many NP-hard problems. Therefore, computing the pathwidth and associated path-decomposition of graphs has both a theoretical and practical interest. In this paper, we design a Branch and Bound algorithm that computes the exact pathwidth of graphs and a corresponding path-decomposition. Our main contribution consists of several non-trivial techniques to reduce the size of the input graph (pre-processing) and to cut the exploration space during the search phase of the algorithm. We evaluate experimentally our algorithm by comparing it to existing algorithms of the literature. It appears from the simulations that our algorithm offers a significative gain with respect to previous work. In particular, it is able to compute the exact pathwidth of any graph with less than 60 nodes in a reasonable running-time ($\leq$ 10 min.). Moreover, our algorithm achieves good performance when used as a heuristic (i.e., when returning best result found within bounded time-limit). Our algorithm is not restricted to undirected graphs since it actually computes the *vertex-separation* of digraphs (which coincides with the pathwidth in case of undirected graphs).

1 Introduction

Because of their well known algorithmic interest, a lot of work has been devoted to the computation of *treewidth* and *tree-decompositions* of graphs [8]. On the theoretical side, exact exponential algorithms [2], Fixed Parameter Tractable (FPT) algorithms [4] and approximation algorithms have been designed. Unfortunately, most of these algorithms are impractical. For instance, the algorithm in [2] is both exponential in time and space and therefore cannot be used for graphs with more than 32 nodes. Another example is the FPT algorithms whose time-complexity is at least exponential in the treewidth: there exists no efficient implementation of the algorithm in [4] even for graphs with treewidth at most 4. On the positive side, efficient algorithms exist for computing the treewidth of particular graph classes,

---

[*] This work has been partially supported by European Project FP7 EULER, ANR project Stint (ANR-13-BS02-0007), the associated Inria team AlDyNet and the project ECOS-Sud Chile.

J. Gudmundsson and J. Katajainen (Eds.): SEA 2014, LNCS 8504, pp. 46–58, 2014.

e.g., graphs with treewidth at most 4 [10]. Many heuristics for computing lower or upper bounds on the treewidth have also been designed [5].

Surprisingly, much less work has been devoted to the computation of *pathwidth* and *path-decompositions* of graphs [8]. Indeed, the pathwidth of any n-node graph is at most $O(\log n)$ times its treewidth. Hence, providing an efficient algorithm to compute pathwidth immediately leads to an efficient approximation algorithm for computing the treewidth. In this paper, we design an algorithm that computes the pathwidth and a corresponding path-decomposition of graphs. We then provide an experimental analysis of our algorithm that presents a significative improvement with respect to existing algorithms.

1.1 Practical Computation of Pathwidth

Any *path-decomposition* of a n-node graph $G = (V, E)$ corresponds to a *layout* (i.e., an ordering) $(v_1, \cdots, v_n)$ of its vertices. The *width* of a path-decomposition is the maximum size of the *border*[1] of the subgraphs induced by $\{v_1, \cdots, v_i\}$, $i \leq n$. The *pathwidth* of a graph is equal to the minimum width of its path-decompositions. A motivation for computing path-decompositions comes from algorithmic applications. Indeed, given a path-decomposition of a graph, dynamic programming algorithms (whose time-complexity depends exponentially on the width of the path-decomposition) can be designed for many combinatorial optimization problems. Other motivations arise from the close relationship between pathwidth and other graph invariants (related to node layouts), e.g., the search numbers [8] and the process number [6]. Computing the pathwidth of graphs is NP-hard [8] in gereral. An FPT algorithm has been proposed in [4] and some kernelization reduction rules are provided in [3]. On the practical side, to the best of our knowledge, very few implementations of algorithms for computing the pathwidth (exact or bounds) have been proposed.

Exact Algorithms. Solano and Pioro proposed in [14] an exact Branch and Bound algorithm to compute the pathwidth of graphs, that checks all possible layouts of the nodes and keeps the best one. A mixed integer linear programming formulation (MILP) has been proposed in [1, 14] and a dynamic programming algorithm (exponential both in time and space) is described in [2]. None of these methods handle graphs with more than 30 nodes. As far as we know, the best practical exact algorithm for computing the pathwidth of graphs with more than 30 nodes is based on a SAT formulation of the problem that is solved using Constraint Programming solver [1].

Heuristics. A polynomial-time heuristic for vertex-separation is proposed in [6]. It aims at computing a layout of the nodes in a greedy way . At each step, the next node of the layout is chosen using a *flow circulation* method. A heuristic based on a Branch and Bound algorithm has been designed in [14]. Recently, a heuristic has been proposed that is based on the combination of a **Shake** function

---

[1] For any graph $G = (V, E)$ and $X \subseteq V$, the border of X is the set of nodes in X that have a neighbor in $V \setminus X$.

and `Local_Search` [9]: it consists in sequentially improving a layout by switching the nodes until a local optimum is achieved.

1.2   Contributions and Organization of the Paper

We design a Branch and Bound algorithm that computes the exact pathwidth of graphs and a corresponding path-decomposition. Basically, our algorithm explores the set of all possible layouts of the vertex-set, returning a layout with smallest width. Our main contribution consists of several non-trivial techniques to reduce the size of the input graph (pre-processing) and to cut the exploration space during the search phase of the algorithm. Note that, our algorithm is not restricted to undirected graphs since it actually computes the *vertex-separation* of digraphs (which coincides with the pathwidth in case of undirected graphs). In Section 2, we prove several technical lemmas that allow us to prove the correctness of the pre-processing phase (Section 2.2) and of the pruning procedures (Section 2.3). We present and prove our algorithm in Section 3. Finally, in Section 4, we evaluate experimentally our algorithm by comparing it to existing algorithms of the literature. It appears from the simulations that our algorithm offers a significative gain with respect to previous work. It is able to compute the exact pathwidth of any graph with less than 60 nodes in a reasonable running-time (≤ 10 min.). Moreover, our algorithm achieves good performance when used as a heuristic (i.e., when returning best result found within bounded time-limit).

2   Preliminaries

In this section, we formally define the notion of vertex-separation. Then, we give some technical lemmas that are used to prove the correctness of our algorithm.

2.1   Definitions and Notations

All graphs and digraphs considered in this paper are connected and loopless. Let $G = (V, E)$ be a graph. For any set $S \subseteq V$, let $N_G(S)$ be the set of nodes in $V \setminus S$ that have a neighbor in S. Similarly, for any digraph $D = (V, A)$ and any subset $S \subseteq V$, let $N_D^+(S)$ (resp., $N_D^-(S)$) be the set of nodes in $V \setminus S$ that have an in-neighbor (resp., an out-neighbor) in S. We omit the subscript when there is no ambiguity. For any digraph $D = (V, A)$ and $a = uv \in A$, let D/a be the graph obtained from D by contracting a. Let $D \setminus u$ be the graph obtained from D by removing u and its incident arcs.

Layouts and Vertex-Separation. Given a set S, a *layout* of S is any ordering of the elements of S. Let $\mathcal{L}(S)$ denote the set of all layouts of S. Let $S' \subseteq S$ and P, Q be two layouts of S' and $S \setminus S'$ respectively. Let $P \odot Q$ be the layout of S obtained by concatenating P and Q. Moreover, let $\mathcal{L}_P(S) = \{L \in \mathcal{L}(S) \mid L = P \odot Q, Q \in \mathcal{L}(S \setminus S')\}$, i.e., $\mathcal{L}_P(S)$ is the set of all layouts of S with prefix P.

Let $D = (V, A)$ be a digraph and let $L = (v_1, \cdots, v_{|S|}) \in \mathcal{L}(S)$ be a layout of $S \subseteq V$. For any $1 \leq i \leq |S|$, let $\nu(L, i) = |N_D^+(\{v_1, \cdots, v_i\})|$ and $\nu(L) =$

$\max_{i \leq |S|} \nu(L, i)$. The *vertex-separation* of D, denoted by $vs(D)$, is equal to the minimum $\nu(L)$ among all layouts L of V, i.e., $vs(D) = \min_{L \in \mathcal{L}(V)} \nu(L)$.

A digraph $D = (V, A)$ is *symmetric* if for each arc $uv \in A$, there is $vu \in A$. For any undirected graph G, let $vs(G)$ denote the vertex-separation of the corresponding symmetric digraph obtained from G by replacing each edge $\{u, v\}$ by two arcs uv and vu. It is well known that the vertex-separation of a undirected graph G equals its pathwidth $pw(G)$ [8].

2.2 Technical Lemmas for Preprocessing

In this section, we prove few technical lemmas that will be useful to prove the correctness of the preprocessing part of our algorithm. Due to lack of space, the proofs are omitted and can be found in [7]. Note that Lemma 2 extends to digraphs a known result in undirected graphs.

Lemma 1. *[Folklore] Let D be any connected digraph and let $SCC(D)$ be the set of strongly connected components of D. Then, $vs(D) = \max_{D' \in SCC(D)} vs(D')$.*

We show that we can contract some well chosen arcs of a digraph without modifying its vertex-separation. It is well known that, in undirected graphs, edge-contraction cannot increase the pathwidth. However, this is not true anymore in digraphs as shown in the following example. Let $D = (\{a, b, c, d\}, \{ab, cb, cd, da\})$. D is acyclic and therefore, $vs(D) = 0$. On the other hand, D/cb is a directed cycle with 3 nodes and $vs(D/cb) = 1$. We also show that under some conditions, the vertex-separation does not decrease after arc-contraction. Altogether, we get:

Theorem 1. *Let $D = (V, A)$ be a n-node digraph and $uv \in A$ such that, $vu \notin A$, and either $(N^-(v) = \{u\}$ and $N^+(u) \cap N^+(v) = \emptyset)$ or $(N^+(u) = \{v\}$ and $N^-(u) \cap N^-(v) = \emptyset)$. Then, $vs(D) = vs(D/uv)$. Moreover, an optimal layout for D can be obtained in linear time from an optimal layout of D/uv.*

Lemma 2. *Let $D = (V, A)$ be a n-node digraph and let $a, b, c \in V$ be three nodes with $N_D^+(b) = N_D^-(b) = \{a, c\}$, $N_D^+(a) = N_D^-(a) = \{b, x\}$, $N_D^+(c) = N_D^-(c) = \{b, y\}$, and $x \neq c$. Then, $vs(D) = vs(D/bc)$.*

2.3 Technical Lemmas for the *Pruning* Part

In this section, we prove few technical lemmas that will be useful to prove the correctness of the *Pruning* part of our *Branch & Bound* algorithm.

When looking for a good layout of the nodes of a digraph, our algorithm will, at some step, consider a layout P of some subset $S \subset V$ and look for the best layout L of V starting with P. Next lemma gives some conditions on a node $v \in V \setminus S$ to ensure that the best solution starting with $P \odot v$ is as good as L.

Lemma 3. *Let $D = (V, A)$ be a n-node digraph, $S \subset V$, and $P \in \mathcal{L}(S)$. If there exists $v \in V \setminus S$ such that either $N^+(v) \subseteq (S \cup N^+(S))$, or $v \in N^+(S)$ and $N^+(v) \setminus (S \cup N^+(S)) = \{w\}$. Then, $\min_{L \in \mathcal{L}_P(V)} \nu(L) = \min_{L \in \mathcal{L}_{P \odot \{v\}}(V)} \nu(L)$.*

Let L be the best layout of V having a prefix P. We express some conditions under which no layout of V starting with a permutation of P is better than L.

Lemma 4. *Let* $D = (V, A)$ *be a* n-*node digraph,* $S \subset V$ *and let* $P, P' \in \mathcal{L}(S)$ *be two layouts of* S. *If* $\nu(P) < \min_{L \in \mathcal{L}_P(V)} \nu(L)$ *or* $\nu(P) \leq \nu(P')$, *then* $\min_{L \in \mathcal{L}_P(V)} \nu(L) \leq \min_{L \in \mathcal{L}_{P'}(V)} \nu(L)$.

3   The Algorithm

In this section, we present our exact exponential-time algorithm for computing $vs(D)$ for any n-node digraph $D = (V, A)$. To ease the presentation, we assume that D is strongly connected. If it is not the case, we apply all the steps of our algorithm to each of its strongly connected components, and then use the construction presented in the proof of Lemma 1 to deduce in linear time the layout L of D such that $\nu(L) = vs(D)$.

Our algorithm is based on a *Branch & Bound* procedure that considers all possible layouts of V and keeps a layout L minimizing $\nu(L)$. Since the worst case time-complexity of such an algorithm is $O(n!)$, we first do a pre-processing to reduce the size of the input.

3.1   Pre-processing Phase

The following reduction rules are applied while it is possible. The rules that have been proposed for undirected graphs can be used only when D is a symmetric digraph. Note that, the digraph D^* obtained after applying once one of these rules is strongly connected, and that given the layout L^* such that $\nu(L^*) = vs(D^*)$, one can deduce in linear time a layout L for D such that $\nu(L) = vs(D)$.

Rule A: If there is $uv \in A$ such that, $vu \notin A$, and either $(N^-(v) = \{u\}$ and $N^+(u) \cap N^+(v) = \emptyset)$, or $(N^+(u) = \{v\}$ and $N^-(u) \cap N^-(v) = \emptyset)$, then let $D^* = D/uv$ (By Theorem 1, $vs(D^*) = vs(D)$).

Rule B: If there are three nodes $a, b, c \in V$ with $N_D^+(b) = N_D^-(b) = \{a, c\}$, $N_D^+(a) = N_D^-(a) = \{b, x\}$, $N_D^+(c) = N_D^-(c) = \{b, y\}$, and $x \neq c$, then let $D^* = D/bc$ (By Lemma 2, $vs(D^*) = vs(D)$).

Rule C: If D is symmetric, let G be the underlying undirected graph. It is shown in [3] that $vs(D^*) = vs(D)$ when D^* is obtained as follows.

    C.1: If (in G) two degree-one vertices u and v share their neighbor, then $D^* = D \setminus u$.

    C.2: If (in G) v, w are two vertices of degree two, and suppose x and y are the common neighbors of v and w, then $D^* = D/vx$.

Note that, each time that one of the above rules is applied, the number of nodes decreases by one and therefore, the worst case time-complexity is divided by n. When no reduction rules can be applied anymore, the algorithm $B\&B$ is applied as explained below.

3.2   Branch & Bound Phase

We now describe our *Branch & Bound*, that we call $B\&B$. The main contribution of our work consists of the way we cut the exploration of all layouts of V. Intuitively, let $S \subset V$ and P be a layout of S that Procedure $B\&B$ is testing, i.e., let us consider a step when $B\&B$ is considering all layouts of V with prefix P. Moreover, let L_{UB} be the best layout of D obtained so far by $B\&B$. We use the two following pruning rules:

1. Decide, using Lemma 4, if it is useful to explore layouts with prefix P. To this end, we maintain a table $\mathcal{P}$ that contains, among others, some subsets of V. If there is an entry in $\mathcal{P}$ for $S \subset V$ and satisfying some properties, we can decide that the best layout starting with the nodes in S cannot be better than L_{UB} and therefore, we do not test further layouts with prefix P.
2. Greedily extend the current prefix P with a vertex $v \in V \setminus S$, with $S = V(P)$, if v satisfies the conditions of Lemma 3. This allows us to restrict the exploration of all the layouts with prefix P to the layouts with prefix $P \odot v$.

As shown in the next section, these two pruning rules allow our algorithm to achieve better performances than existing ones for computing the vertex-separation of digraphs. Let us now describe the recursive procedure $B\&B$ (Algorithm 3 in [7]) more formally and prove its correctness. It takes as inputs:

- $D = (V, A)$, the considered digraph.
- L_{UB} and UB such that L_{UB} is the best layout of V obtained so far and $UB = \nu(L_{UB})$. Note that UB is an upper bound for $vs(D)$.
- A layout P of a subset $S \subseteq V$. Intuitively, P is the prefix of the layouts that will be tested by this execution of $B\&B$. That is, either Procedure $B\&B$ will find a layout $L = P \odot Q$ of V such that $\nu(L) < \nu(L_{UB})$ or it decides that $\nu(L) \geq \nu(L_{UB})$ for any $L \in \mathcal{L}_P(V)$.
- $\mathcal{P}$ is a set of triples $(S_i, \nu_i, b_i)_{i \leq \rho}$, where $S_i \subset V$, ν_i is an integer, and b_i is a boolean for any $i \leq \rho$. Intuitively, $(S_i, \nu_i, b_i) \in \mathcal{P}$ means that a layout P_i of S_i has already been checked, and, if $b_i = 1$, then it is useless to test any other layout of V starting with the nodes in S_i. Moreover, $\nu_i = \nu(P_i)$ and $b_i = 0$ if $\nu_i = \min_{L \in \mathcal{L}_{P_i}(V)} \nu(L)$.

The initial values for the inputs of $B\&B$ are: $P = \emptyset$, L_{UB} is any layout L of V and $UB = \nu(L) < |V|$, and $\mathcal{P} = \emptyset$. Let $\mathcal{P} = (S_i, \nu_i, b_i)_{i \leq \rho}$ and (L_{UB}, UB) be the current values of the global variables at some step of the execution of the algorithm. Then, executing $B\&B(D, P, UB, L_{UB}, \mathcal{P})$, with $P = (v_1, \cdots, v_k)$ being a layout of $S = \{v_1, \cdots, v_k\}$, proceeds as follows.

- if $\nu(P) \geq UB$, then $B\&B(D, P, UB, L_{UB}, \mathcal{P})$ does nothing, i.e., the exploration of the layouts starting by P stops. Indeed, $\min_{L \in \mathcal{L}_P(V)} \nu(L) \geq \nu(P)$ by definition. Therefore, $\min_{L \in \mathcal{L}_P(V)} \nu(L) \geq UB$ and no layout of V starting with P can be better than L_{UB}.

– else, if (S, x, b) belongs to $\mathcal{P}$, then this means that a layout P' of S has been already tested, $x = \nu(P')$ and $UB \leq \min_{L \in \mathcal{L}_{P'}(V)} \nu(L)$. If $b = 1$ or $\nu(P) \geq \nu(P')$, then $B\&B(D, P, UB, L_{UB}, \mathcal{P})$ does nothing, i.e., the exploration of the layouts starting by P stops. Indeed, either $x < \min_{L \in \mathcal{L}_{P'}(V)} \nu(L)$ (if $b = 1$) or $\nu(P) \geq \nu(P')$. By Lemma 4, $\min_{L \in \mathcal{L}_{P'}(V)} \nu(L) \leq \min_{L \in \mathcal{L}_P(V)} \nu(L)$. Again, no layout of V starting with P can be better than L_{UB}.

– otherwise, $B\&B(D, P, UB, L_{UB}, \mathcal{P})$ applies the sub-procedure $Greedy(D, P)$ (Algorithm 1 in [7]) that returns a prefix P' extending P (i.e., P is a prefix of P') and $\min_{L \in \mathcal{L}_P(V)} \nu(L) = \min_{L \in \mathcal{L}_{P'}(V)} \nu(L)$. Then, it calls $B\&B(D, P' \odot v, UB, L_{UB}, \mathcal{P})$ for all $v \in V \setminus V(P')$ such that $\nu(P' \odot v) < UB$, starting from the most promising vertices (i.e., by increasing value of $\nu(P' \odot v)$).

At the end of the exploration of the layouts with prefix P, we update the table $\mathcal{P}$ (Algorithm 2 in [7]). That is,

• if there was no triple $(S, *, *)$ in $\mathcal{P}$, then, $(S, \nu(P), b)$ is added to $\mathcal{P}$ with $b = 0$ if and only if $\nu(P) = \nu(L_{UB})$.

• else, i.e., if there is $(S, x, 0) \in \mathcal{P}$ with $\nu(P) < x$ and $\nu(P) < UB$, then $(S, \nu(P), b)$ replaces $(S, x, 0)$ in $\mathcal{P}$ with $b = 0$ iff $\nu(P) = \nu(L_{UB})$.

4   Simulations and Interpretation of Results

In this section, we evaluate the performance of our algorithm. In particular, we compare it to other exact algorithms and heuristics of the literature. We also analyze the impact of the three optimization phases of our algorithm : pre-processing, greedy steps, and pruning using prefixes. It appears that our algorithm is able to compute the vertex-separation of all graphs with at most 60 nodes but also for some graphs up to 250 nodes. Not only does our algorithm outperform the performance of existing exact algorithms but it can also be used as a good heuristic to obtain upper bounds on the vertex-separation.

4.1   Implementation

We have implemented several variants of our Branch and Bound algorithm in order to analyze the impact of each of our optimization sub-procedures. Briefly, **BAB** is the basic version of the Branch and Bound algorithm (as already implemented in [14]). It takes as an input a digraph D and is applied sequentially on each of the strongly connected components of D. Then, in **BAB-G** we add the greedy-step process (Algorithm 1 in [7]) to **BAB**, in **BAB-GP** we add the pre-processing phase to **BAB-G**, and finally, our main algorithm is **BAB-GPP** in which we add to **BAB-GP** the pruning-process by storing some subsets of nodes a layout of which has already been tested as prefix (Algorithm 2 in [7]).

Our algorithms are implemented in Cython using the *Sage* open-source mathematical software [15]. To speed-up critical operations on neighborhoods (union, intersection, size, etc.), we store out-neighborhoods using bitsets (in particular, a single integer when $n \leq 64$). This enables us to use bitwise operations (OR for union, AND for intersection, etc.). For **BAB-GPP**, the prefixes are stored in a tree structure offering fast access to already tested prefixes. We parameterized the maximum length of a prefix (which corresponds to the tree depth) and set it by default to $\min\{n/3, 50\}$. The total number of stored prefixes is also parameterized to limit memory usage. By default we store at most 10^6 prefixes. Finally, we used a timer to limit computation time per (di)graph. When the time is up, we return the best solution found so far and record it as an upper-bound on $vs(D)$. Following [1], we set this limit to 10 min. All computations have been done on a computer equipped with a Intel Xeon CPU, 3.20GHz, 64GB of RAM.

Previous exact algorithms. In addition to the basic Branch and Bound **BAB** of Solano [14], we are aware of the following implementations of algorithms to compute vertex-separation of graphs.

MILP. Some mixed integer linear programming formulations for the vertex-separation (MILP) have been proposed in [1,14]. A similar formulation has been implemented in $Sage^2$ and it uses the CPLEX solver. We use this function for purpose of comparison.

DYNPROG. This is a dynamic programming algorithm that computes the vertex-separation of a digraph D [2]. An implementation of DYNPROG is available in $Sage^2$. Since the worst case time and space complexities of DYNPROG is $O(2^n)$, it can be used only for graphs with strictly less than 32 nodes.

SAT. In [1], an algorithm to compute the pathwidth of graphs is designed by formulating the PATHWIDTH problem as a Boolean satisfiability testing (SAT) instance and solving this SAT instance using the $MiniSat$ solver. No implementations are provided but the algorithm has been tested on the *Rome graphs* dataset [12]. The SAT algorithm have been able to compute the pathwidth of 17% of the instances in the *Rome graphs* dataset.

4.2   Evaluation in Random Digraphs

In this section, we use the available implementations of MILP and DYN-PROG in *Sage* to compare their performance with our algorithms in random directed graphs. Random directed graphs with N nodes are generated using the `graphs.RandomDirectedGNM(N,M)` method of *Sage* to generate them, where $M = $ density $* N(N - 1)$ is the number of arcs. For several network sizes ($N \in \{20, 25, 30, 40, 50\}$), we execute the algorithms for various densities ($0 < density < 1$). For any N and each density, we run the algorithms on 1 000 instances. The same instances are used for all algorithms. The average running times are depicted on Fig. 1.

Small Graphs ($N = 20$). Fig. 1(a) shows that the MILP Algorithm is outperformed by all other algorithms even for $N = 20$ (two weeks of computations have been required to generate Fig. 1(a) because of MILP). Fig. 1(a) also shows that, for $N = 20$, Algorithm DYNPROG is slightly faster than all variants of

---

^{2} Module `sage.graphs.graph_decompositions.vertex_separation`

BAB. In particular, the running time of **BAB-GPP** is a bit larger than for other variants. This is probably due to the fact that the cuts on the exploration space done by the optimization procedures executed by **BAB-GPP** are negligible compared with the time needed to apply the optimization procedures.

Graphs with $N \in \{25, 30\}$. Figs. 1(b) and 1(c) show that, for graphs with at least 25 nodes, our algorithms become competitive with respect to DYNPROG. In particular, for graphs with 30 nodes, our algorithms are significantly faster (recall that DYNPROG can only be used for graphs with at most 31 nodes).

For $N \geq 30$, the basic variant of **BAB** behaves as well as **BAB-G** for *density* ≥ 0.5, but for smaller densities the computation time is not competitive (several hours). Therefore, we did not report on its performances on the plots.

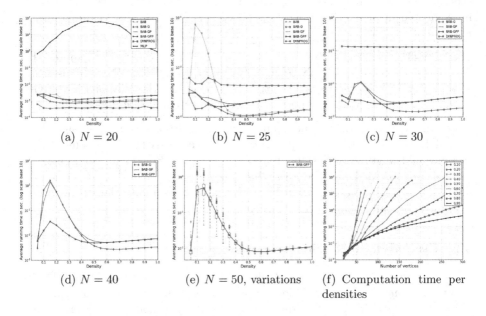

(a) $N = 20$ (b) $N = 25$ (c) $N = 30$

(d) $N = 40$ (e) $N = 50$, variations   (f) Computation time per densities

Fig. 1. Average running time of the algorithms on random digraphs

Impact of Optimization Phases. In Figs. 1(b) to 1(e), we observe that for densities ≥ 0.4, all variants of the Branch and Bound algorithm are very fast with negligeable differences (≤ 0.01 sec.). However, we observe large variations for smaller densities. More precisely, we observe in Fig. 1(b) the significant speed-up offered by the Greedy steps w.r.t. the basic Branch and Bound proposed in [14]. The benefits of the pre-processing phases are however difficult to observe on such small random digraphs since the conditions required to contract arcs rarely occur, and so we observe an improvement in Fig. 1(d) only when the density is 0.05. However, Fig. 1(d) reports large speed-up when using prefixes to cut the search space. In Fig. 1(e), we point out that the running time of **BAB-GPP** varies a lot depending on the graphs (especially for graphs with small density). We report on the variations in running time when $N = 50$ and using the best settings of our algorithm. For densities below than 0.4, the running time varies by up to two orders of magnitude, while for larger densities, the range of variations is very small. Last, we have reported in Fig. 1(f) the evolution of the running time of **BAB-GPP** for different densities. This confirms that the higher the density, the larger the size of the graphs we are able to solve. Note that we have observed the same behaviors for all algorithms with undirected graphs.

4.3   Comparison of BAB-GPP with SAT on *Rome Graphs*

The *Rome graphs* dataset [12] consists of 11 529 undirected n-node graphs with $10 \le n \le 100$. In [1], the performance of the SAT algorithm has been evaluated using this benchmark[3]. With 10 min. time limit per graph, the algorithm of [1] has been able to compute the pathwidth for 17.0% of the Rome graphs. In particular, it is stated that "We note that almost all small graphs ($n + m <$ 45) could be solved within the given timeout, however, for larger graphs, the percentage of solved instances rapidly drops [. . .] Almost no graphs with $n+m >$ 70 were solved." [1] (where m is the number of edges of the considered graphs).

In contrast, our algorithm has computed the pathwidth of 95.6% of the graphs in the *Rome dataset*, with same time limit of 10 min. Note that, in particular, we solved all instances with at most 82 vertices. We report in Tab. 1 the repartition of (un)solved instances.

Table 1. Repartition of (un)solved instances

N	≤ 82	83	84	85	86	87	88	89	90	91	92	93	94	95	96	97	98	99	100	Total
Nb graphs	9586	86	78	73	70	68	69	63	59	116	119	134	154	139	148	139	143	144	141	11 529
Solved	9586	85	77	68	65	62	65	59	50	96	91	104	113	91	110	80	77	70	78	11 027
Unsolved	0	1	1	5	5	6	4	4	9	20	28	30	41	48	38	59	66	74	63	502

k	1	2	3	4	5	6	7	8
$|V(M_k)|$	1	2	5	11	23	47	95	191
$pw(M_k)$	0	1	2	5	10	20	≤ 38	≤ 72

Max. prefix length	Time (in sec.)	Visited nodes	Stored prefix
1	2226.45	1 256 780 074	44
5	144.93	82 417 700	16 386
8	5.81	3 319 482	47 756
10	1.07	664 677	61 466
15	0.77	496 482	65 252
20	0.78	496 482	65 253

Fig. 2. Pathwidth of some Mycielski graphs

Fig. 3. Running time of **BAB-GPP** for the Mycielski graph M_6

---

[3] Computations in [1] have been performed on a computer equipped with a AMD Opteron 6172 processor operating at 2.1GHz and 256GB of RAM.

4.4 Impact of Prefix Length

Our main algorithm **BAB-GPP** is parameterized by the maximum length of the prefixes that we store and that allow us to cut the search space during the Branch and Bound process. We have evaluated the impact of this parameter when our algorithm is executed on some specific graphs. In particular, the *Mycielski graph* is considered as a hard instance for integer programming formulations for pathwidth [11]. The Mycielski graph M_k is a triangle-free graph with chromatic number k having the smallest possible number of vertices. Note that, $|V(M_2)| = 2$ and $|V(M_k)| = 2|V(M_{k-1})| + 1$ for any $k > 2$. We ran our algorithm on the Mycielski graph M_k for $k \leq 8$. Our results are depicted in Tab. 2 (for $k \in \{7, 8\}$ we only got upper bounds).

Tab. 3 presents the running time of **BAB-GPP** for the Mycielski graph M_6 and different bounds on the size of the stored prefixes. In Tab. 3, the column *Visited nodes* refers to the number of nodes of the Branch and Bound exploration space that are actually considered. Tab. 3 shows that storing larger prefixes allows for an impressive reduction of the computation time. However, we have no improvement here when allowing length of 20 instead of 15, and larger values behave similarly. This is probably due to the fact that in this experiment, because of the greedy steps, the algorithm stores only one prefix of length ≥ 15.

This suggests that the combination of the greedy steps and of the storage of the prefixes allows good performance even with bounded length of the prefixes. Thus, memory-space seems not to be an issue for graphs of reasonable size (even though the algorithm might potentially store an exponential number of prefixes).

4.5 Using BAB-GPP as an Heuristic

Some research has been devoted to design heuristic algorithms for this problem. [9] proposed the following: starting from some layout of the nodes of D, the solution is improved by switching pairs of nodes in the layout until no further improvement can be obtained. The VSPLIB [16] has been designed for benchmarking this heuristic [9]. VSPLIB [16] contains 173 instances: 50 $\gamma \times \gamma$ grids with $5 \leq \gamma \leq 54$; 50 trees with respectively 22, 67, and 202 nodes and pathwidth 3, 4, and 5; a set of 73 graphs of order n, called HB, with $10 \leq n \leq 960$ (see [9] for more details). In particular, the pathwidth of $\gamma \times \gamma$ grids equals γ and the pathwidth of trees can be computed in linear time [8]. Therefore, this benchmark allows to evaluate the performance of heuristic. We tested the Algorithm **BAB-GPP** as a heuristic on the graphs of VSPLIB. If the time is up before the end of **BAB-GPP**'s execution, the algorithm returns the value computed so far (i.e., an upper bound on the pathwidth of the graph).

We were able to compute the exact pathwidth for all grids with side $\gamma \leq 13$, trees with $n \leq 67$, and 26 of the HB graphs including one with 957 nodes. For all grids and trees of the VSPLIB, the final value returned by **BAB-GPP** equals the exact pathwidth. That is, our algorithm always finds quickly an optimal layout and most of the execution time is devoted to prove its optimality.

In particular for a $\gamma \times \gamma$ grid, the first solution found is always the pathwidth. This is due to the order in which we add the vertices in the layouts: starting from a node with smallest degree and always adding a node that minimizes the increase of the size of the border. Proceeding that way will always give an optimal layout in square grids.

For trees, the first layout tested by **BAB-GPP** is not always the optimal one, but an optimal one is found very quickly. It would be interesting to understand why our Branch and Bound algorithm performs well in trees[4].

5   Conclusion

In this paper, we have presented a new Branch and Bound algorithm for computing the pathwidth of graphs and the vertex-separation of digraphs. Our approach is more promizing than previous proposals, based on ILP or SAT, for solving large instances. Indeed, the drawbacks of ILP and SAT formulations for layout problems are both in the large number of symmetries of the problems, and in the time needed to fill the optimality gap (i.e., distance between the lower bound based on the fractional relaxation of the formulation and the best integral solution). Our next steps are on one hand to propose our code for inclusion into future releases of *Sage*. On the other hand, we will search for new pruning rules to further reduce computation time. In particular, looking for good lower bounds for pathwidth of graphs is a theoretical issue that received few attention. It would moreover speed up our algorithm since the main computation time seems dedicated to prove the optimality of the best value computed.

References

1. Biedl, T.C., Bläsius, T., Niedermann, B., Nöllenburg, M., Prutkin, R., Rutter, I.: Using ILP/SAT to determine pathwidth, visibility representations, and other grid-based graph drawings. In: Wismath, S., Wolff, A. (eds.) GD 2013. LNCS, vol. 8242, pp. 460–471. Springer, Heidelberg (2013)
2. Bodlaender, H.L., Fomin, F.V., Koster, A.M., Kratsch, D., Thilikos, D.M.: A note on exact algorithms for vertex ordering problems on graphs. Theory Comput. Syst. 50(3), 420–432 (2012)
3. Bodlaender, H.L., Jansen, B.M.P., Kratsch, S.: Kernel bounds for structural parameterizations of pathwidth. In: Fomin, F.V., Kaski, P. (eds.) SWAT 2012. LNCS, vol. 7357, pp. 352–363. Springer, Heidelberg (2012)
4. Bodlaender, H.L., Kloks, T.: Efficient and constructive algorithms for the pathwidth and treewidth of graphs. J. Algorithms 21(2), 358–402 (1996)
5. Bodlaender, H.L., Koster, A.M.C.A.: Treewidth computations ii. lower bounds. Inf. Comput. 209(7), 1103–1119 (2011)
6. Coudert, D., Huc, F., Mazauric, D., Nisse, N., Sereni, J.-S.: Reconfiguration of the routing in WDM networks with two classes of services. In: Optical Network Design and Modeling (ONDM), pp. 1–6. IEEE (2009)
7. Coudert, D., Mazauric, D., Nisse, N.: Experimental evaluation of a branch and bound algorithm for computing pathwidth. Tech. Rep. RR-8470. Inria (February 2014)

---

[4] Note that optimal path-decompositions are computable in linear time in trees [13].

8. Díaz, J., Petit, J., Serna, M.: A survey on graph layout problems. ACM Comput. Surveys 34(3), 313–356 (2002)
9. Duarte, A., Escudero, L.F., Martí, R., Mladenovic, N., Pantrigo, J.J., Sánchez-Oro, J.: Variable neighborhood search for the vertex separation problem. Computers & OR 39(12), 3247–3255 (2012)
10. Hein, A., Koster, A.M.C.A.: An experimental evaluation of treewidth at most four reductions. In: Pardalos, P.M., Rebennack, S. (eds.) SEA 2011. LNCS, vol. 6630, pp. 218–229. Springer, Heidelberg (2011)
11. MIPLIB - mixed integer problem library, http://miplib.zib.de/.
12. Rome graphs, http://www.graphdrawing.org/download/rome-graphml.tgz
13. Skodinis, K.: Construction of linear tree-layouts which are optimal with respect to vertex separation in linear time. J. Algorithms 47(1), 40–59 (2003)
14. Solano, F., Pióro, M.: Lightpath reconfiguration in WDM networks. IEEE/OSA J. Opt. Commun. Netw. 2(12), 1010–1021 (2010)
15. Stein, W., et al.: Sage Mathematics Software (Version 6.0). The Sage Development Team (2013), http://www.sagemath.org
16. VSPLIB (2012), http://www.optsicom.es/vsp/

An Exact Algorithm for the Discrete Chromatic Art Gallery Problem⋆

Maurício J.O. Zambon, Pedro J. de Rezende, and Cid C. de Souza

Institute of Computing, University of Campinas
Campinas, Brazil

Abstract. Recently introduced in [1], the Chromatic Art Gallery Problem (CAGP) is related to the well known Art Gallery Problem to the extent that given a polygon P, a set of guards within P that satisfies a particular property is sought. Define a proper coloring of a guard set that covers the polygon as a color assignment in which any two guards receive different colors whenever their visibility regions intersect. The CAGP aims to find among the sets of guards that cover P one that admits a proper coloring of smallest cardinality. In this paper, we present an Integer Programming formulation for a discrete version of the CAGP and introduce techniques that allow us to attain proven optimal solutions for a variety of instances of the problem. Experiments were conducted considering vertex guards, but the method clearly works for any discrete set of guard candidates that guarantee full polygon coverage. Having been absent from the literature so far, experimental results are also discussed showing that this approach is of practical value.

1 Introduction

Consider an application where a mobile robot (such as an industrial operator, or a mobile agent on a secure area) finds itself in an environment whose floor plan has a boundary described as a simple polygon (possibly with holes). With the objective of guiding the robot, landmarks (or visual emitters) may be placed in this environment so that the robot can distinguish between them and perform motion actions based on their locations. Possible motions include: move in the direction of a landmark, move away from a landmark or move around it. For the purpose of formulating this problem, consider that the landmarks, distinguishable only by their colors, as well as the robot, are idealized as (zero dimensional) points. A motion instruction to the robot is unfeasible if it does not see any landmarks and ambiguous if it sees two landmarks of the same color simultaneously. We say that a landmark covers a point of the environment if a robot is capable of seeing that landmark while placed at that point. In this way, given an environment description, it becomes necessary to determine the position of a set of landmarks that guarantee coverage of all points in the environment and

---

⋆ This research was supported by grants from Capes #P-26329-2012, CNPq #477692/2012-5, #302804/2010-2, and Fapesp #2007/52015-0, #2012/24993-6.

J. Gudmundsson and J. Katajainen (Eds.): SEA 2014, LNCS 8504, pp. 59–73, 2014.

to find a coloring of those landmarks so that no point will be covered by two landmarks of the same color, called a *proper coloring*. Among all possible sets of landmarks covering the environment, we are asked to find one that admits a proper coloring with the least number of colors.

One can easily detect a connection here to the widely studied Art Gallery Problem (AGP) on which the Computational Geometry literature has included many publications for nearly 30 years [2,3,4,5,6]. As a result, the present problem has become known as the Chromatic Art Gallery Problem (CAGP). To contextualize this relation, consider a simple polygon P (possibly with holes) and a finite set of points (guards) $S \subset P$. If for each $p \in P$ there exists $g \in S$ so that the straight line segment gp is fully contained in P, S is said to *cover* P. Obviously, in this case, we can easily derive a proper coloring of S with $|S|$ colors. The search for a minimal proper coloring is justified, in the aforementioned application context, since the fewer the colors the robot has to identify, the more precise the classification of the landmarks will be. After all, depending on illumination conditions and the quality of the robot's sensors, the more similar shades there are, the greater the chance of misclassifications.

Many interesting questions may be raised and a straightforward one is: given P and a finite set $S \subset P$ that covers P, determine a proper coloring of S with the smallest number of colors. Much more interesting in view of an actual field application [7] is the broader question called the *Discrete Chromatic Art Gallery Problem* (DCAGP): given P and a finite set of points $G \subseteq P$, determine a covering $\Gamma \subseteq G$, for P, that admits a proper coloring of least cardinality among all possible coverings from G. The number of colors used in this coloring is called the *chromatic number* of P (from G), denoted $\chi_G(P)$.

Many variations of the art gallery problem have long been known to be NP-hard [8,9]. Moreover, it follows from a result by Broden et al. in [10] that the CAGP is APX-hard. However, the first results on the NP-hardness of the CAGP have only recently been established [7,11]. In [1], besides the introduction of the problem, Erickson and LaValle provide upper bounds for staircase and spiral polygons, $\chi_G(P) \leq 3$ and $\chi_G(P) \leq 2$, respectively. They also show how to construct polygons with n vertices that require $\Omega(n)$ colors, and monotonic and orthogonal polygons that require $\Omega(\sqrt{n})$ colors. In [12], the same authors discuss the case of monotonic polygons covered by guards that are *strongly connected* (i.e., the visibility graph between guards is connected), providing upper bounds and lower bounds for this case; furthermore, they also describe an algorithm that uses $n/3 + 12$ colors to color guards that cover a given monotonic polygon of n vertices.

More recently, another variation of the CAGP, called Conflict-Free Chromatic Art Gallery Problem, received some attention. In this variation, the coloring restriction is relaxed in the sense that it is no longer required that all conflicting guards (those whose visibility regions intersect) receive different colors. The binding requirement becomes that at least one guard be distinguishable at all points within the polygon. In [13], Bärtschi and Suri provide upper bounds for the conflict-free chromatic number of orthogonal and of monotone polygons at $O(\log n)$, and of arbitrary simple polygons at $O(\log^2 n)$.

In the last seven years, besides work on *heuristics* [14], a number of papers that discuss *exact* solutions for the art gallery problem have been published (see [15,16,17,18], as well as [19] and references therein). Some of these works model the AGP as a special case of the Set Covering Problem, which is then solved to optimality through Integer Programming (IP).

Our Contribution. Following an approach similar to the latter, in this paper, we introduce an IP model for the DCAGP and devise the first exact algorithm for the problem presented in the literature. We report on extensive tests conducted on three classes of instances comprised of simple and orthogonal polygons of up to 2500 vertices, which demonstrates that the proposed algorithm is an effective and efficient way for solving the problem to optimality.

Organization of the Text. In Section 2, basic terminology is reviewed, so that the IP formulation presented in Section 3 can easily be understood. Implementation features are explained in Section 4, while test results are detailed in Section 5. We close in Section 6 with some conclusions.

2   Terminology and Problem Statement

Let P be a simple polygon (see [2], Chapter 5, for basic definitions). Since the boundary of P, ∂P, is a Jordan (polygonal) curve, $\mathbb{R}^2 - \partial P$ consists of two disjoint regions: a bounded one (the interior of P) and an unbounded one (the exterior of P). For simplicity, ∂P is written as the sequence of vertices of P. A common notion that generalizes (simple) polygons is that of a (simple) polygon *with holes*. A *hole* in a polygon is merely a (smaller) simple polygon fully contained in its interior. The interior of a hole is then regarded as part of the exterior to the original polygon. Whenever P is a polygon with holes, ∂P denotes the list of the vertices of P's outer boundary along with a list of the vertices of all its holes.

Two points p, $q \in P$ are said to be *(mutually) visible* if the closed segment $\overline{pq}$ does not intersect the exterior of P, and we say that p *sees* q. The *visibility polygon* of a point $p \in P$ is defined as $\mathrm{Vis}(p) = \{q \in P \mid p \text{ sees } q\}$. A finite set of points $\Gamma \subset P$ is called a *guard set* for P if $\bigcup_{g \in \Gamma} \mathrm{Vis}(g) = P$. A pair of guards g_1, $g_2 \in \Gamma$ is said to *conflict* if $\mathrm{Vis}(g_1) \cap \mathrm{Vis}(g_2) \neq \emptyset$. Denote by $C(\Gamma)$ the minimum number of colors sufficient to color Γ so that no pair of conflicting guards receives the same color.

We may now formalize the Discrete Art Gallery Problem (DCAGP).

Let P be a simple polygon (possibly with holes) and $G \subset P$ a (finite) guard set for P. Determine:

$$\chi_G(P) = \min_{\Gamma}\{|C(\Gamma)| \ \text{s.t.} \ \Gamma \subset G \text{ is a guard set for } P\}.$$

Since we seek to find a guard set $\Gamma \subset G$ that minimizes $|C(\Gamma)|$, we refer to the elements of G as *guard candidates*.

Consider a guard set G of points in P and let $A(G)$ be the arrangement induced (in P) by the union of the edges of the visibility polygons of all points in G. This arrangement gives rise to a subdivision of P whose faces are called *Atomic Visibility Polygons* (AVPs). Denote by F the set of AVPs of this subdivision. Incidentally, note that the atomicity of these faces comes from the fact that if a point $g \in G$ sees a point inside an AVP, it necessarily sees the whole AVP. Some of the atomic visibility polygons play an important role in our solution to the DCAGP presented in Section 4. In the spirit of definitions given in [20], denote by $G_f \subset G$ the set of points in G that cover a face $f \in F$. Define a partial order $\prec$ on F as follows. Given $f, f' \in F$ we say that $f \prec f'$ if and only if $G_f \subset G_{f'}$. We call $f \in F$ a *shadow* (*light*) AVP if f is minimal (maximal) with respect to $\prec$. By taking a single point (*a witness*) from each shadow AVP, we can form a *witness set* $W(G)$ for P. It has been shown (see [19]) that a subset of a guard set G that covers $W(G)$ is also guaranteed to cover the whole polygon P.[1] Whenever the set G is understood, we will denote $W(G)$ simply by W.

Notice that the DCAGP can be cast as a graph problem as follows. Let $G_C = (G, E_G)$ be the graph whose vertices correspond to the guard candidates in G and an edge $(g_u, g_v) \in E_G$ iff $\mathrm{Vis}(g_u) \cap \mathrm{Vis}(g_v) \neq \emptyset$, i.e., guards u and v conflict. G_C is called the *2-link-visibility graph* (see [7]). Let $G_W = \{G \cup W, E_W\}$, be the bipartite graph whose vertices correspond to the guard candidates in G along with the witnesses in W and an edge $(g_u, w_v) \in E_W$ iff g_u covers w_v. We call G_W, the *covering graph*, as it describes the visibility relation between guards and witnesses. Finally, let $G_T = G_C \cup G_W$. Given G_T, the DCAGP can be formulated as a graph-theoretic dominating set problem: find a vertex set $C_W \subseteq G$ dominating the vertices in the witness set W, such that, C_W admits a minimum proper coloring among all subsets of G dominating W.

An important distinction between the traditional art gallery problem and the chromatic art gallery problem is that a solution to an instance of the CAGP is not necessarily attained with the minimum number of guards. As an example, consider the polygon P depicted in Figure 1 and the guard set $G = \partial P$. Clearly, P has a minimum covering, $\Gamma_1 \subset G$ of k guards that require k colors to achieve a proper coloring (actually, it is possible to show that all proper colorings of *all* minimum coverings of P require k colors). However, there is a covering, $\Gamma_2 \subset G$ with $k+1$ guards that admits a proper coloring of only *two* colors. This example establishes that the ratio between the smallest proper coloring of a minimum covering of P and $\chi_G(P)$ can be as large as $n/8$.

3   An Integer Programming Formulation for the DCAGP

In the previous section, we described the DCAGP in terms of a graph coloring problem. Based on that transformation, we now present an IP model for the DCAGP.

---

[1] We should alert the insightful reader that the existence of a witness set $W(G)$ for P does not require P to be a *witnessable polygon* in the sense defined by Chwa et al. [21] since, here, the guard set G is given.

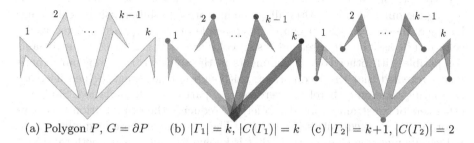

(a) Polygon P, $G = \partial P$ (b) $|\Gamma_1| = k$, $|C(\Gamma_1)| = k$ (c) $|\Gamma_2| = k+1$, $|C(\Gamma_2)| = 2$

Fig. 1. Example of a polygon P, with G given by the vertices of P, for which no guard set Γ of minimum size, k, is 2-colorable, while the chromatic number $\chi_G(P)$ is two

Consider again the graph $G_T = (G \cup W, E_G \cup E_W)$ defined in Section 2, where G represent a set of guard candidates and W is a witness set of G for P. Moreover, suppose that an upper bound K on the optimum of the DCAGP is known (see next section) and assume the guard colors are numbered 1 to K.

As in the natural model for the classical graph coloring problem, our IP formulation contains a binary variable $x_{g\kappa}$ for each pair formed by a guard candidate and a color. Accordingly, $x_{g\kappa} = 1$ iff the guard candidate $g \in G$ is assigned color $\kappa \in \{1, \ldots, K\}$ in an optimal solution. The remaining binary variables c_κ are used to specify the usage of the available colors. In this way, for each color $\kappa \in \{1, \ldots, K\}$, we set $c_\kappa = 1$ iff the color κ is used.

Thus, the DCAGP can be modeled as follows:

$$z = \min \sum_{\kappa \in K} c_\kappa \quad \text{s.t.} \tag{1}$$

$$\sum_{g \in G | w \in \text{Vis}(g)} \sum_{\kappa \in K} x_{g\kappa} \geq 1 \quad \forall\, w \in W \tag{2}$$

$$x_{u\kappa} + x_{v\kappa} - c_\kappa \leq 0 \quad \forall\, (u, v) \in E_G, \kappa \in \{1, \ldots, K\} \tag{3}$$

$$\sum_{\kappa \in K} x_{g\kappa} \leq 1 \quad \forall\, g \in G \tag{4}$$

$$c_\kappa - c_{\kappa+1} \geq 0 \quad \forall\, \kappa \in \{1, \ldots, |K| - 1\} \tag{5}$$

$$\sum_{g \in G} x_{g\kappa} - \sum_{g \in G} x_{g\kappa+1} \geq 0 \quad \forall\, \kappa \in \{1, \ldots, |K| - 1\} \tag{6}$$

$$x_{g\kappa} \in \mathbb{B} \quad \forall\, g \in G, \kappa \in \{1, \ldots, K\} \tag{7}$$

$$c_\kappa \in \mathbb{B} \quad \forall\, \kappa \in \{1, \ldots, K\} \tag{8}$$

where $\mathbb{B} = \{0, 1\}$. The objective function (1) minimizes the number of colors used. The constraint set (2) ensures that all witnesses are covered by at least one guard, implying that the entire polygon is covered. Constraints (3) avoid that any two conflicting guards receive the same color, while guaranteeing that, if some guard is assigned a color, that color must be used in the optimal solution. Constraints (4) impose that each guard can be assigned to at most one color.

Strictly speaking, these constraints can be dropped from the model without affecting the optimal value. After all, given an optimal solution using z colors and having at most one color per guard, it is possible that some guard that received a color can be assigned to a second (used) color and that this assignment remains compatible with other color assignments in the solution. However, our experiments revealed that these constraints, although not necessary, helped decrease computing time and, therefore, were kept as part of the formulation. Similarly, other sets of constraints intended solely on reducing the computation times are incorporated by inequalities (5) and (6). The purpose of these inequalities is to reduce symmetry, a characteristic that is known to cause loss of performance.

One of the keys to the success of integer programming when tackling difficult optimization problems lies on the strength of the linear relaxation of the IP model. One way to strengthen a given IP formulation, is to replace existing inequalities by others that dominate them. In the case of the DCAGP model, this can be done by substituting the *edge* inequalities in (3) by the so-called *clique inequalities*. The concept is simple. For instance, if the vertex set $\{u, v, w\}$ induces a triangle in G_T, each of the three inequalities in (3) relative to the edges joining these vertices are clearly less restrictive than the clique inequality $x_{u\kappa} + x_{v\kappa} + x_{w\kappa} - c_\kappa \leq 0$. We say that the edge inequalities are *dominated* by the clique ones. It is immediate to generalize this idea to larger cliques and it is not difficult to show that the only non-dominated inequalities of that form are those associated to maximal cliques. However, the number of maximal cliques in a graph increases exponentially with the number of vertices. This makes it impractical to use all of them at once. One way to overcome this limitation, while keeping a reasonably strong model, is to consider only inequalities that correspond to a set H of maximal cliques that covers all edges of the graph, in our case, the edges in E_G. Once the set H had been computed, the constraints in (3) can be replaced by the inequalities (9) below, where we have one constraint for each maximal clique η belonging to the edge clique cover H:

$$\sum_{g \in \eta} x_{g\kappa} \leq c_\kappa \qquad \forall\, \eta \in H, \kappa \in \{1, \ldots, K\} \qquad (9)$$

It is easy to check that the binary vectors (x, c) satisfying constraints (3) are precisely the same for which the inequalities in (9) hold. On the other hand, any feasible solution of the linear relaxation of the new model is also feasible for the linear relaxation of the previous model, even though the reverse is usually not true. As a result, the replacement of the edge inequalities by the clique ones lead to a stronger IP model.

4   Implementation Details

In this section, we detail the implementation routines required prior to the construction of the IP model.

Initial Upper Bound. Obtaining a tight upper bound, denoted earlier by K, has a significant impact on the size of our IP formulation since it affects both the number of variables and the number of constraints. The development of a good heuristic for the DCAGP turned out to be quite challenging, which is not surprising since this problem embeds two hard sub-problems, namely, graph coloring and the set covering. On the other hand, any additional color required by the heuristic solution results in a search space much larger than necessary, which is unfavorable for computing times. This is particularly significant in the case of the DCAGP because the optimal chromatic number is usually quite small.

Algorithm 1. DCAGP Greedy Initial Solution

1: **function** GREEDYDCAGP(G, W, G_W)
2:     $S \leftarrow \emptyset$ ▷ Solution (vector of independent sets)
3:     $U \leftarrow W$ ▷ Uncovered witnesses
4:     **repeat**
5:         **for all** $g \in G$ **do**
6:             $w_g \leftarrow |W'|$, where $w \in W'$ iff $w \in \mathrm{Vis}(g)$ and $w \in U$
7:         $x \leftarrow$ solution of the model (10)–(12)
8:         $T \leftarrow \emptyset$ ▷ New guard independent set
9:         **for all** $g \in G$ **do**
10:            **if** $x_g = 1$ **then**
11:                $T \leftarrow T \cup \{g\}$
12:                $U \leftarrow U \setminus \{w \in \mathrm{Vis}(g)\}$
13:        $S \leftarrow S \cup T$
14:     **until** $U = \emptyset$
15:     **return** S
16: **end function**

Taking into consideration the aspects highlighted in the paragraph above, we devised a greedy heuristic whose basic steps are outlined in Algorithm 1. To understand how this heuristic works, keep in mind that the vertices assigned to a given color on any proper coloring of a graph must form an *independent set* (a subset of vertices whose induced subgraph has no edges). After declaring (Line 3) all witnesses to be uncovered, the heuristic proceeds iteratively computing independent sets in G and assigning a different color to each of them. The current independent set is built on a given iteration as follows. Every vertex of G receives a weight corresponding to the number of uncovered witnesses that is seen by the corresponding guard candidate (Line 6). Then, a maximum weighted independent set is computed and its vertices are colored with a new color (Line 7). The set of uncovered witnesses is updated and the heuristic iterates until all witnesses have been covered (Line 12). Since Line 7 of the algorithm requires the solution of a maximum weighted independent set problem, we employ for this task the following IP model[2]:

$$z = \max \sum_{g \in G} w_g x_g \quad \text{s.t.} \tag{10}$$

$$x_u + x_v \leq 1 \quad \forall\, (u,v) \in E_G \tag{11}$$

$$x_g \in \mathbb{B} \quad \forall\, g \in G. \tag{12}$$

When this model has been solved, we have an independent set that covers the maximum number of uncovered witnesses, in compliance with the greedy strategy. This property follows from the calculation of the guard weights and because, by construction, for any independent set in G, no witness is covered more than once. Despite being a hard problem, this IP model is usually solved extremely fast in practice, for the instances in our benchmark. Our experiments shows that this heuristic yields fairly good upper bounds for most instances of the DCAGP leading to an average gap from the optimum of only 1.77 for the solved instances.

Clique Edge Covers. As mentioned in Section 3, the IP model can be strengthened by replacing the constraints in (3) by a set of (maximal) clique inequalities (9). To produce a correct formulation, the cliques in this set must form a clique covering of the edges in E_G. In principle, any clique edge cover could be used for that purpose, and although we could use the same edge clique cover for each one of the K colors, we chose to generate K such sets, one for each color, as this can help decrease the symmetry of the model even further with respect to the linear relaxation.

We could generate the K different clique covers in a random fashion. However, in preliminary experiments, we realized that, with this method, the solver behaved erratically, sometimes reaching the optimum of a given instance in a few seconds and sometimes only after hours of computation. These enormous variations in performance are most likely due to decisions made by the solver's internal routines (heuristics, node/branch selection, etc.), which end users have no control over, rather than to the different input models. Characterizing good clique (edge) covers remains an interesting subject for further investigation.

In the end, we opted for a deterministic approach to produce the K edge covers. The procedure we developed to accomplish this task is given in Algorithm 2. Let v and $e = (u,v)$ be, respectively, a vertex and an edge of the 2-link-visibility graph G_C. Denote the degree of v by $\delta(v)$ and the extremities of e by $e.u$ and $e.v$. Finally, the adjacency of v is represented by $N(v)$.

The initially empty set Ψ (Line 2) is used to return the K clique covers. The first clique of each cover is initialized with a different pair of vertices corresponding to the extremities of K edges forming a matching in G_C. Such a matching is built in a greedy fashion (Line 3) by selecting edges in increasing order of weights given by the sum of the degrees of their endpoints' neighbors. Edges in this matching are said to be covered; otherwise, they are uncovered. Next, the

---

[2] Alternatively, one could utilize an effective heuristic to solve the weighted independent set problem, but this was unnecessary for our testbed.

algorithm generates one maximal clique starting from the corresponding pair of vertices (Line 5). Once a maximal clique has been built, a new one is started from an edge of minimum weight. As before, the edge weight is calculated in terms of the sum of the degrees of the endpoints' neighbors of the edge in the subgraph of G_C containing only the uncovered edges (Line 9). This process is repeated until all edges of G_C are covered by some clique in H (Line 13).

The cliques are constructed greedily (Lines 17 to 24). The process starts with a single edge e given in the input (Line 18). The next vertex n added to the current clique η is the one that maximizes the number of uncovered edges having one extreme on n and the other on a vertex of η (obviously, n must be adjacent to all vertices of the clique).

Algorithm 2. Generating K Clique Edge Covers

1: **function** KCEE(G_C, K)
2:     $\Psi \leftarrow \emptyset$ ▷ Set of edge clique covers
3:     Construct a matching M, composed by the K edges with minimum
$$\sum_{t \in N(u)} \delta(t) + \sum_{z \in N(v)} \delta(z), \text{ where } (u,v) \in E_G$$
4:     **for all** $i \in \{1, \ldots, K\}$ **do**
5:         $\eta \leftarrow$ BUILD CLIQUE($M[i]$)   ▷ $M[i]$: i-th edge in M
6:         $G' \leftarrow G_C \backslash \{e \in \eta\}$ ▷ Remove edges covered by η
7:         $H \leftarrow \eta$ ▷ Initialize the i-th clique edge cover
8:         **repeat**
9:             $\mu \leftarrow e \in G'$ such that $\displaystyle\sum_{t \in N_{G'}(e.u)} \delta(t) + \sum_{z \in N_{G'}(e.v)} \delta(z)$ is minimum
10:             $\eta \leftarrow$ BUILD CLIQUE(μ)
11:             $G' \leftarrow G' \backslash \{e \in \eta\}$ ▷ Remove the covered edges from G'
12:             $H \leftarrow H \cup \{\eta\}$
13:         **until** all edges are covered
14:         $\Psi \leftarrow \Psi \cup \{H\}$
15:     **return** Ψ
16: **end function**

17: **function** BUILD CLIQUE(e)   ▷ Greedy routine to build cliques
18:     $\eta \leftarrow \{e.u, e.v\}$ ▷ η: vertices of the clique
19:     $C \leftarrow N(e.u) \cap N(e.v)$ ▷ C: vertices candidates to extend the clique
20:     **repeat**
21:         $n \leftarrow \gamma \in C$, where γ adds more uncovered edges to the clique
                than any other vertex in C
22:         $C \leftarrow C \cap N(n)$ ▷ Update the set of candidate vertices
23:         $\eta \leftarrow \eta \cup \{n\}$ ▷ Add n to the current clique
24:     **until** ($C = \emptyset$) or (there is no vertex with uncovered adjacent edges)
25:     **return** η
26: **end function**

Lazy Constraints. For a set of $|G|$ guard candidates, the number of shadow AVPs and, consequently, of witnesses in W is $O(|G|^2)$ (see [22]). Since the number of inequalities in (2) is $|W|$, the constraint set of our IP model grows rapidly, making the amount of computer memory available a concern. To overcome this situation, most state-of-the-art IP solvers offer the possibility of applying a *lazy constraint* procedure. According to this strategy, a subset of the inequalities can be chosen to not being included in the model from scratch, but, instead, added as they are found to be violated by a feasible solution. We do so aiming at ending the computation without being forced to consider all the witnesses but only a small subset of them. Notice that this ideal situation may occur in practice as the guards chosen to cover the witnesses considered so far may also cover the remaining ones. However, this stratagem is worthy to be applied only if we can find a small subset of initial witnesses whose coverage (almost) ensure the coverage of the whole polygon. As expected, this is not a trivial task. Nevertheless, we devise a simple procedure that selects in W the $|G|$ witnesses that are covered by the least number of vertices. After that, during the optimization process, every time the solver finds a feasible solution, we check for uncovered witnesses and, if they exist, the corresponding cover constraint is added to the model to ensure their coverage when the process restart.

5   Computational Results

We now present the results obtained by our algorithm for a large set of polygons. The tests were run on a PC running Ubuntu 12.04, featuring an Intel® Xeon® CPU E3-1230 V2 3.30GHz, 4 cores, 8 threads and 32GB of RAM. The code, implemented in C++, used the CGAL 4.3 library for geometric routines and the BOOST 1.54 library for basic algorithms.For each input instance, a time limit of 1800 seconds was enforced for the execution of the IP solver, in this case, CPLEX 12.5.1. The experiments were divided into two groups of instances.

The first set of polygons is comprised of random simple polygons and random orthogonal polygons, both without holes. The tests were run on polygons of sizes (number of vertices) 200, 400, 600, 800, 1500 and 2500, and on 30 instances per size and type. These instances were obtained from [23].

On all tests presented in this section, we opted for considering guard candidates on all vertices since such set always forms a guard set, even though our method would also work with any set of candidates that cover the whole polygon.

Table 1 summarizes the results obtained with this first set of tests. Notice that the algorithm was able to compute the optima for *all* instances in this dataset. Even more surprising is the fact that, for all polygon classes and sizes, the

Table 1. Instances solved for simple and orthogonal polygons (without holes)

	Simple						Orthogonal					
$|G|$	200	400	600	800	1500	2500	200	400	600	800	1500	2500
% of inst. solved	100	100	100	100	100	100	100	100	100	100	100	100
Solver time (s)	0.17	1.17	2.29	3.24	8.48	28.03	0.16	0.40	0.79	1.19	2.08	4.59
Graph density	0.15	0.08	0.05	0.04	0.02	0.01	0.11	0.05	0.04	0.03	0.02	0.01

average time spent by the solver to accomplish the task remained well below one minute. The execution times suggest that simple polygons are harder to solve than orthogonal ones of equivalent sizes. Upon inspecting the last row of Table 1, one notices that, for each given number of vertices, the 2-link-visibility graphs for simple polygons present higher densities than for their orthogonal counterparts. This observation suggests that this measure is relevant to assess the difficulty that our algorithm faces when solving a given instance. Experiments on the next group of instances reinforce this conclusion, even though the hardness of an instance cannot be solely attributed to the density of its 2-link-visibility graph.

We now consider the second group of instances on our dataset, which is composed of three classes of polygons: **VK** random von Koch polygons, **SS** random simple polygons with simple holes and **OO** random orthogonal polygons with orthogonal holes. These classes are fully described in [19].

We tested each polygon class on four different sizes (i.e., number of vertices including those on the holes): 200, 400, 600 and 800. In each case, we considered 30 instances obtained from the benchmarks [23], [24] and [25].

In Table 2, we show the percentage of instances solved and the corresponding average density of the 2-link-visibility graph. Compared to the first group of instances, this new dataset clearly contains instances that are harder to handle. As one would expect, the effect of instance size on the algorithm's performance becomes noticeable since, for all classes of polygons, the percentage of instances solved within the time limit decreases as the number of vertices increases.

Table 2. Instances solved for VK, SS and OO polygons

	SS				OO				VK			
$|G|$	200	400	600	800	200	400	600	800	200	400	600	800
% of inst. solved	100	96.7	76.7	60	100	96.7	86.7	76.7	100	83.3	50	30
Graph density	0.27	0.15	0.11	0.08	0.24	0.13	0.09	0.07	0.69	0.62	0.59	0.57

Observe that the VK instances present much denser graphs (3 to 4 times) than SS and OO polygons of the same size. As discussed before, higher density could help explain why VK polygons are harder to solve, and similarly that the inclusion of holes makes instances more difficult. As is the case with hole-free polygons, the density diminishes as the instance size increases. However, this time, we notice a decline on the percentage of instances solved. This is an evidence that density alone is not enough to determine how hard an instance is.

Since not all instances were solved to optimality within the 1800 seconds time limit, we noted that the average gap between the best upper and lower bounds on all 72 instances was just 1.139. Further investigation is required to determine how often the solution found already reached the optimal.

Table 3 compares solver times for the instances where optimality was reached. Recall that the time limit for the solver was set to 1800 seconds. In that table, the following statistics for the execution times are given: minimum, maximum, average and standard deviation. From these data, it is clear that the solver time varies widely even for instances within the same polygon class having identical number of vertices. Furthermore, there is strong indication that other features, mostly of geometric nature, affect the performance of the algorithm to a larger extent than simply the size of the instance. As an example, from the OO class, we notice that the polygon of 400 vertices that demanded the longest computing time actually required ~3100 times longer to be solved than some OO polygons of twice that size.

Table 3. Solver time, t, (in seconds) for the fully solved VK, SS and OO instances

	SS				OO				VK			
$|G|$	200	400	600	800	200	400	600	800	200	400	600	800
min	2.31	11.64	117.24	0.11	1.57	4.85	4.93	0.46	0.43	15.54	128.74	382.61
max	275.7	1436.8	1624.2	327.7	1074.9	1423.9	1494	49.4	79.7	348.3	1303.2	1690.7
$\bar{t}$	47.9	407.0	384.0	21.2	65.7	108.3	295.6	4.8	8.9	77.8	533.4	1123.6
σ	59.64	454.99	386.21	65.66	196.21	273.74	417.18	9.71	17.63	84.27	350.97	452.47

We now turn our attention to data that allow an assessment of the IP model and of the strategies we adopted for using the CPLEX solver. We start by analyzing the absolute duality gap between the optimum and the linear relaxation value for the instances solved to exactness. The average gaps are displayed in Table 4 for the second group of instances. It is quite remarkable that, in all but one case (VK-600), they never went beyond one unit. This shows the strength of our formulation for the DCAGP.

Table 4. Statistics on the absolute gap, δ, between the number of colors on the exact solution and on the linear relaxation for the optimally solved instances

	SS				OO				VK			
$|G|$	200	400	600	800	200	400	600	800	200	400	600	800
$\bar{\delta}$	0.70	0.66	0.74	0.78	0.60	0.79	0.73	0.52	0.77	0.88	1.07	1.00
max	1.00	1.00	1.00	2.00	2.00	1.00	1.00	1.00	1.00	1.00	2.00	1.00
σ	0.48	0.45	0.43	0.53	0.49	0.45	0.51	0.50	0.43	0.33	0.26	0.00

As seen above, the quality of the model is key for the success of our approach. However, the size of this formulation becomes a drawback when the complexity of the visibility arrangement increases, resulting in a very large number of witnesses to be be covered. As each witness corresponds to one inequality of type (2), a scheme to address this pitfall is to bring these constraints into the model as they become necessary during the computation. This can be done with CPLEX by declaring them as *lazy constraints*. In Table 5, we can evaluate the significance of using this option. On average, less than half of the witnesses were needed to complete the optimization process for all instances solved to optimality. We have found that the use of lazy constraints benefits instances for which memory

Table 5. Percentage of witnesses, w, required to solve the instances using lazy constraints

	SS				OO				VK			
$|G|$	200	400	600	800	200	400	600	800	200	400	600	800
min	26.33	23.36	21.10	23.88	27.62	38.58	37.22	33.50	25.79	13.68	9.82	8.65
max	52.02	39.77	36.62	68.35	60.94	57.40	55.58	71.43	45.98	20.06	14.74	10.16
$\overline{w}$	35.12	30.99	29.36	39.54	45.29	45.08	46.44	49.55	32.86	16.68	12.13	9.39
σ	5.43	4.36	3.77	11.62	8.00	5.00	5.09	9.94	5.28	1.82	1.39	0.49

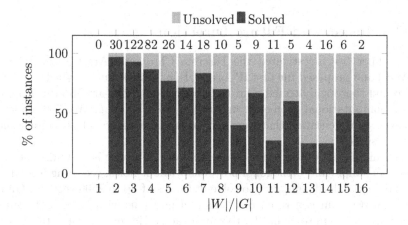

Fig. 2. Significance of $|W|/|G|$ ratio on the percentage of instances solved

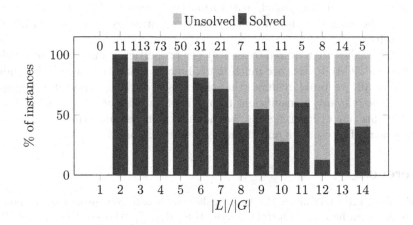

Fig. 3. Significance of $|L|/|G|$ ratio on the percentage of instances solved

usage becomes the bottleneck (even yielding to otherwise unattainable optimal solutions). However, we were unable to conclude that lazy constraints would produce consistent advantage for instances where the the whole model can be loaded and solved within the available memory limit. In an attempt to better understand aspects that are associated to how hard a DCAGP instance from the second group is, we also analyzed the ratio between either the number $|W|$ of witnesses or the number $|L|$ of light AVPs and the number of guard candidates, and their impact. Figures 2 and 3 present the percentage of instances solved and unsolved relative to these ratios. On these graphs, the numeral above each bar indicates the number of instances over which the percentage was calculated. Both sets of results suggest that there is a correlation between how large these fractions are and the difficulty of the corresponding instances.

6   Conclusion and Future Directions

In this paper, we investigated the Discrete Chromatic Art Gallery Problem (DCAGP) and proposed the first IP formulation for its solution, which we implemented and thoroughly tested. Our code uses state-of-the-art libraries from the fields of computational geometry and integer programming. We also report on extensive experimentation that were carried out with several classes of instances of various sizes.

A careful analysis of the results shows that our approach is extremely efficient for random simple and orthogonal polygons without holes, requiring less than a minute on a standard computer to solve instances of up to 2500 vertices. On the other hand, random von Koch polygons and polygons with holes, both simple and orthogonal, seem to be harder to solve exactly. However, even in these cases, optimality was proven for 80% of the instances we tested, with optima being reached within 30 minutes of computation.

The success of our approach arises from the strength of our IP formulation and from the sagacious use of a *lazy constraints* strategy accessible within the CPLEX solver.

Further studies are required to identify the properties, geometric or otherwise, that make a DCAGP instance much more difficult than other seemingly similar ones. After all, we found evidence that the density of the 2-link-visibility graphs alone, albeit relevant, is insufficient to provide a satisfactory explanation. Upon pursuing this line of investigation, we hope to gain new intuition to aid in the development of alternative and stronger IP formulations.

References

1. Erickson, L.H., LaValle, S.M.: An art gallery approach to ensuring that landmarks are distinguishable. In: Durrant-Whyte, H.F., Roy, N., Abbeel, P. (eds.) Robotics: Science and Systems (2011)
2. ORourke, J.: Art Gallery Theorems and Algorithms. Oxford Univ. Press, USA (1987)
3. Ghosh, S.K.: Approximation algorithms for art gallery problems. In: Proc. of the Canadian Information Processing Society Congress, pp. 429–434 (1987)
4. Shermer, T.: Recent results in art galleries. Proc. of IEEE 80(9), 1384–1399 (1992)

5. Urrutia, J.: Art gallery and illumination problems. In: Sack, J.R., Urrutia, J. (eds.) Handbook of Computational Geometry, pp. 973–1027. North-Holland (2000)
6. Ghosh, S.K.: Visibility Algorithms in the Plane. Cambridge U. Press, USA (2007)
7. Erickson, L.H., LaValle, S.M.: How many landmark colors are needed to avoid confusion in a polygon? In: Proc. of the IEEE Int. Conf. on Robotics and Automation (ICRA), pp. 2302–2307. IEEE (2011)
8. Lee, D.T., Lin, A.: Computational complexity of art gallery problems. IEEE Transactions on Information Theory 32(2), 276–282 (1986)
9. Schuchardt, D., Hecker, H.D.: Two NP-hard art-gallery problems for orthopolygons. Mathematical Logic Quarterly 41, 261–267 (1995)
10. Brodn, B., Hammar, M., Nilsson, B.J.: Guarding lines and 2-link polygons is APX-hard. In: Proc. of the 13th Canad. Conf. on Comput. Geom., pp. 45–48 (2001)
11. Fekete, S.P., Friedrichs, S., Friedrichs, S.: Complexity of the general chromatic art gallery problem. In: Proc. of 30th European Workshop on Comput. Geom. (2014)
12. Erickson, L.H., LaValle, S.M.: Navigation among visually connected sets of partially distinguishable landmarks. In: Proc. of the IEEE Int. Conf. on Robotics and Automation (ICRA), pp. 4829–4835. IEEE (2012)
13. Bärtschi, A., Suri, S.: Conflict-free chromatic art gallery coverage. Algorithmica, 1–19 (2013)
14. Amit, Y., Mitchell, J.S.B., Packer, E.: Locating guards for visibility coverage of polygons. Int. J. of Comput. Geom. & Appl. 20(5), 601–630 (2010)
15. Bottino, A., Laurentini, A.: A nearly optimal algorithm for covering the interior of an art gallery. Pattern Recognition 44(5), 1048–1056 (2011)
16. Kröller, A., Baumgartner, T., Fekete, S.P., Schmidt, C.: Exact solutions and bounds for general art gallery problems. J. of Exper. Algorit. 17(1), 2.1–2.23 (2012)
17. Tozoni, D.C., de Rezende, P.J., de Souza, C.C.: The quest for optimal solutions for the art gallery problem: A practical iterative algorithm. In: Bonifaci, V., Demetrescu, C., Marchetti-Spaccamela, A. (eds.) SEA 2013. LNCS, vol. 7933, pp. 320–336. Springer, Heidelberg (2013)
18. Borrmann, D., de Rezende, P.J., de Souza, C.C., Fekete, S.P., Friedrichs, S., Kröller, A., Nüchter, A., Schmidt, C., Tozoni, D.C.: Point guards and point clouds: solving general art gallery problems. In: Proc. of the 29th Annual Symposium on Computational Geometry, SoCG 2013, pp. 347–348. ACM, USA (2013)
19. Couto, M.C., de Rezende, P.J., de Souza, C.C.: An exact algorithm for minimizing vertex guards on art galleries. Int. Trans. in Operat. Res. 18(4), 425–448 (2011)
20. Crepaldi, B., de Rezende, P.J., de Souza, C.C.: An efficient exact algorithm for the natural wireless localization problem. In: Proc. of the 25th Canad. Conf. on Comput. Geom. (2013)
21. Chwa, K., Jo, B., Knauer, C., Moet, E., van Oostrum, R., Shin, C.: Guarding art galleries by guarding witnesses. Int. J. of Comp. Geom. & Appl. 16, 205–226 (2006)
22. Bose, P., Lubiw, A., Munro, J.I.: Efficient visibility queries in simple polygons. Comput. Geom. Theory Appl. 23(3), 313–335 (2002)
23. Couto, M.C., de Rezende, P.J., de Souza, C.C.: Instances for the Art Gallery Problem (2009),
www.ic.unicamp.br/~cid/Problem-instances/Art-Gallery/AGPVG
24. Tozoni, D.C., de Rezende, P.J., de Souza, C.C.: The Art Gallery Problem Project (2013),
http://www.ic.unicamp.br/~cid/Problem-instances/Art-Gallery/AGPPG
25. Crepaldi, B.E., de Rezende, P.J., de Souza, C.C.: Instances for the Natural Wireless Localization Problem (2013),
http://www.ic.unicamp.br/~cid/Problem-instances/Wireless-Localization

Implementation of the Iterative Relaxation Algorithm for the Minimum Bounded-Degree Spanning Tree Problem *

Attila Bernáth, Krzysztof Ciebiera, Piotr Godlewski, and Piotr Sankowski

Institute of Informatics, University of Warsaw, ul. Banacha 2, 02-097 Warsaw, Poland
{athos,ciebie,pgodlewski,sank}@mimuw.edu.pl

Abstract. In the Minimum Bounded-Degree Spanning Tree Problem we want to find a minimum cost spanning tree that satisfies given degree bounds. For this problem a very good quality solution can be found using the iterative relaxation technique of Singh and Lau STOC'07: the cost will not be worse than the cost of the optimal solution, and the degree bounds will be violated by at most one. This paper reports on the experimental comparison of this state-of-art approximation algorithm with standard, although well-tuned meta-heuristics. We have implemented the Iterative Relaxation algorithm of Singh and Lau and speeded it up using several heuristics including row generation and combinatorial LP pivoting. On the other hand, as the heuristic point of reference we have chosen local search techniques in a Simulated Annealing framework, where we allow the violation of degree bounds by one. In such setting there are two natural objectives for comparison: the cost of the solution, and the number of violated degree bounds. If we keep the number of violated constraints fixed in both algorithms then Iterative Rounding usually outperforms Simulated Annealing by several percents.

1 Introduction

This paper is a report on the implementation and experimental verification of the Iterative Relaxation method for the Minimum Bounded-Degree Spanning Tree Problem (MBDSPT Problem). In this problem we are given an undirected graph $G = (V, E)$, costs for the edges $c : E \to \mathbb{R}_+$ and degree bounds $b : V \to \mathbb{Z}_+$. We are looking for the spanning tree T of $G = (V, E)$ that minimizes the total cost $c(T) = \sum_{t \in T} c(t)$ and satisfies degree constrains given by b: namely the number $d_T(v)$ of tree edges incident to a node v should not exceed $b(v)$ at any v. This problem has many applications in situations where we want to connect nodes, but we have physical constraints on the number of connections (wires) we can attach to a node. In particular, this problem arised for the first time in the context of backplane wiring among pins where no more than a fixed number of wire-ends could be wrapped around any pin on the wiring panel [4]. Similar problem is encountered in VLSI design, where the limit on the number of

---

* Research was supported by the ERC StG project PAAl no. 259515.

J. Gudmundsson and J. Katajainen (Eds.): SEA 2014, LNCS 8504, pp. 74–86, 2014.

transistors that can be driven by the output current of a transistor is the degree bound for VLSI routing trees [2]. Another type of applications is the design of a reliable communication network, since the maximum degree in a spanning tree is a measure of vulnerability to single-point failures [14].

Observe that setting all the degree bounds to 2 gives a reformulation of the Minimum Cost Hamiltonian Path Problem. The problem is hopeless to approximate, unless the cost function satisfies some additional property, like triangle inequality (the inapproximability can be generalized to degree bounds larger than two [14]). Hence, two questions naturally arise. First is theoretical: can we propose any theoretically good solution? The second is practical: would such solution be of practical importance, or would meta-heuristic approaches outperform it? There is vast amount of work that try to answer both questions.

From theoretical point of view, after taking into account the mentioned hardness results, the best possible solution has been found in [15] – there is no penalty in cost but some of the degree bounds can be violated. There was a long line of research of this type [14,6,7], ending with the ultimate paper by Singh and Lau [15] that shows that we can find a spanning tree T with cost no more than the cost of the optimal solution of the original problem, and the price we pay is that some of the degree bounds might be violated by one, that is $d_T(v)$ might be $b(v) + 1$ for some nodes. The technique used is the so-called Iterative Relaxation Technique.[1] On a high level the algorithm works as follows: we formulate a natural LP relaxation of the problem; we find a basic optimal solution of this relaxation; we remove edges with zero value in the LP solution; and we remove a degree bound at some node v where the number of remaining edges is not more than $b(v) + 1$. We iterate the above procedure till the solution found is integral. The main technical ingredient of the method is to prove that properties of basic solutions assure that there always exists a removable degree bound.

On the other hand, the literature on experimental solution to the MBDSPT Problem is rather vast. The first implemented algorithm seems to be the branch-and-bound one given by Narula and Ho [12]. Since then almost every metaheuristic framework (e.g., ant-colony, simulated annealing, or genetic algorithms) has been implemented and tested for this problem. In this paper, we concentrate mostly on the implementation issues of the Iterative Relaxation (IR) Technique, and due to space limitation of this submission we refer the reader for the review on the existing work to the two recent papers on local search techniques [16] and Lagrangian relaxation [1].

We have implemented the IR Algorithm of Lau and Singh [15] and optimized it so that it can solve instances of reasonable size. There are several non-trivial issues here that need to be solved. The main issue is that the natural LP relaxation contains an exponential number of constraints (there exists a polynomial sized version of this LP, but it is still too large for LP solvers). We used a row generation technique to overcome this problem, and we tried several heuristics to speed it up. We also implemented a version that solves the LP relaxation only once, and for subsequent iterations it uses a "combinatorial pivoting" subroutine.

---

[1] This technique will be explained in more detail in Section 2.

The implemented IR algorithm is compared with some local search techniques in a simulated annealing (SA) framework. We make both algorithms work with the same assumptions, i.e., we allow the SA algorithm to violate every degree bound by one. We note that our SA algorithm is well tuned to the problem and is able to compete with state-of-art techniques [9,16,1]. Actually, both IR and SA perform much better than solutions implemented in [9,16] and in many cases deliver optimal solutions. You should however, keep in mind that a fair comparison here is hard, as there are two candidate objectives to consider when comparing existing approaches. First, the most natural one is the cost. Second, on the other extreme is the number of violated constraints (e.g., when we are looking for a Hamiltonian path in an unweighted graph, then we don't care about the cost, but we do care about the number of violated constraints). The problem in this comparison is that standard IR does not allow to control the number of violated constraints. Although it is allowed to violate all constraints by one, it actually produces solutions that violate many fewer constraints then SA. Hence, fair comparison of both method requires to develop a method that allows to control the number of violations in IR. We achieve this by rounding different number of variables in each round of the IR. On the other hand, in SA one can control the number of violated constraints by simply adding this requirement to the objective function. Equipped with this methods we observe experimentally that IR performs visibly better then SA when few violations are allowed. In such case the solution cost is lower by 10-20%. When we allow more violations the difference between the methods becomes smaller, but generally IR slightly outperforms SA.

We close the section by introducing some notation. For a graph $G = (V, E)$ and a subset $S \subseteq V$ we write $\delta_G(S)$ for the set of edges with exactly one endnode in S, and $I_G(S)$ for the set of edges with both endnodes in S (we note that this notation differs from the one used in [15]). Let furthermore $d_G(S) = |\delta_G(S)|$, and we will apply the notation $\delta_F(S)$ and $d_F(S)$ as well even if $F \subseteq E$ is only some subset of edges. We will omit the subscript and write $I(S)$, $\delta(S)$, etc. when it causes no confusion. We furthermore simplify $d_G(\{v\})$ to $d_G(v)$. For a vector $x : E \to \mathbb{R}$ and a set $F \subseteq E$ we let $x(F) = \sum_{f \in F} x(f)$. In a directed graph $D = (V, A)$ let $\delta^{in}(S)$ ($\delta^{out}(S)$) denote the set of arcs entering (leaving, resp.) a set of nodes S. An **arborescence** is a directed spanning tree in which every node is entered by at most 1 arc (that is, a spanning tree oriented out of some node r). For nodes $s, t \in V$, a subset S with $s \in S \subseteq V - t$ is called an $s\bar{t}$-**set**.

2   The Iterative Relaxation Algorithm

In this section we briefly recall the Iterative Relaxation Algorithm of Singh and Lau [15] for the Minimum Bounded-Degree Spanning Tree Problem. A nice account of this method (with many other similar iterative algorithms) can be found in the book of Lau, Ravi and Singh [10]. Let us start with some preliminaries.

Theorem 1 (Edmonds [5]). *Given a graph $G = (V, E)$, the convex hull of incidence vectors of spanning trees is described by the following system of linear inequalities.*

$$x \in \mathbb{R}^E, x \geq 0, \qquad (1)$$
$$x(I(S)) \leq |S| - 1 \ for \ any \ non\text{-}empty \ S \subsetneq V, \qquad (2)$$
$$x(E) = |V| - 1. \qquad (3)$$

Consider the following IP formulation of the MBDSPT Problem: we want to find an integer vector $x \in \mathbb{Z}^E$ satisfying (1)-(3) and $x(\delta(v)) \leq b(v)$ for every $v \in V$, and we want to minimize $\sum_{e \in E} c(e)x(e)$. If we drop the integrality constraints, we obtain an LP relaxation called **the Subtour LP**.

The Subtour LP: Assume that we are given an MBDSPT Problem with graph $G = (V, E)$, edge costs $c : E \to \mathbb{R}_+$, and degree bounds $b : V \to \mathbb{Z}_+$. Let us introduce the following polyhedron for some $W \subseteq V$ and $F \subseteq E$ (W is the set of nodes where the degree bound is not yet removed, while F is the subset of edges not yet fixed to zero).

$$SubP(W, F) = \{x \in \mathbb{R}^E : x \geq 0, \qquad (4)$$
$$x(I(S)) \leq |S| - 1 \ \text{for any non-empty} \ S \subsetneq V, \qquad (5)$$
$$x(E) = |V| - 1, \qquad (6)$$
$$x(\delta(v)) \leq b(v) \ \text{for every} \ v \in W, \qquad (7)$$
$$x(e) = 0 \ \text{for} \ e \notin F\}. \qquad (8)$$

Note that $x(e) \leq 1$ is implied by the inequalities (5) above for any edge $e = uv \in E$ (take $S = \{u, v\}$). Note that we could get rid of the dependence on F by simply deleting the edges of $E - F$ as in [10]: we decided not to do so, because this is closer to our implementation (however, we will often ignore the reference to F and simply write $SubP(W)$ instead of $SubP(W, F)$). Given a vector $x \in SubP(W, F)$, a constraint in the system above that holds with equality for x will be called x-**tight** (or simply **tight**, if no confusion can arise). The constraints of form (5) and (6) will be called **set constraints**; those of form (7) will be called **degree constraints**, whereas the first ones (4) are called **non-negativity constraints**. Let $SubP = SubP(\emptyset, E)$ be the polyhedron without the degree bounds (that is, the convex hull of spanning trees by Theorem 1), and let $SubP_b = SubP(V, E)$ be the polyhedron with all the degree constraints.

The linear program below will be called the **Subtour LP** and will be denoted by $SubLP(W, F)$, where $W \subseteq V$ and $F \subseteq E$ (we will often ignore the dependence on F and write $SubLP(W)$ only).

$$\min\{\sum_{e \in E} c(e)x(e) : \ x \in SubP(W, F)\}$$

Clearly, the optimum value of $SubLP(V, E)$ is a lower bound on the optimum of the MBDSPT Problem. The following theorem gives a useful property of extreme-point solutions of this LP.

Theorem 2 (Singh and Lau, [15]). *Assume that x^* is an extreme-point solution of $SubLP(W, F)$ and let $E^* = \{e \in F : x^*(e) > 0\} \subseteq F$ be the support of x^*. Then there exists at least one node $v \in W$ such that $d_{E^*}(v) \leq b(v) + 1$.*

Such a degree bound is called **removable**. The IR Algorithm makes use of this fact by iteratively fixing more and more edges to zero, and relaxing the degree bounds that become removable.

2.1   The Algorithm

The Iterative Relaxation Algorithm of Singh and Lau [15] for the MBDSPT Problem is as follows.

Algorithm IR_MBDSPT(G, c, b)

begin

      INPUT: A graph $G = (V, E)$, edge costs c, and degree bounds $b(v)$ for every $v \in V$.

      OUTPUT: A spanning tree T of G with cost at most the optimum value of $SubLP(V, E)$ and satisfying $d_T(v) \leq b(v) + 1$ at every node v.

1.1. Initialize W to V and F to E.

1.2. **SolveLP:** Find an optimal extreme-point solution x^* of $SubLP(W, F)$.

1.3. While x^* is not an integer vector

1.4.   Let $E^* = \{e \in F : x^*(e) > 0\}$ be the support of x^*.

1.5.   **RelaxLP:** Let W' be W minus **some** of the nodes $v \in W$ with $d_{E^*}(v) \leq b(v) + 1$.

1.6.   **ResolveLP:** Find an extreme-point solution x' of $SubLP(W', E^*)$ satisfying $c^T x' \leq c^T x^*$ (such solution exists, since x^* is feasible to $SubLP(W', E^*)$).

1.7.   Set $W = W'$, $F = E^*$, and $x^* = x'$.

1.8. End while

1.9. Return x^*.

end

Note that in the Relaxation Step we can remove an arbitrary subset of removable degree constraints, but we have to relax at least one of them in order to make progress (i.e., $x' = x^*$ otherwise). We used this observation for finding different solutions for single instances: see details in Section 4.2. Note also that the Resolve Step can simply resolve the LP, but the algorithm also works if x' is not an optimal solution of $SubLP(W', E^*)$. This can be used to speed up the algorithm. However, choosing a suboptimal x' can degrade the quality ($\sim$ cost) of the final solution returned by the algorithm.

The main technical difficulty is Step 1.2: finding an optimal extreme point solution of an LP with exponentially many constraints. We tried many approaches to handle this issue. One possibility is using a polynomial sized LP instead, e.g., the Extended Bidirected LP. However, this LP is still too large to allow for efficient solutions. Instead we used row generation method together with exponential size Subtour LP. Algorithm obtained this way is not guaranteed to run in polynomial time, but performs very well in practice. We have tested the following implementation variants of Algorithm IR_MBDSPT

1. Find_Most_Violated: variant with row generation for the Subtour LP done by choosing the most violated constraint,
2. Find_First_Violated: variant with row generation for the Subtour LP done by choosing the first violated constraint from a random source,

3. Combinatorial_Pivoting: variant with row generation for the Subtour LP done by choosing the first violated constraint from a random source and solving the LP only once using the combinatorial pivoting.

Let t_1, t_2 and t_3 be the average running times (on our benchmark instances) of the respective variants. Row generation (finding violated constraints) for the Subtour LP was the bottleneck of Find_Most_Violated. Changing the row generation method in Find_First_Violated improved the average running time considerably: $t_2 \approx 0.1 \cdot t_1$.

Solving the LP for the first time (Step 1.2 of Algorithm IR_MBDSPT) was the bottleneck of Find_First_Violated. 85% of running time t_2 was spent on that step, and only the the remaining 15% was needed for the loop in Steps 1.3 - 1.8. The third variant: Combinatorial_Pivoting improved the running time of this loop by 60%. It did not, however, have a significant effect on the total running time of the algorithm, as this loop was not the bottleneck of the implementation (that is, the average running time of Combinatorial_Pivoting became $t_3 \approx 0.85 \cdot t_2 + 0.4 \cdot 0.15 \cdot t_2 \approx 0.9 \cdot t_2$).

Combinatorial_Pivoting did however decrease the number of violated degree bounds by a factor of up to 10, while only increasing the solution cost by 2%. This fastest version of the Iterative Relaxation Algorithm solved the largest problems in our benchmarks (cliques of 700 nodes) in approximately 15 minutes.

3   Local Search

We have tried a few approaches to local search and we come to the conclusions that Simulated Annealing [8] gives the best results in our case. We limit the running time of our Local Search to 10s: we have observed that there is no substantial improvement in the cost of the solution after this time limit. Our local search algorithm works as follows. An initial spanning tree T of G is selected. We used a BFS tree rooted at a random node, or a Minimum Cost Spanning Tree. However, this initial choice has no impact on the obtained solution. Then we try to improve T by finding two edges $e_1 \in T$ and $e_2 \in G - T$ such, that $T_{new} = T - e_1 + e_2$ is a spanning tree of G. The penalty of the tree T is defined as follows.

$$P(T) = \sum_{e \in T} c(e) + c_{max} \cdot \sum_{v \in V} \max(d_T(v) - b(v) - 1, 0), \qquad (9)$$

where c_{max} is the maximum cost of edge in G. In other words the penalty is equal to the sum of costs of edges in T plus the penalty for each unsatisfied degree bounds. For every unsatisfied degree bound we add penalty as big as the biggest edge weight times the height of the violation.

The standard hill climbing method tries to find T_{new} with lower penalty than T, whereas in the SA method the penalty can increase with the following probability

$$p(T, T_{new}) = \exp\left(-\frac{P(T_{new}) - P(T)}{t}\right),$$

where t is "temperature" of the process that continually decreases during the execution of the algorithm. We use exponential cooling scheme where the temperature of the next step is equal to αt for a constant $\alpha < 1$.

We have tried two methods of finding T_{new}. They both start by selecting a random non-tree edge $e_1 \in G - E(T)$ and finding a fundamental cycle C in $T + e_1$. The first one chooses e_2 to be a random edge in $C - e_1$ whereas the second one sets e_2 to be the edge on the cycle C that decreases the penalty to the largest extent. In the worst case, this selection takes $O(d)$ time, where d is the diameter of the tree T which is upper bounded by $|V|$. In our benchmarks this implementation ran fast enough, so that the SA process always finished in 10 seconds, and longer cooling schemes did not give better quality solutions.

We have run experiments on both hill climbing and simulated annealing. The latter works visibly better when the degree bounds are small. Moreover, we have tried different penalties for violations of degree bounds and we have found out that setting penalties as in (9) gave the best results. In Figure 1 the results obtained using three implementations of the local search technique are compared using TSPLIB data

- the hill climbing (HC),
- the simulated annealing with random choice of edges e_1 and e_2 (SA 1),
- the simulated annealing with random choice of edge e_1 and respective best choice of edge e_2 (SA 2). Later comparisons will compare SA 2 with IR., since SA 2 gave the best results of three tested algorithms.

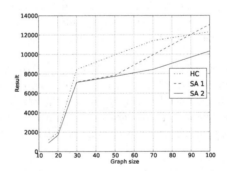

Fig. 1. Comparison of different implementations of local search techniques. Graph size is the number of vertices. The y axis shows the cost of a DBST found by hill-climbing (HC) and two variants of Simulated Annealing (SA 1, SA 2).

4   Experimental Results

We implemented the algorithms described above in C++ (gcc-4.6). As a part of this project a generic framework for implementing IR methods has been developed and became a part of the PAAl library [13]. We also used the LEMON library [11] that proved especially useful for its interface towards LP solvers – we used the CPLEX LP solver [3]. All of our programs were compiled with -O3 optimization option and run on Intel Xeon CPU E5649@2.53GHz machine.

4.1   Test Cases

We tested the algorithms on the following data sets.

TSPLIB Instances: We have taken symmetric TSP instances from the TSPLIB library. We interpreted these as MBDSPT Problem instances by setting the degree bounds identically to 2.

Instances From [9]: We compared the results provided by the algorithms with those found by the heuristics of [9] in terms of the cost function. Using SA we obtained better quality results, e.g., in 16 out of 31 cases we improved heuristically obtained solutions of [9]. On the other hand, in all tests with degree bounds 3, 4 or 5 (except tests *str1009* and *sym704* with degree bound = 3), IR found an optimal integral solution for the initial LP – without violating any degree bounds. For 39 out of 118 tests with degree bounds = 2 the IR also found an optimal (integral) solution for the initial LP.

Instances From [16]: We also compared the results of our algorithms on instances generated using the algorithm described in [16]. For all of the instances generated that way the IR found an optimal (integral) solution for the initial LP. Our SA implementation with limiting of the allowed number of constraint violations to 0 also found results with the same costs as the IR.

Random Generators: We used four random generators. In all generators we generate connected graphs with a given number of nodes and edges, we assign edge costs and degree bounds so that the problem is feasible. Let $U(N)$ denote a random integer variable uniformly distributed in the range $[1, N]$. The generators produce the following types of graphs:

- Generator 1: we generate a random spanning tree T and set the edge costs independently from $U(200)$ (the random spanning tree is generated as follows: add the nodes one by one and connect the new node with a randomly chosen previous one). Then we set the degree bounds to be $d_T(v)$, so that T becomes a feasible solution and $c(T)$ is an upper bound on the cost of an optimal feasible solution. Then we add more edges between nodes not connected so far, with costs from $100 + U(200)$ (thus we don't destroy the solution T very much). Our observations show that the randomly generated tree T rarely has nodes with high degrees.
- Generator 2 uses the ideas from [2]: we generate a graph that has a unique Minimum Cost Spanning Tree (MST), but this tree has high vertex degrees. We set the degree bounds so that the MST is not a feasible solution, but we make sure that there exists a feasible solution. This is achived in the following way. First we generate a random spanning tree $T = (V, E)$ (as in the previous generator), with costs from $100 + U(100)$, and we set the degree bounds to be $d_T(v)$ for every v (again $c(T)$ is an upper bound on the cost of an optimal feasible solution). Then we form the MST by choosing node disjoint stars (each of size roughly $\sqrt{|V|}$), and connecting them in a random way to get a spanning tree. The costs of the edges in this spanning tree are

chosen from $U(100)$, therefore indeed this tree will be the unique MST (if we happen to use an edge created previously, we decrease its cost, so the cost of T might decrease). Finally we add more edges between nodes not connected so far, with costs from $100 + U(100)$.

- Generator 3: first we generate a Hamiltonian path with edge costs from $100 + U(100)$, and set the degree bounds to be equal 2 for every node. Next we add a unique (non-feasible) MST with high degrees, as detailed in the previous generator (edge costs from $U(100)$). Finally we add more edges between nodes not connected so far, and assign the edge costs from $100 + U(100)$.

- Generator 4: we generate a clique of k nodes. Edge costs are chosen from $100 + U(100)$ for all edges except for 3 arbitrary edges incident with an arbitrary node (we call this node the center of the clique), which have costs form $U(100)$. We repeat the described procedure k times to generate k independent graphs. Next we treat those graphs as nodes of a new clique and repeat the above procedure recursively. We add the new edges between the cliques so that each node that was not the center of a clique in the previous recursive step gets exactly one new edge to some other clique. We set the degree bounds to be equal 2 for every node (the generated instance has got a Hamiltonian path, but it also has a non-feasible MST using all 3 low-cost edges in some cliques).

4.2   Comparison of IR and SA

First we performed the vanilla test where both IR and SA are allowed to violate all constraints by one. The comparison of the two methods on TSPLIB instances

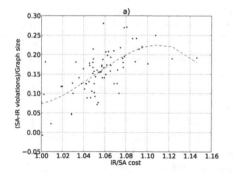
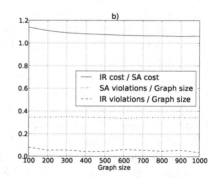

Fig. 2. Comparison of costs and numbers of violated constraints found by IR and SA: (a) for TSPLIB instances, each point represents a single instance, x axis shows cost ratio (SA always outperforms IR), y axis shows difference in the number of violations ratio (IR ouperforms SA in all but one cases) divided by graph size, dashed line is spline interpolation; (b) for random graphs, x axis shows the number of vertices; SA outperforms IR in terms of costs but it violates more constraints.

is shown in Figure 2 (a). Similarly, we have tested the algorithms on random graphs obtained from our 4 generators for sizes from 25 to 1100 nodes. The results for all four graph types are qualitatively the same – Figure 2 (b) gives results for the first generator. We found out that SA almost always found a lower cost spanning tree than IR on those four sets of graphs. On the other hand, IR violates many fewer degree constraints. These results do not give clear answer which method is better. Hence, in order to obtain more decisive results we decided to compare the methods when both of them are allowed the same number of violations. For this comparison we used the Find_First_Violated version of Algorithm IR_MBDSPT (as described in Section 2.1). While it was 10% slower than the fastest version, it had the advantage that in each iteration we could relax more than one constraint. Setting different upper limits on the number of constraints removed in each Relaxation Step of Algorithm IR_MBDSPT gave different trees as the result of the algorithm. Figures 3 and 4 illustrate the comparison of numbers of violated constraints and spanning tree costs in selected cases of random graphs and graphs from TSPLIB. This freedom allows as to compare both algorithms in case of when the number of violated constrains is fixed to be the same. We observed that in most of the cases IR finds solutions

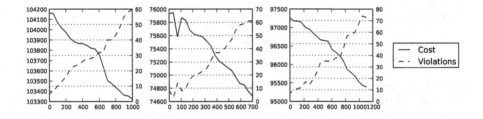

Fig. 3. Comparison of costs and number of violated constraints of three random graphs using IR setting different limits on number of removed constraints (x axis); the left y axis shows the solution cost and the right y axis the number of violations

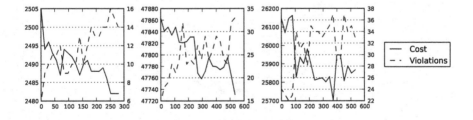

Fig. 4. Comparison of costs and number of violated constraints of three TSPLIB graphs using IR setting different limits on number of removed constraints (x axis); the left y axis shows the solution cost and the right y axis the number of violations

having lower cost than SA with the same number of violations. Figures 5 and 6 show results of such comparison using the same set of graphs as in Figures 3

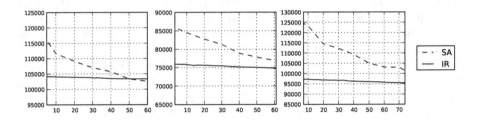

Fig. 5. Comparison of IR and SA algorithm on three random graphs. The x axis shows the number of violated constraints. The y axis shows the solution cost.

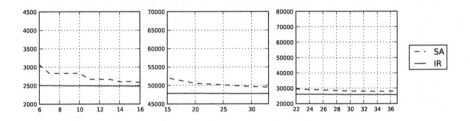

Fig. 6. Comparison of IR and SA algorithm on three TSPLIB graphs. The x axis shows the number of violated constraints. The y axis shows the solution cost.

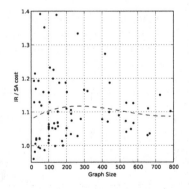

Fig. 7. Comparison of IR and SA algorithm on all TSPLIB graphs when the number of violated constrains in SA is fixed to be the same as violated by IR. The axis x shows number of vertices. The y axis shows ratio of costs. SA finds better solution only in case of small graphs. Dashed line is spline interpolation.

and 4. These figures demonstrate that when the number of violations is small the IR outperforms SA usually by a good 10%. On the other hand, when more violations are allowed this difference decreases, but is still visible in most of the cases. These results on TSPLIB instances are aggregated on Figure 7. It shows that SA was able to find only a few solutions that were better then the ones found by IR. This happened only when graph sizes were small – number of vertices less or equal to 136. The above results give clear evidence that IR delivers betters solutions than SA. We note that, although there is a method to control the number of violated constraints in IR, this method is far from perfect as it is visible on Figure 4. The results are not monotone and so in order to get the right number one needs to go through all the possibilities.

Acknowledgements. The authors would like to thank Tamás Király, Marcin Mucha, Mohit Singh, Łukasz Sznuk, Piotr Wygocki and the developers of the LEMON Library for their help.

References

1. Andrade, R., Lucena, A., Maculan, N.: Using Lagrangian dual information to generate degree constrained spanning trees. Discrete Appl. Math. 154(5), 703–717 (2006), http://www.sciencedirect.com/science/article/pii/S0166218X0500301X
2. Boldon, B., Deo, N., Kumar, N.: Minimum-weight degree-constrained spanning tree problem: Heuristics and implementation on an SIMD parallel machine. Parallel Comput. 22(3), 369–382 (1996)
3. CPLEX, I.I.: High performance mathematical programming engine, http://www-01.ibm.com/software/integration/optimization/cplex-optimizer
4. Deo, N., Hakimi, S.: The shortest generalized hamiltonian tree. In: Proceedings of the 6th Annual Allerton Conference, pp. 879–888. University of Illinois, Illinois (1968)
5. Edmonds, J.: Matroids and the greedy algorithm. Math. Program. 1(1), 127–136 (1971)
6. Furer, M., Raghavachari, B.: Approximating the minimum-degree Steiner tree to within one of optimal. J. Algorithm 17(3), 409–423 (1994)
7. Goemans, M.X.: Minimum bounded degree spanning trees. In: FOCS 2006, pp. 273–282. IEEE Computer Society, Los Alamitos (2006)
8. Kirkpatrick, S., Gelatt, C.D., Vecchi, M.P.: Optimization by simulated annealing. Science 220, 671–680 (1983)
9. Krishnamoorthy, M., Ernst, A.T., Sharaiha, Y.M.: Comparison of algorithms for the degree constrained minimum spanning tree. J. Heuristics 7(6), 587–611 (2001)
10. Lau, L.C., Ravi, R., Singh, M.: Iterative methods in combinatorial optimization. Cambridge University Press, Cambridge (2011)
11. Library for Efficient Modeling and Optimization in Networks (LEMON), http://lemon.cs.elte.hu
12. Narula, S., Ho, C.: Degree-constrained minimum spanning tree. Comput. Oper. Res. 7, 239–249 (1980)

13. Practical Approximation Algorithms Library (PAAL),
    http://paal.mimuw.edu.pl
14. Ravi, R., Marathe, M.V., Ravi, S.S., Rosenkrantz, D.J., Hunt III, H.B.: Many
    birds with one stone: Multi-objective approximation algorithms. In: STOC 1993,
    pp. 438–447. ACM, New York (1993),
    http://doi.acm.org/10.1145/167088.167209
15. Singh, M., Lau, L.C.: Approximating minimum bounded degree spanning trees to
    within one of optimal. In: STOC 2007, pp. 661–670. ACM, New York (2007)
16. Zahrani, M.S., Loomes, M.J., Malcolm, J.A., Albrecht, A.A.: A local search heuris-
    tic for bounded-degree minimum spanning trees. Eng. Optimiz. 40(12), 1115–1135
    (2008), http://www.tandfonline.com/doi/abs/10.1080/03052150802317440

Relocation in Carsharing Systems
Using Flows in Time-Expanded Networks

Sven O. Krumke[1], Alain Quilliot[2], Annegret K. Wagler[2],
and Jan-Thierry Wegener[2*]

[1] University of Kaiserslautern (Department of Mathematics)
Kaiserslautern, Germany
`krumke@mathematik.uni-kl.de`
[2] Université Blaise Pascal (Clermont-Ferrand II)
Laboratoire d'Informatique, de Modélisation et d'Optimisation des Systèmes
BP 10125, 63173 Aubière Cedex, France
`{quilliot,wagler,wegener}@isima.fr`

Abstract. In a carsharing system, a fleet of cars is distributed at stations in an urban area, customers can take and return cars at any time and station. For operating such a system in a satisfactory way, the stations have to keep a good ratio between the total number of places and cars in each station, in order to refuse as few customer requests as possible. This leads to the problem of relocating cars between stations, which can be modeled by means of coupled flows of cars in convoys in a time-expanded network. We present an integer programming formulation for a max-profit flow problem to balance the profit of accepted customer requests against the cost for relocation. This enables us to compute optimal offline solutions only since the computation times are too high for the online situation. We, therefore, devise a heuristic approach based on the flow formulation and show experimentally that it computes reasonable solutions in short time, so that it suits for the online situation.

1 Introduction

Carsharing is a modern way of car rental that contributes to sustainable transport as less car intensive means of urban transport, and an increasing number of cities all over the world establish(ed) such services. Hereby, a fleet of cars is distributed at specified stations in an urban area, customers can take a car at any time and station and return it at any time and station, provided that there is a car available at the start station and a free place at the final station. To ensure the latter, customers have to book their demands in advance.

For operating such a system in a satisfactory way, the stations have to keep a good ratio between the total number of places and the number of cars in each station, in order to refuse as few customer requests as possible. This leads to the problem of balancing the load of the stations, called *Relocation Problem*: an

---

* This work was founded by the French National Research Agency, the European Commission (Feder funds) and the Région Auvergne within the LabEx IMobS3.

J. Gudmundsson and J. Katajainen (Eds.): SEA 2014, LNCS 8504, pp. 87–98, 2014.

operator has to monitor the load situations of the stations and to decide when and how to move cars from "overfull" stations to "underfull" ones.

Balancing problems of this type occur for any car- or bikesharing system, but the scale of the instances, the homogeneity of the fleet, the time delay for prebookings and the possibility to move one or more vehicles in balancing steps differ. We consider an innovative carsharing system, where the cars are partly autonomous, which allows to build wireless convoys of cars leaded by a special vehicle, such that the whole convoy is moved by only one driver (cf. [7]). This setting is similar to bikesharing, where trucks can simultaneously move several bikes during the relocation process [3,4,6]. However, the users of bikesharing systems do not book their requests in advance such that the main goal is to guarantee a balanced system during working hours (dynamic situation as in [6]) or to set up an appropriate initial state for the next morning (static situation as in [4]). In our case, customers book their requests in advance and we can benefit from this forecast of future (im)balances.

For such a carsharing system, the Relocation Problem can be understood as Pickup and Delivery Problem (PDP) in a metric space encoding the considered urban area [11]. Hereby, a set of routes has to be constructed for the convoys, in order to satisfy transportation requests between "overfull" and "underfull" stations, see Section 2 for details. Problems of this type are known to be $\mathcal{NP}$-hard, see e.g. [1,2,12]. In [11], a heuristic approach is presented for the static version of the Relocation Problem that firstly solves a matching problem to generate transport requests, subsequently solves a PDP, and iteratively augments the transport requests and resulting routes.

The Relocation Problem combines two opposite objectives: on the one hand, to maximize the profit of the accepted customer requests, and on the other hand, to minimize the costs incurred by moving cars/drivers during the relocation process. It is natural to assume that the company operating the carsharing system has specified "rental rates" for the customers such that for every request the profit equals the rate. Then, knowing the cost for relocation by means of drivers, the two objectives can be combined as a simple linear combination which specifies the net profit for a specific solution.

Here, we treat the online version of the Relocation Problem in order to capture its dynamic evolution over time. It is natural to interpret the Relocation Problem by means of flows in a time-expanded network [8,10] as, e.g., proposed by [6] for bikesharing systems. In this work, we model the Relocation Problem by coupled flows of cars in convoys in a time-expanded network, taking in addition prebooked customer requests into account, which allows us to solve a max-profit flow problem to balance the profit of accepted customer requests against the cost for relocation, see Section 3. Hereby, considering flows of cars in convoys, i.e., simultaneous flows of cars and convoy drivers in the same network, means that the two flows are not independent or share arc capacities as in the case of multicommodity flows, but are coupled in this sense that the flow of cars is dependent from the flow of drivers (since cars can only be moved in convoys). The coupling constraints for the two flows reflect the complexity of the studied problem (and the constraint matrix of the resulting integer programming formulations is not totally unimodular as in the case of normal flows). This enables us to compute optimal offline solutions only since the computation times are too high for the online situation. We, therefore, devise a heuristic approach based on the integer programming formulation and show experimentally that it computes reasonable solutions in short time, so that it suits for the online situation, see Section 4.

We close with computational results and a discussion on the pros and cons of our approach and future lines of research.

2   Problem Description and Model

We model the Relocation Problem in the framework of a metric task system.

By [11], the studied carsharing system can be understood as a discrete event-based system, where the system components are the stations $v_1, \ldots, v_n$, each having an individual capacity $cap(v_i)$, a system state $z^t \in \mathbb{Z}^n$ specifies for each station v_i its load z_i^t at a time point $t \leq T$ within a time horizon $[0, T]$ and is changed when customers or convoy drivers take or return cars at stations.

To balance the load of the stations, an operator has to monitor the system states and to decide when and how to move cars from overfull stations v_i with $z_i^t > cap(v_i)$ to underfull stations v_j with $z_j^t < 0$, in order to fulfill all pre-booked demands (Relocation Problem). The operator monitors the evolution of system states over time, detects future infeasible states and creates tasks to move cars out of overfull stations and into underfull ones. A *task* is defined by $\tau = (v, rel, due, x)$ where $x \in \mathbb{Z} \setminus \{0\}$ cars are to pickup (if $x > 0$) or to deliver (if $x < 0$) at station v within a time-window $[rel, due] \subseteq [0, T]$ of the release date *rel* and the due date *due*.

To fulfill these tasks, routes have to be created for the convoys in order to perform the desired relocation process. For that, it is suitable to encode the urban area where the carsharing system is running as a *metric space* $M = (V, d)$ induced by a weighted graph $G = (V, E, cap, \ell)$, where the nodes correspond to stations, edges to their physical links in the urban area, node weights to the station's capacities, and the edge weights $\ell : E \to \mathbb{R}_+$ determine the distance d between two points $v, w \in V$ as length of a shortest path from v to w.

This together yields a *metric task system*, a pair $(M, \mathcal{T})$ where $M = (V, d)$ is the above metric space and $\mathcal{T}$ a set of tasks, as suitable framework to embed the routes for the convoys. A driver able to lead a convoy plays the role of a server, the number of available drivers will be denoted by k. Each server has capacity L, corresponding to the maximum possible number of cars per convoy; several servers are necessary to serve a task τ if $x(\tau) > L$ holds.

More precisely, we define the following. An *action* for driver j is a 4-tuple $a = (j, v, t, x)$, where $j \in \{1, \ldots, k\}$ specifies the driver $driv(a)$ performing the action, $v \in V$ specifies the station $loc(a)$, $t \in [0, T]$ is the time $t(a)$ when the action is performed, and $x \in \mathbb{Z}$ the number of cars $\Delta x(a)$ to be loaded (if $x > 0$) or unloaded (if $x < 0$). Hereby, the capacity of the convoy must not be exceeded, i.e., we have $|x| \leq L$. We say that an action is *performed* (by a driver) if he loads (resp. unloads) $|x|$ cars at v.

For technical reasons, the vector $z^t \in \mathbb{N}^{|V|}$ represents the number of cars in a station v at time t before any action is performed. A *move* from one station to another is $m = (j, v, t^v, w, t^w, P, x_m)$, where $j \in \{1, \ldots, k\}$ specifies the driver $driv(m)$ that has to move from the origin station $orig(m) = v \in V$ starting at time $dep(m) = t^v$ to destination station $dest(m) = w \in V$ arriving at time $arr(m) = t^w$, a load of $load(m) = x_m$ cars in the convoy moving along the path $path(m) = P$. Hereby, we require that $0 \leq load(m) \leq L$, P is a shortest (v, w)-path, and that from $orig(m) \neq dest(m)$ follows $arr(m) = dep(m) + d(orig(m), dest(m))$.

A *tour* is an alternating sequence $\Gamma = (m^1, a^1, m^2, a^2, \ldots, a^{n-1}, m^n)$ of moves and actions with

- $driv(m^1) = driv(a^1) = \cdots = driv(a^{n-1}) = driv(m^n)$,
- $dest(m^i) = loc(a^i) = orig(m^{i+1})$,
- $arr(m^i) = t(a^i) = dep(m^{i+1})$, and
- $load(m^{i+1}) = load(m^i) + \Delta x(a^i)$.

Several tours are composed to a transportation schedule. A collection of tours $\{\Gamma^1, \ldots, \Gamma^k\}$ is a *feasible transportation schedule* for $(M, \mathcal{T})$ if every driver has exactly one tour, each task $\tau \in \mathcal{T}$ is served (i.e., for every task $\tau = (v, rel, due, x)$, the $\Delta x(a)$ of all actions a performed at station v sum up to x), and all system states z^t are feasible during the whole time horizon $[0, T]$.

The Relocation Problem consists in creating tasks from given customer requests (by detecting overfull and underfull stations where cars have to be moved out or in) and constructing a transportation schedule from the resulting metric task system $(M, \mathcal{T})$. The approach in [11] solves the static version of the problem in three steps: generate firstly tasks, then transport requests between overfull and underfull stations and solve a PDP to obtain a transportation schedule. We propose an approach for the online version of the problem in one single step by a max-profit flow problem that takes directly the customer requests as input.

3   Flows in Time-Expanded Networks

We propose a way to solve the Relocation Problem in one step by defining a time-expanded network with two coupled flows: a car and a driver flow.

The input for the Relocation Problem consists of the following data:

- a weighted graph $G = (V, E, cap, \ell)$, where the nodes correspond to stations, edges to their links, node weights to the station's capacities, and the edge weights $\ell : E \to \mathbb{R}_+$ determine the driving times between two points $v, w \in V$ as length of a shortest path from v to w;
- the total number k of drivers, the maximum number $L \in \mathbb{N}$ of cars which can be simultaneously moved in one convoy, the initial quantities $z^c(v) \leq cap(v)$ and $z^d(v)$ of cars and drivers located at v at the start time $t = 0$;

- a sequence $R = \{r_1, \ldots, r_h\}$ of customer requests arriving over time, where each request has the form $r_j = (p_j, v_j, q_j, w_j)$ where the pickup time $p_j \in \mathbb{N}$ indicates when a car is requested to be picked up at station $v_j \in V$ and the drop time $q_j \in \mathbb{N}$ indicates when it will be dropped at station $w_j \in V$;
- per unit costs cost^c and cost^d for moving cars and drivers within the network, and per customer request a profit $p(r)$ for serving the request r.

The output of the Relocation Problem is a transportation schedule for a metric task system, whose tasks are directly induced by the customer requests.

We consider a max-profit flow problem which rejects customer requests from R whose profit is smaller than the relocation cost to satisfy them. For that, we build a directed graph $G_T = (V_T, A_T)$ as a time-expanded version of the original network G which includes arcs corresponding to the customer requests in R. The cars and drivers form two flows f and F through G_T which are coupled in the sense that on so-called relocation arcs $a \in A_L \subset A_T$ (see Section 3.1) we have the condition $f(a) \leq L \cdot F(a)$ reflecting the dependencies between the two flows. Note that the tasks are directly derived from the sequence R of customer requests (if an accepted request causes an infeasible system state, the task to balance this state is implicitly generated, see Section 3.2).

3.1 Time-Expanded Networks

We build a time-expanded version $G_T = (V_T, A_T)$ of the original network G.

The node set V_T is constructed as follows. Let $\mathcal{T} = \{0, \ldots, T\}$ be a finite set of points in time, where the time horizon T is the maximum drop time of a (yet known) request in R. For each station $v \in V$ and each time point $t \in \mathcal{T}$, there is a node $(v, t) \in V_T$ which represents station v at time t as a capacitated station where convoys can simply pass or cars can be picked up, delivered and exchanged by drivers.

The arc set $A_T = A_H \cup A_L \cup A_R$ of G_T is composed of several subsets:

- A_H contains, for each station $v \in V$ of the original network and each $t \in \{0, 1, \ldots, T-1\}$, the holdover arc connecting (v, t) to $(v, t+1)$.
- A_L contains, for each edge (v, w) of G and each point in time $t \in \mathcal{T}$ such that $t + d(v, w) \leq T$, the relocation arc from (v, t) to $(w, t + d(v, w))$.
- A_R contains, for each customer request $r = (v, t, w, t') \in R$ a request arc from (v, t) to (w, t').

Note, that the time-expanded network G_T is acyclic by construction.

Remark 1. We note that the above construction of the time-expanded network generalizes in a straightforward way to the case that M is not metric but only *quasi-metric*, which happens for instance if the original graph G is directed. Then, the time needed for driving from v to w may be different from the time needed for driving in opposite direction, i.e., from w to v.

Moreover, although for the scope of this paper we limit the attention to constant driving times, the construction also generalizes in a natural way to time-dependent driving times cf. [9].

3.2   A Max-Profit Flow Problem

On the time-expanded network G_T, we define two different flows, the car flow f and the driver flow F, to encode the relocation of cars in convoys.

Note that a flow on a relocation arc corresponds to a (sub)move in a tour, i.e., some cars are moved by drivers in a convoy from a station v to another station w. Thus, a relocation arc from (v, t) to $(w, t + d(v, w))$ has infinite capacity for the drivers, but to ensure that cars can be moved only in convoys and at most L cars per driver, we require that $f(a) \leq L \cdot F(a)$ for all $a \in A_R$ holds (see (1g)). Since drivers use their "own" cars, the driver flow is not bounded on relocation arcs. Thus, the capacities for f on the relocation arcs are not given by constants but by a function. Note that due to these flow coupling constraints, the constraint matrix of the network is not totally unimodular (as in the case of uncoupled flows) and therefore integrality constraints for both flows are required (1j)) and solving the problem becomes $\mathcal{NP}$-hard.

Flows on holdover arcs correspond to cars/drivers remaining at the station in the time interval $[t, t + 1]$. Thus, the capacity of all holdover arcs for flow f is set to $cap(v)$ (see (1f)), whereas there is no capacity constraint for F on such arcs. Moreover, a car flow on a customer request arc corresponds to an accepted request (see (1h)), whereas driver flow is forbidden on such arcs (see (1i)).

We consider a max-profit flow problem to decide which customer requests can be satisfied without spending more costs in the relocation process than gaining profit by satisfying them. Accordingly, our objective function (1a) considers profits $p(a)$ for the car flow f on all $a \in A_R$ and costs

$$c(a) := \text{cost}^c \cdot d(v, w) \quad \text{and} \quad C(a) := \text{cost}^d \cdot d(v, w)$$

on all relocation arcs $a = ((v, t), (w, t + d(v, w)))$ corresponding to an edge (v, w) in G, whereas all other arcs have zero profits and costs.

To correctly initialize the system, we use the nodes $(v, 0) \in V_T$ as sources for both flows and set their balances accordingly to the initial numbers of cars and drivers at station v and time 0 in $z^c(v)$ and $z^d(v)$ (see (1b) and (1c)). For all internal nodes $(v, t) \in V_T$ with $t > 0$, we use normal flow conservation constraints (which is possible due to the fact that the entire flow of cars is modeled by taking parked cars, convoy moves and customer actions into account), see (1d) and (1e).

This leads to a Max-Profit Relocation Problem, whose output is a subset of accepted customer requests $R' \subseteq R$ and a transportation schedule for a metric task system, whose tasks are induced by the decision which customer requests are accepted. The corresponding integer linear program is as follows:

$$\max \sum_{a \in A_R} p(a)f(a) - \sum_{a \in A_L} c(a)f(a) - \sum_{a \in A_L} C(a)F(a) \tag{1a}$$

$$\sum_{a \in \delta^-(v,0)} f(a) = z^c(v) \qquad \forall\, (v, 0) \in V_T \tag{1b}$$

$$\sum_{a \in \delta^-(v,0)} F(a) = z^d(v) \qquad \forall\, (v, 0) \in V_T \tag{1c}$$

$$\sum_{a\in\delta^-(v,t)} f(a) = \sum_{a\in\delta^+(v,t)} f(a) \quad \forall\ (v,t)\in V_T, t>0 \tag{1d}$$

$$\sum_{a\in\delta^-(v,t)} F(a) = \sum_{a\in\delta^+(v,t)} F(a) \quad \forall\ (v,t)\in V_T, t>0 \tag{1e}$$

$$0 \le f(a) \le cap(v) \qquad\qquad \forall\ a = [(v,t),(v,t+1)] \in A_H \tag{1f}$$

$$f(a) \le L \cdot F(a) \qquad\qquad \forall\ a \in A_L \tag{1g}$$

$$f(a) \le 1 \qquad\qquad \forall\ a \in A_R \tag{1h}$$

$$F(a) = 0 \qquad\qquad \forall\ a \in A_R \tag{1i}$$

$$f, F \text{ integer}, \tag{1j}$$

where $\delta^-(v,t)$ denotes the set of outgoing arcs of (v,t), and $\delta^+(v,t)$ denotes the set of incoming arcs of (v,t).

The Max-Profit Relocation Problem solves the offline version of the Relocation Problem (where the whole sequence R of customer requests is known at time $t = 0$) to optimality. Note that the Offline Max-Profit Relocation Problem is always feasible (since rejecting all customer requests is a solution).

To handle the online situation (where the customer requests in R arrive during a time horizon $[0, T]$), we propose to solve a sequence of Max-Profit Relocation Problems for certain time intervals within $[0, T]$. To take the dynamic evolution of the customer requests into account, we partition R into two subsets R^A of previously accepted customer requests and R^N of customer requests released within the new time interval. Accordingly, we refine the arc set A_R of G_T by considering two subsets $A_{R^A} \cup A_{R^N}$ for to the two request subsets R^A and R^N and set $f(a) = 1$ for all $a \in A_{R^A}$, to ensure that previously accepted customer requests are fulfilled. For every newly released customer request, we bound the car flow on the corresponding arc $a \in A_{R^N}$ by $f(a) \le 1$ to allow that new customer requests can be rejected. The resulting Online Max-Profit Relocation Problem is feasible if and only if the Relocation Problem has a feasible solution with R^A as the input sequence of customer requests. The latter is ensured in the online situation, since the previous solution of a Max-Profit Relocation Problem did not only return a set R^A of accepted customer requests, but also a transportation schedule to solve the resulting Relocation Problem (and rejecting all newly released customer requests in $a \in A_{R^A}$ is possible).

Remark 2. It is also possible to formulate a min-cost flow problem to serve all given customer requests at minimal cost, if possible (to capture the quality-of-service aspect of the Relocation Problem). This can be seen as special case of the above problem with $p(r) = 0\ \forall r \in R$ and $R^A = R$. However, this problem is not always feasible, and detecting a maximal feasible subset $R^A \subseteq R$ is not easy, so that considering the max-profit version of the problem is advantageous.

4   A Flow-Based Heuristic

Solving the Relocation Problem on a complete time-expanded network G_T from Section 3 is generally very slow and, thus, not applicable in practice. However, the runtime can be improved by reducing the number of arcs and nodes in the time-expanded network, which corresponds to a reduction of variables in the corresponding integer linear program. As experiments have shown that only a small percentage of arcs in G_T is used in the optimal solution, the idea is to reduce G_T to a network containing only arcs which are taken in the optimal solution with high probability.

In this section we present the resulting heuristic that solves the Online Max-Profit Relocation Problem in two phases. In the first phase, we compute those arcs which are likely to be used in an optimal solution. In the second phase, we construct a reduced time-expanded network G'_T, where we keep only the previously computed arcs and discard all others; afterwards, we compute an optimal solution on this reduced network G'_T. This does not lead to a globally optimal solution on G_T, but provides reasonable solutions in short time.

4.1   Preprocessing (Phase 1)

This phase itself is performed in two steps. Firstly we compute only a car flow and, secondly, a driver flow that "covers" the car flow.

Car Flow Model and its Linear Program. In this paragraph, we specify the capacities as well as the profits and the costs for each arc with respect to the car flow f_1 on the original time-expanded network G_T. Hereby, we mention only the differences to the time-expanded network from Section 3.1. Finally, we give a linear program in order to compute this car flow.

Unlike to the exact approach, in the preprocessing step, the car flow on a relocation arc is not coupled to another flow but has infinite capacity. The costs with respect to the flow f_1 are set to 1 for each relocation arc.

On customer request arcs we set an upper bound of 1. Customer request arcs have no costs but a profit. In order to ensure that there is positive car flow on a customer request arc $a \in A_R$ whenever possible, the profit must be selected high enough, compared to the costs of the relocation arcs, e.g., $p(a) = |V|$.

In order to compute the car flow f_1 we consider

$$\max \sum_{a \in A_R} p(a) f_1(a) - \sum_{a \in A_L} c(a) f_1(a),$$

with subject to the constraints (1b), (1f), (1h) and f_1 real.

Constraining the flow to be integer is not necessary since the constraint matrix is totally unimodular. Thus, the solution contains integer values only.

Note, from the solution of the car flow, it is possible to directly compute an upper bound on the maximal profit as well as the maximal number of customer requests which can be theoretically served.

Driver Flow Model and its Linear Program. Next, we define a driver flow F_1 on the time-expanded network G_T. We specify the capacities as well as the profits and the costs for each arc with respect to F_1. Finally, we give a linear program in order to compute the driver flow F_1.

The driver flow is influenced by the car flow in such a way that we try to "cover" the car flow on the relocation arcs by giving a profit on the relocation arcs having a positive car flow. All other relocation arcs have costs of 1. Note, that the profits for the relocation arcs must be chosen high enough, e.g., $p(a) = |V|$, $a \in A_L$. The capacity for a relocation arc is infinity.

In order to compute the driver flow F_1 in $G_T = (V_T, A_T)$, we consider

$$\max \sum_{a \in A_L^+} p(a) F_1(a) - \sum_{a \in A_L^-} F_1(a)$$

where $A_L^- = \{a \in A_L \mid f_1(a) = 0\}$ and $A_L^+ = \{a \in A_L \mid f_1(a) > 0\}$, with subject to the (in)equalities (1c), (1e), (1i), and F_1 real.

Like in the previous step, the flow does not need to be constrained to integer values, since the constraint matrix is totally unimodular.

4.2   Computing a Transportation Schedule (Phase 2)

Unlike in the exact approach, the two flows are not coupled in the first phase. This implies that each flow can be rapidly computed, but the computed solution is in general not a feasible transportation schedule. In this section, we describe the construction of a reduced version $G_T' = (V_T', A_H' \cup A_L' \cup A_R)$ of the original time-expanded network $G_T = (V_T, A_H \cup A_L \cup A_R)$ based on the flows computed in the preprocessing (Phase 1). Hereby, we reduce the total number of nodes as well as of holdover and relocation arcs. The set of customer request arcs is not changed. Afterwards, we compute an optimal solution on G_T' providing a feasible transportation schedule.

Constructing the Reduced Time-Expanded Network. The reduced network G_T' is constructed as follows. First, we add for each station $v \in V$ the nodes $(v, 0), (v, T)$ to V_T', and for all customer request arcs $[(v, t), (w, t')] \in A_R$ we add the nodes $(v, t), (w, t')$ to V_T'.

Only the relocation arcs $a = [(v, t), (w, t')] \in A_L$ with $f_1(a) > 0$ or $F_1(a) > 0$ remain in A_L', and we add the nodes $(v, t), (w, t')$ to V_T'.

Next, for each station $v \in V$ we add holdover arcs between two successive nodes on the time line of v. The set of request arcs is taken unchanged from G_T.

Computing a Feasible Transportation Schedule. A feasible transportation schedule is computed by solving the integer linear program from Section 3.2 on the reduced time-expanded network G_T'.

The problem is always feasible due to the following reason: In the Relocation Problem, every customer request can be rejected. Due to the holdover arcs, for every station $v \in V$ there is a path from the source node $(v, 0)$ to the node (v, T) for the car and driver flows. Thus, we can directly conclude that the flow-based heuristic always computes a feasible solution for the Relocation Problem.

4.3 Improving the Solution

Although, the flow-based heuristic already computes a feasible solution for the Relocation Problem, the solution might be suboptimal. Due to the reduction of the relocation arcs, it might be impossible to find a tour for the drivers on the reduced network $G'_T = (V'_T, A'_H \cup A'_L \cup A_R)$ so that they can serve (some) customer request. Adding more relocation arcs to A'_L than described in Section 4.2 can increase the number of served customer requests. However, this generally also results in higher computation times. Therefore, one has to carefully select which relocation arcs shall be added.

We now briefly describe a variation where further relocation arcs are added to A'_L in order to improve the solution. For that, let f_1 be the car flow and let F_1 be the driver flow from the preprocessing step. When the car flow f_1 on a relocation arc $a = [(v, t), (w, t')] \in A_L$ is greater than 0, but the driver flow F_1 on this arc is 0, it is likely that it is impossible to find tours for the drivers in G'_T which transfer cars on the corresponding relocation arc $a' \in A'_L$. In order to increase the probability of the existence of a tour which transfers cars on a, we add all relocation arcs from $a' \in A_L$ where (v, t) is the end node of a' to A'_L.

5 Computational Results and Conclusions

In this paper, we considered the Relocation Problem. Hereby, tours for k drivers have to be computed in a graph G, so that the profit for serving given customer requests is maximized against the costs for relocation. In order to have an exact solution we construct a time-expanded network G_T from the original network G and compute two coupled flows (a car and a driver flow) on this network with an integer linear program (ILP). Due to the coupling constraints, the constraint matrix of the network is not totally unimodular (as in the case of uncoupled flows) and, therefore, solving such problems becomes $\mathcal{NP}$-hard.

We presented a flow-based heuristic approach to solve the Relocation Problem in two phases. Based on the uncoupled car and driver flows (Section 4.1), we compute a reduced time-expanded network containing only those arcs which are likely to be used in the optimal solution. Afterwards, we solve the problem on this reduced network (Section 4.2). Finally, we gave a variation of the heuristic, where the solution is computed on a slightly bigger network (Section 4.3).

All approaches have been tested on 1080 randomly generated instances (with 50–99 stations; the number of cars is 10 times the number of stations; convoy capacities of 20; 5–15 drivers; time-horizons between 60 and 240 time units; and 500–8000 customer requests). Note that the size of these instances corresponds to small car- or bikesharing systems or to clusters of larger systems, as in [13]. The approaches are implemented in Python and Gurobi 5.6 is used for solving the ILPs. The tests have been run on a server with an Intel Xeon E7-8870 processors clocked at 2.40 GHz, with 1 TB RAM, the results are summarized in Table 1.

The flow-based heuristic computes solutions within a reasonable time (in average on all instances in less than 4 minutes, the median is 13 seconds), and a reasonable gap (in average about 3%) to the maximal number of customer requests which can be served. The variant described in Section 4.3 slightly improves the number of served customer requests (in average the gap to the upper bound is less than 3%). However, the average computation time increases to 10 minutes, while the median only increases to 14 seconds. The biggest impact on the runtime is the time horizon and the number of customer requests, while the number of stations seems to have only little impact on the runtime. However, in some cases an optimal solution has been faster computed on instances with a greater amount of stations, when the number of customer requests remains equal. The time limit for the ILP solver was set in all cases to 1 hour, the exact optimal solution was only found on 110 instances within this time limit. On 39 instances, (resp. 137 for the variation of Section 4.3), the flow-based heuristic could not find the optimal solution of Phase 2 within the time limit. A feasible solution was almost always found within the time-limit. Only the exact approach could not find a feasible solution on 40 instances. The two flow-based heuristics always found a feasible solution.

Table 1. This table shows the median computational results for several test sets of instances. Each median is derived from eight different networks and different numbers of drivers (5, 10, 15 drivers). The complete list of results is available in [5]. The following results are shown: the flow-based heuristic (FBH) (Sections 4.1 and 4.2), the improved flow-based heuristic (IFBH) from Section 4.3, the exact approach (OPT) limited to 1 hour runtime (Section 3), and an upper bound (UB), namely to the maximal number of requests which can be accepted. This value is derived from the uncoupled car flow from Section 4.1. In this table, the following parameters and results are shown: the total amount of stations $|V|$, the number of released customer requests and the time horizon T (1st column), the median of the absolute number of accepted requests (2nd column), the median of the relative gap to the upper bound (3rd column), the median of the runtime for solving these instances (4th column), and the number of optimal solutions found (on the complete resp. reduced time-expanded network) within the time limit in percent (5th column). With the exact approach, a feasible solution was not always found within 1h. In this case, only those values are considered, where a feasible solution was found. Such values, i.e., those which are computed from only a subset of experiments, are highlighted (underlined) in this table.

$|V|$	req	T	acc. req. (abs.) FBH	IFBH	OPT	UB	UB gap (%) FBH	IFBH	OPT	runtime (s) FBH	IFBH	OPT	sol. within 1h (%) FBH	IFBH	OPT
50	500	60	467	470	486	500	6.3	6.0	2.7	3.0	3.0	3602.0	100	100	40
50	2000	60	769	769	771	773	0.3	0.3	0.2	4.0	4.0	3602.0	100	100	7
50	2000	120	1106	1106	1107	1135	1.7	1.6	1.8	11.0	21.0	3604.0	100	100	0
50	2000	240	1380	1384	1332	1432	2.9	2.5	6.2	305.0	3612.0	3609.0	93	30	0
50	4000	120	1455	1457	1451	1476	1.2	1.1	1.3	15.0	21.0	3604.0	100	100	0
50	4000	240	1903	1908	1827	1961	2.4	2.2	5.6	322.5	3615.0	3609.5	80	33	0
50	8000	240	2558	2562	<u>2389</u>	2608	2.0	1.9	<u>8.2</u>	824.5	3618.0	<u>3609.5</u>	60	27	<u>0</u>
75	500	60	490	490	495	500	2.0	2.0	1.0	5.0	4.0	32.0	100	96	97
75	2000	60	1052	1052	1053	1064	0.9	0.9	0.8	6.0	5.0	3603.0	100	100	7
75	2000	120	1114	1114	1110	1136	1.4	1.4	2.4	11.0	10.0	3606.0	100	100	0
75	2000	240	1451	1455	1345	1503	2.9	2.8	9.9	49.5	116.5	3614.0	100	87	2
75	4000	120	1531	1532	1523	1565	1.7	1.6	2.5	14.0	13.0	3607.0	100	100	0
75	4000	240	2059	2065	1898	2121	2.9	2.6	10.3	126.5	3435.0	3614.0	97	50	0
75	8000	240	2750	2753	<u>58</u>	2846	2.8	2.7	<u>97.9</u>	523.0	3627.0	<u>3614.0</u>	70	30	<u>0</u>
99	500	60	493	494	497	500	1.3	1.1	0.4	5.0	4.0	26.0	100	100	100
99	2000	60	1423	1425	1435	1482	3.6	3.5	3.0	7.0	7.0	3604.0	100	100	3
99	2000	120	1210	1210	1202	1232	1.9	1.9	2.5	13.0	12.0	3609.0	100	100	0
99	2000	240	1525	1525	1429	1598	3.9	3.6	10.5	45.5	84.0	3619.0	100	100	0
99	4000	120	1495	1496	1488	1519	1.4	1.4	1.8	16.0	14.0	3609.0	100	100	0
99	4000	240	2165	2169	<u>17</u>	2249	3.6	3.4	<u>99.3</u>	64.0	249.0	<u>3619.0</u>	100	87	<u>0</u>
99	8000	240	2827	2832	<u>21</u>	2911	2.5	2.4	<u>99.3</u>	91.5	211.0	<u>3618.0</u>	100	100	<u>0</u>

There are several practical and theoretical open questions concerning the Relocation Problem. Improving the runtime and solutions of the flow-based heuristic is certainly one goal. This possibly can be achieved by a better analysis of the arcs which are used to construct the reduced time-expanded network. Furthermore, applying this idea to similar problems (e.g., the static version of the Relocation Problem) is another open problem. Since the presented heuristic relies on the fact that a feasible solution can reject customer requests, applying this heuristic on other problems is generally not a trivial task.

Additional computational experiments with another set of parameters, e.g., different ratios of cars and stations or other distribution functions for the generation of customer requests, probably give better insights on the performance and capabilities of the heuristic.

References

1. Ball, M.O., Magnanti, T.L., Monma, C.L., Nemhauser, G.L. (eds.): Network Models. Handbooks in Operations Research and Management Science, vol. 7. Elsevier Science B.V., Amsterdam (1995)
2. Ball, M.O., Magnanti, T.L., Monma, C.L., Nemhauser, G.L. (eds.): Network Routing. Handbooks in Operations Research and Management Science, vol. 8. Elsevier Science B.V., Amsterdam (1995)
3. Benchimol, M., Benchimol, P., Chappert, B., de la Taille, A., Laroche, F., Meunier, F., Robinet, L.: Balancing the stations of a self service bike hire system. RAIRO - Operations Research 45, 37–61 (2011)
4. Chemla, D., Meunier, F., Calvo, R.W.: Bike sharing systems: Solving the static rebalancing problem. Discrete Optimization, 120–146 (2013)
5. Krumke, S.O., Quilliot, A., Wagler, A.K., Wegener, J.-T.: (2014), http://www.mathematik.uni-kl.de/opt/mitglieder/krumke/research/
6. Contardo, C., Morency, C., Rousseau, L.-M.: Balancing a dynamic public bike-sharing system. Technical Report 9, CIRRELT (2012), https://www.cirrelt.ca/DocumentsTravail/CIRRELT-2012-09.pdf
7. EL-Zaher, M., Dafflon, B., Gechter, F., Contet, J.-M.: Vehicle Platoon Control with Multi-Configuration Ability. Procedia Computer Science 9, 1503–1512 (2012)
8. Groß, M., Skutella, M.: Generalized maximum flows over time. In: Solis-Oba, R., Persiano, G. (eds.) WAOA 2011. LNCS, vol. 7164, pp. 247–260. Springer, Heidelberg (2012)
9. Hamacher, H.W., Tjandra, S.: Earliest arrival flow with time dependent capacity for solving evacuation problems. In: Pedestrian and Evacuation Dynamics, pp. 267–276. Springer (2002)
10. Koch, R., Nasrabadi, E., Skutella, M.: Continuous and discrete flows over time. Mathematical Methods of Operations Research 73(3), 301–337 (2011)
11. Krumke, S.O., Quilliot, A., Wagler, A.K., Wegener, J.-T.: Models and algorithms for carsharing systems and related problems. Electronic Notes in Discrete Mathematics 44(0), 201–206 (2013)
12. Nemhauser, G.L., Rinnooy Kan, A.H.G., Todd, M.J. (eds.): Optimization. Handbooks in Operations Research and Management Science, vol. 1. Elsevier Science B.V., Amsterdam (1989)
13. Schuijbroek, J., Hampshire, R., van Hoeve, W.-J.: Inventory rebalancing and vehicle routing in bike sharing systems. Working paper (2013)

Dynamic Windows Scheduling with Reallocation [*]

Martín Farach-Colton[1], Katia Leal[2], Miguel A. Mosteiro[3], and Christopher Thraves[2]

[1] Dept. of Computer Science, Rutgers University, Piscataway, NJ, USA & Tokutek Inc.
farach@cs.rutgers.edu
[2] GSyC, Universidad Rey Juan Carlos, Madrid, Spain
{katia,cbthraves}@gsyc.es
[3] Dept. of Computer Science, Kean University, Union, NJ, USA
mmosteir@kean.edu

Abstract. We consider the Windows Scheduling problem. The problem is a restricted version of Unit-Fractions Bin Packing, and it is also called Inventory Replenishment in the context of Supply Chain. In brief, the problem is to schedule the use of communication channels that allow at most one transmission per time slot, to clients specified by a maximum delay between consecutive transmissions. We extend previous online models, where decisions are permanent, assuming that clients may be reallocated at some cost. We present three online reallocation algorithms for Windows Scheduling. We analyze one of them and we evaluate experimentally all three showing that, in practice, they achieve constant amortized reallocations with close to optimal channel usage. Our simulations also expose interesting trade-offs between reallocations and channel usage. To the best of our knowledge, this is the first study of Windows Scheduling with reallocation costs.

Keywords: Reallocation Algorithms, Windows Scheduling, Radio Networks, Unit Fractions Bin Packing.

1 Introduction

We study the Windows Scheduling[1] problem (WS) [4,5,8], which is a restricted version of Unit-Fractions Bin Packing (UFBP) [1,2,5,7,10,12], and it is also called Inventory Replenishment in the context of Supply Chain [18]. In brief, the WS problem is to schedule the use of communication channels to clients. Each client c_i is characterized by an active cycle (with arrival and departure times) and a window w_i also called laxity[2] (in the context of job scheduling). During the period of time that any given client c_i is active, there must be at least one transmission from c_i scheduled in any w_i consecutive

---

[*] This work was supported in part by the National Science Foundation (CCF- 0937829, CCF- 1114930) and Kean University UFRI grant.

[1] Were it not for the extensive literature on windows scheduling, we would have chosen a different name. After all, the goal of windows scheduling is not to schedule windows but transmissions. Moreover, those transmissions need not be scheduled within fixed epochs, as the term *window* seems to suggest. Nonetheless, we use the current notation for literature consistency.

[2] Throughout the paper, we will use the term *laxity* instead of *window* since it conveys better the concept of maximum delay between transmissions. Nonetheless, we will also use the term *window* when describing the related literature.

J. Gudmundsson and J. Katajainen (Eds.): SEA 2014, LNCS 8504, pp. 99–110, 2014.
© Springer International Publishing Switzerland 2014

time slots. The optimization criterion is to minimize the number of channels used. The WS problem appears in many areas such as Operations Research, Networks, Streaming, etc. More application details can be found in [4, 5, 8, 18] and the references therein.

Given that even a restricted version of the WS problem was shown to be NP-hard even for one channel [3], practical WS solutions are only approximations. In the WS literature, competitive analysis of the ratio between the number of channels used by an online algorithm with respect to an optimal algorithm has been carried out in two flavors: the maximum ratio for any given time instant (also called *against current load*), and as the ratio of the maxima (also called *against peak load*). In the model of [4, 5] both competitive ratios are the same because clients do not leave the system. In [8, 18] the competitive ratio is against peak load. In the present work, we carry out competitive analysis against current load.

In [4, 5] clients do not leave the system whereas in [8, 18] clients may leave. We further extend the model assuming that clients may be reallocated at some cost. As used in [6] (for Job Scheduling), we call this class of protocols ***Reallocation Algorithms.*** Reallocation algorithms are a middle ground between online algorithms (where assignments are final) and offline algorithms which can be used repeatedly if reallocations are free. Reallocation has been studied previously for Load Balancing [14, 17], and for UFBP in [1, 2] where it was called *semi-online algorithms*. In [12], also for UFBP, the reallocation cost paid is only computation time. Preemptive Job Scheduling has also been studied assuming that a cost for preemption is paid [11, 13, 15] but preempted jobs in this model are left idling. In the present work, we assume that the cost of reallocation is a constant amount. That is, we aim to minimize the number of reallocations.

In this paper, we present three protocols for online WS with reallocation. Namely, Preemptive Reallocation, Lazy Reallocation, and Classified Reallocation. In brief, the first is a repeated (upon each client arrival or departure) application of the offline protocol in [4]. Aiming to minimize channel usage, clients are preemptively reallocated to guarantee the same offline packing. Instead, in Lazy Reallocation, clients are not reallocated as long as a maximum number of channels in use is not exceeded. The idea is to save reallocations taking advantage of all possible channels. Finally, Classified Reallocation is designed to guarantee an amortized constant number of reallocations. The main approach is to classify clients by laxity.

We evaluate experimentally all three protocols. Our simulations show that, in practice, all three achieve constant amortized reallocations with close to optimal channel usage. Our simulations also expose interesting trade-offs between reallocations and channel usage. On the theoretical side, we introduce a new objective function for WS with reallocations, that can be also applied to models where reallocations are not possible. We analyze this metric for Classified Reallocation which, to the best of our knowledge, is the first online WS protocol for dynamic scenarios (clients may leave) with theoretical guarantees (against current load). Using previous results, we also observe bounds on channel usage for Preemptive Reallocation.

The rest of the paper is organized as follows. First we overview the related work in Section 2. Section 3 includes formal definitions of the problem and the model. The details of our contributions are presented in Section 4. We present and analyze reallocation protocols in Sections 5 and 6, and simulations in Section 7.

2   Related Work

WS is a problem with applications to various areas such as communication networks, supply chain, job scheduling, media on demand systems, etc. We overview here the most related work.

In [4] the authors studied WS for broadcast systems where pages have to be allocated to slotted broadcast channels. The allocation comprises a schedule of pages to slots such that the gap between two consecutive slots reserved for page i is at most its window w_i. The model is static in the sense that, once pages arrive to the system, they do not leave. Two optimization criteria are studied: to minimize the number of channels used by a fixed set of clients and to maximize the number of clients allocated to a fixed set of channels. For the former problem they show an offline approximation of $H + O(\ln H)$ where $H = \sum_i 1/w_i$. The approach followed is to round down each window size to the largest number of the form $x2^y$, where x is an odd positive integer and y is a non-negative integer. Then, allocate all clients of the same window size to the same channel creating a new one whenever necessary. For online algorithms, an approximation of $H + O(\sqrt{H})$ channels was shown later in [5]. The approach is similar but the analysis takes into account that, being online, pages are allocated one by one and decisions are final.

A scenario where pages may leave the system was termed WS with Temporary Items in [8]. Bounds in the latter work were proved on the competitive ratio against peak load. That is, the online and the offline channel-usage maxima compared may occur at different times. That is, taking advantage of (rather than overcoming) windows departures. Specifically, the authors showed that any online algorithm has a competitive ratio at least $1 + 1/w_{min} - \epsilon$, for any $\epsilon > 0$, independently of knowledge of departure time. On the positive results side, they show an online algorithm that is $2 + 2/(\lfloor \lfloor w_{min} \rfloor \rfloor - 1)$-competitive, for $w_{min} \geq 2$. Recently [18], this upper bound was improved to $4w_{min}/(3w_{min} - 4)$, for $w_{min} \geq 4$, in the context of supply chain inventory replenishment. The algorithm makes use of a combined first fit policy for assignment.

In [5], WS was studied as a *restricted* version of UFBP. The reason is apparent, the combination of windows in WS is not additive as in UFBP. Hence, UFBP upper bounds do not apply to WS but lower bounds do. A lower bound of $H + \Omega(\ln H)$ on the online approximation of UFBP, which then applies also to WS, was shown in [5]. A dynamic version of UFBP where items leave was studied in [7]. The authors show an upper bound of 3 on the competitive ratio of best fit and worst fit allocation policies, and an upper bound of 2.4942 for first fit. UFBP with reallocation was studied in [1,2] as *semi-online* algorithms with *repacking*. In their model items do not leave the system, hence only lower bounds apply to WS with reallocation, but only upper bounds are presented. In [12], *fully dynamic* UFBP algorithms apply to settings where items may leave, but the reallocation cost is only bounded in terms of computation time.

3   Model

In this Section, we describe the Windows Scheduling problem and introduce the notation used throughout the paper.

We assume that time is discrete and it is determined by a global clock that runs in *time steps*. Consider a set of clients $\mathscr{C}$ that require periodical radio transmissions. Each client $c_i \in \mathscr{C}$ is determined by three parameters: *arrival* time t_i, *departure* time d_i, and *laxity* w_i. We say that client c_i is *active* during the time interval $[t_i, d_i]$. We assume that there is an infinite number of available radio channels $\mathscr{R} = \{R_j\}_{j \in \mathbb{N}}$. During one time step, only one client is able to transmit via a single channel.

A *schedule* for a client $c_i \in \mathscr{C}$ is a set $S(c_i) \triangleq \{(R_k^i, t_k^i)\}_{k \in [1, \ell_i]}$, where, channel R_k^i is the reserved channel for c_i's transmission at time step t_k^i. Time steps t_k^i should satisfy the following conditions:

$$t_1^i - t_i \leq w_i, \tag{1}$$

$$d_i - t_{\ell_i}^i \leq w_i \text{ and} \tag{2}$$

$$t_{k+1}^i - t_k^i \leq w_i \text{ for all } 1 \leq k < \ell_i. \tag{3}$$

Parameter ℓ_i denotes the number of transmissions performed by client c_i. Note that this parameter is independent for each client, it is not set before hand and has to be set by the schedule so that conditions (1), (2) and (3) are satisfied. Thus, conditions (1), (2) and (3) ensure that each client c_i transmits only while it is active and that periodical transmissions are so that the laxity condition w_i for client c_i is satisfied. Since client c_i has to transmit periodically every (at most) w_i time steps, we say that the *load of client* c_i is $1/w_i$.

We say that schedule $S(c_i)$ *reallocates* client c_i each time that $R_k^i \neq R_{k+1}^i$ for $1 \leq k < \ell_i$. We denote by $reall(S(c_i))$ the number of times schedule $S(c_i)$ reallocates client c_i. The total *number of reallocations* of schedule $S(\mathscr{C})$ for the set of clients $\mathscr{C}$ is denoted by $reall(S(\mathscr{C}))$ and is defined as follows:

$$reall(S(\mathscr{C})) \triangleq \sum_{c_i \in \mathscr{C}} reall(S(c_i)).$$

We are interested in the *amortized* number of reallocations, that is, the number of reallocations of a schedule divided by the number of arrivals and departures in the system. For that purpose, let us define a *round* as the set of time slots between events (arrival or departure of a client). In other words, at each round occurs exactly one event, either an arrival or a departure of a client. Hence, if $reall_r(S(\mathscr{C}))$ denotes the number of reallocations up to round r produced by schedule $S(\mathscr{C})$, the *amortized* number of reallocations of schedule $S(\mathscr{C})$ in the set of clients $\mathscr{C}$ at round r is $reall_r(S(\mathscr{C}))/r$.

Naturally, there exist a trade-off between the number of reallocations and the number of channels used by a schedule, the more reallocations the less number of channels used at each round. Indeed, if a schedule can reallocate clients freely, it can achieve the optimal offline number of channels used at each round by simply computing the optimal offline schedule at each round, and reorganizing the schedule via reallocations in order to mimic the optimal offline schedule. Thus, we are interested in understanding this trade-off at each round. For any round r, let $\mathscr{C}(r) \subseteq \mathscr{C}$ be the set of active clients in the system at round r. Hence, we define $H_r \triangleq \sum_{c_i \in \mathscr{C}(r)} 1/w_i$ as the *load of the system* at round r. The value $\lceil H_r \rceil$ is a lower bound on the optimal number of used channels at round r. Hence, $\mathscr{R}(S(\mathscr{C}))_r / \lceil H_r \rceil$ is an upper bound on the competitive ratio of the number of channels used at round r, where $\mathscr{R}(S(\mathscr{C}))_r$ denotes the number of channels used at round r by schedule $S(\mathscr{C})$.

Using this notation, we define the online WS problem with reallocations as follows:

Definition 1. *Let $\mathscr{C}$ be a set of clients revealed online, that is, arrivals and departures of clients are revealed one by one. The **Online Windows Scheduling with Reallocations** problem is to determine a schedule $S(\mathscr{C})$, possibly with reallocations, so that the following sum is minimum.*

$$\frac{reall_r(S(\mathscr{C}))}{r} + \frac{\mathscr{R}(S(\mathscr{C}))_r}{\lceil H_r \rceil}$$

Notice that this metric is against current load and applies to any given round r. Should we instead be interested in the maximum competitive ratio, we could simply plug the maximum up to round r instead of the current. Likewise, should the reallocation cost be any known value c, it could be simply introduced in the expression above multiplying the first term.

4 Our Contributions

The main contributions of this paper follow. It should be noticed that our upper bounds cannot be compared with previous theoretical work on WS because either their models assume that clients do not leave [4,5], or the bounds on channel usage are proved against peak load [8, 18], or they apply to the less restrictive UFBP problem [1, 2, 5, 7, 12].

- The presentation of three protocols for Online Windows Scheduling with Reallocation called Preemptive Reallocation, Lazy Reallocation, and Classified Reallocation. Preemptive Reallocation guarantees low channel-usage because clients are preemptively reallocated to achieve the offline packing of [4]. Attempting to save reallocations while keeping the channel usage bounded, in Lazy Reallocation clients are not reallocated as long as a maximum number of channels in use is not exceeded. In Classified Reallocation, the main approach is to classify clients by laxity, except for "large" laxities that are allocated in one special channel to maintain the channel-usage overhead below an additive logarithmic factor. To the best of our knowledge, Classified Reallocation is the first online WS protocol for dynamic scenarios (clients may leave) with theoretical guarantees (against current load).
- The experimental evaluation of all three protocols showing that, in practice, all of them achieve constant amortized reallocations with close to optimal channel usage. Our simulations also expose interesting trade-offs between reallocations and channel usage.
- The introduction of a new objective function for Online Windows Scheduling with Reallocations, that can be also applied to models where reallocations are not possible. This metric combines linearly the effect of reallocations amortized over rounds with the number of channels used in contrast with the optimal UFBP, which is a lower bound on the WS optimal allocation.

- Using previous results [4], for Preemptive Reallocation we observe an upper bound of $\lceil 2H_r \rceil$ on channel usage for every round r. Given that in Lazy Reallocation the maximum number of channels is a parameter, its channel usage depends on the implementation.
- We prove that, for any round r, the number of reallocations required by Classified Reallocation is at most $3r/2$, and the number of channels used is at most $OPT(r) + 1 + \log \left(\min \left\{ w_{\max}(r), \lceil \lceil n(r) \rceil \rceil \right\} / w_{\min}(r) \right)$ [3], where $OPT(r)$ is the optimal number of channels required, $w_{\max}(r)$ and $w_{\min}(r)$ are respectively the maximum and minimum laxities in the system, and $n(r)$ is the number of clients in the system, all for round r. We apply these bounds to our objective function in Definition 1.

5   Preemptive Reallocation and Lazy Reallocation Algorithms

The first two algorithms use the concept of a *broadcast tree* to represent the schedule corresponding to each channel.

Broadcast Tree to Represent a Schedule. Similarly to the the work by Bar-Noy et al. [4] and Chan et al. [8], we represent a schedule for a channel with a tree. In particular, we use a binary tree where all nodes have exactly zero or two children. Each leaf of the binary tree has assigned one client (different for each leaf) or the leaf is empty. Each complete binary tree with such an assignment represents uniquely one schedule for a channel. Given a complete binary tree, the schedule picks one leaf at each time step so that either the assigned client transmits or no transmission is produced if the leaf is empty. In order to pick one leaf, walk down from the root up to one leaf following the next recursive rules: the first time a bifurcation is visited go to the left child. The i-th time a bifurcation is visited go to the child that was not visited in the previous visit. Such trees are called **broadcast trees**.

A used channel will be fully described by its broadcast tree in the following two algorithms. Thus, the following two algorithms are determined by *(i)* the procedure to assign clients to a leaf in a broadcast tree when a client arrives and *(ii)* the procedure to reallocate clients if required when a client leaves the system.

Greedy Construction of a Broadcast Tree. Consider an empty channel as a single node, the root of its corresponding tree. Define every root as **available**. For the first client c_1 that arrives, let v_1 be the nonnegative integer such that $2^{v_1} \leq w_1 < 2^{v_1+1}$. Append to any root a binary tree of height v_1 in which for each depth from 1 to $v_1 - 1$, there is a single leaf and for depth v_1 there are two leaves. Set every new leaf as available except for one leaf at depth v which is assigned to c_1. For the i-th client c_i that arrives, let v_i be the nonnegative integer such that $2^{v_i} \leq w_i < 2^{v_i+1}$. If there is an available leaf at depth v_i, assign c_i to that leaf. Otherwise, let $0 < u < v_i$ be any value such that there is an open leaf at depth u in some used broadcast tree. Append a binary tree of height $v_i - u$ to that leaf in which for each depth from u to $v_1 - u - 1$, there is a single leaf and for depth v_1 there are two leaves. Set every new leaf as available except for one

---

[3] Throughout, log means $\log_2$ unless otherwise stated. We define $\lceil \lceil n(r) \rceil \rceil$ as the smallest power of 2 that is not smaller than $n(r)$.

leaf at depth v_i which is assigned to c_i. If no such u exists, append the described tree to an available root. In this case consider $u = 0$.

Preemptive Reallocation. The *Preemptive Reallocation* procedure maintains the following invariant: for each depth there is at most one broadcast tree with a leaf available. If the invariant is violated after some departure, it means that there are two broadcast trees with an available leaf in the same depth. Then, the branches that hang from the twin node of the available leaf can be hanged from the same tree to reinstate the invariant. The Preemptive Reallocation procedure does so minimizing the total number of reallocations, that is, moving the branch that hangs from the broadcast tree with less clients assigned.

Lazy Reallocation. The *Lazy Reallocation* procedure reallocates only when the fraction $\mathscr{R}(S(\mathscr{C}))_r/\lceil H_r \rceil$ exceeds a threshold T after one departure. It exhaustively reallocates clients until no more reallocations can be made. Reallocations are done according to the invariant used by the Preemptive Reallocation procedure merging the smallest depth with two available leaves at the same depth.

The *Preemptive Reallocation* and *Lazy Reallocation* algorithms use the greedy construction of broadcast trees when clients arrive and preemptive and lazy reallocations procedures when clients leave the system, respectively. Lemma 5 of Bar-Noy et al. [4] implies that the Preemptive Reallocation algorithm guarantees $\mathscr{R}(S(\mathscr{C}))_r < \lceil 2H_r \rceil$ for every round r. Hence, $\mathscr{R}(S(\mathscr{C}))_r/\lceil H_r \rceil < 2$ for every round r when $S(\mathscr{C})$ is constructed according to the Preemptive Reallocation algorithm. On the other hand, by definition of the Lazy Reallocation procedure, the Lazy Reallocation algorithm guarantees that $\mathscr{R}(S(\mathscr{C}))_r/\lceil H_r \rceil \leq T$ when $S(\mathscr{C})$ is constructed according to the Lazy Reallocation algorithm. In Section 7, we study via experiments the behavior of $reall_r(S(\mathscr{C}))/r$ for the Preemptive Reallocation and Lazy Reallocation algorithms. In the particular implementation of the Lazy Reallocation algorithm included in this work, we set the threshold T equal to $4\sqrt{H_r}$. Hence, in that case it holds $\mathscr{R}(S(\mathscr{C}))_r/\lceil H_r \rceil \leq 4\sqrt{H_r}$. For both algorithms, Preemptive Reallocation and Lazy Reallocation, we show experimentally that $reall_r(S(\mathscr{C}))/r \leq 1$.

6   Classified Reallocation

In this section, we present a reallocation algorithm that guarantees $O(1)$ reallocations amortized on rounds. For convenience, for any number x, we define the *hyperceiling* of x, denoted as $\lceil \lceil x \rceil \rceil$, to be the smallest power of 2 that is not smaller than x. For this algorithm, we restrict the input to laxities that are powers of 2. The study of inputs with arbitrary laxities is left for future work.

The intuition of the protocol is the following. When a new client c_i arrives and there are already $n - 1$ clients in the system, for $n \geq 1$, we distinguish two cases. If the laxity of c_i is at least $2\lceil \lceil n \rceil \rceil$, assign c_i to a special channel called *big channel*. All clients allocated to the big channel transmit with period $\lceil \lceil n \rceil \rceil$ so, because there are n clients in the system, one big channel is enough. All the other channels being used are called *small channels*. Otherwise, if the laxity is $w_i < 2\lceil \lceil n \rceil \rceil$, assign client c_i to a channel reserved for laxities w_i, we call it w_i-*channel*. If such channel does not exist or all

w_i-channels are full, reserve a new one. For any laxity w_i, all clients allocated to a w_i-channel transmit with period w_i. That is, a maximum of w_i clients can be allocated to a w_i-channel. When a client c_j of laxity w_j leaves a channel C, if C is the big channel do nothing. Otherwise, reallocate a client from the w_j-channel of minimum load (if any other) to the slot left by c_j.

With each arrival or departure the number of clients n change. If, upon an arrival, $\lceil\lceil n\rceil\rceil$ becomes larger than the laxity of some clients allocated to a big channel, reallocate those clients to other channels according to laxity, reserving new channels if necessary. Because n was doubled since the allocation of these clients, these reallocations are amortized by the arrivals that doubled n. If, upon a departure, $2\lceil\lceil n\rceil\rceil$ becomes smaller than the laxity of some clients, reallocate those clients to a big channel, releasing the reservation of the channels that become empty. Because n was halved since the allocation of these clients, these reallocations are amortized by the departures that halved n. Precise details of the Classified Reallocation algorithm are left to the full version of this paper in [9] due to space constraints.

The following theorem bounds the number of reallocations and the number of channels used by the Classified Reallocation algorithm. The proof is left to the full version of this paper in [9] due to space constraints.

Theorem 1. *Given a set of clients $\mathscr{C}$, the schedule $S(\mathscr{C})$ obtained by the Classified Reallocation algorithm requires at most $3r/2$ reallocations up to round r. Additionally, for any round r such that $\mathscr{C}(r) \neq \emptyset$, the number of reserved channels is at most*

$$OPT(r) + 1 + \log \frac{\min\left\{\max_{i\in\mathscr{C}(r)}\{w_i, \lceil\lceil|\mathscr{C}(r)|\rceil\rceil\}\right\}}{\min_{i\in\mathscr{C}(r)} w_i},$$

where $OPT(r)$ is the minimum number of channels required to allocate the clients in $\mathscr{C}(r)$.

The following corollary is a direct consequence of Theorem 1 and the fact that $OPT(r) \geq \lceil\sum_{i\in\mathscr{C}(r)} 1/w_i\rceil = \lceil H_r\rceil$.

Corollary 1. *Given a set of clients $\mathscr{C}$, the schedule $S(\mathscr{C})$ obtained by the Classified Reallocation algorithm achieves, for any round r, is $reall_r(S(\mathscr{C}))/r + \mathscr{R}(S(\mathscr{C}))_r/\lceil H_r\rceil \leq 5/2 + (1 + \log(\min\{\max_{i\in\mathscr{C}(r)} w_i, \lceil\lceil|\mathscr{C}(r)|\rceil\rceil\}/\min_{i\in\mathscr{C}(r)} w_i))/\lceil H_r\rceil$.*

7   Simulations

Model. The deployment of the proposed algorithms on a real environment will require involvement of a large number of active users and resources, which is very hard to coordinate and build, and would prevent repeatability of results. Thus, simulation appears to be the easiest way to analyze the different proposed online reallocation strategies. Based on the simulation results, we can later encourage or discourage the deployment on a real production environment. Next, we present the simulation model implemented for evaluating the performance of the previously proposed reallocation policies, named *CommunicationChannelsSim* (CCSim). The simulated communication channels have been performed by means of SimJava 2.0 [16]. SimJava is a discrete event, process oriented, simulation package. It is an API that augments Java with building blocks for defining and running simulations.

All the simulations are performed from the point of view of one user. This user submits *clients* to the scheduler. The clients are loaded from an XML input file. For our experiments we have created three different input files each containing 4000 clients with different window sizes. These scenarios represent common situations on which clients with different window sizes arrive to the system and departure in an instant of time that depends directly on the laxity demanded by the client. The better the laxity, the sooner he will leave the system. In the same way, the scheduler performs one of the proposed reallocation strategies and acts accordingly when a client arrives or leaves the system. We have implemented three versions of CCSim that only differ in the reallocation policy implemented. As a result, we isolate the reallocation strategy as the only factor that can cause number of channels and number of reallocations variations between these three CCSim versions.

As we mentioned before, we have created our own input files. We have defined the structure of the files using the XML language. Thus, each input file contains 4000 clients with the following characteristics:

* id: each client has its own identifier to differentiate it from the rest.
* t_arrive: arrival simulation time of the client in seconds, for example, 899 seconds.
* size: laxity of the client as a real value, for example, 6.628461669685978. The laxity can be calculated using a Gaussian (normally) distributed double value or it can be chosen uniformly. Thus, we can generate input files following Normal, Uniform or mixed laxity distributions.
* w_size: the laxity is later rounded down to the larger power of 2, for our example, its value would be 4.
* t_leave: simulation time instant in seconds at which the client leaves the system. This value is calculated based on the laxity. Thus, if the client has a laxity less than or equal to 30, t_leave will be t_arrive plus a random number generated uniformly between 500 and 1000. Conversely, if the client has a laxity greater than 30, t_leave will be t_arrive plus a random number generated uniformly between 1000 and 1500. As a consequence, clients with a smaller laxity will remain less time in the system. For instance, the client of the example will leave the system at second 1737.0.

Explanations for the main CCSim participating entities and its simulation setup follow.

CCSim represents the complete simulation, and is responsible for the creation of the main simulated entities: Scheduler and User. When the simulation starts, CCSim creates 1 User and 1 Scheduler.

The **Scheduler** entity represents a generic scheduler implementing the corresponding reallocation algorithm. When a Client *arrives*, the Scheduler allocates it in a Broadcast Tree Node. This operation may involve creating a new Tree. Also, when a Client leaves, the Scheduler removes it from the corresponding Tree Node. This operation may imply reallocating Clients and deleting Trees.

The **User** models a user that submits Clients to the Scheduler. The User is responsible of sending the only two system events: Client arrival and Client departure by using Client's `t_arrive` and `t_leave` times.

The **Client** entity represents a generic Client submitted to the Scheduler. This entity provides specific information about each client as defined in the XML input file.

The Broadcast **Tree** entity, as described in Section 5, represents a channel with up to five different levels representing five Client laxities (2, 4, 8, 16, 32). Thus, at each level there are at most 2^i Nodes, with i being a number between 1 and 5. Clients are allocated in their corresponding Node level according to their rounded down laxity `w_size`. Each tree is characterized by a root Node.

The **Node** entity represents a Tree Node in which a Client can be allocated. Each node knows its level or window size, if it is a root node, a left node or a right node. Also, each node points to its left subtree and to its right subtree.

Discussion. Our simulations show similar behaviors for normal and uniform input distributions. All three protocols achieve constant amortized reallocations with close to optimal channel usage. Figures 1 illustrates the performance of the algorithms along rounds for a normally distributed input. We can observe in Figure 1(a) the overall load of the system for a representative input. The load was increased over 1000 rounds

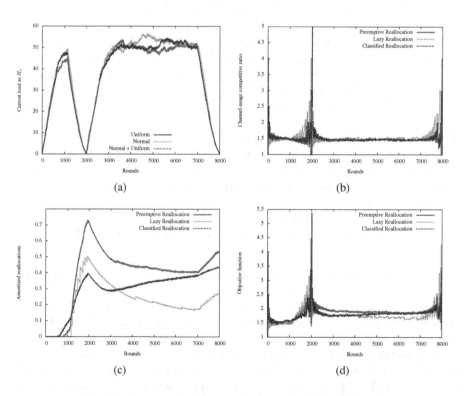

(a)

(b)

(c)

(d)

Fig. 1. Performance along rounds for normally distributed inputs

until reaching a peak and decreased during the following 1000 rounds until reaching low load. Afterwards, the increase/decrease procedure was repeated but now maintaining the system loaded for approximately 4000 rounds. We can see in Figure 1(c) that Preemptive Reallocation incurs in more reallocations than Lazy Reallocation and Classified Reallocation, but still for all three the amortized reallocations are below 1. When the system is loaded, we observe in Figure 1(b) that the competitive ratio stays around 1.5 for all algorithms. When the system has low load, Preemptive Reallocation still maintains the competitive ratio below 2, whereas Lazy Reallocation increases up to the parametric maximum channel usage. For these simulations, that maximum was $H_r + 4\sqrt{H_r}$, that is, the online maximum channel usage of [5] for clients that do not leave. With respect to Classified Reallocation with low load, the competitive ratio is higher because, for simplicity, the simulation was carried out without using a big channel (refer to the Classified Reallocation algorithm). Preemptive Reallocation and Lazy Reallocation reflect clearly the trade-off between reallocations and channel usage. While the former incurs in more reallocations than the latter, the reverse is true with respect to channel usage. This trade-off becomes more dramatic if the maximum number of channels allowed in Lazy Reallocation is increased. Classified Reallocation, on the other hand, which out of the three is the only algorithm providing theoretical guarantees, has low amortized reallocations (less than 0.5) while maintaining a low competitive ratio when the system is loaded. For systems with frequent low load, the big channel should be implemented. Figure 1(d) illustrates the combination of these factors in our objective function. Figure 2 illustrates the trade-offs between channel

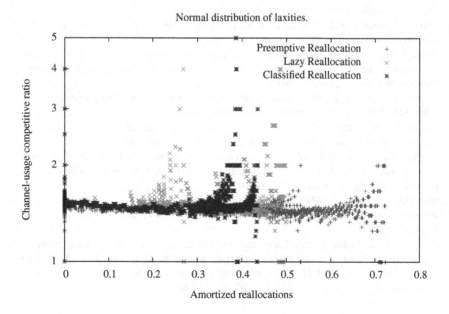

Fig. 2. Channel usage vs. reallocations

usage and reallocations for a normally distributed input. The experiments for uniformly distributed inputs have produced similar results.

References

1. Balogh, J., Békési, J.: Semi-on-line bin packing: A short overview and a new lower bound. Central European Journal of Operations Research, 1–14 (2012)
2. Balogh, J., Békési, J., Galambos, G., Reinelt, G.: On-line bin packing with restricted repacking. Journal of Combinatorial Optimization, 1–17 (2012)
3. Bar-Noy, A., Bhatia, R., Naor, J., Schieber, B.: Minimizing service and operation costs of periodic scheduling. In: Proceedings of the 9th Annual ACM-SIAM Symposium on Discrete Algorithms, pp. 11–20 (1998)
4. Bar-Noy, A., Ladner, R.E.: Windows scheduling problems for broadcast systems. SIAM Journal on Computing 32(4), 1091–1113 (2003)
5. Bar-Noy, A., Ladner, R.E., Tamir, T.: Windows scheduling as a restricted version of bin packing. ACM Transactions on Algorithms (TALG) 3(3), 28 (2007)
6. Bender, M.A., Farach-Colton, M., Fekete, S.P., Fineman, J.T., Gilbert, S.: Reallocation problems in scheduling. In: 25th ACM Symposium on Parallelism in Algorithms and Architectures, SPAA 2013, pp. 271–279 (2013)
7. Chan, J.W.-T., Lam, T.-W., Wong, P.W.H.: Dynamic bin packing of unit fractions items. Theoretical Computer Science 409(3), 521–529 (2008)
8. Chan, W.-T., Wong, P.W.H.: On-line windows scheduling of temporary items. In: Fleischer, R., Trippen, G. (eds.) ISAAC 2004. LNCS, vol. 3341, pp. 259–270. Springer, Heidelberg (2004)
9. M. Farach-Colton, K. Leal, M.A. Mosteiro, C. Thraves. Dynamic windows scheduling with reallocation. arXiv:1404.1087 (April 2014)
10. Han, X., Peng, C., Ye, D., Zhang, D., Lan, Y.: Dynamic bin packing with unit fraction items revisited. Information Processing Letters 110(23), 1049–1054 (2010)
11. Heydari, M., Sadjadi, S.J., Mohammadi, E.: Minimizing total flow time subject to preemption penalties in online scheduling. The International Journal of Advanced Manufacturing Technology 47(1-4), 227–236 (2010)
12. Ivkovic, Z., Lloyd, E.L.: Fully dynamic algorithms for bin packing: Being (mostly) myopic helps. SIAM Journal on Computing 28(2), 574–611 (1998)
13. Liu, Z., Edwin Cheng, T.C.: Minimizing total completion time subject to job release dates and preemption penalties. Journal of Scheduling 7(4), 313–327 (2004)
14. Sanders, P., Sivadasan, N., Skutella, M.: Online scheduling with bounded migration. In: Díaz, J., Karhumäki, J., Lepistö, A., Sannella, D. (eds.) ICALP 2004. LNCS, vol. 3142, pp. 1111–1122. Springer, Heidelberg (2004)
15. Shachnai, H., Tamir, T., Woeginger, G.J.: Minimizing makespan and preemption costs on a system of uniform machines. Algorithmica 42(3-4), 309–334 (2005)
16. SimJava (2006),
   http://www.icsa.inf.ed.ac.uk/research/groups/hase/simjava/
17. WestBrook, J.: Load balancing for response time. Journal of Algorithms 35(1), 1–16 (2000)
18. Yu, H., Xu, Y., Wu, T.: Online inventory replenishment scheduling of temporary orders. Information Processing Letters 113(5), 188–192 (2013)

Parallel Bi-objective Shortest Paths Using Weight-Balanced B-trees with Bulk Updates

Stephan Erb, Moritz Kobitzsch, and Peter Sanders

Karlsruhe Institute of Technology, Am Fasanengarten 5, 76131 Karlsruhe, Germany
stephan.erb@blue-yonder.com, {kobitzsch,sanders}@kit.edu

Abstract. We present a practical parallel algorithm for finding shortest paths in the presence of two objective functions. The algorithm builds on a recent theoretical result that on the first glance looks impractical. We address the problem of significant constant factor overheads due to numerous prefix sum computations by carefully re-engineering the algorithm for moderate parallelism. In addition, we develop a parallel weight-balanced B-tree data structure that cache efficiently supports bulk updates. This result might be of independent interest and closes the gap between the full-blown search tree data structure required by the theoretical result over the simple priority queue for the sequential algorithm. Comparing our implementation against a highly tuned sequential bi-objective search, we achieve speedups of 8 on 16 cores.

1 Introduction

Finding "good" paths through networks is one of the most important problems for graph algorithms with many applications. There is a lot of work for the case that the shortest path with respect to a single objective function is considered. However, often one wants more flexibility and is interested in all paths that are *Pareto optimal* with respect to several objective functions, i.e., paths that are not dominated by any other path (worse with respect to all objective functions). Examples could be cost versus travel time in road networks or hops versus energy consumption in radio networks. Although the problem is NP-hard, it can often be solved in practice for medium sized networks. Of course, to push the size of the instances that can be handled, parallelization is crucial. Nevertheless, there is surprisingly little work on parallel multi-objective path search, although already in 2001 Guerriero and Musmanno state that "parallel computing [...] represents the main goal for future developments" [7]. One might think that the reason for this lack of results is an inherent difficulty of parallelizing shortest path. This is indeed a problem for the single-objective case. However, Sanders and Mandow show that in multi-objective search, at least for the bi-objective case [19], all additional work is parallelizable. Unfortunately, while conceptually simple, their paPaSearch-algorithm looks complicated and forbiddingly expensive. On the one hand, their approach repeatedly uses load distribution based on prefix sums that threatens to destroy locality. Furthermore, paPaSearch replaces the simple priority queue controlling the sequential algorithm by a more

J. Gudmundsson and J. Katajainen (Eds.): SEA 2014, LNCS 8504, pp. 111–122, 2014.
© Springer International Publishing Switzerland 2014

complex *Pareto queue* data structure based on binary search trees – introducing another significant overhead factor.

In the following, we address these problems and develop an efficient implementation of paPaSearch for the bi-objective case. In Section 2, we introduce the results we are building on. Section 3.1 presents a B-tree variant that supports efficient parallel bulk-updates and can be used for retrieving Pareto optimal entries in parallel. This is much more cache efficient than comparable solutions based on binary trees. Based on this, Section 3.2 introduces a streamlined implementation of paPaSearch that reduces the number of processing phases compared to [19] and uses more coarse grained parallelism. The experimental evaluation in Section 4 demonstrates that an implementation using the Intel Threading Building Blocks achieves speedups up to 8 using 16 cores. This speedup is relative to our own sequential implementation that significantly outperforms previous implementations available to us.

2   Preliminaries and Related Work

We consider a directed graph $G = (V, E)$ with $|V| = n$ and $|E| = m$ with 2-dimensional weight function $c : E \to \mathbb{R}^2_{\geq 0}$. a_x, a_y denote the components of a vector a. If a is clear from context, we might choose to omit the subscript and just use x, y. a dominates b if and only if $a_x \leq b_x \wedge a_y \leq b_y$ where $a \neq b$. Given a set of points M, elements of M that are not dominated by any other element in M are called *Pareto optimal*. A set consisting solely of Pareto optimal points is a *Pareto set*. Pareto optimality is always relative to a set of labels. We distinguish between *local* (restricted to the set of a vertex) and *global* optimality (with respect to all *unsettled* labels).

2.1   Sequential Multi-objective Search

Given a source vertex s, we are looking for Pareto optimal paths from s to all other vertices $v \in V$ (*one-to-all* problem), i.e., paths whose length is not dominated by any other path from s to v. (Bi-objective) shortest path search algorithms are a generalization of single-objective algorithms. The search computes labels of vertices in a similar fashion as Dijkstra's algorithm computes tentative distances in the single-objective case. A label $\ell \in \mathbb{R}^2_{\geq 0}$ of a vertex v represents an s–v-path with length vector ℓ. Labels are updated by *edge relaxations*: Given a vertex u with label ℓ and an edge (u, v) with weight w, an edge relaxation computes a candidate label $\ell' = \ell + w$ for vertex v. Furthermore, the relaxation checks whether ℓ' is new and not dominated by any previously computed labels for v. In this case, ℓ' is added to the label set $L[v]$ of v and labels of v dominated by ℓ' are removed (pruned). Often, edge relaxations are grouped into label *scanning* operations where all edges out of u are relaxed.

A generalization of Dijkstra's algorithm then maintains a queue Q of unscanned labels that are Pareto optimal at their respective vertex. In each iteration, it scans one label from Q that is globally Pareto optimal. This choice guarantees that the algorithm has the *label-setting* property, i.e., all scanned labels represent a globally Pareto-optimal path in the final output [9,13,22,15,10]. The queue Q can be efficiently represented by a priority queue where the priority function is a lexicographic ordering or a linear combination of the objective functions with positive coefficients. Several papers look at the performance impact of ordering functions and priority queue implementations [4,15,16]. There are also comparisons with *label-correcting* algorithms [7] that also scan non-Pareto-optimal labels. While this allows a faster and simpler implementation of Q, it also increases the number of label scanning operations. Overall, label-correcting algorithms are at best slightly faster than label-setting algorithms but they may be considerably slower in the worst case.

2.2   Parallel Search Trees

Concurrent access to search trees in data bases is traditionally done using locks but for in-memory parallel processing this is too slow for our purposes. Sewall et al. present parallel bulk updates for B+-Trees [21] to solve this problem. They achieve up to 40 million updates per second on large trees. Our implementation is about a factor of three faster (see Section 4 for details). Their approach also suffers from a massive sequential bottleneck in the worst case, when all updates go to the same leaf. Frias and Singler show how to do parallel bulk insertion into binary search trees [6]. This algorithm is scalable by splitting the tree into pieces such that each piece receives an equal number of insertions. Most of the work can then be done in an embarrassingly parallel way. The final balanced result is computed by concatenating the resulting trees which is possible in logarithmic time. In [19] this result is generalized to accommodate the operations needed for paPaSearch. In principle, this approach can be extended to B-trees resulting in better cache-efficiency than [6,8,11] Here we go one step further by using weight-balanced B-trees [3]. This way we get easily predictable highly coarse grained re-balancing by partial rebuilding and we naturally integrate dynamic load balancing by work stealing. This makes our solution more elegant and robust than [19] since [6,19] uses both static explicit load balancing for bulk updates and dynamic load balancing for finding Pareto optima.

3   Parellel Bi-objective Search

Covering both the theoretical results of [19] and our own contribution in sufficient detail is beyond the scope of this work. For that reason, we only focus on our implementation, giving as numerous details as possible for the many parts in which we divert from the original result; see the Appendix for further details.

3.1   A B-tree Based Pareto Queue

The central component for the parallel bi-objective search is the so called *Pareto queue*. It replaces the heap in Dijkstra's algorithm and has to support bulk-operations for modification (add,remove). In every step of the algorithm, all globally Pareto optimal labels are extracted from the queue Q, new candidate labels are generated and Q is updated accordingly. For our implementation of this Pareto queue, we chose *weight-balanced* B-trees [3] over red-black trees [19]. B-trees have been used to allow for bulk updates before (e.g. [21]) and offer good performance on different levels of the hierarchy found in modern memory architectures [2], especially, as linear access patterns (scanning) are faster than random accesses to memory [1]. Saikkonen shows cache sensitive B-trees [18] to outperform even cache sensitive layouts of binary search trees. Weight balanced B-trees augment the classic definition of a B-tree by allowing for some well controlled imbalances. These imbalances depend on lower/upper bounds to a *weight function* that describes the size of a subtree rooted at a given node.

For our cause, weight-balanced B-trees offer multiple benefits; not only do they allow for a simple cache aware layout if we configure the tree-nodes to be of multiple cache lines in size, but they allow to handle *difficult update sequences* in an efficient way, too. In general, an update sequence to a Pareto queue consists of the request to add a given set of labels and remove a different set of (dominated/globally optimal) labels. We apply such a sequence to our Pareto queue by splitting the sequence according to the router keys of a node and processing the different subsequences recursively in parallel. For this approach to work efficiently in a bulk update scenario, one could argue that we rely on a good distribution of labels and that heavily skewed update sequences will result in mostly sequential execution, especially as we usually process a leaf as an atomic unit. Such difficult sequences, however, can be handled elegantly by performing *partial tree reconstruction*. To do so, we utilize the fact that the weight function gives bounds on the number of labels within a given subtree and skewed sequences result in imbalances; this is due to the limited amount of labels within a given leaf/subtree and the by extension also limited amount of updates that can be applied to it without violating the constraints. If we encounter a sequence of updates that would violate these bounds, instead of performing complex balancing operations, we initiate partial reconstruction at the respective subtree and merge the tree and the updates into a new tree. To be able to do so, we have to detect imbalances at the root of the respective trees. This can be done in constant time after an initial prefix sum computation if we assign +1 to new labels and -1 to any request that deletes a label and calculate a prefix sum over these values. The router keys of a node split the lexicographically ordered update sequence into multiple subsequences, according to the affected subtrees. The difference between the according entries of the prefix sum now directly translates into the effect on the weight function of the respective tree and therefore directly indicates potential imbalances. Accordingly, we call such a difference between two entries weight delta. Similar ideas are utilized in [2] with applications to I/O-efficient bulk loading.

These deltas also prove beneficial in the efficient *partial reconstruction* of the (sub)tree. Given two router keys/labels, the according weight delta enables us to instantly calculate the number of surviving labels in between. Since the structure of a perfectly balanced B-tree is known by definition, this information enables us to directly construct an optimal tree (in parallel) and merge the existing labels and the updates into their correct positions. Since we might deal with skewed sequences in this case, we divert from the interpretation of a leaf as an atomic unit when necessary. If the amount of updates at a given leaf exceeds a threshold, we split the update sequence and locate the respective bounds within the leaf by binary search, each sequence to be processed in parallel.

Overall, this process avoids any form of synchronization and communication between tasks, is completely lock free and enables efficient load balancing via work stealing. In our experiments, we show a strong improvement of our technique over the proposed red-black tree approach.

Similar to a red-black tree implementation, we can extract Pareto-minima from this tree recursively and in parallel. Keeping track of the minimal y value within each subtree, we only descend into nodes for which the minimal y value is lower than the minimal y value to its left; otherwise, the respective labels in the tree are dominated.

3.2   Parallel Bi-objective Search

Similar to Dijkstra's algorithm, our modified parallel bi-objective search operates in rounds. We keep a Pareto queue Q, holding all unsettled labels as well as a separate set of tentative labels at every node. During each iteration, we extract all globally Pareto-optimal labels from Q and relax the respective arcs. The resulting candidate labels are merged into label sets of the affected vertices, potentially dominating other tentative labels. Afterwards, we update the Pareto queue to reflect the changes we made by removing settled/dominated labels and inserting new locally ParetThe steps operate as follows:

Minimum Extraction and Candidate Generation: We combine the generation of candidates and the extraction of Pareto optimal labels. During the process we descend into the tree recursively in parallel. By keeping track of the minimal y value within a given subtree, we can prune parts of the tree directly (due to the lexicographic ordering) if we detect that they do not contain a single Pareto optimal element. Reaching a leaf, we directly generate new candidate labels by relaxing the appropriate arcs for all Pareto optimal labels encountered, allowing us to reuse the load balancing already employed in the traversal process; we, however, do not directly remove candidate labels from the tree but rather schedule an update request to be executed later on in combination with the insertion of newly created, non dominated labels. This combination allows new labels to potentially compensate for removed labels and may prevent unnecessary reconstruction. Candidate labels are collected into a shared memory space which is allocated by the different threads in fixed size chunks (buffers) via fetch_and_add for synchronization; update requests to remove globally optimal labels are handled in the same way. In a final effort of this step, the candidates are sorted lexicographically by vertex id x and y-coordinate.

Updating Label Sets: For every vertex, we maintain a *label set* (compare tentative distance value) that we update sequentially but parallel over all affected vertices. Since we usually expect a large label set and a small number of candidates, it is reasonable to filter out candidates as best as possible before interacting with the label set. We do so in an *interleaved* fashion. The lexicographic ordering among the candidate labels of a vertex implies that the first label is Pareto optimal among all candidates. It may, however, be dominated by a previously existing label of the vertices label set. We locate the appropriate position among the existing labels via binary search. In case the label is not dominated, we merge it with the existing labels, proceeding to the next label candidate not dominated by the label just processed. If the label is dominated, however, we use the label dominating the candidate to check for further dominance among the remaining candidates (compare Figure 1).

This interleaved dominance check improves the filtering rate (depending on the input of course) by an order of magnitude without resulting in additional work when compared to an initial filtering step, only filtering dominated candidate labels among each other, followed by a separate merging process. For any candidate that survives the insertion process we generate an update request to our Pareto queue, as well as for any existing label that ends up being dominated by a new candidate (deletion). We accumulate these updates in the same way as we do for the candidates during the candidate generation. The respective buffers are reused from the update generation during the minimum extraction.

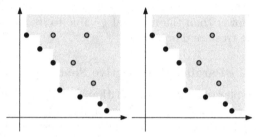

(a) Separate Filtering (b) Interleaved filtering (proposed in [19])

Fig. 1. Comparison of filtering methods. Candidate labels given as solid disks, dominated candidates are grayed out. The area in which candidate labels are deemed dominated is marked. Current labels, already existing at the vertex, are given as crosses. Look-ups within the existing labels are performed for all non-dominated candidates.

Queue Update: In a final step, we sort (shared memory parallel) the update requests lexicographically and compute the prefix sum as described in Section 3.1. The updates are applied to the queue in a combined effort, allowing new labels to compensate for removed labels as much as possible.

4   Experimental Evaluation

This Section consists of two parts. In a first effort, we show that our adapted weight-balanced B-tree forms a viable baseline to act as a Pareto queue and in fact does outperform the implementation of Frias and Singler [6]. Following up on these results, we discuss the baseline for our comparison and present a thorough evaluation of paPaSearch for the bi-objective case.

Unless otherwise stated, we compiled the code using GCC 4.7.2 and the flags -O3 -march=native, -std=c++11. For test hardware we used a dual Octa-Core Intel Xeon E5-2670 (*Sandy Bridge*) clocked at 2.6 GHz (2 NUMA nodes, 20MB shared L3, 256KB/32+32KB private L2/L1-Cache, 4 memory channels). Due to interference effects between the two nodes when it comes to memory access, we mostly present experiments running on a single one of these nodes, operating on 8 cores. The system is operating under Suse Linux Enterprise 11 on kernel 3.0.42.

We configured our B-tree implementation for dense updates ($\alpha = 600, \beta = 32$) for the Pareto queue and for sparse updates ($\alpha = 64, \beta = 32$) for the label sets. We collect our results using fetch_and_add in buckets of size 128.

4.1 Pareto Queue Evaluation

The parallel red-black tree in [6] provides, to the best of our knowledge, the only publicly available competitor to our task parallel implementation of a Pareto queue. They describe an extension to the C++ library data structure std::set[1]. We adopted their latest implementation (MCSTL[2] 0.8.0-beta) and compiled it using GCC version 4.2.4[3] and flags -O3 -march=native. The implementation from the MCSTL features bulk construction and bulk insertion. In terms of the experimental evaluation methodology, we follow the experimental structure of [6]. For measuring the performance of inserting elements into the tree, [6] add elements into a tree 100 times larger than the update count. Since we discovered a tree 10 times larger to be more relevant for our case, we only show plots for a relative difference of a factor of 10 here. We compare both a uniform distribution of the elements as well as skewed updates (targeting the first percent of the tree).

Already in sequential execution, our implementation largely outperforms both the std::set and the mcstl::set implementation which both exhibit a nearly identical runtime behavior. Figure 2 shows our algorithm to be competitive to the mcstl::set implementation running on 8 cores, even when executed sequentially. Depending on the size of the considered queue and insertion sequence, our implementation can outperform the mcstl::set by at least a factor of two. As operations on the queue contribute more than 50 % of our running time, these improvements are a vital factor to the success of our implementation.

Moreover, our implementation is not only faster than the bulk-operations on an mcstl::set, our implementation also exhibits a more uniform behavior. We experience the reasonable speedups, as presented in Figure 2, also for trees 100 times larger than the inserted sequence. The relative performance difference to Frias and Singlers implementation gets a bit smaller, however, as leaves of our tree are hit less often.

The implementation of Frias and Singler does not support bulk removal. Our implementation, however, exhibits nearly identical behavior for the process of removing elements from the tree as observed for inserting elements.

---

[1] Red-black balanced search tree, part of the standard template library.

[2] Multi-Core STL, http://algo2.iti.kit.edu/singler/mcstl

[3] Due to an incompatibility of the mcstl 0.8.0-beta with later GCC versions.

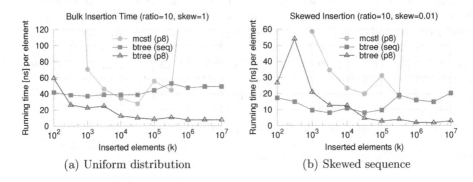

Fig. 2. Parallel insertion of a sequence of elements into a tree 10 times larger than the sequence

Note that we also greatly outperform the results presented in [21] for which 40M updates a second on a tree of size 128M are reported (on two Intel Xeon X5680, 6 cores each at 3.3 GHz). Our implementation manages to insert 10M elements into a tree of size 100M in 0.07 seconds averaging out on 140M inserts a second on a tree of size 100M. The results, however, are not directly comparable, as the implementation presents updates applied in batches of 8K elements and the algorithm focuses on correct streaming behavior of queries (execution order). Nevertheless, we believe that our measured values show the potential of our implementation and further justify our design decision.

4.2   Parallel Bi-objective Search

In the following, we compare our algorithms both to its sequential execution using only a single core, as well as to a tuned sequential algorithm.

A sequential competitor: To show that our competitor is a fair base for comparison, we compare our algorithm to the results of Raith and Ehrgott [17] who present C implementations of multiple approaches for the bi-objective shortest path problem. While we tried different implementations for the competing algorithm, we only report numbers for our best implementation.

Our implementation (LSetClassic) builds upon unbounded arrays to store the label sets, a tuned addressable binary heap as a priority queue and uses lexicographical ordering. The latter choice seems to contradict previous work which favors linear combinations (e.g. [10,15]). However, for one-to-all search, lexicographic ordering allows for higher locality in pruning. More precisely, the lexicographical ordering results in access patterns that mostly operate on the upper end of the label set, as label sets themselves are ordered lexicographically. A direct comparison to the last element often suffices.

Figure 3 shows that, even though running on a roughly comparable machine, our implementation outperforms the implementation of Raith and Ehrgott. Note

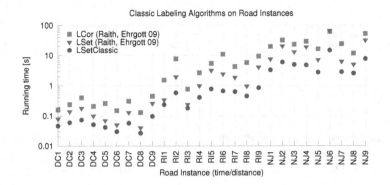

Fig. 3. Baseline algorithm compared to implementation of Raith and Ehrgott [17] on roughly similar machines

that we only depict the road instances from [17] in Figure 3 but experience a similar behavior for the proposed grid instances as well.

Evaluation of Parallel Bi-Objective Search. We implemented our algorithm in C++ using Intel® Threading Building Blocks (Intel® TBB) 4.1[4]).

Correlation and Instance Size: Often (e.g. [17,15]), grid graphs form an instance of choice when evaluating algorithms for the bi-objective shortest path problem; they allow for straightforward variation of input size and arc weight correlation.

Our implementation (see Figure 4) of paPaSearch outperforms our sequential competitor in almost any case, except for highly correlated grids of small dimensions. Figure 4(d), however, shows that already small variations in the correlation can tip the scales in our favor, even for relatively small instances. In general, our implementation shows a very similar running time behavior to our sequential competitor when run single-threaded.

Other Instances: To test the effect of the small weight range to our algorithm, we did some additional experiments with larger maximum arc weight (1 000 instead of 10). While slowing the overall process down, the relative query times for our implementation and the sequential competitor remain the same.

For further evaluation, we also tested the implementation on road instances based on the New York road network, provided by Machuca and Mandow [12]. The networks utilize a cost function consisting out of travel time and economic cost, forming a negatively correlated network. The results resemble the numbers presented for negatively correlated grids of larger size.

To investigate the effect of the rather limiting vertex degree in both grid and road graphs, we also evaluated our algorithm on synthetic sensor networks[5].

---

[4] http://threadingbuildingblocks.org/

[5] Unit disc graphs, containing some holes. For detailed description see [20].

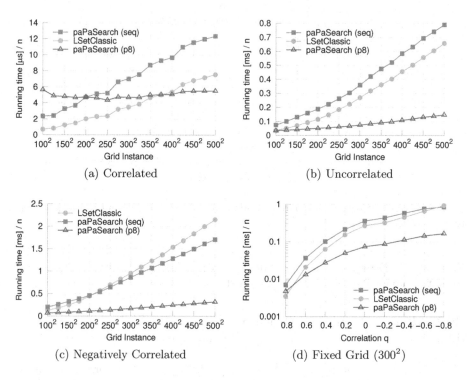

Fig. 4. Running time for selected correlations (q) and varying grid sizes ($a_{\{x,y\}} \leq 10$)

The graphs consist of 100 000 vertices and offer vertex degrees between 5 and 50. For arc weights we chose Euclidean distance and an uncorrelated random integer in the range of $[1, 10\,000]$. The graphs exhibit a near identical running time behavior as uncorrelated grid graphs (Plots can be found in [5]).

5   Conclusion

We have presented the first practical parallel implementation of multi-objective shortest path search. On reasonably sized instances with uncorrelated or even negatively correlated edge weights, we observe speedups larger than 8 on 16 cores. These results seem stable over a variety of different test instances, ranging from road networks over grids to artificial sensor networks. These speedups are made possible by our highly efficient Pareto queue implementation.

Several improvements and generalizations suggest themselves for future work. When only paths to a single target vertex are sought for, this can be exploited by additionally checking whether certain labels plus a precomputed lower bound are dominated by known upper bounds for the target vertex. Since this can be done independently for every label, this is also well parallelizable [19]. What remains to

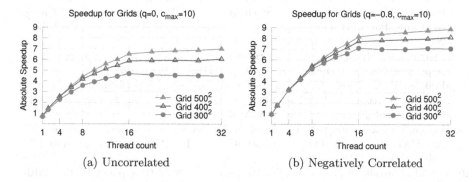

Fig. 5. Speedups relative to the best of either the sequential version of paPaSearch or our implementation of LSetClassic, depending on which algorithm performs better

be seen is whether the lazy filtering approach proposed there or more aggressive schemes perform better in practice.

In [19] it is also explained how to accommodate more than two objective functions by projecting them to two objective functions used in the Pareto queue. The main challenge here is to find out what projection strategies expose most parallelism for practical instances.

An interesting potential improvement is to expose more parallelism by going to label-correcting algorithms thus achieving speedup also for smaller instances and perhaps using more cores for large instances. For example, it might be possible to generalize Δ stepping [14] for a bi-objective setting.

Acknowledgments. We would like to thank Andrea Raith, Matthias Ehrgott, Enrique Machuca, and Lawrence Mandow for providing us with their road networks, and Dennis Schieferdecker for the generation of the sensor networks. We thank Timo Bingmann for his B-tree implementation panthema.net/2007/stx-btree.

References

1. Abraham, I., Delling, D., Goldberg, A.V., Werneck, R.F.: A hub-based labeling algorithm for shortest paths in road networks. In: Pardalos, P.M., Rebennack, S. (eds.) SEA 2011. LNCS, vol. 6630, pp. 230–241. Springer, Heidelberg (2011)
2. Achakeev, D., Seeger, B.: Efficient bulk updates on multiversion B-trees. PVLDB 6(14), 1834–1845 (2013)
3. Arge, L., Vitter, J.S.: Optimal external memory interval management. SIAM J. Comput. 32(6), 1488–1508 (2003)
4. Cherkassky, B.V., Goldberg, A.V., Radzik, T.: Shortest paths algorithms: Theory and experimental evaluation. Math. Program. 73, 129–174 (1996)
5. Erb, S.: Engineering Parallel Bi-Criteria Shortest Path Search. Master's thesis. Karlsruhe Institute of Technology (2013)

6. Frias, L., Singler, J.: Parallelization of bulk operations for STL dictionaries. In: Bougé, L., Forsell, M., Träff, J.L., Streit, A., Ziegler, W., Alexander, M., Childs, S. (eds.) Euro-Par Workshops 2007. LNCS, vol. 4854, pp. 49–58. Springer, Heidelberg (2008)

7. Guerriero, F., Musmanno, R.: Label correcting methods to solve multicriteria shortest path problems. J. Optim. Theory Appl. 111(3), 589–613 (2001)

8. Hankins, R.A., Patel, J.M.: Effect of node size on the performance of cache-conscious B^+-trees. In: Harchol-Balter, M., Douceur, J.R., Xu, J. (eds.) SIGMETRICS, pp. 283–294. ACM Press, New York (2003)

9. Hansen, P.: Bicriterion path problems. In: Fandel, G., Gal, T. (eds.) Multiple Criteria Decision Making Theory and Application. LNEMS, vol. 177, pp. 109–127. Springer, Berlin (1980)

10. Iori, M., Martello, S., Pretolani, D.: An aggregate label setting policy for the multi-objective shortest path problem. Eur. J. Oper. Res. 207(3), 1489–1496 (2010)

11. Levandoski, J.J., Lomet, D.B., Sengupta, S.: The Bw-tree: A B-tree for new hardware platforms. In: Jensen, C.S., Jermaine, C.M., Zhou, X. (eds.) ICDE, pp. 302–313. IEEE Computer Society, Washington (2013)

12. Machuca, E., Mandow, L.: Multiobjective heuristic search in road maps. Expert Syst. Appl. 39(7), 6435–6445 (2012)

13. Martins, E.: On a multicriteria shortest path problem. Eur. J Oper. Res. 16(2), 236–245 (1984)

14. Meyer, U., Sanders, P.: Δ-stepping: A parallelizable shortest path algorithm. J. Algorithms 49(1), 114–152 (2003)

15. Paixáo, J.M., Santos, J.: Labeling methods for the general case of the multi-objective shortest path problem—-A computational study. In: Madureira, A., Reis, C., Marques, V. (eds.) Computational Intelligence and Decision Making, Intell. Syst. Control Autom. Sci. Eng., vol. 61, pp. 489–502. Springer, Dordrecht (2013)

16. Raith, A.: Speed-up of labelling algorithms for biobjective shortest path problems. In: Ehrgott, M., Mason, A., O'Sullivan, M., Raith, A., Walker, C., Zakeri, G. (eds.) ORSNZ, pp. 313–322. Operations Research Society of New Zealand, Auckland (2010)

17. Raith, A., Ehrgott, M.: A comparison of solution strategies for biobjective shortest path problems. Computers & OR 36(4), 1299–1331 (2009)

18. Saikkonen, R.: Bulk Updates and Cache Sensitivity in Search Trees. Ph.D. thesis. University of Jyväskylä (2009)

19. Sanders, P., Mandow, L.: Parallel label-setting multi-objective shortest path search. In: IPDPS, pp. 215–224. IEEE Computer Society, Washington (2013)

20. Schieferdecker, D., Völker, M., Wagner, D.: Efficient algorithms for distributed detection of holes and boundaries in wireless networks. CoRR abs/1103.1771 (2011)

21. Sewall, J., Chhugani, J., Kim, C., Satish, N., Dubey, P.: PALM: Parallel architecture-friendly latch-free modifications to B+ trees on many-core processors. PVLDB 4(11), 795–806 (2011)

22. Tung, C.T., Chew, K.L.: A multicriteria Pareto-optimal path algorithm. Eur. J. Oper. Res. 62(2), 203–209 (1992)

Beyond Synchronous: New Techniques for External-Memory Graph Connectivity and Minimum Spanning Forest

Aapo Kyrola, Julian Shun, and Guy Blelloch

Carnegie Mellon University
5000 Forbes Avenue, Pittsburgh, PA 15213, USA
{akyrola,jshun,guyb}@cs.cmu.edu

Abstract. GraphChi [16] is a recent high-performance system for external memory (disk-based) graph computations. It uses the Parallel Sliding Windows (PSW) algorithm which is based on the so-called *Gauss-Seidel* type of iterative computation, in which updates to values are immediately visible within the iteration. In contrast, previous external memory graph algorithms are based on the *synchronous* model where computation can only observe values from previous iterations. In this work, we study implementations of connected components and minimum spanning forest on PSW and show that they have a competitive I/O bound of $O(\text{sort}(E) \log(V/M))$ and also work well in practice. We also show that our MSF implementation is competitive with a specialized algorithm proposed by Dementiev et al. [10] while being much simpler.

1 Introduction

Research on external-memory graph algorithms was an active field in the 1990s and early 2000s, however not much work has been done on external-memory algorithms for fundamental graph problems since then. Recently, fueled by the interest to study large social networks and other massive graphs, there has been renewed interest for large-scale disk-based graph computation. In 2012, Kyrola et al. proposed GraphChi [16], which uses the *Parallel Sliding Windows* (PSW) algorithm for external-memory graph computation with the vertex-centric programming model. More recently, alternative solutions have been proposed, most notably X-Stream [20] and TurboGraph [12]. TurboGraph works efficiently only on modern Solid State Disks (SSD) that can support hundreds of thousands of random disk accesses per second, while GraphChi and X-Stream work efficiently even on traditional spinning disks.

In this paper, we study how GraphChi's PSW technique can be used efficiently to solve classic problems of graph connectivity and finding the minimum spanning forest (MSF) of a graph. Analyzing PSW is particularly interesting, since it expresses iterative computation using the so called Gauss-Seidel model (abbreviated G-S)[1] of computation in contrast to the Bulk Synchronous Parallel

---

[1] We borrow terminology from the study of iterative linear system solvers.

J. Gudmundsson and J. Katajainen (Eds.): SEA 2014, LNCS 8504, pp. 123–137, 2014.

(BSP) model of X-Stream. While in the BSP model program execution can only observe values computed on the previous iteration, in the G-S model the most recent value for any item is available for computation. We show by simulations and theoretical analysis that the G-S model can reduce the number of costly passes over the graph data significantly compared to BSP.

We show how to use PSW to write simple implementations of external-memory (EM) algorithms for connected components and MSF, in contrast to many previous algorithms that are difficult to implement. These algorithms take advantage of the G-S form of computation. We describe a problem called *minimum label propagation* (MLP) in which vertices update its identifier with the minimum of its identifier and its neighbors' identifiers. We show that a G-S implementation of this problem gives speedup over the traditional synchronous implementation, and use it to compute connected components and MSF. Our MSF algorithm is a graph contraction-based algorithm which repeatedly applies a single step of MLP. We prove that it terminates in a logarithmic number of iterations, and has an expected I/O bound of $O(\text{sort}(E)\log(V/M))$. We also show experimentally that it is competitive with the only available external-memory MSF implementation, while being much simpler. For connected components, we present two algorithms—a simple one based on MLP that requires a number of iterations proportional to the graph diameter, and one based on graph contraction (with the same I/O bound as for MSF). We show experimentally that for low-diameter graphs, the label propagation algorithm is competitive, while the contraction-based algorithm is efficient on general graphs.

2   Related Work

Chiang et al. [8] describe the first external-memory (deterministic) algorithm for MSF, and it has an I/O complexity bound of $O(\min(\text{sort}(V^2), \log(V/M)\text{sort}(E)))$. Kumar and Schwabe [14] give an improved deterministic algorithm with a bound of $O(\text{sort}(E)\log(B) + \log(V)\text{scan}(E))$. Arge et al. [3] give a deterministic algorithm requiring $O(\text{sort}(E)\log\log(B))$. The best I/O bound is for a randomized algorithm by Abello et al. [1] using $O(\text{sort}(E))$ I/O's with high probability. As MSF can be used for connected components, these bounds apply for external-memory connected components as well. There has been work on many other external-memory graph algorithms as well (see [13] for a survey).

As far as we know, the only experimental work on external-memory connected components is by Lambert and Sibeyn [17] and Sibeyn [21]. The implementations by Lambert and Sibeyn [17] do not have theoretical guarantees on general graphs and the implementation by Sibeyn [21] is only provably efficient for random graphs. For external-memory MSF, Dementiev et al. [10] present an implementation which requires $O(\text{sort}(E)\log(V/M))$ I/O's in expectation.

Convergence of iterative asynchronous computation has been studied in other settings, for example by Bertsekas and Tsitsiklis [5] for parallel linear system solving and Gonzalez et al. [11] in the context of probabilistic graphical models.

3   Preliminaries

A graph is denoted by $G = (V, E)$ where V is the set of vertices and E is a set of tuples (u, v) such that $u, v \in V$. When clear from the context, we also use V and E to refer to the number of vertices and number of edges in G, respectively.

I/O Model. Our analysis is based on the *I/O model* introduced by Aggarwal and Vitter [2]. The cost of a computation is the number of block transfers from external memory (disk) to main memory (RAM) or vice versa, and any computation that is done with data in RAM is assumed to be free. When modeling the I/O complexity, the following parameters are defined: N is the number of items in the problem instance, M is the number of items that can fit into main memory, and B is the number of items per disk block transfer. Fundamental primitives for I/O efficient algorithms are *scan* and *sort* (generalized prefix-sum). Their respective I/O complexities were derived by Aggarwal and Vitter [2]: $\text{scan}(N) = O(N/B)$ and $\text{sort}(N) = O((N/B) \log_{(M/B)}(N/B))$.

Jacobi (synchronous) and Gauss-Seidel (asynchronous) Computation. We now define formally the semantics of the two different models of computation. We borrow terminology from the numerical linear algebra literature since the commonly used terms "synchronous" and "asynchronous" have various other meanings in the context of numerical computation. In the following definitions, we denote by $x_i(t) \in A$ to be the value of a variable x_i (associated with an edge or vertex indexed by i) after iteration $t \geq 1$; $x(0)$ is the initial value of x and A is the domain of the variables.

Definition 1. Jacobi (Synchronous) Model: *A function F that computes $x_i(t)$ depends only on values from the previous iteration, i.e. $F(v) := F(\{x_j(t - 1)\}_{\forall j}) \to A$.*

To define the Gauss-Seidel computation, we define **schedule**, which determines the order of vertex "updates" in a vertex-centric computation.

Definition 2. *Let $\pi := V \to \{1, 2, ... |V|\}$ be a bijective function. Then $\pi(v)$ defines an update **schedule** so that if $\pi(u) < \pi(v)$, then vertex u will be updated before vertex v.*

The simplest schedule $\pi(v) = v$ updates vertices in the order they are labeled and is the default schedule used in GraphChi. In this paper, we study random schedules where π is a uniformly random permutation. We now give a definition of G-S computation that extends the traditional definition with a schedule:

Definition 3. Gauss-Seidel (Asynchronous) Model: *A function F that computes $x_i(t)$ uses the most recent values of its dependent variables. That is, $F(v) := F(\{x_i(t)\}_{i|\pi(i)<\pi(v)} \cup \{x_j(t - 1)\}_{j|\pi(j)>\pi(v)}) \to A$.*

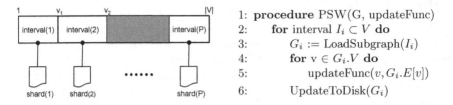

```
1:  procedure PSW(G, updateFunc)
2:      for interval I_i ⊂ V do
3:          G_i := LoadSubgraph(I_i)
4:          for v ∈ G_i.V do
5:              updateFunc(v, G_i.E[v])
6:          UpdateToDisk(G_i)
```

Fig. 1. Left: The vertices of graph (V, E) are divided into P intervals. Each interval is associated with a shard, which stores all edges that have destination vertex in that interval. Right: Pseudo-code for the main loop of Parallel Sliding Windows. Note that both for-loops can iterate in random order.

Parallel Sliding Windows. We now introduce the framework used to implement the algorithms in this paper. Parallel Sliding Windows (PSW) is based on the *vertex-centric* model of computation [16]. The state of the computation is encapsulated in the graph $G = (V, E)$, where $V = \{0, 1, \ldots, |V| - 1\}$, E is a set of ordered[2] tuples (src, dst) such that $src, dst \in V$. We associate a value (data) with each vertex and edge, denoted by d_v and d_e respectively. PSW executes programs that are presented as imperative vertex *update-functions* with the form: updateFunc$(v, E[v])$. This function is passed a vertex v, and arrays of the in- and out-edges of the vertex (denoted as $E[v]$, where $E[v] = \{(src, dst) \mid src \in V \vee dst \in V\}$). The vertex data and the data for its incident edges are accessible via pointer. Values of other vertices cannot be accessed. The update function is executed on each vertex in turn (under some schedule), with Gauss-Seidel semantics, i.e. changes to edge values are immediately visible to subsequent updates.

PSW executes the programs on a sequence of (partial) subgraphs $G_i \subset G$, $i \in \{1, .., P\}$, where each subgraph fits into memory. Each subgraph contains vertices of a continuous interval I_i of vertices: $I_1 = \{1, \ldots, a_1\}, \{a_1 + 1, \ldots, a_2\}, \ldots, I_P = \{a_{P-1}, \ldots, N\}$ so that $\bigcup_{i=1..P} I_i = V$ and $\forall i \neq j$, $I_i \cap I_j = \emptyset$. In addition to vertex values, each subgraph contains all the edges of those vertices[3]. After loading a subgraph, PSW executes the update function on the vertices and then writes the changes to vertex and edge values back to disk (see the pseudo-code in Fig. 1). Note that the order of processing the subgraphs, as well as the order that the update function is invoked on the vertices of the subgraph, can be arbitrary.

We now describe how PSW stores the edges on disk, and how the vertex intervals I_i are defined. Each interval I_1 is associated with a file, called shard(i). shard(i) stores all the *in-edges* (and their associated values) of vertices in interval I_i (see Fig. 1). Moreover, the edges in a shard are stored in sorted order based on their source vertex. The size of a shard must be less than the available memory M, and is typically set to $M/4$. To create the shards, we first sort all the edges based on their destination vertex ID, with an I/O cost of sort(E).

---

[2] For undirected graphs, we simply ignore the direction and it can be chosen arbitrarily.

[3] The subgraph is partial since it does not include vertex values for neighbors outside of the interval.

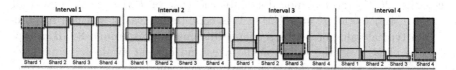

Fig. 2. Visualization of one iteration of PSW. In this example, vertices are divided into four intervals, each associated with a shard. The computation proceeds by constructing a subgraph of vertices one interval at a time. In-edges for the vertices are read from the corresponding shard (in dark color) while out-edges are read from all of the shards.

Then we create one shard at a time by scanning the sorted edges from the beginning, and add edges to a new shard until it reaches its maximum size (after which the shards are sorted in-memory prior to storing on disk). The vertex intervals and P are thus defined dynamically during the preprocessing phase. The second phase has an I/O cost of scan(E).

Each shard is split into two components: an immutable *adjacency shard* and a mutable *data shard*. Edges are stored in the adjacency shard as follows: for each vertex u that has out-edges stored in the shard, we write u followed by its number of edges. For each edge we store the neighbor vertex ID. The edge values are stored in the data-shard as a flat array A so that $A[i]$ has the value of ith edge. For each interval we also store a flat array on disk containing the vertex values for the vertices in the interval.

We now describe how the edges for a subgraph g_i are loaded from disk. First, the in-edges (and their values) for vertices in interval I_i are all contained in shard(i). Thus, that shard is loaded completely into memory. Secondly, the out-edges of the subgraph are contained in contiguous blocks in each of the shards, since we had sorted the edges by their source ID in the preprocessing phase. Note that we can easily store in memory the boundaries for the out-edges in each shard. After loading the blocks containing the edge adjacencies and values, we construct the subgraph in memory so that each edge is associated with a pointer to the location in a block that was loaded from disk, so all modifications are made directly to the data blocks. We can then execute the update function on each of the vertices in interval I_i. After finishing the updates, the edge data blocks are rewritten back to disk replacing the old data. Thus, all changes are immediately visible to subsequent updates (for the next subgraph), giving us Gauss-Seidel semantics (note that updates within each subgraph are done in a G-S manner in memory). This process is illustrated in Fig. 2.

Theoretical Properties of PSW. PSW can process any graph that fits on disk. The original GraphChi paper [16] states that there must be enough memory to store any one vertex and its edges, but we later describe in Section 5 how to get around this limitation. As described in [16], one iteration (pass) over the graph, in which both in- and out-edges of vertices are updated, has cost $PSW(G)$, where $2(E/B) + (V/B) \leq PSW(G) \leq 4(E/B) + (V/B) + \Theta(P^2)$.

For the $\Theta(P^2)$ term to not dominate the cost of PSW, we assume that $E < M^2/(4B)$ (recall $P = 4E/M$). This condition is easily satisfied in practice as a typical value of M for a commodity machine is around 8 GB, a typical value of B is on the order of kilobytes and current available graph sizes are less than a petabyte. Many algorithms only modify out-edges, and read in-edges in which case the I/O complexity is only $PSW(G)/2$. PSW also requires a preprocessing step for creating the shards, which has an I/O cost of sort(E). Note that due to the assumption $E < M^2/(4B)$, we have that sort(E) = $O((E/B)\log_{M/B}(M^2/(4B))) = O(E/B)$.

Graph Contraction with PSW. We can implement graph contraction under the PSW framework by allowing update-functions to write edges to a file (in a buffered manner). After the computation is finished, we create the contracted graph from the emitted edges and remove duplicate edges in the process. This has same cost as the preprocessing of the graph, which is sort(E). We can then continue executing PSW on the newly created contracted graph.

4   Minimum Label Propagation

We use *minimum label propagation (MLP)* as a subroutine in our minimum spanning forest algorithm. It can also be used directly to compute the connected components of a graph. The algorithm works by initializing each vertex with a label equal to its ID, and on each iteration, updating (in random order) the vertices' labels with the smallest of its neighbors' labels and its own labels.

The pseudo-code for computing connected components with MLP is shown in Algorithm 1. With PSW, and in the external-memory setting, vertices must communicate their labels via edges. Each edge (u, v) has two fields, *leftLabel* and *rightLabel*, where leftLabel contains the label of the smaller of u and v, and rightLabel contains the label of the other vertex. The functions in lines 1–6 implement this logic. The vertex values are initialized with the vertex IDs on line 16. We note that if MLP is run until convergence (i.e. until all edges "agree"), vertices in the same connected component will have the same label (in fact, if MLP is run until convergence, each edge can store only one field as both vertices will eventually have the same label).

Our algorithm executes the MLPUpdate update function (lines 8–14) using PSW as long as any vertex changes its label during an iteration (lines 18–20). MLPUpdate is passed a vertex and its edges. On line 11, it finds the minimum label written to its incident edges by its neighbors. If that label is smaller than the previous minimum label, it is written to all adjacent edges on line 14.

I/O Complexity. We analyze the number of iterations of MLP required until all connected vertices in each connected component of a graph share the same label. We refer to the the **distance** dist(u, v) between vertices $u, v \in V$ as the length of the path from u to v with the fewest number of edges. If u and v are not connected, then dist(u, v) is undefined. The **diameter** of a graph, denoted as D_G, is the maximum dist(u, v) between any two connected vertices $u, v \in V$.

Algorithm 1. Minimum-label Propagation (MLP)

1: **function** GETNEIGHBORLABEL(vertex, edge)
2:    **if** vertex.ID < edge.ID **then** return edge.rightLabel
3:    **else** return edge.leftLabel

4: **function** SETMYLABEL(vertex, edge, newLabel)
5:    **if** vertex.ID < edge.ID **then** edge.leftLabel = newLabel
6:    **else** edge.rightLabel = newLabel

7: **global** var changed
8: **procedure** MLPUPDATE(vertex, vertexedges)
9:    var minLabel = vertex.label
10:    **for** edge ∈ vertexedges **do**
11:       minLabel = min(minLabel, GetNeighborLabel(vertex, edge))
12:    **if** vertex.label ≠ minLabel **then** changed = *true*
13:       vertex.label = minLabel
14:       **for** edge ∈ vertexedges **do** SetMyLabel(vertex, edge, minLabel)

15: **procedure** RUNMLP(G)
16:    **for** vertex ∈ G.V **do** vertex.label = vertex.ID
17:    changed = *true*
18:    **while** changed = true **do**
19:       changed = false
20:       PSW(G, MLPUpdate)

The following lemma states the number of iterations a synchronous computation requires for convergence, and its proof is straightforward.

Lemma 1. *Let the vertex with the minimum identifier be v_{min}. Then the Jacobi (synchronous) computation of minimum label propagation requires exactly $\max_{v \in V} dist(v, v_{min}) \leq D_G$ iterations to converge.*

Clearly, the Gauss-Seidel model requires at most as many iterations as the synchronous model of computation so the I/O complexity is at most $D_G \times PSW(G) = O(D_G(V + E)/B)$. But with G-S, the computation can converge in fewer iterations because on each iteration the minimum label can propagate over multiple edges. We can study this analytically on a simple chain graph: Let C_n be a chain graph with n vertices $V_n = \{1, 2, ..., n\}$ and $n - 1$ edges $E_n = \{(1, 2), (2, 3), ..., (n - 1, n)\}$.

Theorem 1. *On a chain graph C_n, the expected number of iterations required for the Gauss-Seidel computation for convergence of MLP is $(n - 1)/(e - 1) \approx 0.582(n - 1)$. The synchronous computation requires exactly $n - 1$ iterations.*

Proof. The smallest label is 1 ("minimum label"), and on each iteration the label "advances" one or more steps towards the end of the chain. In the beginning of an iteration, let u be the vertex furthest from vertex 1 (the beginning of the chain) that has already assigned label 1. Vertex $u + 1$ will receive label 1 after it is updated. Now, if $\pi(u + 2) > \pi(u + 1)$, also vertex $u + 2$ will receive label 1 during the same iteration. Similarly, if $\pi(u + 3) > \pi(u + 2) > \pi(u + 1)$, the label reaches $u + 3$, and so on. We see that the probability that the minimum label advances exactly k steps is the probability of a permutation $\pi(u+1), \ldots, \pi(u+k), \pi(u+k+1)$ where the permutation is ascending from $\pi(u+1)$ to $\pi(u+k)$ but $\pi(u+k+1) < \pi(u+k)$. The probability of such a permutation is $\frac{k}{(k+1)!}$. Let X be the random variable denoting the number of steps the minimum label advances in the chain during one iteration. Then for large n, $\mathbb{E}[X]$ approaches $\sum_{k=1}^{\infty} \frac{k^2}{(k+1)!} = e - 1$. The theorem follows from this and Lemma 1.

5   Minimum Spanning Forest and Graph Contraction

Previous algorithms for computing the minimum spanning forest (MSF) in the external-memory setting use different variations of graph contraction to recursively solve the problem. We implement a variation of Boruvka's algorithm [7] on PSW, based on the MLP algorithm. On each iteration, Boruvka's algorithm selects the minimum weight edge of each vertex. These minimum edges are surely part of the minimum spanning forest, and the induced graph consisting only of these minimum edges is a forest. In Boruvka's algorithm, each tree is contracted into one vertex, edges are relabeled accordingly and the computation is repeated on the contracted graph. Each edge in the contracted graph contains information of its identity in the original graph so that the MSF edges can be identified.

Min-Label Contraction (MLC) Algorithm. The MLP algorithm described in previous section can be used to implement graph contraction: Let (x, y) be the labels stored on edge $e = (u, v)$ after one or more iterations of MLP. Then, we output edge (x, y) for the contracted graph, unless $x = y$. If there are multiple copies of an edge (x, y) that are output for the new graph, they are merged into one edge. The number of vertices in the new graph is equal to the number of distinct labels at the end of last MLP iteration. See the description at the end of Sec. 3 for details on how the contraction step is implemented.

MSF: One-Iteration Min-Label Contraction on a Forest. The pseudocode for our MSF algorithm is shown in Algorithm 2. A super-step of the algorithm (lines 19–23) consists of three invocations of PSW. The first PSW executes the CHOOSEMINIMUM update function (lines 2–4), which marks the minimum weighted edge of each vertex by setting the inMSF field of the edge data. These minimum edges are part of the MSF, and induce a collection of subtrees (a forest) on the graph. The second PSW execution is similar to the MLP algorithm (Algorithm 1), but the minimum labels are selected only among edges that have inMSF set to true. On line 19, we initialize a new graph and then output relabeled edges to the new graph. The third PSW execution contracts the graph and writes it to file by calling the update function CONTRACTIONSTEP (lines 11–16). Finally, on line 23 we preprocess the new graph into shards that can be used for PSW on the next iteration. Note that we have omitted details on keeping track of the original identity of each edge.

In contrast to Boruvka's algorithm, our algorithm runs only one iteration of propagation and contraction per super-step. This does not guarantee that the

---

Algorithm 2. Minimum Spanning Forest using PSW

1: **global** var outfile
2: **procedure** CHOOSEMINIMUM(vertex, vertexedges)
3:     var minEdge = [find minimum weighted edge of vertexedges]
4:     minEdge.inMSF = true
5: **procedure** MINIMUMLABELPROPONEITER(vertex, vertexedges)
6:     var minLabel = vertex.value
7:     **for** edge ∈ vertexedges **do**
8:         **if** edge.inMSF **then**
9:             minLabel = min(minLabel,GetNeighborLabel(vertex, edge))
10:     **for** edge ∈ vertexedges **do** SetMyLabel(vertex, edge,vertex.label)
11: **procedure** CONTRACTIONSTEP(vertex, vertexedges)
12:     **for** e ∈ vertexedges **do**
13:         **if** e.dst = vertex.ID **then**
14:             **if** e.leftLabel ≠ e.rightLabel **then**
15:                 writeEdge(outfile, e.leftLabel, e.rightLabel, e.value)
16:             **if** e.inMSF **then** outputToMSFFile(e)
17: **procedure** RUNMSF(G)
18:     **while** $|G.E| > 0$ **do**
19:         PSW(G, CHOOSEMINIMUM)
20:         PSW(G, MINIMUMLABELPROPONEITER)
21:         outfile = [initialize empty file]
22:         PSW(G, CONTRACTIONSTEP)
23:         G = PreprocessNewGraph(outfile)

---

trees will be completely contracted, but we will derive a lower bound on the number of vertices contracted on each step. In the G-S setting we assume that the unique labels are adversarial but the schedule of the vertices is random. Denote the label of a vertex v at the beginning of an iteration by $l(v)$. We only need to consider the min-label contraction problem on a tree. We want to show that a constant number of vertices will be contracted in each iteration, which allows us to bound the number of iterations of MSF by $O(\log(V/M))$.

Fact 1. *The number of degree-one (leaves) and degree-two vertices in a tree of V nodes is at least $V/3$.*

By Fact 1, we only need to consider the expected number of leaves and degree-two vertices contracted. In our algorithm, all vertices with the same label will be contracted into one. If a vertex ends up with the same label as a neighbor, at least one of the two will be contracted. Among the vertices with the same label, the vertex that is contracted can be chosen randomly. So on average, a vertex with the same label as a neighbor is contracted with at least $1/2$ probability.

Lemma 2. *For a tree with V_1 leaves, the expected number of leaves contracted is at least $V_1/4$.*

Proof. Consider the ID of a leaf v and its neighbor w. There are two cases: (1) $l(w) < l(v)$ and (2) $l(v) < l(w)$. In case (1), if $\pi(w) < \pi(v)$ then v will get the same label as w. There is a $1/2$ probability of the event $\pi(w) < \pi(v)$. In case (2), fix the ordering with respect to π for all vertices except v and w. Let $\pi(x) = \max(\pi(w), \pi(v))$. Consider the permutation from the start to $\pi(x)$ excluding $\pi(w)$ and $\pi(v)$. There are two sub-cases: (2a) the permutation causes the ID of w to become smaller than $l(v)$ after w is executed, and (2b) the permutation does not cause the ID of w to become smaller than $l(v)$ after w is executed. In case (2a), if $\pi(w) < \pi(v)$ then v will end up with the same label as w. In case (2b), if $\pi(v) < \pi(w)$ then w will end up with the same label as v. This is true because we know that all vertices before w do not reduce w's label to below $l(v)$. The probability of the desired ordering of $\pi(w)$ and $\pi(v)$ in either case (2a) or (2b) is $1/2$ as the events $\pi(w) < \pi(v)$ and $\pi(v) < \pi(w)$ are equally likely. Thus for any initial labeling of the vertices, a leaf must fall into either case (1) or case (2), and have a $1/2$ probability of having the same label as its neighbor. A leaf with the same label as its neighbor is contracted with at least $1/2$ probability. By linearity of expectations, at least $V_1/4$ leaves are contracted.

Lemma 3. *For a tree with V_2 degree-2 vertices, the expected number of degree-2 vertices contracted is at least $V_2/6$.*

Proof. Consider a degree-2 vertex v with neighbors u and w. If $\pi(v) < \pi(u) < \pi(w)$ or $\pi(u) < \pi(v) < \pi(w)$ then w will not affect the resulting labels of v or u. Similarly if $\pi(v) < \pi(w) < \pi(u)$ or $\pi(w) < \pi(v) < \pi(u)$ then u will not affect the resulting labels of v or w. In either of these cases, we can consider u as a leaf and use the analysis of Lemma 2 because the neighbor that is after v in π does not affect whether v will be contracted. One of these two cases will happen with $2/3$ probability. In the other orderings of u, v and w according to π we pessimistically assume that v is not contracted. Thus a degree-2 vertex will be contracted with at least $(2/3) \cdot (1/4) = 1/6$ probability. By linearity of expectations, the expected number of degree-2 vertices contracted is at least $V_2/6$.

By applying Fact 1 and Lemmas 2 and 3, we have the following theorem:

Theorem 2. *For a tree with V vertices, the number of vertices contracted in one iteration of min-label contraction is at least $V/18$.*

We note that our analysis applies for any (adversarial) labeling of the vertices.

Corollary 1. *The I/O complexity of our MSF algorithm is $O(sort(E) \log(V/M))$.*

Proof. By Theorem 2, after at most $\log_{18} V - \log_{18} M = O(\log(V/M))$ iterations, the number of vertices remaining will be at most M, at which point we can switch to a semi-external algorithm. Each iteration of the MSF algorithm requires $O(sort(E))$ I/O's. The result follows.

Our I/O complexity is worse than the $O(sort(E))$ bound of Abello et al. [1], but matches the bound of the only available external-memory MSF implementation by Dementiev et al. [10].

Dealing with Very High Degree Vertices. The original version of PSW requires any vertex and its edges to fit in memory. With the contraction procedure described earlier, it is possible that a vertex in the contracted graph gets many edges, possibly more than $O(M)$. We can address this issue by storing such high-degree vertices in their own shards. To find the minimum weighted edge or a neighbor ID, the order of the edges does not matter, so we can process such shards in parts, such that each part fits in memory. Since such shards only store edges for one vertex, this does not affect G-S semantics, and the I/O complexity bounds remain unchanged. Note that using this procedure, we can even remove the degree restriction on vertices in the *original* graph.

Connected Components. We can use the one-iteration MLC algorithm to compute connected components also. Instead of choosing the edge with minimum weight, each vertex chooses the neighbor with the minimum ID. During contraction, for each contracted vertex we keep pairs $(v, p(v))$ where $p(v)$ is the ID of its neighbor. On the way back up from the recursion we can relabel each vertex to be the same as its neighbor's label. Since the labels for the remaining are computed recursively we also have a list of pairs for them. This can be done by sorting the contracted vertex pairs by $p(v)$ and the remaining vertex pairs by v, and scanning them in parallel as done in [1]. The cost is $O(\text{sort}(V) + \text{scan}(V))$ per iteration, which is within our complexity bounds stated in Corollary 1.

6   Experiments

We use the following real-world and synthetic graphs in our experiments. *twitter_rv* is a graph of the Twitter social network with 41.7M vertices and 1.47B edges [15]. *uk-2007-05* is a subset of the UK WWW-network with 105M vertices and 3.8B edges [6]. The *web-Google* graph is a small web-graph with 0.5M vertices and 5M edges [18]. The *live-journal* graph is a social network graph with 5M vertices and 68M edges [4]. The first two are among the largest real-world graphs publicly available. We use synthetic k-grid graphs—each of the k-dimensional grids contain 100^k vertices, where each vertex has an edge to each of its 2^k neighbors. The *chain* graph is a 1-D grid graph.

Min-label Propagation Simulations. To our knowledge, it remains an open question to obtain a closed-form expression for the number of Gauss-Seidel iterations for MLP convergence on general graphs. Fortunately, it is simple to run simulations on arbitrary graphs and in Fig. 3, we show results on our input graphs. For the G-S computation, we randomized the schedule.

We see that the G-S computation always requires fewer iterations than the synchronous computation. The highest speedup in terms of number of iterations G-S achieves over Jacobi iterations is a 10-fold speedup on the 4d-grid. We note that on the grids, the advantage of G-S improves when the dimensionality (and thus the average number of edges) increases. Our intuition for this phenomenon is that between any pair of vertices, the number of possible paths for the minimum label to propagate increases rapidly with the average degree of vertices.

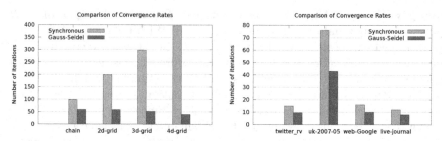

Fig. 3. Number of iterations required for convergence for synchronous and Gauss-Seidel label propagation on various graphs

Graph Contraction Simulations. In Theorem 2 we proved that at least a $V/18$ fraction of the vertices contract on each iteration. However, we found that the rate of contraction is much higher in practice. Although for MSF we only need to apply the one-iteration min-label contraction on trees, we conjecture that our contraction technique also works well on general graphs. We confirm the efficiency of the contraction technique by simulation. We compare our algorithm to Reif's random mate technique [19], where in each iteration vertices flip coins, and vertices that flipped "tails" pick a neighbor that flipped "heads" (if any) to contract with. Random mate gives the same I/O complexity as our algorithm since it contracts a constant fraction of the vertices per iteration. The simulation results are shown in Fig. 4. We see that for all input graphs, Gauss-Seidel under a randomized schedule achieves a better contraction rate (60–80%) rate than the synchronous version. The contraction rate observed is much higher than the bound indicated in Theorem 2. We see that both the synchronous and G-S versions of our algorithm achieve a higher contraction rate than random mate.

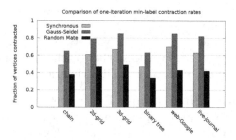

Fig. 4. The fraction of vertices contracted in one-iteration min-label contraction on various graphs. The binary tree has 4000 vertices.

Connected Components. We implemented two version of connected components on GraphChi (PSW). The first version runs the MLP algorithm until convergence. The implementation contains a simple optimization: if all vertices in a subgraph have converged, it is not reprocessed. This optimization requires only $O(P)$ memory. The second version is based on the one-iteration min-label contraction technique described in Section 5. One optimization we used was that instead of inducing a collection of subtrees per iteration (as in MSF), we did propagation over all of the edges. We found this to work much better in practice. The timing results are shown in Fig. 5 (left). All experiments were run on a MacMini (2011) with an Intel I5 CPU, 8 GB of RAM, and a 1 TB rotational hard drive. The results clearly confirm that on low-diameter real-world graphs the MLP version works well, but with the grid that has very high diameter, the contraction-based algorithm is orders of magnitudes faster. On real-world graphs, the contraction algorithm is also competitive, and actually outperforms the MLP version on the large *uk-2007-05* web graph. Unfortunately, we could not compare with the implementation of Sibeyn [21] as it is unavailable.

Minimum Spanning Forest. Perhaps due to the difficulty of realizing the algorithms, the only other implemented external memory MSF algorithm is due to Dementiev et al. [10], which is available as an open-source implementation using STXXL [9]. In Fig. 5 (right) we show timing results on various graphs, using a 8-CPU AMD server with 32 GB of RAM and a rotational SCSI hard drive (we could not compile Dementiev's algorithm on the Mac Mini.). Based on the results, we see that neither algorithm is the clear winner, but the relative difference varies strongly between different graphs. This is not surprising since they employ different graph contraction techniques. This shows that our algorithm using PSW is competitive with a special-purpose MSF implementation. The advantage of our algorithm is that it is much simpler (requiring only tens of lines of simple code), being part of a general purpose framework.

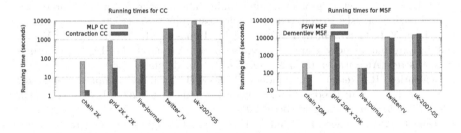

Fig. 5. Left: Timings for the MLP CC algorithm vs. contraction-based CC algorithm. Right: Timings for the PSW MSF algorithm vs. the Dementiev et al. [10] implementation. The numbers are averages—the standard deviations were less than 10%.

7   Conclusion

We have presented simple external-memory algorithms for connected components and minimum spanning forest implemented using GraphChi, and have proven an I/O complexity bound competitive with the only previous implementations for the problems. Our algorithms take advantage of the Gauss-Seidel type of computation, which leads to an improvement in convergence rate over the synchronous type of computation used in previous external-memory algorithms for these problems. Parallel Sliding Windows is an exciting development in the research of external-memory graph algorithms as they provide a generic framework for designing and implementing simple and practical algorithms.

Acknowledgements. This work is supported by the National Science Foundation under grant number CCF-1314590. Kyrola is supported by a VMware Graduate Fellowship. Shun is supported by a Facebook Graduate Fellowship.

References

1. Abello, J., Buchsbaum, A.L., Westbrook, J.R.: A functional approach to external graph algorithms. Algorithmica 32(3), 437–458 (2002)
2. Aggarwal, A., Vitter, J., et al.: The input/output complexity of sorting and related problems. Commun. ACM 31(9), 1116–1127 (1988)
3. Arge, L., Brodal, G.S., Toma, L.: On external-memory MST, SSSP and multi-way planar graph separation. J. Algorithms 53(2), 186–206 (2004)
4. Backstrom, L., Huttenlocher, D., Kleinberg, J., Lan, X.: Group formation in large social networks: Membership, growth, and evolution. In: 12th ACM SIGKDD International Conference on Knowledge Discovery and Data Mining, pp. 44–54. ACM, New York (2006)
5. Bertsekas, D.P., Tsitsiklis, J.N.: Parallel and Distributed Computation: Numerical Methods. Prentice-Hall, Upper Saddle River (1989)
6. Boldi, P., Santini, M., Vigna, S.: A large time-aware web graph. ACM SIGIR Forum 42(2), 33–38 (2008)
7. Boruvka, O.: O jistem problemu minimalnim (about a certain minimal problem). In: Prace, Moravske Prirodovedecke Spolecnosti, pp. 37–58 (1926)
8. Chiang, Y.J., Goodrich, M.T., Grove, E.F., Tamassia, R., Vengroff, D.E., Vitter, J.S.: External-memory graph algorithms. In: 6th Annual ACM-SIAM Symposium on Discrete Algorithms, pp. 139–149. SIAM, Philadelphia (1995)
9. Dementiev, R., Kettner, L., Sanders, P.: STXXL: Standard template library for XXL data sets. In: Brodal, G.S., Leonardi, S. (eds.) ESA 2005. LNCS, vol. 3669, pp. 640–651. Springer, Heidelberg (2005)
10. Dementiev, R., Sanders, P., Schultes, D., Sibeyn, J.: Engineering an external memory minimum spanning tree algorithm. In: Levy, J.-J., Mayr, E.W., Mitchell, J.C. (eds.) Exploring New Frontiers of Theoretical Informatics. IFIP, vol. 155, pp. 195–208. Springer, Heidelberg (2004)
11. Gonzalez, J., Low, Y., Guestrin, C.: Residual splash for optimally parallelizing belief propagation. In: International Conference on Artificial Intelligence and Statistics. pp. 177–184. JMLR (2009)
12. Han, W.S., Lee, S., Park, K., Lee, J.H., Kim, M.S., Kim, J., Yu, H.: TurboGraph: A fast parallel graph engine handling billion-scale graphs in a single PC. In: 19th ACM SIGKDD International Conference on Knowledge Discovery and Data Mining, pp. 77–85. ACM, New York (2013)
13. Katriel, I., Meyer, U.: Elementary graph algorithms in external memory. In: Meyer, U., Sanders, P., Sibeyn, J. (eds.) Algorithms for Memory Hierarchies. LNCS, vol. 2625, pp. 62–84. Springer, Heidelberg (2003)

14. Kumar, V., Schwabe, E.J.: Improved algorithms and data structures for solving graph problems in external memory. In: 8th IEEE Symposium on Parallel and Distributed Processing, pp. 169–176. IEEE Press, New York (1996)
15. Kwak, H., Lee, C., Park, H., Moon, S.: What is Twitter, a social network or a news media? In: 19th International Conference on World Wide Web, pp. 591–600. ACM, New York (2010)
16. Kyrola, A., Blelloch, G., Guestrin, C.: GraphChi: Large-scale graph computation on just a PC. In: 10th USENIX Symposium on Operating Systems Design and Implementation, vol. 8, pp. 31–46. USENIX (2012)
17. Lambert, O., Sibeyn, J.F., Stadtwald, I.: Parallel and external list ranking and connected components. In: International Conference of Parallel and Distributed Computing and Systems, pp. 454–460. IASTED (1999)
18. Leskovec, J., Lang, K.J., Dasgupta, A., Mahoney, M.W.: Community structure in large networks: Natural cluster sizes and the absence of large well-defined clusters. Internet Mathematics 6(1), 29–123 (2009)
19. Reif, J.H.: Synthesis of Parallel Algorithms. Morgan Kaufmann, San Francisco (1993)
20. Roy, A., Mihailovic, I., Zwaenepoel, W.: X-Stream: edge-centric graph processing using streaming partitions. In: 24th ACM Symposium on Operating Systems Principles, pp. 472–488. ACM, New York (2013)
21. Sibeyn, J.F.: External connected components. In: Hagerup, T., Katajainen, J. (eds.) SWAT 2004. LNCS, vol. 3111, pp. 468–479. Springer, Heidelberg (2004)

Retrieval and Perfect Hashing
Using Fingerprinting

Ingo Müller[1,2], Peter Sanders[1], Robert Schulze[2], and Wei Zhou[1,2]

[1] Karlsruhe Institute of Technology, Karlsruhe, Germany
{ingo.mueller,sanders}@kit.edu, wei.zhou@student.kit.edu
[2] SAP AG, Walldorf, Germany
robert.schulze@sap.com

Abstract. Recent work has shown that perfect hashing and retrieval of data values associated with a key can be done in such a way that there is no need to store the keys and that only a few bits of additional space per element are needed. We present FiRe – a new, very simple approach to such data structures. FiRe allows very fast construction and better cache efficiency. The main idea is to substitute keys by small fingerprints. Collisions between fingerprints are resolved by recursively handling those elements in an overflow data structure. FiRe is dynamizable, easily parallelizable and allows distributed implementation without communicating keys. Depending on implementation choices, queries may require close to a single access to a cache line or the data structure needs as low as 2.58 bits of additional space per element.

1 Introduction

Consider a set S of n keys from some universe U. Often we want to map S to unique integer IDs from a small range. A mapping with this property is called a *perfect hash function*. Similarly, we often want to store data values associated with the keys. This is known as a *retrieval data structure*. These two problems are closely interrelated. In particular, perfect hash functions can be used to implement a retrieval data structure by indexing an array of values. The classical way to implement these data structures uses hash tables storing the keys and/or values. However, it turns out that it is not necessary to store the key values. If the keys are big, this optimization can be important. For example, suppose that S is a set of URLs and we want to store one out of a small number of categories for each element of S. Another application example is storing flags for graph exploration in a large implicitly defined graph where the keys are quite large state descriptions of a finite automaton [9]. For further applications refer to [3]. We also encountered the retrieval problem in context of a the SAP HANA main memory column oriented data base [12]. In such a *column store* DB, each attribute of a relation is stored as a separate column. In order to keep large data sets in main memory, data compression is important in column stores. Perhaps the most important compression technique replaces elements from a large universe U (e.g., strings) by an ID that can be encoded with a few

J. Gudmundsson and J. Katajainen (Eds.): SEA 2014, LNCS 8504, pp. 138–149, 2014.

bits in many cases (dictionary compression). A retrieval data structure can be used to provide efficient mapping from U to IDs. SAP was interested in variants with very high query performance even at the price of somewhat larger space consumption than previous work. Since many data structures are accessed at the same time in HANA, an additional aspect was that we cannot assume lookup tables to remain in cache between accesses. Hence a small "cache footprint" of the data structure was an important design consideration.

Previous retrieval data structures in one way or the other cause multiple cache faults for each query. Refer to Section 3 for details. Here we explore the possibility to achieve higher performance by using a single hash function evaluation leading to access of a single cache line which yields the desired result. Our basic approach is quite simple minded and builds on traditional hash table data structures, in particular those for external memory: Keys are mapped to *buckets* capable of storing several values. The new aspect is that we do not store the keys themselves but only small *fingerprints* based on another hash function. On the first glance, this idea does not work since several elements in the same bucket may have the same fingerprint making them indistinguishable. This problem is solved by moving *all* colliding elements to an overflow data structure. Similarly, surplus elements from overfull buckets are moved to the overflow data structure. The overflow data structure can be based on the same principle using a fresh hash function. Elements not fitting there are moved to a secondary overflow data structure and so on. We may stop the recursion once the number of elements is small enough to use a more expensive data structure. This yields constant worst case access time and the expected number of cache faults can be close to one. Section 4 describes the Fingerprint Retrieval approach (FiRe) in more detail. In particular, we explain how using compression of the set of fingerprints in a bucket, the required space can become close to the space needed just for storing the function values. It turned out that the FiRe approach has additional advantages that may be even more important for some applications. In particular we can dynamize the data structure allowing insertions and deletions in expected constant time. In Section 5 we explain how the FiRe approach can be adapted to perfect hash functions. In Section 6 we report on experiments with a performance oriented implementation of FiRe. FiRe significantly outperforms competing solutions with respect to construction time and query time. The price is sometimes but not always higher space overhead which is nonetheless much smaller than for ordinary hashing. Section 7 summarizes the results and discusses possible future work.

2   Preliminaries

We use $i..j$ as a shorthand for $\{i, \ldots, j\}$. Let $n = |S|$. An obvious lower bound for the space consumption of a retrieval data structure is rn bits. For the analysis we assume that the used hash functions $h : U \to 1..m$ behave like truly random functions, i.e., we assume that they are drawn uniformly at random from the set of mappings from U to $1..m$. This can be justified theoretically using a "splitting trick" [7]. A *perfect hash function* $h : U \to 1..m$ is an injective mapping from S to $1..m$. h is a *minimal perfect hash function* if $m = n$. In this paper, $\log x$ denotes the base two logarithm. $f(n) \sim g(n)$ expresses that f converges to g as $n \to \infty$. A pair $(s, t) \in M \subseteq \mathbb{R}^2$ is *Pareto optimal* with respect to M if no other element $(s', t') \in M$ *dominates* x, i.e., $s' \leq s$ and $t' \leq t$.

3 Related Work

Fingerprinting is a well known technique [17] for indexing data structures but it has not been applied to perfect hashing or retrieval so far. Most applications use fairly large fingerprints in order to avoid collisions. The cuckoo filter [11] uses small fingerprints to obtain an approximate dictionary. Unique features of FiRe are its simplicity and that it is able to repair collisions.

Theoretical solutions for perfect hashing with a constant or even optimal number of bits per element have been known for a long time [15]. However, practical solutions have emerged only recently – raising significant interest in the topic. The compressed hash-and-displace algorithm [1] (CHD) uses a primary hash function to identify the index of a secondary hash function. This index is geometrically distributed with constant expectation and with clever compression [13] needs only a constant number of bits. CHD has relatively expensive queries since it performs select operations on large sparse bit vectors. CHD is also inherently sequential since it relies on a greedy construction algorithm. The BPZ algorithm by Botelho, Pagh and Ziviani [2] computes the hash function value based of three random table lookups. Hence, using BPZ for retrieval implies about four cache faults for each access for large inputs. Computing the table is based on an inherently sequential greedy algorithm for ordering the edges of a 3-regular random hypergraph. The EPH algorithm [3] uses the splitting trick to compute minimal perfect hash functions. The main difference to FiRe is that the resulting buckets in EPH have variable size and use no fingerprints but the BPZ algorithm to build bucket local perfect hash functions. Another external hash table represents buckets using *entropy coded tries* (ECT) [18] storing the longest distinguishing prefix of a hash value. These can be viewed as "perfect" fingerprints leading to a single stage lookup desirable for external memory but also introducing complication and computational overhead not appropriate for our high performance setting.

The retrieval problem can also be solved directly. The CHM algorithm [8] uses yet another greedy algorithm to compute a table of $m = \mathcal{O}(n)$ r-bit values so that the retrieved value is the xor of $k \geq 2$ values at hashed table positions.

These results can be viewed as show cases of algorithm engineering since they combine interesting ideas and highly nontrivial theoretical analysis in such a way that one gets surprisingly good results that are practically useful. FiRe is different in that it starts from a simple-minded idea and has a very simple analysis and implementation. The surprising part is that one gets competitive and in some aspects superior results this way.

4 Retrieval Using Fingerprint Hashing

4.1 The FiRe Data Structure

The first level of a FiRe data structure consists of an array B of $m = n/b$ buckets. A bucket is an array of a values with r bits each. For each stored value, the bucket also stores a fingerprint in the range $1..k$. An element $s \in S$ is mapped to a bucket by a hash function $h_B : U \rightarrow 1..m$. Its fingerprint is obtained by a hash function $h_f : U \rightarrow 1..k$. We can equivalently assume that we have a single hash function $h \rightarrow 1..km$ defining both bucket position and fingerprint. The first level can only hold values for elements with a unique value of h, i.e., no two elements of S stored in the first level may be mapped to the same bucket *and* have the same fingerprint – a fingerprint collision. If more then a such eligible elements are mapped to the same bucket, any a of them can be chosen.

The first level is constructed by mapping elements to their buckets, removing elements involved in fingerprint collisions, and then removing elements from overloaded buckets. Elements not stored in the first level are moved on to the second level which is built in an analogous way. This process is repeated until a maximum number of levels L is reached. Layer $L + 1$ is a *fallback* data structure which stores the remaining elements and guarantees constant worst case access time using any of the previous techniques. We can also use $L = \infty$, eliminating the fallback data structure. Figure 1 summarizes the structure of FiRe. As long as a constant fraction of the remaining elements considered can be stored in each iteration, the overall construction time is linear.

A query for key u checks whether an element with fingerprint $h_f(u)$ is stored in bucket $h_B(u)$ of level 1. In the positive case, the associated data element is returned. Otherwise, the next level is queried. As long as L and a are constants,

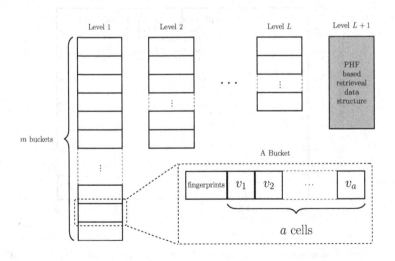

Fig. 1. Schematic diagram of a FiRe data structure

this yields constant worst case query time. For $L = \infty$ the worst case query time can be made logarithmic by restarting the construction process of a layer whenever it is much smaller than expected.

Representing Fingerprints. There are several ways to represent fingerprint information with different trade-offs between space consumption and query time. Refer to the full paper for details. Most of the time we will consider bit vectors using k bits for representing all fingerprints in a bucket. Also interesting are information theoretically optimal representations requiring $\left\lceil \log \sum_{i=0}^{a} \binom{k}{i} \right\rceil$ bits.

Asymptotic Analysis. For details refer to the full paper. Here we only outline the basic ideas of an analysis as $m \to \infty$ assuming that $L = \infty$ and the parameters a, b, and k are the same on all levels. We first show that the probability of a particular fingerprint value to represent exactly one element is $p_1 \sim \frac{b}{k} e^{-b/k}$. We then argue that the number of non-colliding elements allocated to a bucket is approximately $B(k, p_1)$ binomially distributed which implies that the expected number of empty cells in a bucket is $a_0 \sim \sum_{i=0}^{a-1} (a-i) \binom{k}{i} p_1^i (1-p_1)^{k-i}$. Since the *overall* number of empty cells is sharply concentrated around its expectation, we can use a_0 to estimate the space overhead per element as $s \sim \frac{ra_0 + s_f}{a - a_0}$ bits per element where s_f denotes the number of bits needed to store the fingerprints of a bucket. The expected number of accessed levels is $\ell \sim \frac{b}{a - a_0}$.

4.2   Choosing Parameters

The performance of the FiRe data structure with respect to space consumption and query time depends on the parameters a, b, k, on implementation choices,

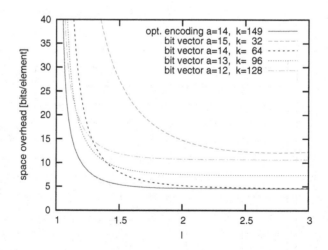

Fig. 2. Expected number of levels accessed versus space overhead for $r = 32$ and bit-vector encoding of fingerprints

in particular for representing fingerprints, on r, and on hardware parameters like the cache line size. Hence, it is a complex problem how to actually set the parameters. Moreover, in most situations there is not *one* optimal choice but a trade-off between space and time. Hence, we are interested in a set of parameter settings representing Pareto optimal solutions with respect to space and time while excluding suboptimal choices that are dominated by other choices. We propose to attack this problem by starting from hardware and implementation constraints generating a small set of reasonable choices for setting a, k, and the fingerprint representations. For each of these choices, b will be the only remaining free parameter. We then have to find out which of these choices yield Pareto optimal solutions for some values of b.

We exemplify this methodology for an example oriented at the column store application mentioned in the introduction. We consider a machine with cache line size 64 and we want to store values with $r = 32$ bits. We are interested in very fast access and thus choose the bit vector representation for fingerprints. For the same reason, we want buckets to fit perfectly into a cache line. This implies that k should be a multiple of 32 and that $a = 16 - k/32$. Figure 2 plots the resulting trade-off for $k \in \{32, 64, 96, 128\}$. These plots were obtained by computing the pairs (ℓ, s) for $b \in 1..100$. The case $k = 32$ is not useful since it leads to too many fingerprint collisions. The case $k = 64$, $a = 14$ is perhaps the most useful one since it is good for a wide range of trade-offs. In particular, its space overhead becomes as low as 4.586 bits per element for $b = 64$. If we are willing to spend 5 bits per element, $\ell = 2.1$ expected level access suffice (at $b = 29$). To achieve $\ell < 1.3$, it is better to use $k = 96$, $a = 13$. To achieve $\ell < 1.15$, it is better to use $k = 128$, $a = 12$. The corresponding ranges of b for which the respective cases are Pareto optimal are 1..11 for $k = 128$, 12..16 for $k = 96$, and 17..64 for $k = 64$. For comparison, the solid curve shows the values for $a = 14$, $k = 129$ assuming an information theoretically optimal encoding of the fingerprints. This curve dominates all the other curves. However, note that the query time for this case is likely to be quite large anyway due to overhead for decoding the fingerprints. Hence, this curve should rather be viewed as an optimistic estimate for the performance of some very clever implementation that offers a combination of space efficient encoding and fast decoding.

External, Parallel, and Distributed Processing. In the full paper we explain how to construct a FiRe data structure in external memory and on a parallel machine. Due to the decomposition into independent buckets this is very easy and efficient. Perhaps more interestingly, FiRe can be implemented in various distributed settings so that communication volume is very low. In particular, there is no need to communicate keys.

4.3 Dynamization

FiRe directly supports an update operation – changing the value associated with a key. The same holds for retrieval data structures based on perfect hashing but other data structures such as CHM [8] do not allow updates.

An advantage of the FiRe is that it can be augmented to allow *modifications* (insertions and deletions) in expected constant time. Existing superlinear lower space bounds [5,10] make this appear difficult without access to the key information. However, we will now argue that additional information only needs to be available during modifications. This setting does not save memory but has useful applications anyway. One example is when the FiRe data structure fits into fast memory (e.g., L3 cache) and the augmented information fits into the next level of the memory hierarchy (e.g., main memory). In this situation, the dynamized FiRe will be faster than alternative solutions when there are much more queries than modifications. Another scenario is distributed computing where the key information is available on one site A and another site B only stores the data needed for queries. A modification will then be done on A which sends only the information required to change the FiRe data structure at site B.

In order to support insertions, we need to store two kinds of additional information. First, we need to know the keys of the elements stored in a bucket. When a newly inserted element x suffers a fingerprint collision with an element y, both x and y need to be moved on to the next level. When x collides with a fingerprint that already suffered a collision previously, this also needs to be known. We call such a fingerprint "blocked". To insert an element x, we inspect the bucket $i = h_B(x)$. If the fingerprint $f = h_f(x)$ is unused, and block i is not full, the value for x is inserted into block i and f is marked as used. The dynamic part of the data structure remembers x. In all other cases, f becomes blocked for block i and the insertion attempt moves on to the next level. If x had to move on because of another element y stored there (i.e., f was used but not blocked previously), element y is also moved to the next level.

Note that in the worst case, a single insertion can cause a chain reaction leading to a number of element moves exponential in the number of levels L. In the full paper we analyze this effect using branching processes and show that the situation remains stable as long as $k > b$. When this condition becomes violated, the data structure should be rebuilt.

The easiest way to handle deletions is to ignore them – the specification of retrieval data structures does not specify anything about the result of a query for an element outside S. The stability condition for the branching process should then define n (and, as a consequence $b = n/m$) as the number of stored elements plus the number of inserted elements. We can also trigger rebuilding when this value of n differs too much from the number of non-deleted elements. Actual deletion of an element stored in a bucket is easy. We just remove it from the cell it previously occupied and set its finger print from used to unused.

If we insist on keeping the retrieval part of the FiRe data structure identical to what we would get in the static case, we additionally need to be able to unblock blocked fingerprints. For this we need to count the number of times a fingerprint is used. When this count goes down to one, we have to find the element which wants to move there, delete it from the subsequent layers and move it to the current layer. A similar case applies when an element is deleted from a full bucket. Then the block has room for an element associated with a blocked fingerprint with count one.

5   Fingerprint Based Perfect Hashing (FiPHa)

Perfect hashing can be viewed as FiRe with $r = 0$, i.e., there is no need to store any associated information – we only store fingerprint information in the buckets. The fingerprint information can be used to define the injective function $h_p(x) := ah_b(x) - a + \text{rank}_B(x)$ if $h_f(x)$ is a valid fingerprint in bucket $B[h_b(x)]$ and where $\text{rank}_B(x)$ counts the number of one-bits in the fingerprint bit vector of bucket $B[h_b(x)]$ up to position $h_f(x)$. If $h_f(x)$ is not a valid fingerprint in bucket $B[h_b(x)]$, we return the value of $h_p(x)$ for the next layer and add the offset am. The analysis of FiPHa is analogous to the analysis of FiRe. We get expected space overhead of $s = ns_f/(a - a_0)$ bits and as before $\ell = b/(a - a_0)$. The expected range of the perfect hash function is $n/(1 - \frac{a_0}{a})$. This can be seen as follows: In layer 1, we will consume range am. In expectation, $am - a_0 m$ elements will be mapped to this range. The resulting ratio is $am/(am - a_0 m) = 1/(1 - \frac{a_0}{a})$. Since the same ratio will be observed on all levels, the overall expected range is $n/(1 - \frac{a_0}{a})$.

For example, assuming information theoretically optimal representation of fingerprints, cache line size 64 bytes, and optimizing for space, we set $k = b = 543$ and $a = 200$. We get expected space $s = 2.61$ bits per element, $\ell = 2.77$ expected layer accesses and the perfect hash function has expected range $1.026n$. In this case, compression does not actually help a lot. Consider uncompressed bit vector representation, $k = b = 512$ and $a = 188$. We then get space 2.79 bits, 2.78 layer accesses, and range $1.023n$.

FiPHa is dynamizable in a way analogous to FiRe.

Minimal Perfect Hashing. We can get a minimal perfect hash function by using only a single large bucket (i.e., $b = n$, $m = 1$, $a = \infty$) and by setting the bucket size a to the actual number of elements stored in that bucket (i.e., all those elements not moved on to the next layer). The price we pay for this conceptual simplification is that now the rank function has to work on an input of size $\mathcal{O}(n)$. Fortunately, it is well known how to do this with constant query time and information theoretically optimally up to lower order terms. There are even practical implementation, e.g., [19]. The asymptotic analysis is also greatly simplified since there is no need to account for empty cells. In the full paper we argue that the expected space consumption per stored element is $\frac{H(p_1)}{p_1}$ where H denotes the entropy function and p_1 is the probability that a fingerprint is used exactly once in a bucket. The expected number of accessed levels is $e^{n/k}$. Optimizing for space consumption we get $p_1 = 1/e$ at $k = n$. This value yields $H(1/e) \approx 2.58$ bit of space per element and expected number of accessed levels e. If we are willing to spend 3 bits per element, we get about 1.58 expected level accesses. For 4 bits per element the same figure becomes 1.21 expected level access.

6 Experiments

We show results for 10^8 32 bit integers (data set **INT**) and 10^8 3-grams (sequences of 3 words) randomly choosen from the $1.33 \cdot 10^8$ 3-grams from Google Books [14] starting with **n** (data set **NGRAM**). We evaluate the retrieval data structures on both data sets and with $r \in \{8, 32, 64\}$ bits for data values.

We have implemented the FiRe and FiPHa data structures using bit vector representation of fingerprints. Refer to the bachelor thesis of Wei Zhou [21] for more details. Buckets are aligned to cache lines. We use Jenkins [16] fast and simple hash function (we also tried the newer SpookyHash with similar results). The implementation uses GNU C++ 4.8.1 with compilation options `-O3 -m64 -msse4.2 -fopenmp -std=c++11 -march=native`. Elements are assigned to buckets with the fast parallel radix sort algorithm from [20]. All experiments have been performed using a single core of a machine with 48 GByte RAM and 2 Intel Xeon X5650 hexa-core processors with 2.66 GHz clock frequency and 12 MByte L3 cache. The cache line size of this processor is 64 byte. For each of value of r we have configured FiRe with three parameter settings that achieve ℓ close to 1.05, 1.25, and 1.5 respectively. In the full paper we list these configurations. Here we denote them FiRe5, FiRe25, and FiRe50 respectively. All these variants use $L = 8$ levels. The fallback data structure uses our own implementation of the BPZ algorithm [2]. FiPHa is configured as introduced in Section 5.

We compare FiRe and FiPHa with five state of the art implementations from the CMPH library [4]. All these algorithms use the Jenkins hash function [16]. CHM-x refers to the CHM algorithm [8] where x indicates how many values are xor-ed. CHD-α stands for the CHD algorithm [1] where α denotes the load factor used in [1] – 0.99 means that a nearly minimal perfect hash function is generated. We have also made measurements using the STL `unordered_map` hash table (using Jenkins hash function and preallocating as many buckets as elements). STL not only needs much more space but is also about three times slower than FiRe both with respect to construction time and query time.

Figure 3 visualizes the remaining results. The full paper gives the corresponding numeric values. With respect to construction time FiRe is four times faster than the fastest competitors (CHD-0.5) and 17 times faster than the slowest one (CHD-0.99). FiPHa is considerably slower to construct, but still faster than all competitors. With respect to space consumption, FiRe50 is competitive with the other implementations for $r \geq 32$. Only FiPHa and CHD-0.99, which compute a near minimal perfect hash function, beat all the other codes significantly, with a small advantage for CHD-0.99. However, this comes at the price of a much larger construction time. For $r = 8$ or for FiRe5 and FiRe25, FiRe needs significantly more space then the other codes. However this is still much less than an ordinary hash table. However, more space efficient implementations of fingerprint sets may improve this in the future, in particular for small r.

As expected, FiRe has the best query times. The only competitor with comparable performance, CHM-2, actually needs a similar amount of space as FiRe but is clearly beaten with respect to construction time. Also recall that CHM does not support updates of the retrieved information. As expected FiPHa has

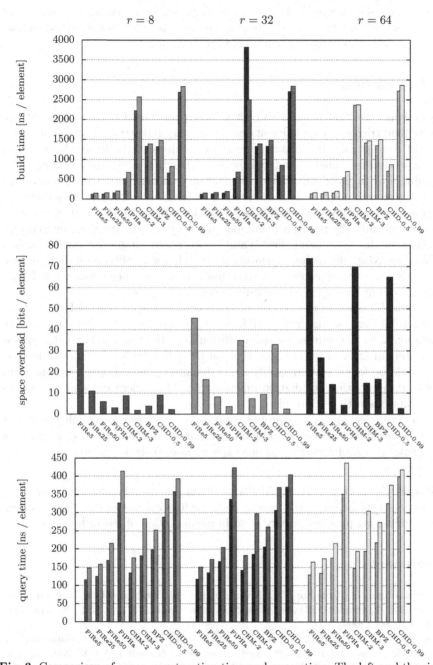

Fig. 3. Comparison of space, construction time and query time. The left and the right bar show the values of the INT and the NGRAM data set respectively.

worse query performance than FiRe but remains competitive with CHD-0.99, the only competitor with a similar space overhead, while having a considerably lower construction time. Overall, we also see a significant performance advantage compared to competitors of comparable functionality.

7   Conclusion

In retrospect, we find it surprising that fingerprint based hashing was not the first method tried for space efficient perfect hashing and retrieval data structures. At least conceptually it looks simpler than previous methods. We have shown that it also allows faster (and parallel) construction and faster queries. The advantages of fingerprinting for dynamization and distributed implementation seem even more fundamental. Since so many results follow from the fingerprinting approach, it looks interesting to look for further applications and refinement.

We could radically reduce the number of empty cells by mapping overflowing elements to the same level. For example, we could adapt the bucket-cuckoo hashing approach [6] to fingerprinting as in [11] for approximate dictionaries. We pay with more expensive construction and we will need somewhat larger fingerprints but for large r this should overall save space.

It might be possible to find better trade-offs between space and query time by using different parameters on different levels of the FiRe data structure. It looks promising to optimize for space efficiency on the first level and change the parameters in favor of lower query time on the following levels. Good combinations could be found systematically using dynamic programming – we maintain a set of Pareto optimal configurations using i levels of hierarchy and use this to build solutions with $i + 1$ levels.

In order to get a good practical compromise between space efficiency and speed, it would be interesting to look for a representation of the fingerprint sets that allows fast rank-queries and need space close to the lower bound.

Acknowledgements. We would like to thank Martin Dietzfelbinger for valuable suggestions including the idea to combine fingerprints with cuckoo hashing. Sebastian Schlag provided a very good implementation of the radix sorter.

References

1. Belazzougui, D., Botelho, F.C., Dietzfelbinger, M.: Hash, displace, and compress. In: Fiat, A., Sanders, P. (eds.) ESA 2009. LNCS, vol. 5757, pp. 682–693. Springer, Heidelberg (2009)
2. Botelho, F.C., Pagh, R., Ziviani, N.: Simple and space-efficient minimal perfect hash functions. In: Dehne, F., Sack, J.-R., Zeh, N. (eds.) WADS 2007. LNCS, vol. 4619, pp. 139–150. Springer, Heidelberg (2007)
3. Botelho, F.C., Ziviani, N.: External perfect hashing for very large key sets. In: 16th ACM Conference on Information and Knowledge Management, pp. 653–662. ACM, New York (2007)

4. de Castro Reis, D., Belazzougui, D., Botelho, F.C., Ziviani, N.: CMPH – C Minimal Perfect Hashing Library, http://cmph.sf.net
5. Demaine, E.D., der Heide, F.M.A., Pagh, R., Pătraşcu, M.: De dictionariis dynamicis pauco spatio utentibus. In: Correa, J.R., Hevia, A., Kiwi, M. (eds.) LATIN 2006. LNCS, vol. 3887, pp. 349–361. Springer, Heidelberg (2006)
6. Dietzfelbinger, M., Weidling, C.: Balanced allocation and dictionaries with tightly packed constant size bins. Theoret. Comput. Sci. 380(1-2), 47–68 (2007), http://dx.doi.org/10.1016/j.tcs.2007.02.054
7. Dietzfelbinger, M.: Design strategies for minimal perfect hash functions. In: Hromkovič, J., Královič, R., Nunkesser, M., Widmayer, P. (eds.) SAGA 2007. LNCS, vol. 4665, pp. 2–17. Springer, Heidelberg (2007), http://dx.doi.org/10.1007/978-3-540-74871-7_2
8. Dietzfelbinger, M., Pagh, R.: Succinct data structures for retrieval and approximate membership (Extended abstract). In: Aceto, L., Damgård, I., Goldberg, L.A., Halldórsson, M.M., Ingólfsdóttir, A., Walukiewicz, I. (eds.) ICALP 2008, Part I. LNCS, vol. 5125, pp. 385–396. Springer, Heidelberg (2008)
9. Edelkamp, S., Sanders, P., Šimeček, P.: Semi-external LTL model checking. In: Gupta, A., Malik, S. (eds.) CAV 2008. LNCS, vol. 5123, pp. 530–542. Springer, Heidelberg (2008)
10. Eppstein, D., Goodrich, M.: Straggler identification in round-trip data streams via newton's identities and invertible Bloom filters. IEEE Trans. Knowl. Data Eng. 23(2), 297–306 (2011)
11. Fan, B., Andersen, D.G., Kaminsky, M.: Cuckoo filter: Better than bloom. Login 38(4) (2013)
12. Färber, F., et al.: SAP HANA Database: Data management for modern business applications. SIGMOD Rec. 40(4), 45–51 (2012), http://doi.acm.org/10.1145/2094114.2094126
13. Fredriksson, K., Nikitin, F.: Simple compression code supporting random access and fast string matching. In: Demetrescu, C. (ed.) WEA 2007. LNCS, vol. 4525, pp. 203–216. Springer, Heidelberg (2007)
14. Google: Google books Ngram Viewer, http://storage.googleapis.com/books/ngrams/books/datasetsv2.html
15. Hagerup, T., Tholey, T.: Efficient minimal perfect hashing in nearly minimal space. In: Ferreira, A., Reichel, H. (eds.) STACS 2001. LNCS, vol. 2010, pp. 317–326. Springer, Heidelberg (2001)
16. Jenkins, B.: Algorithm alley: Hash functions. Dr. Dobb's Journal (1997)
17. Karp, R.M., Rabin, M.O.: Efficient randomized pattern-matching algorithms. IBM J. Res. Dev. 31(2), 249–260 (1987)
18. Lim, H., Andersen, D.G., Kaminsky, M.: Practical batch-updatable external hashing with sorting. In: ALENEX, pp. 173–182. SIAM, Philadelphia (2013)
19. Navarro, G., Providel, E.: Fast, small, simple rank/select on bitmaps. In: Klasing, R. (ed.) SEA 2012. LNCS, vol. 7276, pp. 295–306. Springer, Heidelberg (2012)
20. Wassenberg, J., Sanders, P.: Engineering a multi-core radix sort. In: Jeannot, E., Namyst, R., Roman, J. (eds.) Euro-Par 2011, Part II. LNCS, vol. 6853, pp. 160–169. Springer, Heidelberg (2011)
21. Zhou, W.: A Compact Cache-Efficient Function Store with Constant Evaluation Time. Bachelor thesis, KIT and SAP (2013)

Efficient Wavelet Tree Construction and Querying for Multicore Architectures[*]

José Fuentes-Sepúlveda, Erick Elejalde, Leo Ferres, and Diego Seco

Universidad de Concepción. Edmundo Larenas 219, Concepción, Chile
{jfuentess,eelejalde,lferres,dseco}@udec.cl

Abstract. Wavelet trees have become very useful to handle large data sequences efficiently. By the same token, in the last decade, multicore architectures have become ubiquitous, and parallelism in general has become extremely important in order to gain performance. This paper introduces two practical multicore algorithms for wavelet tree construction that run in $O(n)$ time using $\lg \sigma$ processors, where n is the size of the input and σ the alphabet size. Both algorithms have efficient memory consumption. We also present a querying technique based on batch processing that improves on simple domain-decomposition techniques.

1 Introduction and Motivation

After their introduction in the mid-2000s, *multicore* computers —computers with a shared main memory and more than one processing unit— have become pervasive. In fact, it is hard nowadays to find a single-processor desktop, let alone a high-end server. The argument for multicore systems is simple [20,22]: thermodynamic and material considerations prevent chip manufacturers from increasing clock frequencies beyond 4GHz. Since 2005, clock frequencies have stagnated at around 3.75GHz for commodity computers, and even in 2013, 4GHz computers are rare. Thus, one possible next step in performance is to take advantage of multicore computers. To do this, algorithms and data structures will have to be modified to make them behave well in parallel architectures.

In the past few years, much has been written about compressed data structures. One such structure that has benefited from thorough research is the *wavelet tree* [12], henceforth *wtree*. Although the *wtree* was original devised as a data structure for encoding a reordering of the elements of a sequence[12,9], it has been successfully used in many applications. For example, it has been used to index documents [24], grids [19] and even sets of rectangles [3], to name a few applications (w.r.t. [18,16] for comprehensive surveys).

Our contributions in this paper are as follows: we first propose two linear $O(n)$ time parallel algorithms for the most expensive operation on wavelet trees, construction, using $\lg \sigma$ processors[1] (see Sect. 3). We report experiments which show the algorithms to be practical for large datasets, achieving close to perfectly

---

[*] This work was supported in part by CONICYT FONDECYT 11130377.
[1] We use $\lg x = \log_2 x$.

linear speedup (Sect. 4). In order to exploit multicore architectures, we also investigated techniques to speed up range queries and propose BQA (*batch-query-answering*), a hybrid domain-decomposition/parallel batch processing technique (see Sect. 14) that exploits both multicore architectures and cache data locality effects in hierarchical memory systems. We empirically achieve almost perfect linear throughput by augmenting the number of cores (see Sect. 4). To the best of our knowledge, this is the first proposal of a parallel wavelet tree (in the dynamic multithreading model, see Sect. 2).

2   Preliminaries

Wavelet Trees: For the purpose of this paper, a wavelet tree is a data structure that maintains a sequence of n symbols $S = s_1, s_2, \ldots, s_n$ over an alphabet $\Sigma = [1..\sigma]$ under the following operations: $access(S, i)$, which returns the symbol at position i in S; $rank_c(S, i)$, which counts the times symbol c appears up to position i in S; and $select_c(S, j)$, which returns the position in S of the j-th appearance of symbol c. Wavelet trees can be stored in space bounded by different measures of the entropy of the underlying data, thus enabling compression. In addition, they can be implemented efficiently [4] and perform well in practice.

The wavelet tree is a balanced binary tree. We identify the two children of a node as left and right. Each node represents a range $R \subseteq [1, \sigma]$ of the alphabet Σ, its left child represents a subset R_l, which corresponds with the first half of R, and its right child a subset R_r, which corresponds with the second half of R. Every node virtually represents a subsequence S' of S composed of symbols whose value lies in R. This subsequence is stored as a bitmap in which a 0 bit means that position i belongs to R_l and a 1 bit means that it belongs to R_r.

In its simplest form, this structure requires $n\lceil \lg \sigma \rceil + o(n \lg \sigma)$ bits for the data, plus $O(\sigma \lg n)$ bits to store the topology of the tree, and supports aforementioned queries in $O(\lg \sigma)$ time by traversing the tree using $O(1)$-time *rank/select* operations on bitmaps [21]. Its construction takes $O(n \lg \sigma)$ time (we do not consider space-efficient construction algorithms [6,23]). As mentioned before, the space required by the structure can be reduced: the data can be compressed and stored in space bounded by its entropy (via compressed encodings of bitmaps and modifications on the shape of the tree), and the $O(\sigma \lg n)$ bits of the topology can be removed [15], which is important for large alphabets. We focus on the simple form, though our results can be extended to other encodings and tree shapes.

The *wtrees* support more complex queries than the primitives described above. For example, Mäkinen and Navarro [15] showed its connection with a classical two-dimensional range-search data structure. They showed how to solve range queries in a wavelet tree and its applications in *position-restricted searching*. In [13], the authors represent posting lists in a *wtree* and solve ranked AND queries by solving several range queries synchronously. Thus, solving range queries *in parallel* becomes important. As we present a parallel version of these queries, let us define them here. Given $1 \le i \le i' \le n$ and $1 \le j \le j' \le \sigma$, a range query $rq(S, i, i', j, j')$ reports all the symbols s_x such that $x \in [i, i']$ and $s_x \in [j, j']$. The counting version of the problem can be defined analogously.

Some work has been done in parallel processing of wavelet trees. In [1], the authors explore the use of wavelet trees in distributed web search engines. They assume a distributed memory model and propose partition techniques to balance the workload of processing wavelet trees. Note that our work is complementary to theirs, as each node in their distributed system can be assumed a multicore computer that can benefit from our algorithms. In [14], the authors explore the use of SIMD instructions to improve the performance of wavelet trees (and other string algorithms, see, for example, [8]). This set of instructions can be considered as low-level parallelism. We can also benefit from their work as it may improve the performance of the sequential parts of our algorithms.

Dynamic Multithreading Model: *Dynamic multithreading* (DYM) [7, Chapter 27] is a model of parallel computation which is faithful to several industry standards such as Intel's CilkPlus (cilkplus.org) and OpenMP Tasks (openmp.org/wp), and Threading Building Blocks (threadingbuildingblocks.org).

We will define a *multithreaded computation* as a directed acyclic graph (DAG) $G = (V, E)$, where the set of vertices V are instructions and $(u, v) \in E$ are dependencies between instructions; whereby in this case, u must be executed before v.[2] In order to signal parallel execution, we will augment sequential pseudocode with three keywords, **spawn**, **sync** and **parfor**. The **spawn** keyword signals that the procedure call that it precedes *may be* executed in parallel with the next instruction in the instance that executes the **spawn**. In turn, the **sync** keyword signals that all spawned procedures must finish before proceeding with the next instruction in the stream. Finally, **parfor** is simply "syntactic sugar" for **spawn**'ing and **sync**'ing ranges of a loop iteration. If a stream of instructions does not contain one of the above keywords, or a **return** (which implicitly **sync**'s) from a procedure, we will group these instruction into a single *strand*. Strands are scheduled onto processors using a *work-stealing* scheduler, which does the load-balancing of the computations. Work-stealing schedulers have been proved to be a factor of 2 away from optimal performance [2].

To measure the efficiency of our parallel wavelet tree algorithms, we will use three metrics: the *work*, the *span* and the number of processors. In accordance to the parallel literature, we will subscript running times by P, so T_P is the running time of an algorithm on P processors. The *work* is the total running time taken by all (unit-time) strands when executing on a *single* processor (i.e., T_1),[3] while the *span*, denoted as T_∞, is the *critical path* (the longest path) of G. In this paper, we are interested in speeding up wavelet tree manipulation and finding out the upper bounds of this speedup. To measure this, we will define *speedup* as $T_1/T_P = O(P)$, where linear speedup $T_1/T_P = \Theta(P)$, is the goal. We also define *parallelism* as the ratio T_1/T_∞, the maximum theoretical speedup that can be achieved on *any* number of processors.

---

[2] Notice that the RAM model is a subset of the DYM model where the outdegree of every vertex $v \in V$ is ≤ 1.

[3] Notice, again, that analyzing the work amounts to finding the running time of the serial algorithm using the RAM model.

3   Multicore Wavelet Tree

Parallel Construction: We focus on binary wavelet trees where the symbols of Σ are contiguous in $[1, \sigma]$. If they are not contiguous, a bitmap is used to remap the sequence to a contiguous alphabet [4]. Under these restrictions, the *wtree* is a balanced binary tree with $\lceil \lg \sigma \rceil$ levels.

In this scenario, a simple recursive algorithm, such as the one implemented in LIBCDS (http://libcds.recoded.cl), can build a *wtree* in $T_1 = O(n \lg \sigma)$ time by a linear processing of the symbols at each node. This algorithm works by halving Σ recursively into binary sub-trees whose left-child are all 0s and the right-child are all 1s, until 1s and 0s mean only one symbol in Σ. Instead, we propose an iterative construction algorithm that performs worse when executed sequentially, but shows nice parallel behavior. The key idea of the algorithm is that we can build any level of the *wtree* independently from the others. Unlike the classical construction, when building a level we cannot assume that a previous step is providing us the correct permutation of the elements of S. Instead, we compute the node at level i for each symbol of the original sequence. The following proposition shows how it can be computed.

Proposition 1. *Given a symbol $s \in S$ and a level i, $0 \le i < l = \lceil \lg \sigma \rceil$, of a wtree, the node at which s is represented at level i can be computed as $s \gg l - i$.*

In other words, if the symbols of Σ are contiguous, then the i most significant bits of the symbol s gives us its corresponding node at level i. In the word-RAM model with word size $\Omega(\lg n)$, this computation takes $O(1)$ time, and thus the following corollary holds:

Corollary 1. *The node at which a symbol s is represented at level i can be computed in $O(1)$ time.*

The iterative parallel construction procedure is shown in Algorithm 1 (the sequential version can be obtained by replacing **parfor** instructions with sequential **for** instructions). The algorithm takes as input a sequence of symbols S, the length n of S, and the length of the alphabet, σ (see Sect. 2). The output is a *wtree WT*, which represents the input data S. We denote the ith level of WT as WT_i, $\forall i, 0 \le i < \lceil \lg \sigma \rceil$ and the jth node in level WT_i as WT_i^j, $\forall j, 0 \le j < 2^i$.

The outer loop (line 3) iterates in parallel over $\lceil \lg \sigma \rceil$ levels. Lines 4 to 13 scan each level performing the following tasks: the first step (lines 4 to 7) calculates the maximum number of nodes for the current level, and traverse it initializing each node and its bitmap. For this initialization we can compute the size of each node with a linear time sweep of the elements in the node. The second step (lines 8 to 13) computes for each symbol in S, the node that represents the alphabet range that holds it at the current level (line 9 show an equivalent representation of the idea in Proposition 1). Then, the algorithm computes whether the symbol belongs to either the first or second half of the part of Σ that represents that node. Notice that *bitmapSetNextBit* needs to keep track of the positions already written in the bitmap and set the value of the next bit. When we reach the last

Input : S, n, σ
Output: A wavelet tree representation WT of S

1 $l \leftarrow \lceil \lg \sigma \rceil$
2 $WT \leftarrow$ Create a new tree with l levels
3 **parfor** $i \leftarrow 0$ **to** $l - 1$ **do**
4    $m \leftarrow 2^i$
5    $WT_i \leftarrow$ Create a new level with m nodes
6   **parfor** $j \leftarrow 1$ **to** m **do**
7      $|$ $WT_i^j \leftarrow$ Initialize a new Node
8   **for** $v \leftarrow 1$ **to** n **do**
9      $u \leftarrow \lfloor s_v / \lfloor \frac{\sigma}{2 \times m} \rfloor \rfloor$
10   **if** u *is odd* **then**
11        $|$ bitmapSetNextBit($WT_i^{u/2}.bitmap$, 1)
12   **else**
13        $|$ bitmapSetNextBit($WT_i^{u/2}.bitmap$, 0)
14 **return** WT

Algorithm 1. Wavelet tree parallel construction (pwt)

element, all the bitmaps contain the necessary information for that level. Finally, the bitmaps in this structure need to support rank/select operations, thus the construction algorithm must create additional structures after the bitmaps are completed. Notice that the **parfor** starting at line 6 scans the nodes of level i initializing them. The number of nodes grows exponentially larger when more levels are created, until we reach n nodes. This brings about a task workload imbalance among the worker threads because any given task may have exponentially more work to do. To prevent this, we also divide the first step into strands that the work-stealing scheduler will balance (see Sect. 2). It is easy to see that a sequential version of this algorithm takes $O(n \lg \sigma)$ time, which matches the time for construction found in the literature for non space-efficient construction algorithms. If **parfor** implements parallelism in a "divide-and-conquer" fashion (as in our model and implementation), then the DAG represents a binary-tree of constant-time division of Σ until it reaches the leaves of said tree, each of which has $O(n)$ weight. The work of pwt is still $T_1 = O(n \lg \sigma)$. All paths in the DAG, however, are the same length, and the same weight: the internal nodes are all $O(1)$, and the leaves are all $O(n)$. Thus, the critical path is $T_\infty = O(n)$ in all cases. In the same vein, parallelism will be $T_1/T_\infty = O(\lg \sigma)$, again for all cases. It follows that having $P = \lg \sigma$ will be enough to obtain optimal speedup.

The main drawback of the pwt algorithm is that it only scales until the number of cores equals the number of levels in the wavelet tree. So, even if we have more cores available, the algorithm will only use up to $\lg \sigma$ processors. In order to make an efficient use of all available cores we followed a different approach. This time we split the input data instead of the steps of the algorithm. In this way, the new algorithm creates as many chunks of the text as available cores, allowing it to take full advantage of the resources at hand. The algorithm then executes the construction for each substring and finally merges all resulting *wtrees* into a single one that represents the entire input text. We called this algorithm dd because of its domain decomposition nature. Although the dd algorithm is easy to follow, the merging part is not trivial. Here the order in which the partial trees are merged is important. In the merge step, the algorithm assigns an entire level i to each thread. Each thread concatenates the corresponding bitarrays of each tree, creating the bitarray for the final tree in the assigned level i. The concatenation takes the range of bits corresponding to each WT_i^j node of the level, in each tree and the result is the range of bits associated to the WT_i^j node in the final *wtree*. Concatenation is done following the same relative order of the substrings with respect to the input text. The procedure is repeated for each node in the level. In this case the thread working with the last level (the leaves) will do most of the work. Although all levels have the same size in bits, the last level is the one with the greatest number of calls to the concatenation function (one for each virtual node), leading to a bigger overhead.

The dd algorithm has the same asymptotic complexity for the total work $T_1 = O(n \lg \sigma)$. When running on P processors the construction of the partial trees takes $O(\frac{n}{P} \log \sigma)$ time. Merge takes $O(\frac{n \log \sigma}{PW})$, where W is the word size of that architecture. Note that merge has a linear speedup while $P \le \log \sigma$. Thus, considering W as a constant, the *span* is $T_\infty = O(n)$ in all cases.

Parallel Querying: We distinguish between two kinds of queries on wavelet trees: *path* and *branch* queries. Path queries are characterized by following just a single path from the root to a leaf and the value in level $i-1$ has to be computed before the value in level i. Examples of this type of queries are *select*, *rank*, and *access*. On the other hand, branch queries may follow more than one path root-to-leaf (indeed they may reach more than one leaf). Each path has the same characteristics as path queries and each path is independent from others paths. Examples of this type of queries are *range count* and *range report* [10].

In a parallel setting, a single path query cannot be parallelized because only one level of the query can be computed at a time. The common alternative is parallelizing several path queries using domain decomposition over queries (i.e., dividing queries over P). For this naïve approach, we obtained near-optimal throughput, defined as the number of processors times sequential throughput. We do not report this experiment for lack of space.

We implemented two branch-query-answering techniques: *individual*-query-answering (IQA) and *batch*-query-answering (BQA). The IQA technique is the obvious query by query processing. The BQA technique involves grouping sets of queries to take advantage of spatial and temporal locality in hierarchical memory architectures. For instance, at each node in the *wtree*, we can evaluate all the queries in a batch reusing the node's bitarray, thus increasing locality.

With little effort, we can parallelize sequential IQA in a domain decomposition fashion (denoted as dd-IQA), achieving near-optimal throughput (more than nine times the throughput for $P = 12$ compared to the sequential IQA).

The parallelBQA algorithm is shown in Algorithm 2. It implements the general BQA technique mentioned above, answering queries in a single batch.

Input : WT_i^j, *batch*, *num_queries*, *states*, *results*
Output: *results*: A collection containing the results for each query

1 *lbatch* ← a new collection with *num_queries* elements
2 *rbatch* ← *batch*
3 *local_states* ← a new collection with *num_queries* elements
4 **for** q ← 1 **to** *num_queries* **do**
5 | *local_states$_q$* ← *states$_q$*
6 | **if** *local_states$_q$* = 1 **then continue**
7
8 | **if** *batch$_q$.x^s* > *batch$_q$.x^e* ∨ *(*isLeaf(WT_i^j) ∩ *batch$_q$.y_range)* = ∅ **then**
9 | | *local_states$_q$* ← 1
10 | | **continue**
11 | **if** isLeaf(WT_i^j) **then**
12 | | *results$_q^j$* ← *batch$_q$.x^e* − *batch$_q$.x^s* + 1   /* *j* is the label */
13 | | **continue**
14 | *lbatch$_q$.x^s* ← **rank$_0$**$(WT_i^j.bitmap, batch_q.x^s − 1) + 1$
15 | *lbatch$_q$.x^e* ← **rank$_0$**$(WT_i^j.bitmap, batch_q.x^e)$
16 | *rbatch$_q$.x^s* ← **rank$_1$**$(WT_i^j.bitmap, batch_q.x^s − 1) + 1$
17 | *rbatch$_q$.x^e* ← **rank$_1$**$(WT_i^j.bitmap, batch_q.x^e − 1)$
18 | *lbatch$_q$.y_range* ← *batch$_q$.y_range*
19 **if** ∀$_q$, *local_states$_q$* = 1 **then return**
20
21 **spawn parallelBQA**$(WT_{i+1}^{2j}$, *lbatch*, *num_queries*, *local_states*, *results*)
22 **parallelBQA**$(WT_{i+1}^{2j+1}$, *rbatch*, *num_queries*, *local_states*, *results*)
23 **return**   /* implicit sync */

Algorithm 2. Parallel batch querying of range report (**parallelBQA**)

In particular, the algorithm portrayed here parallelizes the recursive calls during the traversal of the *wtree* (the **spawn** instruction in line 19). This is what we call "internal" parallelization. If the **spawn** is taken out, we would be left with a sequential batch processing algorithm. In addition, if P is sufficiently large, we can also apply domain decomposition techniques to the list of batches, calling **parallelBQA** also in parallel (denoted here as **dd-parallelBQA**). This means a "double" parallelization, an internal one and an external one. Notice that this technique is also applicable to *range count*.

4   Experimental Results

All algorithms were implemented in the C programming language and compiled using GCC 4.8 using the -O3 optimization flag. The experiments were carried out on a dual-processor Intel Xeon CPU (E5645) with six cores per processor, for a total of 12 physical cores running at 2.40GHz. Hyperthreading was disabled. The computer runs Linux 3.5.0-17-generic, in 64-bit mode. This machine has per-core L1 and L2 caches of sizes 32KB and 256KB, respectively and a per-processor shared L3 cache of 12MB, with a 5,958MB (~6GB) DDR RAM memory. Algorithms were compared in terms of running times using the usual high-resolution (nanosecond) C functions in `<time.h>`.[4]

Construction Experiments: We compared the implementation of our parallel wavelet tree construction algorithms, considering one pointer per node and one pointer per level, against the best current implementations available: LIBCDS and SDSL.[5] Both libraries were compiled with their default options. In particular, LIBCDS was compiled with optimization -O9 and SDSL with optimization -O3. Regarding to the bitarray implementation, in our implementations and in LIBCDS we use the 5%-extra space structure presented in [11]. In SDSL we use its `bit_vector_il` implementation, which yielded the best time performance in our experiments. The "trials" consisted in manipulating different alphabet and input sizes (i.e., σ and n), and the number of processors P recruited to work on the task. Since we are interested in big data, in order to obtain large enough alphabets, we took the `english` corpus of the Pizza & Chili website[6] as a sequence of *words* instead of the characters of the English alphabet. This representation gave us a large initial alphabet Σ of σ=633,816 symbols, which was ordered by their frequency of occurrence in the original `english` text. For experimentation, we generated an alphabet Σ' of size 2^k, taking the top 2^k *most frequent* words in the original Σ, and then assigning to each symbol a random index in the alphabet using a Marsenne Twister [17], with $k \in \{4, 6, 8, 10, 12, 14\}$. To create the input sequence S of n symbols, we searched for each symbol in Σ' in the original `english` text and, when found, appended it to S until it reached the maximum possible size given σ' (~1.5GB, in the case of $\sigma' = 2^{14}$), maintaining the order of the original `english` text. We then either split S until we reached the target sizes, which varied from 2MB (i.e., 2^{21} bytes), 4, 8, 16, 32, 64, 128, 256 up to 512MB (the maximum in the `english` corpus), or concatenated S with initial sub-sequences of itself to reach larger sizes up to 2GB (2^{31} bytes).[7]

We repeated each trial three times. We worked with the minimum time taken by all three repetitions of a trial, assuming that slightly larger values for any given trial are just "noise" from external processes such as operating system and networking tasks. In our experiments, LIBCDS showed better times than SDSL, hence we consider LIBCDS as the baseline to report speedup.

Fig. 1 shows the speedup varying the number of threads. The figure shows that algorithms scale with the number of threads, up to 12 threads, because the hardware has 12 processors. With more processors and enough work for each thread, our algorithms should scale appropriately. Upon reaching 12 threads, the operating system has to schedule the excess tasks on the same processors, slowing

---

[4] This is a reproducible-research-friendly paper, everything needed to replicate these results is available at `www.inf.udec.cl/~leo/sea2014`

[5] `https://github.com/simongog/sdsl-lite`. We thank Simon Gog for the conversations about the use and implementation of SDSL.

[6] `http://pizzachili.dcc.uchile.cl/texts/nlang/`

[7] Notice that since we worked with integers, these numbers expressed in bytes should be divided by 4 (in our architecture) to get the number of elements n of S.

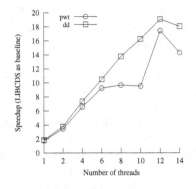

Fig. 1. Speedup over threads with a 2GB text and $\sigma = 2^{12}$

Fig. 2. Time over n, with $\sigma = 2^{12}$ and 12 threads

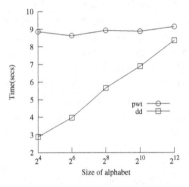

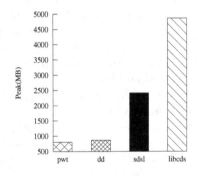

Fig. 3. Time over σ with a 2GB text and 12 threads

Fig. 4. Memory consumption with a 2GB text and $\sigma = 2^{12}$

down the general construction process. The graph shows a maximum speedup greater than the number of cores. However, this should not be considered a *super-linear speedup* (i.e., a speedup $> P$). Speedups shown in Fig. 1 were calculated using LIBCDS, the best current sequential implementation. It is necessary to consider that LIBCDS as a library may have an extra overhead in its performance, but this will affect it up to a constant factor.

The time of both algorithms is dominated by the thread that has the most work to do. In the case of dd, all threads have the same amount of work and curve scales linearly. However, pwt has a different behaviour. Fig. 1 shows the construction of a wavelet tree with 12 levels, the time will be dominated by the thread that will build more levels. In the case of six threads, each thread has to build two levels. In the case of 8 and 10 threads, four and two threads, respectively, will have to build two levels, with a time similar to the case of 6 threads. Finally, with 12 threads, each thread has to build only one level.

Fixing the number of threads to 12 and varying n and σ, we obtain the results of Fig(s). 2 and 3, respectively. Fig. 2 shows that dd has the best performance, followed closely by pwt. Fig. 3 shows an interesting result. When varying the size of the alphabet, the times of pwt are almost constant whereas the times of dd increase. This shows that in domains where σ is small, dd is the best approach, mainly because it has a better behaviour in memory transfers, yielding less cache misses. However, for larger σ, which results in *wtree* structures with more levels, the pwt shows better scalability. This is useful, for example, for gene encoding, grids, natural language vocabularies, etc.

We also report memory consumption in Fig. 4. To measure it, we monitored the memory allocated with *malloc* and released with *free*, taking the peak of consumption. We only considered the memory allocated during the construction, but not the memory allocated to store the text. The pwt and dd implementations do not need to copy the input sequence as LIBCDS and SDSL. Instead, they just read the input sequence. In particular, the memory consumption of dd depends on P, n and σ: $O(P\sigma \lg(n))$, because each thread maintains the topology of the tree to do the merge later. The memory consumption of pwt depends on n and σ, but does not depend on P. In all algorithms tested, we assumed that the input sequence cannot be destroyed. If they could, SDSL and LIBCDS would save space. By the same token, dd and pwt algorithms could be even faster if they were allowed to use more memory resources like LIBCDS and SDSL do. Given the improvement in performance we achieve using parallelism, we explicitly designed our algorithms to save memory and still achieve significant speedups.

Querying Experiments: To generate *branch queries*, ranges over the text were selected with random bounds and the size was fixed at 1%. In order to stress the querying algorithm, we also took $[1, \sigma]$ as the range of the alphabet. This ensured that the query traversal reached the leaves of the *wtree*. To compare sequential IQA and parallel IQA, we randomly generated 10,000 range queries. To test the BQA techniques we took a new set of randomly generated 10,000 queries and grouped them into 100-query batches. As we did for construction experiments, all *branch query* experiments were tested on the 2GB english text, $\sigma' = 2^{14}$ and varying the available processors from 1 to 12 (see Fig. 5). We repeated each experiment three times. We worked with the maximum throughput taken by all three repetitions, similar to Sect. 4.

As discussed in Sect. 14, the BQA technique implies a little more programming effort but improves throughput over the IQA by answering 10% more queries/second in the sequential case and over 23% more queries/second for the domain-decomposition case (denoted here as dd-BQA). As we saw, a consequence of the BQA is that tasks now demand enough work to offset the cost of *internally* parallelizing the batch answering process (see Algorithm 2). By combining domain decomposition with internal batch parallelization, we achieve 34% more throughput (i.e., we answer one third more queries a second) compared to the domain-decompositioned IQA. Throughput scales well over P. Table 1 shows the running times of the different implementations.

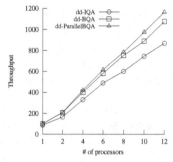

Fig. 5. Throughput over P for 100 batches of 100 queries (10,000 queries)

Table 1. Running times of branch queries (in seconds $\times$ 10,000 queries)

	Threads						
Algorithms	1	2	4	6	8	10	12
IQA	109.2	-	-	-	-	-	-
dd-IQA	113	60.6	30.3	20.5	16.7	13.4	11.5
BQA	99.2	-	-	-	-	-	-
dd-BQA	99.1	48.4	24.9	17.2	13.3	11.2	9.3
parallelBQA	98.5	49.2	26	17.8	14.2	11.5	9.9
dd-parallelBQA	98.4	48	23.8	16.2	12.7	10.3	8.6

5   Conclusion

Despite the vast amount of research done around wavelet trees, very little has been done to optimize them for current multicore architectures. We have shown that it is possible to have practical multicore implementations of wavelet tree construction by exploiting information related to the levels of the *wtree*, achieving $O(n)$-time construction and good use of memory resources. We also have shown a non-trivial parallelization of querying wavelet tree data.

In this paper we focused on the most general representation of a wavelet tree. However, some of our results may apply to other variants of wavelet trees. For example, it would be interesting to study how to extend our results to compressed wavelet trees (e.g., Huffman shaped *wtrees*) and to generalized wavelet trees (i.e., multiary wavelet trees where the fan out of each node is increased from 2 to $O(polylog(n))$). We shall explore also the extension of our results to the Wavelet Matrix [5] (a different level-wise approach to avoid the $O(\sigma \lg n)$ space overhead for the structure of the tree, which turns out to be simpler and faster than the wavelet tree without pointers). We also plan to experiment with real data from other domains such as inverted indices [13], genome information and two-dimensional range searching (useful, for example, in position restricted substring searching) [15]. More future work also involves dynamization, whereby the *wtree* is being modified concurrently by many processes as it is queried.

After the last decades, it is evident that architecture has become relevant again. It is nowadays difficult to find single-core computers. It therefore seems like a waste of resources to stick to sequential algorithms. We believe one natural way to improve performance of important data structures such as wavelet trees is to squeeze every drop of parallelism of modern multicore machines.

References

1. Arroyuelo, D., Costa, V.G., González, S., Marín, M., Oyarzún, M.: Distributed search based on self-indexed compressed text. Inf. Process. Manag. 48(5), 819–827 (2012)
2. Blumofe, R.D., Leiserson, C.E.: Scheduling multithreaded computations by work stealing. J. ACM 46(5), 720–748 (1999)

3. Brisaboa, N.R., Luaces, M.R., Navarro, G., Seco, D.: Space-efficient representations of rectangle datasets supporting orthogonal range querying. Inf. Syst. 38(5), 635–655 (2013)
4. Claude, F., Navarro, G.: Practical rank/select queries over arbitrary sequences. In: Amir, A., Turpin, A., Moffat, A. (eds.) SPIRE 2008. LNCS, vol. 5280, pp. 176–187. Springer, Heidelberg (2008)
5. Claude, F., Navarro, G.: The wavelet matrix. In: Calderón-Benavides, L., González-Caro, C., Chávez, E., Ziviani, N. (eds.) SPIRE 2012. LNCS, vol. 7608, pp. 167–179. Springer, Heidelberg (2012)
6. Claude, F., Nicholson, P.K., Seco, D.: Space efficient wavelet tree construction. In: Grossi, R., Sebastiani, F., Silvestri, F. (eds.) SPIRE 2011. LNCS, vol. 7024, pp. 185–196. Springer, Heidelberg (2011)
7. Cormen, T.H., Leiserson, C.E., Rivest, R.L., Stein, C.: Multithreaded Algorithms. In: Introduction to Algorithms, 3rd edn., chap. pp. 772–812. The MIT Press (2009)
8. Faro, S., Külekci, M.O.: Fast multiple string matching using streaming SIMD extensions technology. In: Calderón-Benavides, L., González-Caro, C., Chávez, E., Ziviani, N. (eds.) SPIRE 2012. LNCS, vol. 7608, pp. 217–228. Springer, Heidelberg (2012)
9. Ferragina, P., Manzini, G., Mäkinen, V., Navarro, G.: Compressed representations of sequences and full-text indexes. ACM Trans. Algorithms 3(2) (2007)
10. Gagie, T., Navarro, G., Puglisi, S.J.: New algorithms on wavelet trees and applications to information retrieval. Theoret. Comput. Sci. 427, 25–41 (2012)
11. González, R., Grabowski, S., Mäkinen, V., Navarro, G.: Practical implementation of rank and select queries. In: WEA, pp. 27–38. CTI Press, Greece (2005)
12. Grossi, R., Gupta, A., Vitter, J.S.: High-order entropy-compressed text indexes. In: SODA, pp. 841–850. Soc. Ind. Appl. Math, Philadelphia (2003)
13. Konow, R., Navarro, G.: Dual-sorted inverted lists in practice. In: Calderón-Benavides, L., González-Caro, C., Chávez, E., Ziviani, N. (eds.) SPIRE 2012. LNCS, vol. 7608, pp. 295–306. Springer, Heidelberg (2012)
14. Ladra, S., Pedreira, O., Duato, J., Brisaboa, N.R.: Exploiting SIMD Instructions in Current Processors to Improve Classical String Algorithms. In: Morzy, T., Härder, T., Wrembel, R. (eds.) ADBIS 2012. LNCS, vol. 7503, pp. 254–267. Springer, Heidelberg (2012)
15. Mäkinen, V., Navarro, G.: Rank and select revisited and extended. Theoret. Comput. Sci. 387(3), 332–347 (2007)
16. Makris, C.: Wavelet trees: A survey. Comput. Sci. Inf. Syst. 9(2), 585–625 (2012)
17. Matsumoto, M., Nishimura, T.: Mersenne twister: a 623-dimensionally equidistributed uniform pseudo-random number generator. ACM Trans. Model. Comput. Simul. 8(1), 3–30 (1998)
18. Navarro, G.: Wavelet trees for all. In: Kärkkäinen, J., Stoye, J. (eds.) CPM 2012. LNCS, vol. 7354, pp. 2–26. Springer, Heidelberg (2012)
19. Navarro, G., Nekrich, Y., Russo, L.M.S.: Space-efficient data-analysis queries on grids. Theoret. Comput. Sci. 482, 60–72 (2013)
20. Otellini, P.: Keynote Speech at Intel Developer Forum (2003), http://www.intel.com/pressroom/archive/speeches/otellini20030916.htm
21. Raman, R., Raman, V., Satti, S.R.: Succinct indexable dictionaries with applications to encoding k-ary trees, prefix sums and multisets. ACM Trans. Algorithms 3(4) (2007)
22. Sutter, H.: The free lunch is over: A fundamental turn toward concurrency in software (2005), http://www.gotw.ca/publications/concurrency-ddj.htm
23. Tischler, G.: On wavelet tree construction. In: Giancarlo, R., Manzini, G. (eds.) CPM 2011. LNCS, vol. 6661, pp. 208–218. Springer, Heidelberg (2011)
24. Välimäki, N., Mäkinen, V.: Space-efficient algorithms for document retrieval. In: Ma, B., Zhang, K. (eds.) CPM 2007. LNCS, vol. 4580, pp. 205–215. Springer, Heidelberg (2007)

Wear Minimization for Cuckoo Hashing:
How Not to Throw a Lot of Eggs into One Basket

David Eppstein[1], Michael T. Goodrich[1], Michael Mitzenmacher[2], and Paweł Pszona[1]

[1] Department of Computer Science, University of California, Irvine, CA, USA
[2] School of Engineering and Applied Sciences, Harvard University, Cambridge, MA, USA

Abstract. We study wear-leveling techniques for cuckoo hashing, showing that it is possible to achieve a memory wear bound of $\log \log n + O(1)$ after the insertion of n items into a table of size Cn for a suitable constant C using cuckoo hashing. Moreover, we study our cuckoo hashing method empirically, showing that it significantly improves on the memory wear performance for classic cuckoo hashing and linear probing in practice.

"I did throw a lot of eggs into one basket, as you do in your teenage years..."

—Dylan Moran [31]

1 Introduction

A *dictionary*, or *lookup table*, is a data structure for maintaining a set, S, of key-value pairs, which are often called *items*, to support the following operations:

- **add**(k, v): insert the item (k, v) into S.
- **remove**(k): remove the item with key k from S.
- **lookup**(k): return the value v associated with k in S (or return **null** if there is no item with key equal to k in S).

The best dictionaries, both in theory and in practice, are implemented using hashing techniques (e.g., see [4,21]), which typically achieve $O(1)$ expected time performance for each of the above operations.

One technique that has garnered much attention of late is *cuckoo hashing* [26], which is a hashing method that supports lookup and remove operations in *worst-case* $O(1)$ time, with insertions running in expected $O(1)$ time. Based on the metaphor of how the European cuckoo bird lays its eggs, inserting an item via cuckoo hashing involves throwing the item into one of two possible cells and evicting any existing item, which in turn moves to its other cell, repeating the process. In this standard formulation, each item in S is stored in one of two possible locations in a table, T, but other variants have also been studied as well (e.g., see [2,11,12,14,20]). For instance, the *cuckoo hashing with a stash* variation sacrifices some of the elegance of the basic algorithm by adding an auxiliary cache to break insertion cycles, greatly reducing the failure probability of the algorithm and reducing the need to rebuild the structure when a failure happens [20]. There have also been improvements to the space complexity for cuckoo hashing [13,22], as well as experimental studies of the practical applications of cuckoo hashing [1,35]. In spite of its elegance, efficiency, and practicality, however, cuckoo hashing has a major drawback with respect to modern memory technologies.

J. Gudmundsson and J. Katajainen (Eds.): SEA 2014, LNCS 8504, pp. 162–173, 2014.

1.1 Wear Leveling in Modern Memories

Phase-change memory (PCM) and flash memory are memory technologies gaining in modern interest, due to their persistence and energy efficiency properties. For instance, they are being used for main memories as well as for external storage (e.g., see [6,27,32,34]). Unfortunately, such modern memory technologies suffer from *memory wear*, a phenomenon caused by charge trapping in the gate oxide [16], in which extensive writes to the same cell in such memories (typically 10,000–100,000 write/erase cycles for flash memory) causes all memory cells to become permanently unusable.

To deal with this drawback, several heuristic techniques for *wear leveling* have been proposed and studied (e.g., see [5,7,8,9,17,28,33]). Such techniques are intended to limit memory wear, but they do not produce high-probability wear guarantees; hence, we are motivated to provide special-purpose solutions for common data structure applications, such as for dictionaries, that have provable wear-leveling guarantees.

In this paper, we focus on a previously-ignored aspect of cuckoo hashing, namely its *maximum wear*—the maximum number of writes to any cell in the hash table, T, during the execution of a sequence of n operations. Unfortunately, in spite of their other nice properties, cuckoo hashing and its variants do not have good wear-leveling properties. For instance, after inserting n items into a hash table, T, implemented using cuckoo hashing, the expected maximum wear of a cell in T is $\Omega(\log n / \log \log n)$—this follows from the well-known balls-in-bins result [25] that throwing n balls into n bins results in the expected maximum number of balls in a bin being $\Theta(\log n / \log \log n)$. Ideally, we would strongly prefer there to be a simple way to implement cuckoo hashing that causes its memory wear to be closer to $O(1)$.

We study a simple but previously unstudied modification to cuckoo hashing that uses $d > 2$ hash functions, and that chooses where to insert or reinsert each item based on the wear counts of the cells it hashes to. We prove that this method achieves, with high probability, maximum wear of $\log \log n + O(1)$ after a sequence of n insertions into a hash table of size Cn, where C is a small constant. Moreover, we show experimentally that this variant achieves significantly reduced wear compared to classical cuckoo hashing in practice, with deletions as well as insertions.

1.2 Related Work

With respect to previous algorithmic work on memory wear leveling, Ben-Aroya and Toledo [5] introduce a memory-wear model consisting of $N \geq 2$ cells, indexed from 0 to $N-1$, such that each is a single memory word or block, and such that there is a known parameter, L, specifying an upper-bound limit on the number of times that any cell of this memory can be rewritten. They study competitive ratios of wear-leveling heuristics from an online algorithms perspective. In addition, in even earlier work, Irani *et al.* [18] study several schemes for performing general and specific algorithms using write-once memories. Qureshi *et al.* [28], on the other hand, describe a wear-leveling strategy, called the *Start-Gap* method, which works well in practice but has not yet been analyzed theoretically. With respect to other previous work involving wear-leveling analysis of specific algorithms and data structures, Chen *et al.* [10] study the wear-leveling performance of methods for performing database index and join operations.

Azar *et al.* [3] show that if one throws n balls into n bins, but with each ball being added to the least-full bin from $k \geq 2$ random choices, then the largest bin will have size $\log\log n + O(1)$ with high probability. Subsequentially, other researchers have discovered further applications of this approach, which has given rise to a paradigm known as the *power of two choices* (e.g., see [24]). This paradigm is exploited further in the work on *cuckoo hashing* by Pagh and Rodler [26], which, as mentioned above, uses two random hash locations for items along with the ability to dynamically move keys to any of its hash locations so as to support worst-case $O(1)$-time lookups and removals. Several variants of cuckoo hashing have been considered, including the use of a small cache (called a "stash") and the use of more than two hash functions (e.g., see [2,11,12,14,20]). Also, as mentioned above, there has also been work improving the space complexity [13,22], as well as experimental analyses of cuckoo hashing [1,35]. Nevertheless, we are not familiar with any previous work on the memory-wear performance of cuckoo hashing, from either a theoretical or experimental viewpoint.

Our approach to improving the memory wear for cuckoo hashing is to use a technique that could be called the *power of three choices*, in that we exploit the additional freedom allowed by implementing cuckoo hashing with at least three hash functions instead of two. Of course, as cited earlier, previous researchers have considered cuckoo hashing with more than two hash functions and the balls-in-bins analysis of Azar *et al.* [3] applies to any number of at least two choices. Nevertheless, previous many-choice cuckoo hashing schemes do not have good memory wear bounds, since, even by the very first random choice for each item, there is an $\Omega(\log n/\log\log n)$ expected maximum memory-wear bound for classic cuckoo hashing with $d \geq 3$ hash functions. In addition, the approach of Azar *et al.* does not seem to lead to a bound of $O(\log\log n)$ for the memory wear of the variant of cuckoo hashing we consider, because items in a cuckoo hashing scheme can move to any of their alternative hash locations during insertions, rather than staying in their "bins" as required in the framework of Azar *et al.* Moreover, there are non-trivial dependencies that exist between the locations where an item is moved and a sequence of insertions that caused those moves, which complicates any theoretical analysis.

1.3 Our Results

In this paper, we consider the *memory wear* of cells of a cuckoo hash table, where by "wear" we mean the number of times a cell is rewritten with a new value. We introduce a new cuckoo hashing insertion rule that essentially involves determining the next location to "throw" an item by considering the existing wear of the possible choices. We provide a theoretical analysis proving that our simple rule has low wear—namely, for a suitably small constant α_d, when inserting αn items into a cuckoo hash table with n cells, d distinct choices per item, and $\alpha < \alpha_d$, the maximum wear on any cell can be bounded by $\log\log n + O(1)$.

Of course, for any reasonable values of n, the above $\log\log n$ term is essentially constant; hence, our theoretical analysis necessarily begs an interesting question:

Does our modified insertion rule for cuckoo hashing lead to improved memory-wear performance in practice?

To answer this question, we performed a suite of experiments which show that, in fact, even for large, but non-astronomical, values of n, our simple insertion rule leads to significant improvements in the memory-wear performance of cuckoo hashing over classic cuckoo hashing and hashing via linear probing.

2   Algorithm

In classical cuckoo hashing, each item of the dynamic set S is hashed to two possible cells of the hash table; these may either be part of a single large table or two separate tables. When an item x is inserted into the table (at the first of its two cells), it displaces any item y already stored in that cell. The displaced item y is reinserted at its other cell, which in turn may displace a third item z, and so on. If n items are inserted into a table whose number of cells is a sufficiently large multiple of n, then with high probability all chains of displaced items eventually terminate within $O(\log n)$ steps, producing a data structure in which each lookup operation may be performed in constant time. The expected time per insertion is $O(1)$. Removals may be performed in the trivial way, by simply removing an item from its cell and leaving all other items in their places. By a standard analysis of balls-and-bins processes, the wear arising just from placing items into their first cells, not even counting the additional wear from re-placing displaced items, is $\Omega(\log n / \log \log n)$.

A standard variant of cuckoo hashing allows more than two choices. In this setting, when an item x is inserted, it first goes through its choices in order to see if any are empty. If none are, one must be displaced. A common approach in this case is to use what is called random walk cuckoo hashing [13,15]; when an item is to be displaced, it is chosen randomly from an item's choices. (Usually, after the first displacement, one does not allow the item that was immediately just placed to be displaced again, as this would be wasteful in terms of moves.) Allowing more choices allows for higher load factors in the hash table and makes the probability of a failure much smaller.

We modify the standard algorithm, using $d \geq 3$ choices for each item, in the following ways:

- Associated with each cell we store a *wear count* of the number of times that cell has been overwritten by our structure.
- When an item is inserted, it is placed into one of its d cells with the lowest wear count, rather than always being placed into the first of its cells.
- When an item y is displaced from a cell, we compare the wear counts of all d of the cells associated with y (including the one it has just been displaced from, causing its wear count to increase by one). We then store y in the cell with the minimum wear count, breaking ties arbitrarily, and displace any item that might already be placed in this cell.

A computational shortcut that makes no difference to the analysis of our algorithm is to take note of a situation in which two items x and y are repeatedly displacing each other from the same cell; in this case, the repeated displacement will terminate when the wear count of the cell reaches a number at least as high as the next smallest wear count of one of the other cells of x or y. Rather than explicitly performing the repeated displacement, the algorithm may calculate the wear count at which this termination will happen, and perform the whole sequence of repeated displacements in constant time.

3 Analysis

As is common in theoretical analyses of cuckoo hashing (e.g., see [2,11,12,14,20,26]), let us view the cuckoo hash table in terms of a random hypergraph, H, where there is a vertex in H for each cell in the cuckoo hash table, and there is a hyperedge in H for each inserted item, with the endpoints of this hyperedge determined by the d cells associated with this item. For the sake of this analysis, we consider a sequence of αn insertions into an initially empty table of size, n, and we disallow deletions. (Although the analysis of mixed sequences of insertions and deletions is beyond the approach of this section, we study such mixed update sequences experimentally later in this paper.) Thus, H has n vertices and αn hyperedges, each of which is a subset of d vertices. In this context, we say that a subgraph of H with s edges and r vertices is a *tree*, if $(d-1)s = r - 1$, and it is *unicyclic*, if $(d-1)s = r$ (e.g., see [19]). Since the d hash functions are random, H is a standard random d-ary (or "d-uniform") hypergraph, with $d \geq 3$ being a fixed constant; hence, we may use the following well-known fact about random hypergraphs.

Lemma 1 (Schmidt-Pruzan and Shamir [29]; Karoński and Łuczak [19]). *When* $\alpha < 1/(d(d-1))$, *with probability* $1 - O(1/n)$ *the maximum connected component size in H is $O(\log n)$, and all components are either trees or unicyclic components.*

Here the constant implicit in the $O(\log n)$ may depend on α and d, and affects the $O(1)$ term in the final $\log \log n + O(1)$ bound on the wear.

We define an orientation on H where each hyperedge in H is oriented to one of its d cells, denoting where the associated item resides. Thus, such an orientation has at most one hyperedge oriented to each vertex, since there is at most one item located in each cell.

Recall our modified cuckoo hashing insertion procedure, which allows us to achieve our bound on wear. We assume that each cell tracks its current wear (the number of times the cell value has been rewritten) in a counter. On insertion for a time, x, if any of the d cells for x is empty, then x is placed in one of its empty cells (randomly chosen). Otherwise, x replaces the item, y, in the cell with the smallest wear out of its d possible choices. The replaced item, y, continues by replacing the item in the cell with the smallest wear out of its d choices, and so on, until all items are placed. Note that a pair of items, x and y, may repeatedly replace each other in a cell until the wear of the cell increases to the point when there is an alternative lower-wear cell for x or y (in practice we would not have perform the repeated replacements, but would update the wear variable accordingly).

For a cell of wear k, we define its *wear-children* to be the $d-1$ other cells corresponding to the item contained in this cell. Our proof of the $O(\log \log n)$ bound on the maximum wear for our modified cuckoo hashing scheme depends, in part, on the following three simple, but crucial, observations.

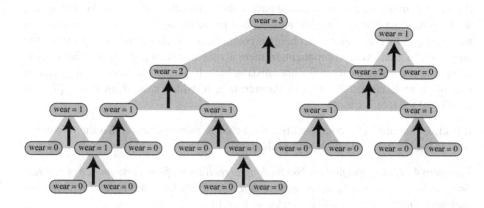

Fig. 1. A tree component in the hypergraph of cells (yellow ovals) and values (blue and pink triangles) for a cuckoo hash table with $d = 3$. The arrows in the value triangles indicate the cell into which each value is placed. The complete binary tree associated with a cell of wear 3 is shown by the pink triangles. Some additional values and cells needed to achieve the depicted wear values are not shown.

Observation 1. *A cell of wear k has $d - 1$ wear-children with wear at least $k - 1$.*

This is because for a cell to obtain wear k all the choice corresponding to the item placed in that cell have to have wear at least $k - 1$ at the time of its placement.

Observation 2. *Consider any tree component in the hypergraph H with a cell of wear k. Then the component contains a $(d-1)$-ary tree consisting of the cell's wear-children, and the wear-children of those wear-children, and so on, of at least $(d-1)^k + 1$ distinct nodes.*

This is because, by Observation 1, a cell of wear k has $d - 1$ wear-children with wear at least $k - 1$, each of which has $d - 1$ of its own children, and we can iterate this argument counting descendants down to wear 0. The nodes must be distinct when the component is a tree. Fig. 1 depicts an example.

If we did not need to consider unicyclic components, we would be done at this point. This is because, by Lemma 1, component sizes are bounded by $O(\log n)$ with high probability, and, if all components are trees, then, by Observation 2, if there is a cell of wear k, then there is a component of size at least $(d - 1)^k$. Thus, for $d \geq 3$, the maximum wear would necessarily be $\log \log n + O(1)$ with high probability. Our next observation, therefore, deals with unicyclic components.

Observation 3. *Consider any unicyclic component in the hypergraph H with a cell of wear k. Then the component contains a $(d - 1)$-ary tree of $(d - 1)^{k-1} + 1$ distinct nodes.*

Proof. We follow the same argument as for a tree component, in which we build a complete $(d - 1)$-ary tree from the given cell. Because the component is not acyclic, this process might find more than one path to the same cell. If this should ever happen, we keep only the first instance of that cell, and prune the tree at the second instance of the same cell (preventing it from being a complete tree). This pruning step breaks the only cycle in a unicyclic component, so there is only one such pruning and the result is a tree that is complete except for one missing branch. The worst place for the pruning step to occur is nearest the root of the tree tree, in which case at most a $1/(d-1)$ fraction of the tree is cut off.   □

It is clear that unicyclic components do not change our conclusion that the maximum wear is $\log\log n + O(1)$ with high probability.

Theorem 4. *If our modified cuckoo hashing algorithm performs a sequence of n insertions, with $d \geq 3$, on a table whose size is a sufficiently large multiple of n, then with high probability the wear will be $\log\log n + O(1)$.*

Proof. We form a d-regular hypergraph whose vertices are the cells of the hash table, and whose hyperedges record the sets of cells associated with each item. By Lemma 1, with high probability, all components of this graph are trees or unicyclic components, with $O(\log n)$ vertices. By Observations 2 and 3, any cell with wear k in one of these components has associated with it a binary tree with $\Omega(2^k)$ vertices. In order to satisfy $2^k = O(\log n)$ (the binary tree cannot be larger than the component containing it) we must have $k \leq \log\log n + O(1)$.   □

It would be of interest to extend this analysis to sequences of operations that mix insertions with deletions, but our current methods handle only insertions.

4   Experiments

We have implemented and tested three hashing algorithms:

- our variant of cuckoo hashing with $d = 3$ hash functions
- standard cuckoo hashing with $d = 3$ hash functions[1]
- standard open-address hashing with linear probing and eager deletion (explained below) [30].

We ran a series of tests to gauge the behavior of maximum wear for the above three algorithms. In all cases, the setup was the same: we start with an initially empty hash table of capacity 30 million items, then perform a number of insertions, until desired usage ratio (fill ratio) is achieved (we tested usage ratios 1/6, 1/3, 1/2 and 2/3). Then, for the main part of the test, we perform 1 billion (10^9) operation pairs, where a single operation pair consists of

---

[1] Using standard cuckoo hashing with $d = 2$ proved counterproductive, as multiple *failures* (i.e., situations where we are unable to insert new item into the table) were observed. This is due to the failure probability being non-negligible, of the order $\Theta(1/n)$, in this version of cuckoo hashing. Of course, such failures can be circumvented by using a *stash* [20] for storing items that failed to be inserted, but a stash necessarily has to be outside of the wear-vulnerable memory, or it has to be moved a lot, since it will have many rewrites. Thus, we did not include comparisons to cuckoo hashing with two hash functions.

1. deleting a randomly selected item that is present in the hash table
2. inserting a new item into the hash table.

This way, once the desired usage ratio is achieved in the first phase, it is kept constant throughout the rest of the test. For the sequence of insertions, we simply used integers in the natural order (i.e., 0, 1, 2, . . .), since the hash functions are random. Different test runs with the same input were parametrized by the use of different hash functions in the algorithm. We ran about 13 tests for each (*usage ratio*, *algorithm*) combination. Our tests were implemented in C++. We used various cryptographic hash functions provided by the mhash library [23].

As discussed above, we store wear information for each cell in the hash table. Each time a new value is written into the cell, we increase its wear count. In the case of linear probing, when an item, a, is erased, all subsequent items in the same probe sequence (until an empty cell is encountered) need to be rehashed. This is implemented as erasing an item and inserting it again using the insertion algorithm. It is easy to implement this in a way such that if a rehashed item ends up in its original position, no physical write is necessary. Therefore we did not count this case as a wear increase.

In addition to the above scenario of mixed insertions and deletions, we also measured the max wear after an insertion of 20 million items into a hash table of capacity 30 million, with no deletions. The results are shown in Fig. 2.

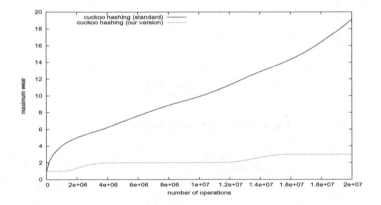

Fig. 2. Average maximum wear during initial insertion sequence (20 million insertions into hash table of capacity 30 million), with no deletions. Linear probing is not shown, as it has maximum wear equal to 1 as long as insertions are the only operations involved.

Even though our theoretical analysis is only valid for sequences of insertions, our algorithm behaves equally well when insertions are interspersed with deletions, as explained in the description of the test setting. The results for different usage ratios are shown in Figures 3, 4, 5 and 6. Table 1 contains average wear (calculated over all cells) when the test has concluded.

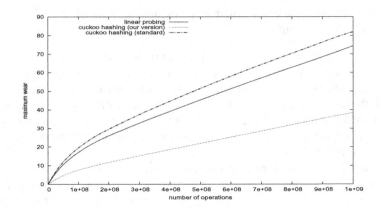

Fig. 3. Maximum wear for usage ratio $\frac{1}{6}$

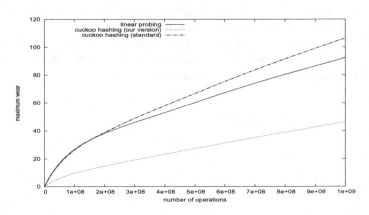

Fig. 4. Maximum wear for usage ratio $\frac{1}{3}$

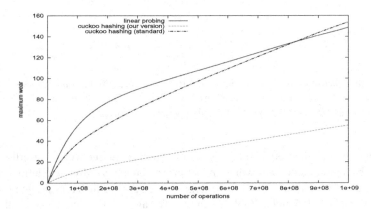

Fig. 5. Maximum wear for usage ratio $\frac{1}{2}$

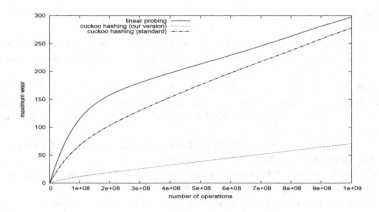

Fig. 6. Maximum wear for usage ratio $\frac{2}{3}$

Table 1. Average wear after sequence of operations

Usage ratio	Cuckoo hashing (ours)	Cuckoo hashing (standard)	Linear probing
1/6	33.92	40.52	37.12
1/3	36.57	52.33	44.41
1/2	44.68	74.51	58.96
2/3	64.52	128.91	90.19

4.1 Discussion

When the only operations performed are insertions, linear probing achieves optimal wear (wear = 1). However, when a mixed sequence of insertions and deletions is performed, our algorithm is clearly the best among those tested in terms of minimizing maximum wear. In this situation, linear probing and standard cuckoo hashing behave in a similar way, while the maximum wear is much smaller for our algorithm. It is evident that the difference grows as the hash table becomes more filled (its usage ratio becomes higher). It is also worth noticing that minimizing maximum wear also results in smaller overall wear of the hash table, as shown in Table 1.

Acknowledgements. This research was supported in part by NSF grants 1011840, 1228639, and ONR grant N00014-08-1-1015. Michael Mitzenmacher was supported in part by NSF grants CCF-1320231, CNS-1228598, IIS-0964473, and CCF-0915922, and part of this work was done while visiting Microsoft Research New England.

References

1. Alcantara, D.A., Sharf, A., Abbasinejad, F., Sengupta, S., Mitzenmacher, M., Owens, J.D., Amenta, N.: Real-time parallel hashing on the GPU. In: Proc. ACM SIGGRAPH Asia 2009, pp. 154:1–154:9(2009), doi:10.1145/1661412.1618500
2. Arbitman, Y., Naor, M., Segev, G.: De-amortized cuckoo hashing: Provable worst-case performance and experimental results. In: Albers, S., Marchetti-Spaccamela, A., Matias, Y., Nikoletseas, S., Thomas, W. (eds.) ICALP 2009, Part I. LNCS, vol. 5555, pp. 107–118. Springer, Heidelberg (2009)
3. Azar, Y., Broder, A., Karlin, A., Upfal, E.: Balanced allocations. SIAM J. Comput. 29(1), 180–200 (1999), doi:10.1137/S0097539795288490
4. Baeza-Yates, R., Poblete, P.V.: Searching. In: Algorithms and Theory of Computation Handbook, pp. 2-1–2-16. Chapman & Hall/CRC (2010)
5. Ben-Aroya, A., Toledo, S.: Competitive analysis of flash-memory algorithms. In: Azar, Y., Erlebach, T. (eds.) ESA 2006. LNCS, vol. 4168, pp. 100–111. Springer, Heidelberg (2006), http://dx.doi.org/10.1007/11841036_12, doi:10.1007/11841036_12
6. Bez, R., Camerlenghi, E., Modelli, A., Visconti, A.: Introduction to flash memory. Proc. IEEE 91(4), 489–502 (2003), doi:10.1109/JPROC.2003.811702
7. Chang, L.-P.: On efficient wear leveling for large-scale flash-memory storage systems. In: Proc. ACM Symp. on Applied Computing, pp. 1126–1130 (2007), doi:10.1145/1244002.1244248
8. Chang, Y.-H., Hsieh, J.-W., Kuo, T.-W.: Endurance enhancement of flash-memory storage, systems: an efficient static wear leveling design. In: Proc. 44th ACM/IEEE Design Automation Conf. (DAC 2007), pp. 212–217 (2007), doi:10.1145/1278480.1278533
9. Chang, Y.-H., Hsieh, J.-W., Kuo, T.-W.: Improving flash wear-leveling by proactively moving static data. IEEE Trans. Comput. 59(1), 53–65 (2010), doi:10.1109/TC.2009.134
10. Chen, S., Gibbons, P.B., Nath, S.: Rethinking database algorithms for phase change memory. In: Proc. 5th Conf. on Innovative Data Systems Research (CIDR), pp. 21–31 (2011), http://www.cidrdb.org/cidr2011/Papers/CIDR11_Paper3.pdf
11. Devroye, L., Morin, P.: Cuckoo hashing: further analysis. Inform. Process. Lett. 86(4), 215–219 (2003), doi:10.1016/S0020-0190(02)00500-8
12. Dietzfelbinger, M., Goerdt, A., Mitzenmacher, M., Montanari, A., Pagh, R., Rink, M.: Tight thresholds for cuckoo hashing via XORSAT. In: Abramsky, S., Gavoille, C., Kirchner, C., Meyer auf der Heide, F., Spirakis, P.G. (eds.) ICALP 2010. LNCS, vol. 6198, pp. 213–225. Springer, Heidelberg (2010), doi:10.1007/978-3-642-14165-2_19
13. Fotakis, D., Pagh, R., Sanders, P., Spirakis, P.G.: Space efficient hash tables with worst case constant access time. Theory of Computing Systems 38(2), 229–248 (2005), doi:10.1007/s00224-004-1195-x
14. Frieze, A.M., Melsted, P., Mitzenmacher, M.: An analysis of random-walk cuckoo hashing. In: Dinur, I., Jansen, K., Naor, J., Rolim, J. (eds.) Approximation, Randomization, and Combinatorial Optimization. Algorithms and Techniques. LNCS, vol. 5687, pp. 490–503. Springer, Heidelberg (2009)
15. Frieze, A.M., Melsted, P., Mitzenmacher, M.: An analysis of random-walk cuckoo hashing. SIAM J. Comput. 40(2), 291–308 (2011), doi:10.1137/090770928
16. Grupp, L.M., Caulfield, A.M., Coburn, J., Swanson, S., Yaakobi, E., Siegel, P.H., Wolf, J.K.: Characterizing flash memory: anomalies, observations, and applications. In: Proc. 42nd IEEE/ACM Int. Symp. on Microarchitecture (MICRO 42), pp. 24–33 (2009), doi:10.1145/1669112.1669118
17. Hunter, A.: A brief introduction to the design of UBIFS. White paper (2008), http://www.linux-mtd.infradead.org/doc/ubifs_whitepaper.pdf

18. Irani, S., Naor, M., Rubinfeld, R.: On the time and space complexity of computation using write-once memory or is pen really much worse than pencil? Mathematical Systems Theory 25(2), 141–159 (1992), doi:10.1007/BF02835833

19. Karoński, M., Łuczak, T.: The phase transition in a random hypergraph. J. Comput. Appl. Math. 142(1), 125–135 (2002), doi:10.1016/S0377-0427(01)00464-2

20. Kirsch, A., Mitzenmacher, M., Wieder, U.: More robust hashing: cuckoo Hashing with a Stash. SIAM J. Comput. 39(4), 1543–1561 (2009), doi:10.1137/080728743

21. Knuth, D.E.: The Art of Computer Programming, 2nd edn. Sorting and Searching, vol. 3. Addison Wesley (1998)

22. Lehman, E., Panigrahy, R.: 3.5-way cuckoo hashing for the price of 2-and-a-bit. In: Fiat, A., Sanders, P. (eds.) ESA 2009. LNCS, vol. 5757, pp. 671–681. Springer, Heidelberg (2009), doi:10.1007/978-3-642-04128-0_60

23. Mavroyanopoulos, N., Schumann, S.: Mhash. Open-source software library, http://mhash.sourceforge.net/

24. Mitzenmacher, M., Richa, A.W., Sitaraman, R.: The power of two random choices: A survey of techniques and results. In: Handbook of Randomized Computing, vol. 1, pp. 255–312. Kluwer Academic Publishers (2000)

25. Mitzenmacher, M., Upfal, E.: Probability and Computing: Randomized Algorithms and Probabilistic Analysis. Cambridge University Press (2005)

26. Pagh, R., Rodler, F.F.: Cuckoo hashing. J. Algorithms 51(2), 122–144 (2003), doi:10.1016/j.jalgor.2003.12.002

27. Pavan, P., Bez, R., Olivo, P., Zanoni, E.: Flash memory cells—an overview. Proc. IEEE 85(8), 1248–1271 (1997), doi:10.1109/5.622505

28. Qureshi, M.K., Karidis, J., Franceschini, M., Srinivasan, V., Lastras, L., Abali, B.: Enhancing lifetime and security of PCM-based main memory with start-gap wear leveling. In: Proc. 42nd IEEE/ACM Int. Symp. on Microarchitecture (MICRO), pp. 14–23 (2009), doi:10.1145/1669112.1669117

29. Schmidt-Pruzan, J., Shamir, E.: Component structure in the evolution of random hypergraphs. Combinatorica 5(1), 81–94 (1985), doi:10.1007/BF02579445

30. Sedgewick, R.: Algorithms in Java, Parts 1–4: Fundamentals, Data Structures, Sorting, and Searching, 3rd edn., pp. 615–619. Addison Wesley (2003)

31. Turner, L.: Dylan Moran Interview: On Music Loved & Loathed. The Quietus (November 24, 2009), http://thequietus.com/articles/03283-dylan-moran-interview-on-music-loved-loathed

32. Wong, H.-S.P., Raoux, S., Kim, S., Liang, J., Reifenberg, J.P., Rajendran, B., Asheghi, M., Goodson, K.E.: Phase change memory. Proc. IEEE 98(12), 2201–2227 (2010), http://dx.doi.org/10.1109/JPROC.2010.2070050, doi:10.1109/JPROC.2010.2070050

33. Woodhouse, D.: JFFS: the journalling flash file system. In: Ottawa Linux Symposium (2001), https://sourceware.org/jffs2/jffs2.pdf

34. Wu, M., Zwaenepoel, W.: eNVy: a non-volatile, main memory storage system. In: Proc. 6th Int. Conf. on Architectural Support for Programming Languages and Operating Systems (ASPLOS), pp. 86–97 (1994), doi:10.1145/195473.195506

35. Zukowski, M., Héman, S., Boncz, P.: Architecture-conscious hashing. In: Proc. 2nd Int. Worksh. on Data Management on New Hardware (DaMoN 2006) (2006), doi:10.1145/1140402.1140410

Loop Nesting Forests, Dominators, and Applications

Loukas Georgiadis[1], Luigi Laura[2], Nikos Parotsidis[1], and Robert E. Tarjan[3]

[1] Department of Computer Science & Engineering, University of Ioannina, Greece
{loukas,nparotsi}@cs.uoi.gr
[2] Department of Computer, Control, and Management Engineering and
Research Centre for Transport and Logistics
"Sapienza" Università di Roma, Italy
laura@dis.uniroma1.it
[3] Department of Computer Science, Princeton University, and Microsoft Research
Silicon Valley, USA
ret@cs.princeton.edu

Abstract. Loop nesting forests and dominator trees are important tools in program optimization and code generation, and they have applications in other diverse areas. In this work we first present carefully engineered implementations of efficient algorithms for computing a loop nesting forest of a given directed graph, including a very efficient algorithm that computes the forest in a single depth-first search. Then we revisit the problem of computing dominators and present efficient implementations of the algorithms recently proposed by Fraczak et al. [12], which include an algorithm for acyclic graphs and an algorithm that computes both the dominator tree and a loop nesting forest. We also propose a new algorithm than combines the algorithm of Fraczak et al. for acyclic graphs with the algorithm of Lengauer and Tarjan. Finally, we provide fast algorithms for the following related problems: computing bridges and testing 2-edge connectivity, verifying dominators and testing 2-vertex connectivity, and computing a low-high order and two independent spanning trees. We exhibit the efficiency of our algorithms experimentally on large graphs taken from a variety of application areas.

1 Introduction

A *flow graph* is a directed graph with a distinguished *start* vertex s such that every vertex is reachable from s. Throughout this paper $G = (V, A, s)$ is a flow graph with vertex set V, arc set A, and start vertex s. We denote the number of vertices by n and the number of arcs by m. Since s reaches all vertices, $m \geq n-1$. Loop nesting forests and dominator trees are two fundamental concepts in flow graphs that we define next. They are important tools in global flow analysis, program optimization and code generation [1, 31], and also have a variety of applications in other areas.

A loop nesting forest represents a hierarchy of strongly connected subgraphs of G. There are several ways to define such a forest [31]. Here we consider the definition that was first presented by Tarjan [34] and later rediscovered by Havlak [26],

J. Gudmundsson and J. Katajainen (Eds.): SEA 2014, LNCS 8504, pp. 174–186, 2014.
© Springer International Publishing Switzerland 2014

which is useful in the applications we study. A loop nesting forest H of G is defined with respect to a depth-first search (dfs) spanning tree T rooted at s. For any vertex u, the *loop* of u, denoted by $loop(u)$, is the set of all descendants x of u in T such that there is a path from x to u containing only descendants of u in T. Vertex u is the *head* of $loop(u)$. For any two vertices u and v, $loop(u)$ and $loop(v)$ are either disjoint or nested (i.e., one contains the other). This property allows us to define the *loop nesting forest* H of G, with respect to T, as the forest in which the parent of any vertex v is the nearest proper ancestor u of v in T such that $v \in loop(u)$ if there is such a vertex u, null otherwise. See Figure 1. Tarjan [34] gave an $O(m\alpha(n, m/n))$-time pointer-machine algorithm to compute a loop nesting forest using disjoint set union, where α is a functional inverse of Ackermann's function [33]. The Gabow-Tarjan static tree disjoint set union algorithm [16] reduces the running time of this algorithm to $O(m)$ on a RAM. Buchsbaum et al. [5] gave an $O(m)$-time pointer-machine algorithm.

A vertex u is a *dominator* of a vertex v (u *dominates* v) if every path from s to v contains u. The dominator relation is reflexive and transitive. Its transitive reduction is a rooted tree, the *dominator tree* D: u dominates v if and only if u is an ancestor of v in D. If $v \neq s$, $d(v)$, the parent of v in D, is the *immediate dominator* of v: it is the unique proper dominator of v that is dominated by all proper dominators of v. Dominators have a variety of important applications, notably in optimizing compilers [2, 9] but also in many other areas. See [23, 25] and the references therein. Lengauer and Tarjan [28] gave two near-linear-time algorithms for computing dominators that run fast in practice and have been used in many of these applications. The simpler of these runs in $O(m \log_{(m/n+1)} n)$ time, and the more sophisticated runs in $O(m\alpha(n, m/n))$ time. Subsequently, more-complicated but truly linear-time algorithms to compute dominators were discovered [3, 5, 6, 21]; these are based on the Lengauer-Tarjan algorithm and achieve linear time by incorporating several other techniques, including the precomputation of answers to small subproblems. Recently, Gabow [15] and Fraczak et al. [12] presented linear-time algorithms that are based on a different approach, and require only simple data structures and a data structure for static tree set union [16]. Gabow's algorithm uses the concept of minimal-set posets [13, 14], while the algorithm of Fraczak et al. uses arc contractions.

Contribution. In Section 2 we present carefully engineered versions of Tarjan's algorithm [34] for computing a loop nesting forest of a flow graph, including a single-pass streamlined version [5, 12]. To the best of our knowledge, this is the first experimental study of Tarjan's loop nesting forest algorithm, and the first implementation of its streamlined version. We revisit the problem of computing dominators in Section 3, and present efficient implementations of the algorithms proposed by Fraczak et al. [12], which include an algorithm for acyclic graphs and an algorithm that computes both the dominator tree and a loop nesting forest. We also propose a new algorithm than combines the algorithm of Fraczak et al. for acyclic graphs with the algorithm of Lengauer and Tarjan. We compare the performance of these algorithms with the efficient implementations of the Lengauer-Tarjan algorithm and the hybrid algorithm SNCA from [25]. Finally, in

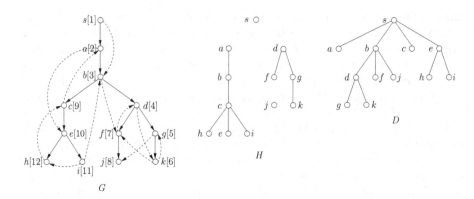

Fig. 1. A flow graph G with a depth-first search spanning tree T shown with solid arcs; non-tree arcs are shown dashed; vertices are numbered in reverse postorder (in brackets). Loop nesting forest H of G with respect to T. Dominator tree D of G.

Section 4, we apply the loop nesting forest and dominator-finding algorithms in order to provide efficient algorithms for the following problems: finding the bridges of a flow graph, testing 2-edge and 2-vertex connectivity of a directed graph, and computing a low-high order and two independent spanning trees of a flow graph. We also present, as a byproduct, a new linear-time algorithm to compute the *derived graph* [36] of a flow graph. We exhibit the efficiency of our algorithms experimentally on large graphs taken from a variety of application areas. Section 5 contains some final remarks.

Related Work. Empirical evaluations of algorithms for computing dominators were presented in [20, 25], where the performance of careful implementations of both versions of the Lengauer-Tarjan algorithm, an iterative algorithm of Cooper, Harvey, and Kennedy [8], and a new hybrid $O(n^2)$-time algorithm [25] were compared. The performance of all these algorithms was similar for graphs of moderate size (with at most a few thousand vertices and edges), but the iterative algorithm was not competitive for larger and more complicated graphs (with millions of vertices and edges). The experimental results also showed that the simple version of the Lengauer-Tarjan algorithm and the hybrid algorithm perform at least as well as the sophisticated version of the Lengauer-Tarjan algorithm, in spite of their larger worst-case time bounds. In [20], efficient implementations of algorithms that compute a low-high order were given, by augmenting the computations of the Lengauer-Tarjan algorithm. Here we consider the implementation of an alternative low-high order algorithm, based on loop nesting forests [23]. Finally, [11] presented, using the WebGraph Java library [38], efficient implementations of algorithms for computing the strong bridges and the strong articulation points of a directed graph [27]. The algorithm for computing strong bridges is based on dominator trees. We obtain an alternative algorithm by using a loop nesting forest instead.

Table 1. Real-world graphs sorted by file size; n is the number of vertices, m the number of edges, and δ_{avg} is the average vertex degree

Graph	n	m	file size	δ_{avg}	type
Rome99	3.3k	8.8k	98k	2.65	road network
s38584	20.7k	34.4k	434k	1.67	circuit
Oracle-16k	15.6k	48.2k	582k	3.08	memory profiling
p2p-gnutella25	22.6k	54.7k	685k	2.41	peer2peer
soc-Epinions1	75.8k	508k	5.9M	6.71	social network
USA-road-NY	264k	733k	11M	2.78	road network
USA-road-BAY	321k	800k	12M	2.49	road network
Amazon0302	262k	1.2M	18M	4.71	product co-purchase
web-NotreDame	325k	1.4M	22M	4.6	web graph
web-Stanford	281k	2.3M	34M	8.2	web graph
Amazon0601	403k	3.4M	49M	8.4	product co-purchase
wiki-Talk	2.3M	5.0M	69M	2.1	social network
web-BerkStan	685k	7.6M	113M	11.09	web graph
SAP-4M	4.1M	12.0M	183M	2.92	memory profiling
Oracle-4M	4.1M	14.6M	246M	3.55	memory profiling
Oracle-11M	10.7M	33.9M	576M	3.18	memory profiling
SAP-11M	11.1M	36.4M	638M	3.27	memory profiling
LiveJournal	4.8M	68.9M	1G	14.23	social network
USA road complete	23.9M	58.3M	1.1G	2.44	road network
SAP-32M	32.3M	81.9M	1.5G	2.53	memory profiling
SAP-47M	47.0M	131.0M	2.2G	2.8	memory profiling
SAP-70M	69.7M	215.7M	3.7G	3.09	memory profiling
Hollywood-2011	2.1M	228.5M	3.7G	104.8	movie co-appearance
Uk-2002	18.5M	285.1M	5.1G	15.34	web graph

Experimental Setting. We wrote our implementations in C++, using g++ v. 4.6.4 with full optimization (flag -O3) to compile the code. The experimental testbed is similar to the one used in [20]; we report the running times on a GNU/Linux machine, with Ubuntu (12.04LTS): a dell PowerEdge R715 server 64-bit NUMA machine with four AMD Opteron 6376 processors and 128GB of RAM memory. Each processor has 8 cores sharing a 16MB L3 cache, and each core has a 2MB private L2 cache and 2300MHz speed. We report CPU times measured with the getrusage function. All the running times reported in our experiments were averaged over ten different runs. To minimize fluctuations due to external factors, we used the machine exclusively for tests, took each measurement three times, and picked the best. (We note that the differences among the three measurements were, in all the cases, almost negligible.) Running times do not include reading the input file, but they do include creating the graph (successor and predecessor lists, as required by each algorithm), and allocating and deallocating the arrays used by each algorithm. The dataset used in our experiments, detailed in Table 1, includes graphs from several distinct application domains: road networks from the 9th DIMACS Implementation Challenge website [10], a circuit from VLSI-testing applications [4], from the ISCAS'89 suite [7], graphs taken from applications of dominators in memory profiling (see e.g., [30]), and graphs from the Stanford Large Network Dataset Collection [29].

2   Loop Nesting Forests

Tarjan's algorithm [34] processes the vertices of the flow graph G in postorder with respect to a dfs spanning tree T. When it processes vertex v, it contracts $loop(v)$ into v. The vertices in $loop(v)$ are found by a backward search from v, and the effect of contraction is achieved with the use of a *disjoint set union* (dsu) data structure [33].

We tested four implementations of Tarjan's algorithm that we refer to as T3, T2, T1, and ST1. Algorithms T3 and T2 are, respectively, a three-pass and a two-pass implementation of Tarjan's original algorithm: T3 computes T, represented by its parent function p, the nearest common ancestors (nca's) in T for all arcs (this is needed so that the backward search from v visits only descendants of v), and finally the loop heads in a separate pass; T2 combines the computation of T with that of the nca's in one pass, and computes loop heads in a second pass. Algorithm T1 implements the single-pass streamlined version of Tarjan's algorithm [5, 12] that avoids the nca computations. Finally, algorithm ST1 gives a simpler version of T1, where we use a disjoint set union data structure with path compression but not balancing.

The experimental results are shown in Figure 2 (top). Algorithm T1 is consistently faster than T3 and T2 by a factor that is in the range from 1.3 to 3.2, and from 1.3 to 2.8, respectively. Next we compare T1 with ST1. Although the use of a simpler dsu data structure increases the worst-case bound of the algorithm to $O(m \log_{(m/n+1)} n)$ [37], with the exception of three graphs, it improved the running time by 5% to 33%. This is improvement is due to the reduced storage space required for the dsu data structure (two fewer arrays of size n), and fewer operations performed by *unite*. Similar behaviour was observed in all the algorithms we implemented that used a dsu data structure. In fact, if we do not need T after the computation of the loop nesting forest, then we can maintain the dsu structure in the same array that stores p; if suffices to have, for each vertex $v \neq s$, a bit indicating if v has been united with $p(v)$.

3   Dominators

Recently, Fraczak et al. [12] presented new algorithms for computing dominators, based on dominator-preserving transformations. These algorithms require only simple data structures and a data structure for static tree set union [16] that is used to contract tree arcs. They run in near-linear or linear time, depending on the implementation of the disjoint set data structure. We evaluate experimentally the performance of three algorithms, AD, GD, and HD, from [12]. Algorithm AD computes dominators in a directed acyclic graph, by performing contractions. It processes the vertices in reverse preorder, with respect to a dfs spanning tree T. A vertex is v is contracted into its parent $p(v)$ in T, thereby deleting v from the current graph, when the last arc into v has been marked. An arc can be marked only if it is a forward arc in the current graph. To keep track of these arcs, we maintain, for each undeleted vertex x, a bag (multiset) $out(x)$ of vertices v such that (x, v)

is a current undeleted arc. These bags are initially empty. When we process a vertex u, we insert the arcs entering u into the appropriate *out* bags. Then we mark the arcs that correspond to the vertices in $out(u)$. The other two algorithms, GD and HD, work for general directed graphs, and require to keep track both of arcs leaving and of arcs entering each undeleted vertex in the current graph, i.e., we maintain both *out* and *in* bags. Algorithm GD requires the computation of nca's in T for all arcs in order to determine when to insert an arc into an *out* and an *in* bag. Algorithm HD avoids the need for nca's by computing a loop nesting forest during the dfs. We refer to [12] for the details. We also implemented a new algorithm, SLT-AD, that combines the semi-dominator computation of SLT with AD. Semi-dominators are computed by the Lengauer-Tarjan algorithm [28] as an initial approximation to immediate dominators, and are defined as follows. Number the vertices in preorder of T, and identify vertices by their numbers. A path from u to v is *high* if all its vertices other than u and v are higher than both u and v. If $v \neq s$, the *semi-dominator* of v, $sd(v)$, is the minimum vertex u such that there is a high path from u to v. Algorithm SLT-AD, first computes semi-dominators, using simplified data structures [25], and then runs a simplified version of AD on an acyclic graph $\hat{G}$, formed by the arcs of T and the additional (forward) arcs $(sd(v), v)$ for all $v \neq s$. Therefore, it has the same asymptotic running time as SLT. This approach is similar to the one used in SNCA. In SNCA, the dominator tree is constructed incrementally by computing nca's of the arcs in $\hat{G}$. This is simpler than running AD but leads to $O(n^2)$ running time in the worst case.

We compare the performance of these algorithms with SLT and SNCA from [25]. In order to test AD we created acyclic versions of the graphs of Table 1, by deleting all back arcs with respect to T. In Figure 2 we can see the experimental results for acyclic graphs (second plot) and general directed graphs (third plot). Algorithm AD outperforms SLT in several instances. In particular, for SAP-11M, it is faster than SLT by a factor more than 2.63. With the exception of two graphs, Rome99 and Hollywood-2011, AD is not slower than SLT by a factor larger than 1.63. For Rome99, AD required twice as much time as SLT, while for Hollywood-2011 AD is slower than SLT by a factor less than 3.57. This behavior is due to the fact that AD maintains dynamically lists of outgoing arcs (*out* bags), which makes it more sensitive to the structure of the graph than SLT and its relatives (SNCA and SLT-AD), especially for graphs with large average vertex degree.

Now we turn to general directed graphs. The new algorithm SLT-AD is slightly faster than SLT in more than half of the instances, but it does not outperform SNCA. On average, algorithms GD and HD are slower than SLT by a factor close to 1.96 and 1.83, respectively. Since HD computes both a loop nesting forest and the dominator tree (which is required in some applications [17, 31]), we compare its performance with an algorithm that combines ST1 with SLT. We refer to this combined algorithm ST1-SLT; it does a dfs of G, that produces a dfs spanning tree T and the corresponding loop nesting forest. Then, it uses T to compute the dominators of G. The experimental results are shown in Figure 2 (bottom). In this respect, HD performs well, being slower than ST1-SLT by a factor close to 1.25 on average.

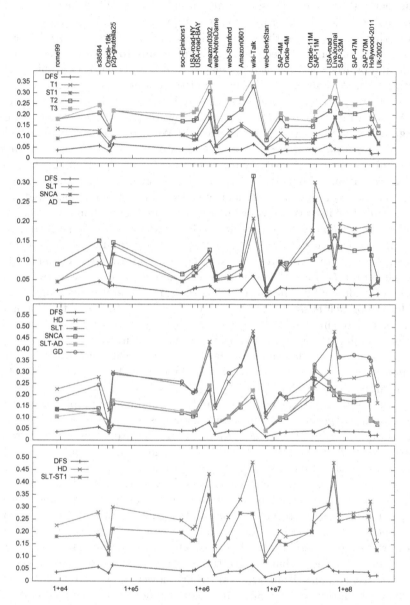

Fig. 2. Top: experimental comparison of versions of Tarjan's loop nesting forest algorithms; second row: experimental comparison of dominator-finding algorithms for directed acyclic graphs; third row: experimental comparison of dominator-finding algorithms for general directed graphs; bottom: experimental comparison of algorithms that compute both a loop nesting forest and the dominator tree of a general directed graph. Running times (in microsecs) normalized to the number of edges (shown in logarithmic scale).

4   Connectivity and Related Applications

In this section we evaluate algorithms that apply a loop nesting forest or a dominators computation in order to get efficient solutions to various problems related to 2-edge and 2-vertex connectivity. A *strong bridge* (resp. *strong articulation point*) of a directed graph G is an arc (vertex) whose removal increases the number of strongly connected components of G [27]. Clearly, a directed graph is 2-edge (2-vertex) connected if and only if it is strongly connected and has no strong bridges (strong articulation points) [18, 27].

Bridges. An arc (u, v) is a *bridge* if all paths from s to v include (u, v). Italiano et al. [27] showed that one can apply a bridge-finding algorithm to compute the strong bridges of G, and thus also to test 2-edge connectivity. As shown in [34], one can use a loop nesting forest to compute bridges efficiently. For each vertex v, we need to find if there is a non-tree arc (x, w) such that $w \in loop(v)$ and x is not a descendant of v. Arc (u, v) is a bridge if and only if no such arc exists. This test is implemented by algorithm ST1-BRIDGES, where we use ST1 to compute the loop nesting forest. An alternative test was given in [32]: Arc (u, v) is a bridge if and only if it is a tree arc and all other arcs (w, v) entering v are such that w is dominated by v. Algorithm SLT-BRIDGES implements this test, where SLT is used to compute the dominator tree. This approach was used in [11] to compute the strong bridges of a directed graph. We compare the two methods in Figure 3. Algorithm ST1-BRIDGES outperforms SLT-BRIDGES in all but four instances. It is faster by a factor that is close to 1.54 on average, and at most 3.64.

2-Connectivity. The strong articulation points of G can be found analogously by computing dominators [27]. This observation implies that testing if a directed graph is 2-vertex connected reduces to testing if the a flow graph has a *flat* dominator tree, that is, $d(v) = s$ for all vertices $v \neq s$. We refer to the corresponding algorithm as SLT-FLAT, which uses SLT to compute the dominator tree.

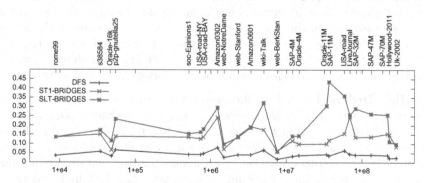

Fig. 3. Experimental comparison of bridges computation using the Lengauer-Tarjan dominators algorithm (SLT-BRIDGES) and the loop nesting forest algorithm (ST1-BRIDGES). Running times (in microsecs) normalized to the number of edges (shown in logarithmic scale).

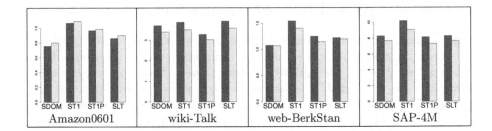

Fig. 4. Experimental comparison of algorithms that test if an input flow graph has a flat dominator tree: SDOM-FLAT, ST1-FLAT, ST1P-FLAT, and SLT-FLAT. Each bar represents the average running time of the corresponding algorithm on 2-vertex connected graphs, obtained from four graphs of Table 1, after applying two distinct preprocessing methods [19]: 1) canceling articulation points (left bars), and 2) adding a bidirectional Hamiltonian path (right bars). Running times are in seconds.

We can also design a simplified algorithm based on the following observation: Suppose $d(v) \neq s$ for some vertex $v \neq s$. Then there is a vertex u such that $sd(u) = d(u) = p(u) \neq s$. Algorithm SDOM-FLAT implements this test, using the simplified computation of semi-dominators. Another way to test if a flow graph has a *flat* dominator tree is with the use of a loop nesting forest, which avoids the need to actually compute dominators [22–24]. To that end, let $S(v)$ be the set of vertices u for which there is an arc (u, w) such that $w \in loop(v)$. Flow graph G has a flat dominator tree if and only if, for every vertex $v \neq s$, $S(v)$ contains s or at least two vertices numbered less than v in reverse postorder. To apply this idea, it suffices to store only a subset $S_2(v)$ of $S(v)$ containing two vertices in $S(v)$ that are minimum in reverse postorder. Then we can compute the S_2-sets in $O(m)$ time in any bottom-up order of T. We refer to the corresponding algorithm as ST1-FLAT. We also implemented a variant of this method, ST1P-FLAT, that uses preorder numbering; instead of storing $S_2(v)$, for each vertex v, we compute the minimum and maximum vertices in preorder that have an arc entering $loop(v)$. In this way, we avoid one pass over the arcs of the graph. We compare the above three methods in Figure 4. As expected, SDOM-FLAT outperforms SLT-FLAT, and ST1P-FLAT outperforms ST1-FLAT, on all instances. On the other hand, there is no clear winner between SDOM-FLAT and ST1P-FLAT.

Low-High Orders. Let G be a flow graph with start vertex s, and let D be the dominator tree of G. A *low-high order* of G is a preorder of D, such that, for all $v \neq s$, $(d(v), v) \in A$ or there are two arcs $(u, v) \in A$, $(w, v) \in A$ such that u is less than v, v is less than w, and w is not a descendant of v in D. A low-high order is a *correctness certificate* for a dominator tree, and its computation makes a dominator-finding algorithm *self-certifying* [23]. Low-high orders are related to the notion of *independent spanning trees*. Given a low-high order it is easy to construct in $O(n)$ time two independent spanning trees, and, conversely, given two independent spanning trees one can construct a low-high order in $O(n)$ time [23, 24]. The above definitions are interesting in their own right, and have applications in other graph problems [23]. Fast algorithms (almost-linear-time or linear-time, depending on the implementation) for computing a *low-high order* were presented in [23, 24]. A key component of these algorithms is the computation of the derived graph that we define next. Then we consider efficient implementations of an algorithm in [23] for computing a low-high order given a flow graph and its dominator tree. The algorithm uses only simple data structures and static tree set union, and computes a loop nesting forest and the derived graph as an intermediary.

Derived graph. Let (v, w) be an arc of G. The properties of D imply that $d(w)$ is an ancestor of v in D [23]. The *derived arc* of (v, w) is null if w is an ancestor of v in D, (v', w) otherwise, where $v' = v$ if $v = d(w)$, v' is the sibling of w that is an ancestor of v if $v \neq d(w)$. The *derived graph* G' is the graph with vertex set V and arc set $A' = \{(v', w) \mid (v', w)$ is the non-null derived arc of some arc in $A\}$. The derived graph can be computed in $O(m)$ time using a three-pass radix sort [23]; we refer to this algorithm as DER-SORT. We also introduce an alternative method, DER-DFS, that computes the derived arcs by executing a dfs of G. This is accomplished by maintaining, during the dfs, a stack of vertices for each level in the dominator tree.

Construction of a low-high order. We test two implementations of the low-high algorithm via a loop nesting forest [23] that differ in the computation of the derived graph: ST1-LH-SORT uses radix sort and ST1-LH-DFS uses dfs. Unlike all the algorithms considered above, these low-high order construction algorithms require not only the flow graph but also its dominator tree to be provided in the input. Figure 5 shows that ST1-LH-DFS is faster than ST1-LH-SORT in all but two instances, by a factor that is at most 1.50. We also tested how the choice between DER-SORT and DER-DFS affects the running times of the low-high order algorithms SLTCERT and SLTCERT-II from [20], which are based on SLT. For these algorithms the speedup was marginal, which is due to the fact

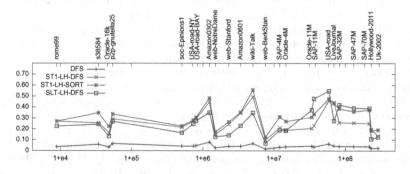

Fig. 5. Experimental comparison of low-high order computation from a loop nesting forest, using radix sort (ST1-LH-SORT) and depth-first search (ST1-LH-DFS) to compute the derived graph. Running times (in microsecs) normalized to the number of edges (shown in logarithmic scale).

that they compute the derived arcs of two spanning trees instead of the whole graph. For completeness, we include in Figure 5 the running times of the variant of SLTCERT that uses DER-DFS, referred to as SLT-LH-DFS.

5 Concluding Remarks

Our experimental results showed that the streamlined version of Tarjan's algorithm performs very well in practice. Based on this implementation, we gave efficient algorithms for a variety of applications related to 2-connectivity. We compared the performance of these algorithms with alternative solutions based on dominator trees. The algorithms based on the computation of a loop nesting forest perform equally well as the dominator-based alternatives, and require only simple data structures and a data structure for disjoint set union. On the other hand, the dominator-finding algorithm of Lengauer and Tarjan uses one conceptually complicated data structure for computing minima on paths in a tree [35]. We also tested implementations of four new dominator-finding algorithms, one introduced here and the others presented in [12]. At least three of them seem promising in practice: AD for acyclic graphs, SLT-AD for general graphs, and HD when both the dominator tree and a loop nesting forest are needed.

References

1. Aho, A.V., Sethi, R., Ullman, J.D.: Compilers: Principles, Techniques, and Tools. Addison-Wesley, Reading (1986)
2. Aho, A.V., Ullman, J.D.: Principles of Compilers Design. Addison-Wesley (1977)
3. Alstrup, S., Harel, D., Lauridsen, P.W., Thorup, M.: Dominators in linear time. SIAM Journal on Computing 28(6), 2117–2132 (1999)
4. Amyeen, M.E., Fuchs, W.K., Pomeranz, I., Boppana, V.: Fault equivalence identification using redundancy information and static and dynamic extraction. In: Proceedings of the 19th IEEE VLSI Test Symposium (March 2001)
5. Buchsbaum, A.L., Georgiadis, L., Kaplan, H., Rogers, A., Tarjan, R.E., Westbrook, J.R.: Linear-time algorithms for dominators and other path-evaluation problems. SIAM Journal on Computing 38(4), 1533–1573 (2008)
6. Buchsbaum, A.L., Kaplan, H., Rogers, A., Westbrook, J.R.: A new, simpler linear-time dominators algorithm. ACM Transactions on Programming Languages and Systems 20(6), 1265–1296 (1998); Corrigendum in 27(3), 383–387 (2005)
7. CAD Benchmarking Lab. ISCAS 1989 benchmark information, http://www.cbl.ncsu.edu/www/CBL_Docs/iscas89.html
8. Cooper, K.D., Harvey, T.J., Kennedy, K.: A simple, fast dominance algorithm. Software Practice & Experience 4, 110 (2001)
9. Cytron, R., Ferrante, J., Rosen, B.K., Wegman, M.N., Zadeck, F.K.: Efficiently computing static single assignment form and the control dependence graph. ACM Transactions on Programming Languages and Systems 13(4), 451–490 (1991)
10. Demetrescu, C., Goldberg, A.V., Johnson, D.S.: 9th DIMACS Implementation Challenge: Shortest Paths (2007), http://www.dis.uniroma1.it/~challenge9/
11. Firmani, D., Italiano, G.F., Laura, L., Orlandi, A., Santaroni, F.: Computing strong articulation points and strong bridges in large scale graphs. In: Klasing, R. (ed.) SEA 2012. LNCS, vol. 7276, pp. 195–207. Springer, Heidelberg (2012)

12. Fraczak, W., Georgiadis, L., Miller, A., Tarjan, R.E.: Finding dominators via disjoint set union. Journal of Discrete Algorithms 23, 2–20 (2013)

13. Gabow, H.N.: Applications of a poset representation to edge connectivity and graph rigidity. In: Proc. 32th IEEE Symp. on Foundations of Computer Science, pp. 812–821 (1991)

14. Gabow, H.N.: The minimal-set poset for edge connectivity (2013) (unpublished manuscript)

15. Gabow, H.N.: A poset approach to dominator computation (2013) (unpublished manuscript 2010, revised unpublished manuscript)

16. Gabow, H.N., Tarjan, R.E.: A linear-time algorithm for a special case of disjoint set union. Journal of Computer and System Sciences 30(2), 209–221 (1985)

17. Georgiadis, L.: Computing frequency dominators and related problems. In: Hong, S.-H., Nagamochi, H., Fukunaga, T. (eds.) ISAAC 2008. LNCS, vol. 5369, pp. 704–715. Springer, Heidelberg (2008)

18. Georgiadis, L.: Testing 2-vertex connectivity and computing pairs of vertex-disjoint s-t paths in digraphs. In: Abramsky, S., Gavoille, C., Kirchner, C., Meyer auf der Heide, F., Spirakis, P.G. (eds.) ICALP 2010. LNCS, vol. 6198, pp. 738–749. Springer, Heidelberg (2010)

19. Georgiadis, L.: Approximating the smallest 2-vertex connected spanning subgraph of a directed graph. In: Demetrescu, C., Halldórsson, M.M. (eds.) ESA 2011. LNCS, vol. 6942, pp. 13–24. Springer, Heidelberg (2011)

20. Georgiadis, L., Laura, L., Parotsidis, N., Tarjan, R.E.: Dominator certification and independent spanning trees: An experimental study. In: Bonifaci, V., Demetrescu, C., Marchetti-Spaccamela, A. (eds.) SEA 2013. LNCS, vol. 7933, pp. 284–295. Springer, Heidelberg (2013)

21. Georgiadis, L., Tarjan, R.E.: Finding dominators revisited. In: Proc. 15th ACM-SIAM Symp. on Discrete Algorithms, pp. 862–871 (2004)

22. Georgiadis, L., Tarjan, R.E.: Dominator tree verification and vertex-disjoint paths. In: Proc. 16th ACM-SIAM Symp. on Discrete Algorithms, pp. 433–442 (2005)

23. Georgiadis, L., Tarjan, R.E.: Dominator tree certification and independent spanning trees. CoRR, abs/1210.8303 (2012)

24. Georgiadis, L., Tarjan, R.E.: Dominators, directed bipolar orders, and independent spanning trees. In: Proc. 39th Int'l. Coll. on Automata, Languages, and Programming, pp. 375–386 (2012)

25. Georgiadis, L., Tarjan, R.E., Werneck, R.F.: Finding dominators in practice. Journal of Graph Algorithms and Applications (JGAA) 10(1), 69–94 (2006)

26. Havlak, P.: Nesting of reducible and irreducible loops. ACM Transactions on Programming Languages and Systems 19(4), 557–567 (1997)

27. Italiano, G.F., Laura, L., Santaroni, F.: Finding strong bridges and strong articulation points in linear time. Theoretical Computer Science 447, 74–84 (2012)

28. Lengauer, T., Tarjan, R.E.: A fast algorithm for finding dominators in a flowgraph. ACM Transactions on Programming Languages and Systems 1(1), 121–141 (1979)

29. Leskovec, J.: Stanford large network dataset collection (2009), http://snap.stanford.edu

30. Mitchell, N.: The runtime structure of object ownership. In: Thomas, D. (ed.) ECOOP 2006. LNCS, vol. 4067, pp. 74–98. Springer, Heidelberg (2006)

31. Ramalingam, G.: On loops, dominators, and dominance frontiers. ACM Transactions on Programming Languages and Systems 24(5), 455–490 (2002)

32. Tarjan, R.E.: Edge-disjoint spanning trees, dominators, and depth-first search. Technical report, Stanford University, Stanford, CA, USA (1974)

33. Tarjan, R.E.: Efficiency of a good but not linear set union algorithm. Journal of the ACM 22(2), 215–225 (1975)
34. Tarjan, R.E.: Edge-disjoint spanning trees and depth-first search. Acta Informatica 6(2), 171–185 (1976)
35. Tarjan, R.E.: Applications of path compression on balanced trees. Journal of the ACM 26(4), 690–715 (1979)
36. Tarjan, R.E.: Fast algorithms for solving path problems. Journal of the ACM 28(3), 594–614 (1981)
37. Tarjan, R.E., van Leeuwen, J.: Worst-case analysis of set union algorithms. Journal of the ACM 31(2), 245–281 (1984)
38. The WebGraph Framework Home Page, http://webgraph.dsi.unimi.it/

DenseZDD:
A Compact and Fast Index for Families of Sets

Shuhei Denzumi[1], Jun Kawahara[2], Koji Tsuda[3,4], Hiroki Arimura[1],
Shin-ichi Minato[1,4], and Kunihiko Sadakane[5]

[1] Graduate School of IST, Hokkaido University, Japan
[2] Nara Institute of Science and Technology (NAIST), Japan
[3] National Institute of Advanced Industrial Science and Technology (AIST), Japan
[4] ERATO MINATO Discrete Structure Manipulation System Project, JST, Japan
[5] National Institute of Informatics (NII), Japan
{denzumi,arim,minato}@ist.hokudai.ac.jp, jkawahara@is.naist.jp,
koji.tsuda@aist.go.jp, sada@nii.ac.jp

Abstract. In many real-life problems, we are often faced with manipulating families of sets. Manipulation of large-scale set families is one of the important fundamental techniques for web information retrieval, integration, and mining. For this purpose, a special type of *binary decision diagrams (BDDs)*, called *Zero-suppressed BDDs (ZDDs)*, is used. However, current techniques for storing ZDDs require a huge amount of memory and membership operations are slow. This paper introduces DenseZDD, a compressed index for static ZDDs. Our technique not only indexes set families compactly but also executes fast member membership operations. We also propose a hybrid method of DenseZDD and ordinary ZDDs to allow for dynamic indices.

1 Introduction

Binary Decision Diagrams (BDD) [1] are a graph-based representation of Boolean functions and widely used in VLSI logic design and verification. A BDD is constructed reducing a binary decision tree, which represents a decision making process through the input variables. If we fix the order of the input variables and apply the following two reduction rules, then we obtain a minimal and canonical form for a given Boolean function:

1. Delete all redundant nodes (whose two children are identical) and
2. Merge all equivalent nodes (having the same index and pair of children).

Among unique canonical representations of Boolean functions, BDDs are smaller than others such as CNF, DNF, and truth tables for many classes of functions. BDDs have the following features:

- Boolean functions are uniquely represented like other representations.
- Multiple functions are stored compactly by sharing common subgraphs.
- Fast logical operations are executed on Boolean functions.

J. Gudmundsson and J. Katajainen (Eds.): SEA 2014, LNCS 8504, pp. 187–198, 2014.
© Springer International Publishing Switzerland 2014

Zero-suppressed Binary Decision Diagrams (ZDDs) [9] are variation of traditional BDDs, used to manipulate families of sets. Using ZDDs, we can implicitly enumerate combinatorial item set data and efficiently compute set operations over the ZDDs. In the rest of this section, we use the term BDD to indicate both the original BDD and the ZDD unless specified because any ZDD is regarded as a BDD representing some function.

Though BDDs are more compact than other representations of Boolean function and set families, they are still large; a node of a BDD uses 20 to 30 bytes depending on implementations [10]. BDDs become inefficient if the graph size is too large to be held in memory. Therefore the aim of this paper is to reduce the size (number of bits) used to represent BDDs. We classify implementations of BDDs into three types:

- Dynamic: The BDD can be modified. New nodes can be added to the BDD.
- Static: The BDD cannot be modified. Only query operations are supported.
- Freeze-dried: All the information of the BDD is stored, but it cannot be used before restoration.

Most of the current implementations of BDDs are dynamic. There is previous work on freeze-dried representations of BDDs by Starkey and Bryant [15] and later, by Mateu and Prades-Nebot [8]. Hansen, Rao and Tiedemann [5] developed a technique to compress BDD and reduce the size of the BDD to 1-2 bits per node. However there is no implementation of BDDs that is specialized for static case.

This paper is the first to propose a static representation of ZDDs, which we call *DenseZDDs*. The size of ZDDs in our representation is much smaller than an existing dynamic representation [10]. Not only compact, DenseZDD supports much faster membership operations than [10]. Experimental results show that DenseZDDs are five times smaller and membership queries are twenty to several hundred times faster, compared to [10]. Note that our technique can be directly applied to compress traditional BDDs too.

2   Preliminaries

Let $e_1, \ldots, e_n$ be items such that $e_1 < e_2 < \cdots < e_n$. Let $S = \{a_1, \ldots, a_c\}$, $c \geq 0$, be a set of items. We denote the *size* of S by $|S| = c$. The empty set is denoted by $\emptyset$. A *family* is a subset of the power set of all items. A finite family F of sets is referred to as a *set family*.[1] The *join* of families F_1 and F_2 is defined as $F_1 \sqcup F_2 = \{ S_1 \cup S_2 \mid S_1 \in F_1, S_2 \in F_2 \}$.

In Appendix A of our technical report [2], we describe existing succinct data structures. The balanced parenthesis sequence (BP), the Fully Indexable Dictionary (FID), and some basic structures used in this paper are reviewed. We also explain operations on the data structures such as $rank_c$, $select_c$, and so on.

---

[1] In the original ZDD paper by Minato, a set is called a combination, and a set family is called a combinatorial set.

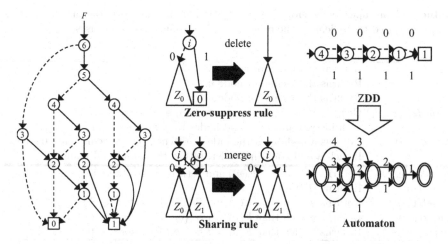

Fig. 1. An example of ZDD

Fig. 2. Reduction rules of ZDDs

Fig. 3. Worst-case example of a straightforward translation

2.1 Zero-Suppressed Binary Decision Diagrams (ZDDs)

A *zero-suppressed binary decision diagram* (a ZDD) [9] is a variant of a binary decision diagram [1], customized to manipulate finite families of set. A ZDD is a directed acyclic graph satisfying the following conditions. A ZDD has two types of nodes, terminal and nonterminal nodes. A *terminal node* v has as attribute a value $value(v) \in \{0, 1\}$, indicating whether it is a *0-terminal node* or a *1-terminal node*, denoted by **0** and **1**, respectively. A *nonterminal node* v has as attributes an integer $index(v) \in \{1, \ldots, n\}$ called the *index*, and two children $zero(v)$ and $one(v)$, called the *0-child* and *1-child*. The edges from nonterminals to their 0-child (1-child resp.) are called *0-edges* (*1-edges* resp.). In the figures, terminal nodes are denoted by squares, and nonterminal nodes are denoted by circles. 0-edges are denoted by dotted arrows, and 1-edges are denoted by solid arrows. We define $triple(v) = \langle index(v), zero(v), one(v) \rangle$, called the *attribute triple* of v. For any nonterminal node v, $index(v)$ is larger than the indices of its children.[2] We define the *size* of the graph, denoted by $|G|$, as the number of its nonterminals.

Definition 1 (set family represented by ZDD). *A ZDD G rooted at a node $v \in V$ represents a finite family of sets $F(v)$ on U_n defined recursively as follows: (1) If v is a terminal node: $F(v) = \{\emptyset\}$ if $value(v) = 1$, and $F(v) = \emptyset$ if $value(v) = 0$. (2) If v is a nonterminal node, then $F(v)$ is the finite family of sets $F(v) = (\{e_{index(v)}\} \sqcup F(one(v))) \cup F(zero(v))$.*

---

[2] In ordinary BDD or ZDD papers, the indices are in ascending order from roots to terminals. For convenience, we employ the opposite ordering in this paper.

Table 1. Main operations supported by ZDD. The first group is the primitive ZDD operations used to implement the others, yet they could have other uses

$index(v)$	Returns the index of node v.
$zero(v)$	Returns the 0-child of node v.
$one(v)$	Returns the 1-child of node v.
$getnode(i, v_0, v_1)$	Generates (or makes a reference to) a node v with index i and two child nodes $v_0 = zero(v)$ and $v_1 = one(v)$.
$topset(v, i)$	Returns a node with the index i reached by traversing only 0-edges. If such a node does not exist, return the 0-terminal node.
$member(v, S)$	Returns $true$ if $S \in F(v)$, and returns $false$ otherwise.
$count(v)$	Returns $|F(v)|$.
$offset(v, i)$	Returns v such that $F(v) = \{\, S \subseteq U_n \mid S \in F, e_i \notin S \,\}$.
$onset(v, i)$	Returns v such that $F(v) = \{\, S \backslash \{e_i\} \subseteq U_n \mid S \in F, e_i \in S \,\}$.
$apply_\diamond(v_1, v_2)$	Returns v such that $F(v) = F(v_1) \diamond F(v_2)$, for $\diamond \in \{\cup, \cap, \backslash, \oplus\}$.

The example in Fig. 1 represents a sets family $F = \{\ \{6, 5, 4, 3\}, \{6, 5, 4, 2\}, \{6, 5, 4, 1\}, \{6, 5, 4\}, \{6, 5, 2\}, \{6, 5, 1\}, \{6, 5\}, \{6, 4, 3, 2\}, \{6, 4, 3, 1\}, \{6, 4, 2, 1\}, \{6, 2, 1\}, \{3, 2, 1\}\ \}$.

A set $S = \{a_1, \ldots, a_c\}$ describes a path in the graph G starting from the root. At each nonterminal node, the path continues to the 0-child if $e_i \notin S$ and to the 1-child if $e_i \in S$. The path eventually reaches the 1-terminal (or 0-terminal resp.), indicating that S is accepted (or rejected resp.).

In ZDD, we employ the following two reduction rules to compress the graph: (a) Zero-suppress rule: A nonterminal node whose 1-child is the 0-terminal node. (b) Sharing rule: Two or more nonterminal nodes having the same attribute triple. By applying above rules, we can reduce the graph without changing its semantics. If we apply the two reduction rules as much as possible, then we obtain a canonical form for a given family of sets.

We can reduce the size of ZDDs by using a type of *attributed edges* [12] named *0-element edges*. Each nonterminal node v has an $\emptyset$-flag $empflag(v)$ on its 1-edge to implement 0-element edges. If $empflag(v) = 1$, the subgraph pointed by the v's 1-edge includes the empty set $\emptyset$ in the family represented by the subgraph. In this paper, effective $\emptyset$-flags are denoted as circles at starting points of 1-edges.

Table 1 summarizes operations of ZDDs. The upper half shows the primitive operations, while the lower half shows other operations which can be implemented by using the primitive operations. The operations $index(v)$, $zero(v)$, $one(v)$, $topset(v, i)$ and $member(v, S)$ do not create new nodes. Therefore they can be done on a static ZDD. The operation $count(v)$ does not create any node; however we need an auxiliary array to memorize which nodes are already visited.

2.2   Problem of Existing ZDDs

Let m be the number of nodes of a given ZDD and n be the number of distinct indices of nodes. Existing ZDD implementations have the following

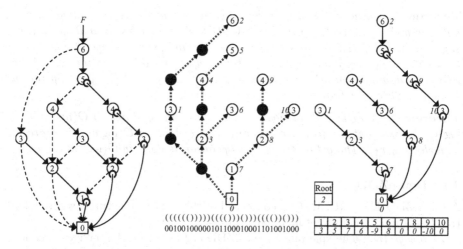

Fig. 4. The ZDD using 0-element edges that is equivalent to the ZDD in Fig. 1

Fig. 5. A zero-edge tree and a dummy node vector obtained from the ZDD in Fig. 4

Fig. 6. A one-child array obtained from the ZDD in Fig. 4

problems. First, they require too much memory to represent a ZDD. Second, the $member(v, S)$ operation is too slow, needing $\Theta(n)$ time in the worst case. In practice, the size of query sets is usually much smaller than n, and so an $\mathcal{O}(|S|)$ time algorithm is desirable. However it is impossible to attain this in the current implementation [10] because the $member(v, S)$ operation is implemented by using the $zero(v)$ operation repeatedly.

For example, we traverse 0-edges 255 times when we search $S = \{e_1\}$ on the ZDD for $F = \{\{e_1\}, \ldots, \{e_{256}\}\}$. If we translate a ZDD to an equivalent automaton by using an array to store pointers (see Fig. 3), we can execute searching in $\mathcal{O}(|S|)$ time. ZDD nodes correspond to labeled edges in the automaton. However, the size of such automaton via a straightforward translation can be $\Theta(n)$ times larger than the original ZDD [3] in the worst case. Therefore, we want to perform $member(v, S)$ operations in $O(|S|)$ time on ZDDs.

Minato proposed Z-Skip Links [11] to accelerate the traversal of ZDDs of large-scale sparse datasets. His method adds one link per node to skip nodes that are concatenated by 0-edges. Therefore the memory requirement of this augmented data structure cannot be smaller than original ZDDs. Z-Skip-Links make membership operations much faster than using conventional ZDD operations when handling large-scale sparse datasets. However, the computation time is probabilistically analyzed only for average case.

3 Data Structure

In this section, we describe our data structure *DenseZDD* which solves the two problems defined in Section 2.2. We obtain the following results.

Theorem 1. *Let u be the size of the ZDD that removes the zero-suppress rule only for nodes pointed to by 0-edges. A ZDD with m nodes on n items can be stored in $2u + m \log m + 3m + o(u)$ bits so that the primitive operations except getnode(i, v_0, v_1) are done in constant time. In other words, u is the size of the ZDD with dummy nodes that are described below. The getnode(i, v_0, v_1) operation is done in $\mathcal{O}(\log m)$ time.*

Theorem 2. *A ZDD with m nodes on n items can be stored in $\mathcal{O}(m(\log m + \log n))$ bits so that the primitive operations are done in $\mathcal{O}(\log m)$ time except getnode(i, v_0, v_1). The getnode(i, v_0, v_1) operation is done in $\mathcal{O}(\log^2 m)$ time.*

3.1   DenseZDD

A DenseZDD $DZ = \langle U, M, I \rangle$ is composed of three data structures: a zero-edge tree U, a dummy node vector M, and a one-child array I.

Zero-edge tree: The spanning tree of ZDD G formed by the 0-edges is called the *zero-edge tree* of G and denoted by T_Z. In a zero-edge tree, all 0-edges are reversed and the 0-terminal node becomes the root of the tree. The preorder rank of each node is used to identify it. Zero-edge trees are based on the same idea as left or right trees by Maruyama *et al.* [7].

An important difference between our structure and theirs is the existence of *dummy nodes*. We call nodes in the original ZDD as *real nodes*. We use the zero-edge tree with dummy nodes, denoted by T'_Z. We create dummy nodes on each 0-edge to guarantee that the depth of every real node v in the zero-edge tree equals $index(v)$. We define the depth of the 0-terminal node, the root of this tree, to be 0. Let U be the BP of T'_Z. The length of U is $\mathcal{O}(mn)$ because we create $n - 1$ dummy nodes for one real node in the worst case. An example of a zero-edge tree and its BP are shown in Fig. 5. Black circles are dummy nodes and the number next to each node is its preorder rank. The 0-terminal node is ignored in the BP because we know the root of a zero-edge tree is always that node.

Dummy node vector: A bit vector of the same length as U is used to distinguish dummy nodes and real nodes. We call it the *dummy node vector* of T'_Z and denote it by B_D. The i-th bit is 1 if and only if the i-th parenthesis of U is '(' and its corresponding node is a real node in T'_Z. An example of a dummy node vector is also shown in Fig. 5. The 0-terminal node is also ignored. Let the FID of B_D be M. Using M, we can determine whether a node is dummy or real, and compute preorder ranks among only real nodes. We can also obtain positions of real nodes on BP from their preorder ranks by the select operation on M.

One-child array: An integer array to represent the 1-child of each node is called the *one-child array* and denoted by C_O. This array contains node preorder ranks of all 1-children in preorder on T_Z. That is, its i-th element is the preorder rank of the 1-child of the nonterminal node whose preorder rank is i. We also require one bit for each element of the one-child array to store the $\emptyset$-flag. If $empflag(v) = 1$ for a nonterminal node v, the corresponding element in the one-child array will be negative. An example of a one-child array is shown in Fig. 6.

Let I be the compressed representation of C_O. In I, one integer is represented by $\lceil \log(m + 1) \rceil + 1$ bits, including one bit for the $\emptyset$-flag.

4 Algorithm

4.1 Conversion of an Ordinary ZDD to a DenseZDD

We show how to construct the DenseZDD. We first build the zero-edge tree from the given ZDD. A pseudo-code is shown in Appendix C.3 in our report [2]. The zero-edge tree consists of all 0-edges of the ZDD, with their directions being reversed. For a nonterminal node v, we say that v is a 0^r-child of $zero(v)$. To make a zero-edge, we use a list $revzero$ in each node, which stores 0^r-children of the node. The lists for all the nodes are computed by a depth-first traversal of the ZDD. This is done in $\mathcal{O}(m)$ time and $\mathcal{O}(m)$ space, since each node is visited at most twice and the total size of $revzero$ is the same as the number of nonterminal nodes.

We obtained a zero-edge tree T, but it is not an ordered tree. We define pre-order rank $prank(v)$ for every node v before traversal. The nodes in $revzero$ are sorted in descending order of their pairs $\langle index, prank \rangle$, that is, $index(revzero[i]) \geq index(revzero[i + 1])$ for $1 \leq i < |revzero(v)|$. Then, nodes with higher indices are visited first. This ordering is useful to reduce the number of dummy nodes and to implement ZDD operations simply. It seems impossible to define visiting order of nodes by preorder rank of their 1-children during computing preorder, but it is possible. Since a ZDD node v satisfies $index(v) > index(zero(v))$ and $index(v) > index(one(v))$, we can decide $prank$ for every node by the pseudo-code of Appendix C.1 in our report [2], which is a BFS algorithm based on $index$ value starting from 0-terminal. To compute $prank$ efficiently, we construct the temporary BP for the zero-edge tree. Using the BP, we can compute the size of each subtree rooted by v in T in constant time and compact space.

Next, we create dummy nodes imaginarily. For a node v, we create $q = \max\{ i \in \{1, \ldots, n\} \,|\, i = index(revzero[j]) - 1, 1 \leq j \leq |revzero(v)| \}$ dummy nodes $d_1, \ldots, d_q$ such that $triple(d_i) = \langle index(v) + i, d_{i-1}, \mathbf{0} \rangle$, and $empflag(d_i) = 0, 1 \leq i \leq q$. For convention, d_0 denotes v.

To sum up, the DenseZDD for the given ZDD is composed of the zero-edge tree, the one-child array, and the dummy node vector. We traverse the zero-edge tree in DFS order as if dummy nodes exist and construct the BP representation U, the dummy node vector M, and the one-child array I. The BP and dummy node vector are constructed for the zero-edge tree with dummy nodes. On the other hand, the one-child array ignores dummy nodes. DenseZDD $DZ = \langle U, M, I \rangle$ is obtained. Pseudo-codes are also given in Appendix C.2 and C.3 in our report [2].

4.2 Primitive ZDD Operations

We show how to implement primitive ZDD operations on DenseZDD $DZ = \langle U, M, I \rangle$ except $getnode$. We give an algorithm for $getnode$ in Section 5.

In the zero-edge tree, there are two types of nodes: real nodes and dummy nodes. Real nodes are those in the ZDD, while dummy nodes have no corresponding ZDD nodes. Real nodes are numbered from 1 to m based on preorders in the tree. Below a node is identified with this number, which we call its *node number*. We can convert between the node number i of a node and the position p in the BP sequence U by $p := select_1(M, i)$ and $i := rank_1(M, p)$. The 0-terminal has node number 0 and nonterminal nodes have positive node numbers. If a node number of a negative value is used, it means a node with an $\emptyset$-flag.

In addition, we consider an additional primitive operation for DenseZDDs: $chkdum(p)$. This operation checks if a node at position p on U is a dummy node or not. If it is a dummy $chkdum$ returns false; otherwise it returns true. This operation is implemented by simply looking at the p-th bit of M. If the bit is 0, then the node is dummy; otherwise it is a real node.

$index(i)$: Since the item of the node is the same as the depth of the node, we can obtain $index(i) := depth(U, select_1(M, i))$.

$one(i)$: Because 1-children are stored in preorder of the parents of nodes, we can obtain $one(i) := I[i]$.

$topset(i, d)$: The node $topset(i, d)$ is the ancestor of node i in the zero-edge tree with index d. A naive solution is to iteratively climb up the zero-edge tree from node i until we reach a node with index d. However, as shown above, the index of a node is identical to its depth. By using the power of the succinct tree data structure, we can directly find the answer by $topset(i, d) := rank_1(M, level\_ancestor(U, select_1(M, i), d))$.

$zero(i)$: Implementing the $zero$ operation requires a more complicated technique. Consider a subtree T of the zero-edge tree consisting of the node i, its real parent node r, all real children of r, and dummy nodes between those nodes. As a pre-condition, the zero-edge tree is constructed by the algorithm of Appendix C.2 of our report [2]. That is, for the children of r, the nodes with higher *index* value have smaller preorder, and the imaginary parents of the children are dummy nodes (or i) that are added on the edge between r and the child having the highest *index* value. Computing $zero(i)$ is equivalent to finding r. Because the children of r are ordered from left to right in descending order of their depths, and dummy nodes are shared as much as possible, the deepest node in T is on the leftmost path from r. Furthermore, the parents of other real children are also on the leftmost path. This property also holds in the original zero-edge tree. The dummy node vector B_D stores flags in the preorder in the zero-edge tree. Then $B_D[p_r] = B_D[p_i] = 1$, where p_r and p_i are positions of nodes r and i in the BP sequence U, and $B_D[j] = 0$ for any $p_r < j < p_i$. Therefore we can find p_r by a rank operation. In summary, $zero(i) := rank_1(M, parent(U, select_1(M, i)))$.

4.3   Compressing the Balanced Parentheses Sequence

The balanced parentheses sequence U is of length $2u$, where u is the number of nodes including dummy nodes. Let a ZDD have m real nodes and the number of items be n, u is mn in the worst case. Here we compress the BP sequence U.

The BP sequence U consists of at most $2m$ runs of identical symbols. To see this, consider the substring of U between the positions for two real nodes. There is a run ')))...' followed by a run '(((...' in the substring. We encode lengths of those runs using some integer encoding scheme such as the delta-code or the gamma-code [4]. An integer $x > 0$ is encoded in $\mathcal{O}(\log x)$ bits. Because the maximum length of a run is n, U can be encoded in $\mathcal{O}(m \log n)$ bits. The range min-max tree of U has $2m/\log m$ leaves. Each leaf corresponds to a substring of U that contains $\log m$ runs. Then any tree operation can be done in $\mathcal{O}(\log m)$ time. The range min-max tree is stored in $\mathcal{O}(m(\log n + \log m)/\log m)$ bits.

We also compress the dummy node vector B_D. Because its length is $2u \leq 2mn$ and there are only m ones, it can be compressed in $m(2 + \log m) + o(u)$ bits by FID. The operations $select_1$ and $rank_1$ take constant time. We can reduce the term $o(u)$ to $o(m)$ by using a sparse array [14]. The operation $select_1$ is done in constant time, while $rank_1$ takes $\mathcal{O}(\log m)$ time. From the discussions above, we can prove Theorem 1 and Theorem 2. For the proof, see Appendix B in our report [2].

5 Hybrid Method

In this section, we show how to implement dynamic operations on DenseZDD. Namely, we need to implement the $getnode(i, v_0, v_1)$ operation. Our approach is to use a hybrid data structure using both the DenseZDD and a conventional dynamic ZDD. Assume that initially all the nodes are in a DenseZDD. Let m_0 be the number of initial nodes. In a dynamic ZDD, the operation $getnode(i, v_0, v_1)$ is implemented by a hash table indexed with the triple $\langle i, v_0, v_1 \rangle$.

We show first how to check whether the node $v := getnode(i, v_0, v_1)$ already exists. That is, we want to find a node v such that $index(v) = i$, $zero(v) = v_0$, $one(v) = v_1$. If v does not exist, we create such a node using the hash table as well as a dynamic ZDD. If it exists, in the zero-edge tree, v is a real child node of v_0. Consider again the subtree of the zero-edge tree rooted at v_0 and having all real children of v_0. All children of v_0 with index i share the common (possible dummy) parent node, say w. Because w is on the leftmost path in the subtree, it is easy to find it. Namely, $w := level\_ancestor(U, select_1(M, rank_1(M, v_0)+1), i)$. The node v is a child of w with $one(v) = v_1$. Because all children of w are sorted in the order of one values by the construction algorithms, we can find v by a binary search. For this, we use $degree$ and $child$ operations on the zero-edge tree.

Theorem 3. *The existence of $getnode(i, v_0, v_1)$ can be checked in $\mathcal{O}(t \log m)$ time, where t is the time complexity of primitive ZDD operations.*

If the BP sequence is not compressed, $getnode$ takes $\mathcal{O}(\log m)$ time. Otherwise it takes $\mathcal{O}(\log^2 m)$ time. We should check the hash table before checking the zero-edge tree if dynamic nodes are already exist. As well as a conventional ZDD, hashing increases constant factors of time bounds significantly and add space bound $\mathcal{O}(x \log x)$ where x is the number of dynamic nodes.

6   Experimental Results

We ran experiments to evaluate the compression, construction, and operation times of DenseZDDs. We implemented the algorithms described in Sec. 3 and 4 in C/C++ languages on top of the SAPPORO BDD package [10]. The package is general implimentation of ZDD with 0-element edges, and uses 30 bytes per ZDD node. We performed experiments on eight quad-core 3.09 GHz AMD Opteron 8393 SE processors and 512 GB DDR2 memory shared among cores running. SUSE 10 Our algorithms use a single core since they are not parallelized.

We show the characteristics of the ZDDs in Table 2. Original ZDDs are denoted by Z. DenseZDDs without/with compression of the balanced parentheses sequences of the zero-edge trees are denoted by DZ/DZ$_c$, respectively.

As real data sets, for $N = 5, 10, 20, 50, 100$, the source ZDD *webviewN* was constructed from the data set BMS-Web-View-2[3] by using mining algorithm LCM over ZDD [13] with minimum support N. For artificial data sets, the ZDD *gridN* represents all self-avoiding paths on an $N \times N$ grid graph from the top

Table 2. Comparison of performance, where δ denotes the dummy node ratio

data set	#items	#nodes	#itemsets	size (bytes)			comp. ratio		δ
			Z	Z	DZ	DZ$_c$	DZ	DZ$_c$	
grid5	40	584	8,512	17,520	2,350	2,196	0.134	0.126	0.28
grid10	180	377,107	4.1×10^{20}	11,313,210	1,347,941	1,265,773	0.119	0.112	0.20
grid15	420	1.5×10^8	2.3×10^{48}	4,342,789,110	678,164,945	647,843,001	0.156	0.149	0.19
webview5	952	2,299	11,928	10,592,760	3,871,679	1,851,889	0.365	0.174	0.93
webview10	1,617	6,060	70,713	281,700	1,034,471	477,299	0.367	0.169	0.93
webview20	2,454	30,413	634,065	912,390	290,661	140,873	0.318	0.154	0.92
webview50	2,905	93,900	4.4×10^6	181,800	44,846	24,596	0.246	0.135	0.88
webview100	3,149	353,092	2.7×10^7	68,970	11,455	7,967	0.166	0.115	0.75
webviewALL	3,149	465,449	3.2×10^7	13,963,470	4,964,303	2,413,625	0.355	0.172	0.92
randjoin128	32,696	6,751	2.5×10^8	202,530	408,149	99,117	2.015	0.489	0.99
randjoin2048	32,768	377,492	1.8×10^{13}	11,324,760	2,415,648	1,511,658	0.213	0.133	0.82
randjoin8192	32,768	1.3×10^6	3.7×10^{15}	38,094,930	5,328,502	4,386,452	0.139	0.115	0.42
randjoin16384	32,768	1.9×10^6	2.8×10^{16}	56,447,280	7,056,418	6,113,910	0.125	0.108	0.14

Table 3. Converting time and random searching time

data set	conversion time (sec)				traverse time (sec)			search time (sec)		
	read	convert	const.	comp.	Z	DZ	DZ$_c$	Z	DZ	DZ$_c$
grid5	0.001	0.001	0.009	0.000	0.000	0.001	0.006	0.029	0.038	0.229
grid10	0.461	0.634	0.449	0.060	0.075	0.247	1.388	0.005	0.013	0.056
grid15	124.887	407.502	112.379	8.186	41.214	102.673	398.397	0.006	0.011	0.064
webview5	0.256	0.690	1.361	0.055	0.066	0.154	0.250	1.966	0.045	0.099
webview10	0.217	0.226	0.564	0.041	0.017	0.042	0.073	1.901	0.043	0.100
webview20	0.066	0.036	0.313	0.022	0.005	0.014	0.027	1.875	0.046	0.101
webview50	0.013	0.019	0.050	0.004	0.002	0.020	0.005	1.314	0.273	0.102
webview100	0.004	0.002	0.017	0.001	0.000	0.004	0.007	0.777	0.129	0.376
webviewALL	0.551	0.927	1.644	0.108	0.091	0.207	0.346	1.706	0.049	0.105
randjoin128	0.004	0.053	0.149	0.008	0.001	0.002	0.003	0.527	0.044	0.095
randjoin2048	0.243	0.742	0.946	0.029	0.093	0.126	0.145	8.071	0.044	0.098
randjoin8192	0.858	2.573	1.259	0.043	0.338	0.240	0.304	15.604	0.039	0.092
randjoin16384	1.270	5.016	1.471	0.070	0.676	0.353	0.447	19.501	0.040	0.093

---

[3] http://fimi.ua.ac.be

left corner to the bottom right corner [6][4]. Finally, *randjoinN* is a ZDD that represents the join $C_1 \sqcup \cdots \sqcup C_4$ of four ZDDs for random families $C_1, \ldots, C_4$ consisting of N sets of size one drawn from the set of $n = 32768$ items.

In Table 2 we show the sizes of the original ZDD, the DenseZDD with/without compression and their compression ratio. We compressed FID for dummy node vector if the dummy node ratio is more than 75%. In almost cases, we observe that the DenseZDD is from 2.5 to 9 times smaller than the original ZDD, and that compressed DenseZDD is from 6 to 9 times smaller than the original ZDD. The compressed DenseZDD is quarter the size of the DenseZDD in the best case, and is half the size of the original ZDD in the worst case. For most of our data sets, the ratio δ of the number of dummy nodes to the size of DenseZDD is roughly 90%, except for *gridN* and *randjoin16384*.

In Table 3, we show the conversion times from ZDDs to DenseZDDs, traversal times, and search times on ZDDs and DenseZDDs. Conversion time is composed of four parts: time to read a file containing a stored ZDD and reconstruct the ZDD, convert it to raw parentheses, bits, and integers, construct succinct representation of them, and compress the BP of the zero-edge tree. The conversion time appears almost linear in the input size showing its scalability for large data. Traverse operation used $zero(v)$ and $one(v)$, while membership operation used $topset(v, i)$ and $one(v)$. We observed that the DenseZDD has almost twice longer traverse time and more than 10 times shorter search time than an original ZDD. These results show the efficiency of our implementation of the $topset(v, i)$ operation on DenseZDD using level-ancestor operations.

From the above results, we conclude that DenseZDDs are more compact than ordinary ZDDs unless the dummy node ratio is extremely high, and the membership operations for DenseZDDs are much faster if the number of items is large or the dummy node ratio is small. We observed that in DenseZDDs, traversal time is approximately double and search time approximately one-tenth compared to the original ZDDs. The traversal is accelerated especially for large-scale sparse datasets because the number of nodes connected by 0-edges grows as large as the index number n. Recently, processing of "Big Data" have attracted a great deal of attention, and we often deal with a large-scale sparse dataset, which has more than ten thousands of items as the columns of a dataset. In the era of Big Data, we expect that DenseZDD will be effective for various real-life applications, such as data mining, system diagnosis, and network analysis.

7   Conclusion

In this paper, we have presented a compressed index for static ZDDs named DenseZDD. We also proposed a hybrid method for dynamic operations on DenseZDD. For future work, the one-child array should be stored in more compact space. Constructing the DenseZDD from the normal ZDD using external memory is an important open problem. We will implement the hybrid method on our ZDD package and convert/update algorithm with less memory. We expect that our technique can be extended to other variants of BDDs.

---

[4] An algorithm animation: http://www.youtube.com/watch?v=Q4gTV4rOzRs

Acknowledgments. The authors would like to thank Prof. R. Grossi, Prof. R. Raman, and Dr. Y. Tabei for their discussions and valuable comments. This work was supported by Grant-in-Aid for JSPS Fellows 25-193700. This research was partly supported by Grant-in-Aid for Scientific Research on Innovative Areas — Exploring the Limits of Computation, MEXT, Japan and ERATO MINATO Discrete Structure Manipulation System Project, JST, Japan. KT is supported by JST CREST and JSPS Kakenhi 25106005. KS is supported by JSPS KAKENHI 23240002.

References

1. Bryant, R.E.: Graph-based algorithms for Boolean function manipulation. IEEE Transactions on Computers C-35(8), 677–691 (1986)
2. Denzumi, S., Kawahara, J., Tsuda, K., Arimura, H., Minato, S., Sadakane, K.: A compact and fast index structure for families of sets. Tech. rep., TCS Technical Report Series A, TCS-TR-A-14-71, Division of Computer Science, Hokkaido University (2014), http://www-alg.ist.hokudai.ac.jp/tra.html
3. Denzumi, S., Yoshinaka, R., Arimura, H., Minato, S.: Notes on sequence binary decision diagrams: Relationship to acyclic automata and complexities of binary set operations. In: Prague Stringology Conference 2011, Prague, pp. 147–161 (2011)
4. Elias, P.: Universal codeword sets and representation of the integers. IEEE Transactions on Information Theory IT-21(2), 194–203 (1975)
5. Hansen, E.R., Rao, S.S., Tiedemann, P.: Compressing binary decision diagrams. In: 18th European Conference on Artificial Intelligence, pp. 799–800 (2008)
6. Knuth, D.E.: Combinatorial Algorithms, part 1, 1st edn. The Art of Computer Programming, vol. 4A. Addison-Wesley Professional, Boston (2011)
7. Maruyama, S., Nakahara, M., Kishiue, N., Sakamoto, H.: ESP-index: A compressed index based on edit-sensitive parsing. Journal of Discrete Algorithms (2013)
8. Mateu-Villarroya, P., Prades-Nebot, J.: Lossless image compression using ordered binary-decision diagrams. Electronics Letters 37, 162–163 (2001)
9. Minato, S.: Zero-suppressed BDDs for set manipulation in combinatorial problems. In: 30th International Design Automation Conference, pp. 272–277 (1993)
10. Minato, S.: SAPPORO BDD package. Division of Computer Science, Hokkaido University (2012) (to be released)
11. Minato, S.: Z-skip-links for fast traversal of zdds representing large-scale sparse datasets. In: Bodlaender, H.L., Italiano, G.F. (eds.) ESA 2013. LNCS, vol. 8125, pp. 731–742. Springer, Heidelberg (2013)
12. Minato, S., Ishiura, N., Yajima, S.: Shared binary decision diagram with attributed edges for efficient Boolean function manipulation. In: 27th International Design Automation Conference, pp. 52–57 (1990)
13. Minato, S., Uno, T., Arimura, H.: LCM over ZBDDs: Fast generation of very large-scale frequent itemsets using a compact graph-based representation. In: Washio, T., Suzuki, E., Ting, K.M., Inokuchi, A. (eds.) PAKDD 2008. LNCS (LNAI), vol. 5012, pp. 234–246. Springer, Heidelberg (2008)
14. Okanohara, D., Sadakane, K.: Practical entropy-compressed rank/select dictionary. In: Ninth Workshop on Algorithm Engineering and Experiments, pp. 60–70 (2007)
15. Starkey, M., Bryant, R.: Using ordered binary-decision diagrams for compressing images and image sequences. Tech. Rep. CMU-CS-95-105, Carnegie Mellon University (1995)

An Evaluation of Dynamic Labeling Schemes for Tree Networks

Noy Rotbart, Marcos Vaz Salles, and Iasonas Zotos

Department of Computer Science, University of Copenhagen (DIKU)
Universitetsparken 5, 2100 Copenhagen, Denmark
{noyro,vmarcos}@diku.dk, tlc736@alumni.ku.dk

Abstract. We present an implementation and evaluation based on simulation of dynamic labeling schemes for tree networks. Two algorithms are studied: a general scheme that converts static labeling schemes to dynamic, and a specialized dynamic distance labeling scheme. Our study shows that theoretical bounds only partially portray the performance of such dynamic labeling schemes in practice. First, we observe order-of-magnitude differences between the gains in label size when compared to the number of messages passed. Second, we demonstrate a significant bottleneck in the tree network, suggesting that the current practice of counting total messages passed in the whole network is insufficient to properly characterize performance of these distributed algorithms. Finally, our experiments provide intuition on the worst case scenarios for the stated algorithms, in particular path tree networks and fully dynamic schemes permitting both node additions and deletions.

1   Introduction

Labeling schemes are methods used to calculate global properties of a graph based solely on local information present at its nodes. They were introduced in a restricted manner by Bruer and Folkman [5], revisited by Kannan, Naor and Rudich [12], and explored by a wealth of subsequent work [2–4, 10, 19, 20]. Due to their main role in the understanding of graph structures and their practical relevance, trees are central in these studies. In more detail, asymptotically tight labeling schemes for trees were found for the functions adjacency [4], sibling [2], ancestry [10], nearest common ancestor (NCA) [3], routing [20], and distance [19]. The papers mentioned aim to find labeling schemes that minimise the maximum label size, with minor attention to efficient encoding and decoding of the labels.

Korman, Peleg and Rodeh [18] studied labeling schemes in the context of distributed systems. In this context, information about labeling is exclusively communicated via messages, and nodes may join or leave the system dynamically. Since communication is required, maximum label size is not the sole criterion for the quality of such labeling schemes, but also metrics related to the number and total size of messages passing in the network. The authors suggested two algorithms that fit this context: a conversion of static labeling schemes, and a

J. Gudmundsson and J. Katajainen (Eds.): SEA 2014, LNCS 8504, pp. 199–210, 2014.

specialized distance labeling scheme. Additional work investigated specialized labeling schemes for routing [15, 16] and various trade-offs for distance [14, 17].

One could attempt to design dynamic labeling schemes without node re-labeling at all. Unfortunately, Cohen, Kaplan and Milo [8] showed that if re-labeling is not permitted, there is a a tight bound of $\Theta(m)$ bit for labels supporting m insertions. The lower bound shows the inapplicability of dynamic labeling schemes without relabeling for most functions.[1] For the above reasons, we focus on dynamic models with re-labeling. In particular, we focus on the function distance, since distance labels are expressive enough to answer all the queries mentioned above.

Previous experimental papers on the topic focused on the performance of static labeling schemes, emphasizing the label size as the main metric. Caminiti et al. [6] experimented on the influence that different tree decompositions have on label size for NCA labeling schemes. Fischer [9] evaluated the label size expected for a variant of NCA, using various coding algorithms. Kaplan et al. [13] presented experiments to support an ancestry labeling scheme trading slightly larger labels for much faster query time. Finally, Cohen et. al [7] performed experiments on a novel technique for labeling schemes for graphs.

These papers reveal that different families of trees present very different label sizes. A natural question is whether dynamic labeling schemes present the same behavior. In addition, it remains unclear how the various communication and label size trade-offs suggested for dynamic labeling schemes behave in practice. Moreover, one may question whether it is sensible to develop dynamic labeling schemes for individual functions, or whether the trade-offs given by general methods are good enough [18]. Furthermore, the price of allowing for deletions, rather than just insertions, still needs to be characterized in representative scenarios. In short, many questions are relevant and interesting for dynamic labeling schemes, but we are unaware of any experimental papers on the topic.

In this paper, we devise a simulation of a tree network and answer those questions for the dynamic labeling schemes proposed by Korman, Peleg and Rodeh [18]. These labeling schemes can be classified along two dimensions: whether the labeling scheme is specialized for distance or supports general static labelers, and whether the labeling scheme supports both insertion and deletions or only insertions. Our findings suggest three main observations:

1. Generally, the amortized complexity analysis of [18] is verified in our experiments. However, for labeling schemes supporting general static labelers, the observed behavior indicates that the complexity of the static labeler is overshadowed by the overhead introduced by the dynamic scheme.

2. In order to achieve an asymptotically tight label size, the specialized distance labeling scheme uses significantly more communication.

3. The message distribution in labeling schemes that support both insertion and deletion is severely skewed. We show an experiment in which the majority of messages in the system is communicated by less than 0.4% of the nodes.

---

[1] More specifically: NCA, routing, distance, flow and center. In contrast, siblings, adjacency and connectivity enjoy simple dynamic labeling schemes of size $2 \log n$.

1.1   Preliminaries

In the remainder of the paper, we call the collection of all n-node unweighted trees $Trees(n)$. We assume trees to be rooted, and denote $\log_2 n$ as $\log n$. We adopt the terminology in [18], which names dynamic encoding algorithms as *distributed online protocols* or simply *protocols*. From here on, all dynamic labeling schemes discussed are for tree networks. The following types of topology changes are considered: *a) Add-Leaf*, where a new degree-one node u is added as a child of an existing node v; and *b) Remove-Leaf*, where a (non-root) leaf v is deleted. Dynamic labeling schemes that support only *Add-Leaf* are denoted *semi-dynamic*, and those that support both operations are denoted *fully-dynamic*.

Metrics for Dynamic Labeling Schemes. In the literature surveyed, labeling schemes aim for smallest possible label. If communication is not accounted for, dynamic labeling schemes can trivially achieve the optimal, static bounds. Therefore, in order to account for the cost of communication, the following metrics are introduced [18]. Let M be a dynamic labeling scheme, with re-labeling allowed, and let $S(n)$ be a sequence of n topological changes.

1. *Label Size $\mathcal{LS}(M, n)$*: the maximum size of a label assigned by M on $S(n)$.
2. *Message Complexity, $\mathcal{MC}(M, n)$*: the maximum number of messages sent in total by M on $S(n)$. Messages are sent exclusively between adjacent nodes.
3. *Bit Complexity, $\mathcal{BC}(M, n)$*: the maximum number of bits sent in total by M on $S(n)$. Note that $\mathcal{BC}(M, n) \leq \mathcal{MC}(M, n) \cdot \mathcal{LS}(M, n)$.

2   Dynamic Labeling Schemes for Tree Networks

Korman et al. [18] presented *semi-dynamic* and *fully-dynamic* labeling schemes that support distance queries and a *semi-dynamic* and *fully-dynamic* labeling scheme that receives a static labeling scheme[2] as input and supports queries of its type. In this section, we give a brief and informal description of both semi-dynamic labeling schemes as well as of the conversion necessary to make these schemes fully-dynamic.

We denote the static distance labeling scheme as *Distance*, its extension to a specialized semi-dynamic mode *SemDL*, and its fully-dynamic specialized extension as *DL*. Korman et al. [18] define the general dynamic labeling schemes *SemGL* and *GL*, which maintain labels on each node of a dynamically changing tree network using a static labeling scheme as a subroutine. The performance of these schemes is tightly coupled to the performance of the static scheme used. Throughout the remainder, we denote the general labeling schemes operating on *Distance* simply as *SemGL* and *GL*, and explicitly mention other distance functions where appropriate. Table 1 reports the different complexities for the aforementioned schemes (the parameter d is explained in Sec. 2.1).

---

[2] The static labeling scheme must respect a set of conditions as mentioned in [18], which to the best of our knowledge are respected by all labeling schemes in the literature.

Table 1. Simplified complexity estimates for $Trees(n)$. Bounds for GL are reported with $d = 2$. See [18] for an elaborate complexity report.

Labeling Scheme	Label Size	$\mathcal{MC}$	$\mathcal{BC}$
$Distance$	$O(\log^2 n)$	$O(n)$	$O(n \log^2 n)$
$SemDL$	$O(\log^2 n)$	$O(n \log^2 n)$	$O(n \log^2 n \log \log n)$
DL	$O(\log^2 n)$	$O(n \log^2 n)$	$O(n \log^2 n \log \log n)$
$SemGL$	$O(\frac{d-1}{\log d} \log^3 n)$	$O(n \frac{\log n}{\log d})$	unreported
GL	$O(\log^3 n)$	$O(n \log^2 n)$	unreported

2.1   Brief Overview

In this section, we provide an informal description for *Distance*, *SemDL*, *SemGL*, and the approach for semi-dynamic to dynamic conversion.

Encoding *Distance* Directly from Heavy-Light Decomposition. For every non-leaf v in a rooted tree T, we denote a child with the heaviest weight[3] as *heavy* and the rest as *light*; we mark the root of the tree as light and its leaves as heavy. The light nodes divide T into disjoint *heavy paths*. For any $v \in T$, the path from the root traverses at most $\log n$ light nodes, and accordingly at most $\log n$ heavy paths. A label of a node can now be defined by interleaving *light sub-labels*, containing the index of a light child, with *heavy sub-labels*, containing only the length of the heavy path. Such labels clearly require $O(\log^2 n)$ bits, which is asymptotically optimal for labels supporting distance on trees. Computing the distance between two nodes can be done by comparing the prefixes of their corresponding labels. Finally, we note that the labels produced by *Distance* can be used to solve most queries in the literature, namely adjacency, sibling, ancestry, routing and NCA.

SemDL. It is not essential for the correctness of the decomposition that the heavy node selected be in fact the heaviest, or say, the 3rd heaviest. It is only essential for the bound on the label size. *semDL* maintains a dynamic version of the *Distance* labels using precisely this observation. Every node maintains an estimate of its weight, and transfers this estimate to its parent, using a previously introduced binning method [1]. When the estimate exceeds a threshold in a node, this implies that a node other than the heavy node is now significantly heavier than the heavy node. Thus, a re-labeling (shuffle, in [18]) is instantiated on the subtree to maintain the $O(\log^2 n)$ label size.

SemGL. The protocol is designed to transform any static labeling scheme *Static* to semi-dynamic, and therefore, does not utilize the heavy-light decomposition directly. Instead, the protocol operates *Static* separately on a cleverly constructed sets of so called bubbles described as follows.

Each node $v \in T$ is included in exactly one induced subtree of T, denoted *bubble* of *order* i ($0 \leq i \leq \log_d n$) that contains at least d^i nodes. The bubbles

---

[3] The weight of a node v is the number of its descendants in a tree.

constitute a *bubble tree*, where the order of a bubble is always less or equal to the order of the parent bubble. In addition, there are no d consecutive bubbles of the same order in any path of the bubble tree. A node added to the tree is assigned the order 0. If this insertion yields d consecutive bubbles of order 0 in the bubble tree, they are merged to a single bubble of order 1, and the condition is checked again for bubbles of order 1 and so on. We denote the parent of the root of bubble b as b_p and the function that *Static* supports as f. Essentially, the label of a node v in a bubble b is a concatenation of the label of b_p with both a local *Static* label of v in b, and the result of $f(v, b_p)$.

Semi-dynamic to Fully-Dynamic Conversion. Both labeling schemes are converted to their counterpart fully-dynamic labeling schemes, *DL* and GL, using an additional simple protocol. The protocol uses the binning method [1] to maintain local weight estimates for each node. The protocol simply ignores the topological changes up to the point in which their number is large, and then re-labels the entire tree.

3 Experimental Framework

We have used simulation as our performance measurement methodology [11]. There are several reasons for this choice. First, distributed systems of realistic size in the order of millions of nodes are unavailable to us. Second, simulation allows us to remain isolated from effects such as network interference. Third, simulation is a sufficient tool to compare the metrics intended for our purposes. All implementations have been realized in C#. The full package is available over the Internet at the URL: `http://www.diku.dk/~noyro/dynamic-labeling.zip`.

Performance Indicators. In order to match the analysis, we use the metrics defined above, namely $\mathcal{MC}$, $\mathcal{BC}$, and $\mathcal{LS}$, denoted *total network messages*, *total network bandwidth*, and *maximum label length* resp. In particular, we are interested in the tradeoff between $\mathcal{LS}$ and $\mathcal{MC}$ in the different systems. Even though we report the bit complexity, message sizes in our setting are bounded by the label sizes, which tend to be very small. Since in networks the start up costs of sending small messages dominate messaging cost, it is sufficient to focus on message complexity alone. In addition, we study the distribution of messages passing through a single node. The latter provides insight into the congestion for worst-case network performance.

The algorithms present slightly improved bounds if the number of topological events n is known in advance. We simulate labeling schemes that are not aware of n. It is beyond the scope of our paper to account for neither the construction time, nor the performance of the decoder.

Selected Static Functions. To test *SemGL* and *GL*, we provide, in addition to *Distance*, the static labeling scheme *Ancestry* [12] with labels of size $2 \log n$.

Test Sets. We consider insertions sequences creating the following trees.
- **Complete K-ary Trees.** K-ary trees are trees that have a maximum number of children allowed per node. A complete K-ary tree is a K-ary tree in

which all leaf nodes reside on the lowest level and all other nodes have K children. The leaves are inserted starting from the leftmost nodes of the tree. We denote a 2-ary tree as *full binary tree*.

- **Skewed Trees: Star and Path.** A star is a rooted tree with children adjacent to the root. In a path tree, all nodes have either one or no children.
- **Random Trees.** We implemented random insertion sequences creating trees of bounded depth δ as well as of unbounded depth. While generating trees that are truly random is challenging, we follow a simple iterative procedure. First, select a node with equal probability among the nodes with depth less than δ. Then, insert a new node under the selected node and iterate.

From [6, 9], we conclude that labels constructed using a heavy-light decomposition yield the largest label size when applied on full binary trees. Since we would like to focus on the performance of the dynamic protocols, we use full binary trees as our main test set.

Counted Bits. The labels constructed have polylogarithmic size, and each is composed of a number of components of variable sizes. Therefore, we must account for a realistic bit encoding. Bit accounting matters for both $\mathcal{LS}$ and $\mathcal{BC}$. All labels are accounted by twice their theoretical label size, so that the encoder could find the beginning and the end of each of the components. The same applies for messages. We use unary encoding to represent the number of non-empty elements in each row of the label matrix [18].

4   Experimental Results

In this section, we summarize our main experimental findings. First, we test *semGL* with two static labeling schemes and outline the role d plays in the protocol (Section 4.1). We then proceed by directly comparing *SemGL* to *SemDL* (Section 4.2). We conclude by observing the performance of *GL* in comparison to *SemGL* and *DL*, as well as comparing the distribution of messages in *GL* and *SemGL* (Section 4.3). All figures presented but Figures 3 and 7 have the number of node additions in their x-axes in a log-scale. Figures describing $\mathcal{MC}$ and $\mathcal{BC}$ have their y-axes in log-scale, and figures describing $\mathcal{LS}$ are in linear scale.

4.1   *SemGL*

We discuss two experiments dealing exclusively with the performance of *semGL*.

Varying Tree Type. For the purposes of this section, we compare *SemGL* under two static labeling schemes, namely *Distance* and *Ancestry*. The test is performed on eight different, deterministically constructed trees, where the full K-ary trees are expanded either in a *breadth-first* or a *depth-first* manner. We fix $d = 4$, as it best illustrates the differences observed.

Overall, we observe similar trends for both static labeling schemes (Figures 1a, 1c, and 1e vs. 1b, 1d, and 1f). However, we observe an apparent anomaly on the largest label sizes produced in Figure 1b. While in general labels for *Ancestry* should be more compact than labels for *Distance*, the opposite effect is evident

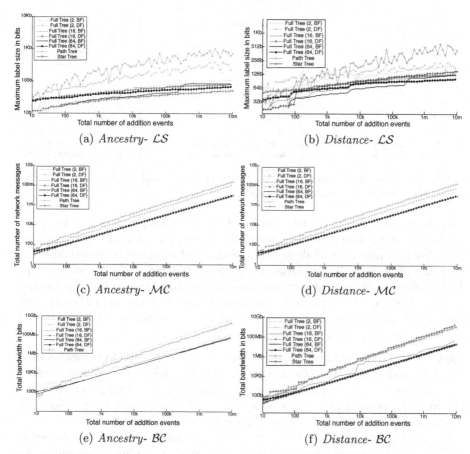

Fig. 1. *SemGL* on *Distance* and *Ancestry* static labeling schemes, for various trees of different order and node insertion modes, with $d = 4$

for paths. This effect can be explained when we recall that on a path, the label size of *Distance* is a straightforward $\log n$ bits, whereas the label size of *Ancestry* is fixed to $2 \log n$.

We observe that the maximum label size of *SemGL* is directly related to the depth of the tree, such that trees with higher depth yield larger labels. In addition, trees with higher depth have more variance in their maximum label size. Both of these properties can be attributed to the *bubble tree* structure. Since the bubble order directly relates to the maximum node depth, higher depth leads to larger *bubble order*, which yields a larger label size. The second property, namely higher variance of maximum label size in trees of higher depth, is attributed to more frequent high-level relabeling. At the noticeable decline points, the encoder relabels large parts of the tree to ensure the logarithmic label size.

A clear evidence for the correlation of label size to the depth of the tree is that for random trees of bounded and unbounded depth, the label size grows

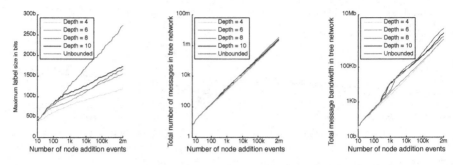

Fig. 2. *SemGL* labels for random trees with variable bounded depth. Each curve represents the average of 10 executions, with $d = 4$.

consistently with depth, as shown in Figure 2. Comparing Figure 2 to Figure 1, the values for random trees lie between the values for path and the remaining trees. We also note that while the total number of messages is essentially identical, the bit complexity is higher as the depth is higher. We deduce that the depth influences the size of each message, and not just the number of messages.

Given these results, in the rest of the experiments we focus on full binary trees traversed in depth-first manner. We have observed in separate experiments that these trees generate similar trends as random trees.

Varying the d Variable. In Figure 3, we vary the value of d in values between 2 and 16, and analyze its influence on *SemGL*. We choose a set of low d values to trigger larger overhead on the message complexity of *SemGL*, such that performance changes are more noticeable. The results are similar for *SemGL* with the *Ancestry* static labeling scheme, and thus omitted from the figure.

We observe two trends in the figures. First, the number of messages passed in the network drops logarithmically as we increase d. The same drop can be observed for the total bandwidth. In addition, the label size increases as d increases. This trend is in line with the expected $\frac{d-1}{\log d}$ expression reported in Table 1.

In addition, we notice a decay in number of messages in Figure 3 which is also in line with the expected $O((\log d)^{-1})$ curve for $\mathcal{MC}(SemGL, n)$. The total bandwidth consumed by tree network reacts to d in the same manner that the total number of messages does. The values stabilize when $d = 12$ for approx. 5 m (million) nodes and 11 for 1 m nodes. Since both complexities are showing similar behavior throughout all experiments, we omit graphs describing the total bandwidth in the remainder for brevity.

4.2   Comparison of *SemGL* and *SemDL*

Figure 4a shows as expected that *SemGL* produces larger labels than *semDL*. Recall that by Table 1 we expect a log factor in the label size between both methods. For trees with 4m nodes, the latter translates into a factor of 20, in contrast to the factor of four seen in the figure. On the other hand, Figure 4b shows that the distance-specialized *semDL* uses up to 270 times more messages

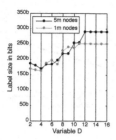
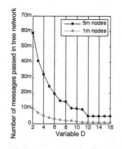
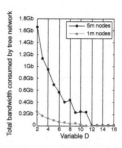

Fig. 3. *SemGL* Variable d experiment for full binary trees expanded in a depth-first manner. Samples are taken for d in [2,16] in trees of size 1m and 5m nodes.

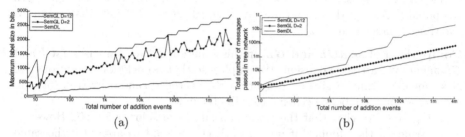

Fig. 4. *SemGL* versus *SemDL* on full binary trees expanded in a depth-first manner: (a) the maximum label size; (b) the number of messages sent

in comparison to *SemGL*, when we would again by Table 1 expect a reverse multiplicative log factor[4]. This suggests that the trade-off between label size and communication expected by the reported complexities is in practice overshadowed by significant constants. The ratio of the factor differences amounts to two orders of magnitude.

Comparing Figures 4a and 4b, *semDL* presents a stable increase in the label size in contrast to the two bumps in its message size. The latter is in contrast to the previously explained phenomena in *SemGL*. The two bumps mark a re-label operation done from the root, and contribute to almost an order-of-magnitude difference in the number of messages passed.

4.3   Fully-Dynamic Labeling Schemes

In this section, we analyze the performance of fully-dynamic protocols, and the performance differences between semi-dynamic and dynamic labeling schemes. We use a full binary tree, generated in a depth-first manner.

Comparison of *DL* and *GL*. In this experiment, we compare *DL* and *GL* for tree expansion, i.e., additions of nodes to the tree. We remark that *SemGL* exhibited a definite advantage in $\mathcal{MC}$ over *semDL* (Section 4.2). Similarly, we note in Figures 5a and 5b that *GL* performs significantly better than *DL* in terms

---

[4] Note that d is a low constant in our experiment.

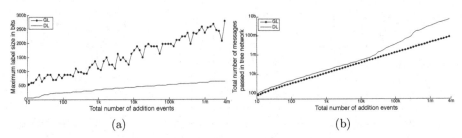

(a)   (b)

Fig. 5. Comparison of *DL* and *GL* (a) is the maximum label size measured in bits and (b) is the total number of messages passed

of network messages. For 4m nodes, *GL* uses roughly 100 times less messages compared to *DL*. At the same time, as expected *DL* yields labels that are up to five times smaller than *GL*.

Comparison of *semGL* and *GL*. In this experiment, we compare *semGL* and *GL* again over a tree expansion. We emphasise that no deletion operations are performed. We aim to give an insight on the overhead caused by transforming the semi-dynamic labeling scheme to fully-dynamic.

In Figure 6, we view that the maximum label size is lower for *GL*. However, as we scale on the number of nodes, both curves tend to approach the same value. *GL*'s number of messages is, however, 100 times larger than *semGL* for 4m nodes. This constitutes a significant overhead in the execution of the fully-dynamic labeling scheme. For perspective, 4.3m messages suffice to label 4m nodes using *semGL*, but only 0.23m nodes using *GL*.

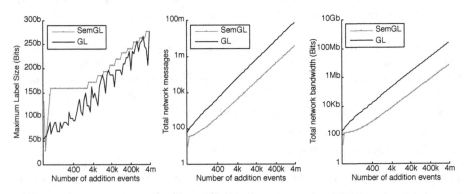

Fig. 6. Comparison of *semGL* ($d = 12$) and *GL*

Message Distribution. We group the nodes according to the number of messages that pass through them throughout the *GL* protocol. Table 2 shows that the distribution of messages is remarkably skewed. Interestingly, 0.00003% of the nodes account for 21.55% of the total number of messages passed. Each of the top 30 nodes in the distribution have either sent or received between 50001 − 500000

Table 2. The distribution of messages per node for *GL*, with $d = 14$

messages (range)	nodes	Percent of nodes	Percent of messages
0 - 5	500018	50.00%	2.6%
5-50	468725	46.87%	23.4%
51-500	27366	2.74%	15.9%
501-5000	3646	0.36%	20.86%
5001-50000	215	0.02%	15.66%
50001-500000	30	0.00003%	21.55%

messages. The top node, in particular, has either sent or received about 0.5m out of a total of 19m messages. We have verified that the top node in the distribution corresponds to the root, and the top 30 nodes are the shallowest nodes in the tree.

In contrast, Figure 7 shows the distribution of messages for *semGL* on a worst case path network with 1m nodes. *semGL* does not incur similar bottlenecks as *GL*. Instead, the figure shows a distribution that resembles a bell curve: Most nodes either send or receive around 20 messages, and the differences in number of messages among nodes are not stark.

This result suggests that *GL* creates a network bottleneck by concentrating the messages passed. This behavior is undesirable in a distributed system.

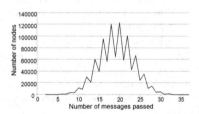

Fig. 7. The distribution of messages per node for *semGL*, with $d = 2$ on a path

5   Conclusions

Overall, the algorithms perform as expected given the amortized complexity analysis. However, the experiments reveal constant factors that clearly affect performance. We observed that fully-dynamic protocols induce a severe performance overhead when compared to semi-dynamic variants. Moreover, as seen in the distribution of messages passed, fully-dynamic protocols also lead to severe bottlenecks in the tree network.

We affirm Korman et al.'s recommendations, namely to reset the d parameter of *GL* on a restart, and for *semGL*, to predict, if plausible, the target tree length n_f in order to set d to a value close to $\log n_f$. On the other hand, Korman et al. argue for trading network communication for improved label size in *DL* [18].

In light of our results, we suggest that in practice, the opposite trade-off is required, namely larger labeling schemes with decreased message complexity.

Korman et al. have developed *DL* with a metric that accumulates network messages, amortizing this value over a number of insertions. Our experiments show that such a metric allows for major bottlenecks and non-distributed system in practice. We suggest that a revised metric that accounts for the total number of messages passing through a single node be investigated as part of future work.

References

[1] Afek, Y., Awerbuch, B., Plotkin, S., Saks, M.: Local management of a global resource in a communication network. JACM 43(1), 1–19 (1996)
[2] Alstrup, S., Bille, P., Rauhe, T.: Labeling schemes for small distances in trees. SIAM J. Discret. Math. 19(2), 448–462 (2005)
[3] Alstrup, S., Halvorsen, E.B., Larsen, K.G.: Near-optimal labeling schemes for nearest common ancestors. arXiv preprint arXiv:1312.4413 (2013)
[4] Alstrup, S., Rauhe, T.: Small induced-universal graphs and compact implicit graph representations. In: FOCS 2002, pp. 53–62 (2002)
[5] Breuer, M.A., Folkman, J.: An unexpected result in coding the vertices of a graph. J. Mathematical Analysis and Applications 20, 583–600 (1967)
[6] Caminiti, S., Finocchi, I., Petreschi, R.: Engineering tree labeling schemes: A case study on least common ancestors. In: Halperin, D., Mehlhorn, K. (eds.) ESA 2008. LNCS, vol. 5193, pp. 234–245. Springer, Heidelberg (2008)
[7] Cohen, E., Halperin, E., Kaplan, H., Zwick, U.: Reachability and distance queries via 2-hop labels. SIAM J. Comp. 32(5), 1338–1355 (2003)
[8] Cohen, E., Kaplan, H., Milo, T.: Labeling dynamic xml trees. SIAM Journal on Computing 39(5), 2048–2074 (2010)
[9] Fischer, J.: Short labels for lowest common ancestors in trees. In: Fiat, A., Sanders, P. (eds.) ESA 2009. LNCS, vol. 5757, pp. 752–763. Springer, Heidelberg (2009)
[10] Fraigniaud, P., Korman, A.: An optimal ancestry scheme and small universal posets. In: STOC 2010, pp. 611–620 (2010)
[11] Jain, R.: The art of computer systems performance analysis, vol. 182. John Wiley & Sons, Chichester (1991)
[12] Kannan, S., Naor, M., Rudich, S.: Implicit representation of graphs. SIAM Journal on Discrete Mathematics, 334–343 (1992)
[13] Kaplan, H., Milo, T., Shabo, R.: A comparison of labeling schemes for ancestor queries. In: SODA 2002, pp. 954–963 (2002)
[14] Korman, A.: General compact labeling schemes for dynamic trees. Distributed Computing 20(3), 179–193 (2007)
[15] Korman, A.: Improved compact routing schemes for dynamic trees. In: PODC 2008, pp. 185–194. ACM (2008)
[16] Korman, A.: Compact routing schemes for dynamic trees in the fixed port model. In: Distributed Computing and Networking, pp. 218–229 (2009)
[17] Korman, A., Peleg, D.: Labeling schemes for weighted dynamic trees. Information and Computation 205(12), 1721–1740 (2007)
[18] Korman, A., Peleg, D., Rodeh, Y.: Labeling schemes for dynamic tree networks. Theory of Computing Systems 37(1), 49–75 (2004)
[19] Peleg, D.: Proximity-preserving labeling schemes. Journal of Graph Theory 33(3), 167–176 (2000)
[20] Thorup, M., Zwick, U.: Compact routing schemes. In: SPAA 2001, pp. 1–10 (2001)

Engineering Color Barcode Algorithms for Mobile Applications*

Donatella Firmani, Giuseppe F. Italiano, and Marco Querini

University of Rome "Tor Vergata", Via Politecnico 1, 00133, Rome, Italy
firmani@ing.uniroma2.it, {italiano,querini}@disp.uniroma2.it

Abstract. The wide availability of on-board cameras in mobile devices and the increasing demand for higher capacity have recently sparked many new color barcode designs. Unfortunately, color barcodes are much more prone to errors than black and white barcodes, due to the chromatic distortions introduced in the printing and scanning process. This is a severe limitation: the higher the expected error rate, the more redundancy is needed for error correction (in order to avoid failures in barcode reading), and thus the lower the actual capacity achieved. Motivated by this, we design, engineer and experiment algorithms for decoding color barcodes with high accuracy. Besides tackling the general trade-off between error correction and data density, we address challenges that are specific to mobile scenarios and that make the problem much more complicated in practice. In particular, correcting chromatic distortions for barcode pictures taken from phone cameras appears to be a great challenge, since pictures taken from phone cameras present a very large variation in light conditions. We propose a new barcode decoding algorithm based on graph drawing methods, which is able to run in few seconds even on low-end computer architectures and to achieve nonetheless high accuracy in the recognition phase. The main idea of our algorithm is to perform color classification using force-directed graph drawing methods: barcode elements which are very close in color will attract each other, while elements that are very far will repulse each other.

Keywords: Graph Drawing, Color Barcodes, Color classification.

1 Introduction

Barcodes are optical machine-readable representations of data, capable of storing digital information. They are currently deployed in many scenarios and are typically printed and scanned by several devices, including desktop scanners and mobile devices. This process is often modeled by a noisy printing and scanning (print-scan) channel. Due to the noise, interference and distortion introduced by this channel, a scanned barcode image may be corrupted, and thus algorithms for

---

* This paper has been partially supported by MIUR, the Italian Ministry of Education, University and Research, under Project AMANDA (Algorithmics for MAssive and Networked DAta).

J. Gudmundsson and J. Katajainen (Eds.): SEA 2014, LNCS 8504, pp. 211–222, 2014.

decoding barcodes must necessarily be able to cope with errors. This is accomplished by adding redundancy, which is a viable method to increase reliability in a noisy channel at the price of a reduction of the information rate. In such a scenario, the trade-off between reliability and data density of barcodes is a significant design consideration: the larger is the expected number of errors to be tolerated, the larger is the amount of redundancy needed in the barcode, and the smaller is the resulting barcode data density.

In order to increase the data density, several 2-dimensional (2D) barcodes have been introduced, such as the Aztec and QR (Quick Response) codes, which are gaining enormous popularity in a wide range of applications. In many cases, 2D barcodes are captured with digital cameras embedded in smartphones and portable devices, which are inherently capable of capturing color information. Both the wide availability of on-board cameras in mobile devices and the increasing demand for higher density barcodes have thus motivated the need for 2D color barcodes [2,3,5,7]. Color barcodes generate each module of the data area with a color selected from 2^k different colors (e.g., 4-color barcodes encode 2 bits per module and 8-color barcodes encode 3 bits per module), where a module (or cell) is the atomic information unit of a 2D barcode. In theory, the data density of a color barcode with 4 (respectively 8) colors can be twice (respectively three times) larger than the data density of the corresponding black and white barcode. In practice, the actual data density of color barcodes is substantially limited by the redundancy added for error correction, due to distortions introduced by the print-scan channel. In particular, colors are more sensitive to chromatic distortions, and thus the error rate of color barcodes may be significantly larger than the error rate of black and white barcodes, all other conditions being equal (i.e., with all barcodes being generated, printed and scanned under same conditions, such as module size, amount of redundancy, printing and scanning resolution and devices). The larger error rates typical in color barcodes could be mitigated by the use of a larger amount of redundancy in the coding. Unfortunately, this approach reduces correspondingly the data density potentially offered by color barcodes, thus weakening substantially their benefits.

In this paper we investigate the applicability of 2D color barcodes to mobile applications. Besides addressing the general trade-off between error correction and data density, there are challenges that are specific to mobile scenarios and that make the problem much more complicated than in traditional desktop environments. Indeed, traditional desktop scanners operate typically under homogeneous light-controlled conditions, while mobile phone cameras induce a strongly non-uniform illumination to the barcode to be decoded, thus introducing more noise and distortion in their print-scan channel. This would motivate the need for more sophisticated decoding algorithms. Unfortunately, in mobile devices computational resources and performance are critical issues, which makes the usage of more sophisticated algorithms largely unpractical. In other words, the main challenge here is to design simple decoding algorithms, which can be implemented and run in few seconds on a low-end computer architecture (such as a smartphone), but are still capable of achieving high decoding accuracy: this is particularly important to decrease the error rates and to consequently reduce the redundancy needed for error correction, so as to fully exploit the full potential in data density offered by the usage of colors.

The main contributions of this paper are the following. First, we perform a thorough experimental study of several known color classifiers and clusterers, in order to identify the most promising approaches for decoding color barcodes in mobile scenarios. Our experiments show that simpler methods are much faster in practice, but appear not to be effective in decoding accuracy (in terms of error rates). On the opposite side, more sophisticated methods are more effective in accuracy, but require higher computational costs and decoding times, and thus are not practical enough to be implemented on low-end computing platforms. Motivated by this, we design and engineer a new decoding algorithm which seems to take the best of the two worlds: i.e., it is fast and efficient as simple methods, but at the same time it has a decoding accuracy similar to the more sophisticated methods. Interestingly enough, our algorithm provides a new and intriguing application of graph drawing methods. For the sake of homogeneity, we report here experiments performed on the same color barcode scheme, i.e., HCC2D [5]. However, we stress that most of the conclusions that can be drawn from our experiments are about color classification, and thus depend only on chromatic issues, rather than on the particular barcode considered. As a result, the main findings of this paper can be extended in general to any color barcode.

2   Experimental Setup

Test Environment. Our codes were mostly written in Java, compiled with JDK 1.7. For performance issues, some critical functions (e.g., image processing) were written in C and compiled with gcc 4.7.3 with optimization flags -O2. All the experiments reported in this paper were run on a low-end PC, equipped with a 1.73 GHz Intel dual core processor, 32KB L1-Cache, 256 KB L2 cache, 3MB L3-cache, 2 GB RAM, running OS Linux Debian 6.0. All our software implementations could not be ported to and timed accurately in mobile devices, such as a smartphone, and thus as a first approximation we took the performance on a low-end PC as a proxy of the performance a modern smartphone.

Data Sets. Our data set consists of 100 different color barcode (HCC2D) scans collected from real-life applications. Those scans were pictures acquired from the on-board cameras of different mobile phones, mainly from the Google Nexus family. Pictures were taken from different angles, under a large variety of light conditions and at various distances (with the only constraint that the distance between the device and the barcode allowed the camera's autofocus to work properly). In particular, each barcode used 4 colors and contained a total of 21,609 (147×147) color cells, each of size 4×4 printer dots. In the HCC2D barcode design, 512 color cells are used for representing 128 color palettes, which are used to display the reference colors in several parts of the barcode, and 1,377 cells are used for control purposes (i.e., for internal orientation calibration and self-alignment markers). The remaining 19,720 color cells (4,930 bytes) are used for data, which included also the space needed for error correction. Each barcode underwent a real printing and scanning process (i.e., barcodes were not distorted artificially) and the printing resolution was set to 600 dpi.

Metrics Used. We define the number of byte errors as the number of incorrectly received bytes (altered due to noise, interference, distortion or synchronization errors in the print-scan channel). We measure errors in the print-scan channel with the *byte error rate* (ByER), defined as the number of byte errors divided by the total number of bytes available for data in the barcode. ByER may be expressed as a percentage.

3 Color Classification in Desktop and Mobile Scenarios

In previous work [8], we performed a thorough experimental study of algorithms for decoding color barcodes in desktop scenarios, i.e., when barcodes were printed and scanned on low-cost color laser multifunction printers/scanners. In particular, we evaluated the practical performance of several state-of-the-art methods for color classification in this framework. The algorithms considered were chosen so as to be representative of general methods, and included minimum distance classifiers based on the Euclidean distance in the color space (Euclidean), clustering based on k-means (K-means) and on the Louvain method [1] (Louvain), decision trees based on Logistic Model Trees (LMT), probabilistic classifiers (Naive Bayes) and algorithms based on support vector machines (SVM). Note that the last three methods (LMT, Naive Bayes and SVM) require an initial preprocessing phase of supervised learning through training examples. We refer the interested reader to [8] for a description of those methods and of the experiments performed in a desktop environment. We only report here that the main experimental findings of [8] showed that the impact of different color classifiers on the error rate achieved in decoding can be quite significant. In particular, the use of more complicated techniques, such as support vector machines, did not seem to pay off, as they did not achieve better accuracy in classifying color barcode cells. The lowest error rates were indeed obtained with K-means (ByER of 4.54% on average) and Naive Bayes (ByER of 6.21% on average). From the computational viewpoint, K-means seemed to be the method of choice, since it was faster in practice than Naive Bayes (and it did not required supervised learning).

We start by illustrating here the results of the same experiment as in [8], but this time on the color barcodes acquired from mobile devices described in Section 2 rather than on the barcodes acquired with desktop scanners. We expect different outcomes in this experiment, since the two data sets (barcodes from desktop scanners and from mobile devices) have rather different underlying characteristics: in particular, since desktop scanners have controlled light intensity, while pictures taken from phone cameras present a much larger variation in light conditions, we expect that it would be more difficult to decode successfully barcodes acquired from mobile devices. The results of this experiment are illustrated in Figures 1(a) and 1(b), which show respectively the Box-and-Whisker plots for

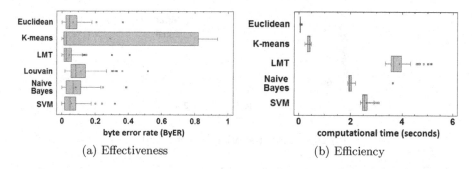

(a) Effectiveness   (b) Efficiency

Fig. 1. Box-and-Whisker plot for (a) ByER data and (b) computational time related to barcode reading by mobile phones. (Viewed better in color).

the byte error rates and the running times of the different methods in this new scenario. We recall here that Box-and-Whisker plots depict the smallest observation (sample minimum), lower quartile (Q1), median (Q2), upper quartile (Q3), largest observation (sample maximum), and the mean. The running times of the Louvain method are not included in Figure 1(b), since they were much higher (several tens of seconds) than the other methods. The first striking difference with the experiments for desktop scanners was that in the new data set K-means was no longer the most effective algorithm: in the new experiment K-means was able to decode successfully only 68% of barcodes acquired from mobile devices, and its accuracy degraded sharply for the remaining 32% of barcodes. Most of those barcodes were suffering from strongly non-uniform illumination conditions, a problem which seems to occur frequently with mobile phone cameras.

To show this phenomenon, Figure 2 illustrates the differences in quality when acquiring the same barcode with a desktop scanner or with the on-board camera of a mobile device. Out of the $21,609$ cells of this barcode, $5,610$ were black, $5,082$ cyan, $5,216$ magenta and $5,701$ white, which yields that out of the approximately $4.66 \cdot 10^8$ (ordered) cell pairs, $1.16 \cdot 10^8$ are of the same color, and $3.50 \cdot 10^8$ are of different colors. If we consider the chromatic distances between cell pairs, in this sample at least ideally $1.16 \cdot 10^8$ cell pairs will have distance equal to 0, and $3.50 \cdot 10^8$ pairs will have distance strictly larger than 0. In particular, there will be 6 different groups with Euclidean distance larger than 0, one for each (unordered) pair with different colors. This situation is illustrated in Figure 2(a). In practice, however, there will be chromatic distortions, and consequently we expect that the chromatic distances between color cells can be quite different from their theoretical values: in particular, two cell pairs of the same color might have distances larger than 0, while two cell pairs of different colors might get much closer in distance. Figures 2(b) and 2(c) illustrate the chromatic distances between color cells when the same barcode is acquired by a desktop scanner (Figure 2(b)) or by the on-board camera of a smartphone (Figure 2(c)). As it can be seen from those figures, it is much easier to recognize colors when the barcode is acquired by a desktop scan: one could just simply assume that any two color cells encode the same color if their chromatic distance is below a

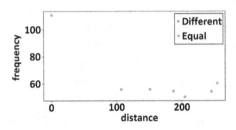

(a) Digital File (no print-scan channel)

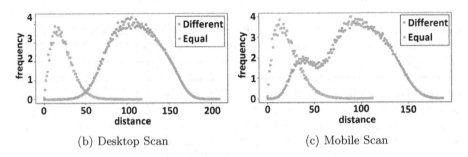

(b) Desktop Scan   (c) Mobile Scan

Fig. 2. Frequency plots (millions of occurrences) of Euclidean distances in ideal, desktop and mobile scenarios. (Viewed better in color).

certain threshold, and they encode different colors otherwise. On the opposite, deciding whether two cells encode the same color or not becomes a much more difficult task when the chromatic distances are distributed as in Figure 2(c).

Turning back to the running times of the different algorithms (Figure 1(b)), the simpler methods (`Euclidean` and `K-means`) were able to run in order of seconds on a low-end computing architecture, while the remaining methods were too slow or too difficult to implement efficiently and required an initial preprocessing for supervised learning. This raises an interesting challenge, since there does not seem to be an efficient and effective method for mobile applications. Indeed, efficient methods (such as `Euclidean` and `K-means`) appear not to be effective (in terms of decoding accuracy and error rates) while the more effective methods (such as `LMT`, `Naive Bayes` and `SVM`) do not seem to be practical enough to be implemented on low-end computing platforms. Ideally, we would like to design methods that can take the best of the two worlds: i.e., simple and efficient as `Euclidean` and `K-means`, but at the same time effective as `LMT`.

4   Force-Directed Graph Drawing Algorithms

Towards this end, we experimented with many other algorithms for color classification. For lack of space, we do not report here all the result of our experiments, but we only mention that relatively promising results were obtained with force-directed graph drawing algorithms, which are mainly used to draw graphs in an aesthetically pleasing way. Their main objective is to position the nodes of a graph in a given space, by assigning forces among the set of edges and the set of nodes, based on their relative positions, and then using these forces either to simulate the motion of the edges and nodes or to minimize some particular energy value. In particular, the algorithm by Fruchterman and Reingold [4]) resulted to be among the most effective approaches in our framework.

In order to exploit graph drawing algorithms in our settings, we build a graph from a color barcode: this graph is referred to as a *barcode graph* and is defined as follows. For each color cell in a barcode there is a node in the barcode graph, and thus a barcode graph contains 21,609 nodes. The barcode graph is a complete graph: there is an edge between any pair of nodes, with its cost being proportional to the chromatic distance between the two corresponding color cells, for a total of more than 246 million edges. The smaller the cost of a given edge, the more likely is that the two barcode cells at its endpoints share the same color, and thus the attraction / repulsion forces between nodes in the barcode graphs can be set according to the edge costs. Clearly, the barcode graph is very large, and several heuristics could be applied to reduce its size (such as considering edges only within certain thresholds of their cost). Unfortunately, force-directed algorithms tend to be slow in practice: in our experiments, they required more than 20 seconds to process even very small barcode graphs, such as graphs corresponding only to smaller portions of a given barcode (e.g., 30 × 30 color cells, corresponding to graphs with hundred nodes and thousand edges).

To circumvent this problem, we tried to combine the effectiveness of a force-directed algorithm with the efficiency of simpler methods (such as `Euclidean`) by designing a 2-phase algorithm which intuitively works as follows. In the first phase, we use `Euclidean` to decode the color cells which are apparently easy to classify. Since it plays a role in our algorithm, we now describe in detail this method. Given a barcode, `Euclidean` first computes reference black, cyan, magenta and white colors by averaging the chromatic components of cells contained in the color palettes. Each color cell is then classified according to its shortest (Euclidean) distance to the reference colors.

In the second phase, we apply a force-directed algorithm only to the remaining color cells that still need to be classified. If we succeed in classifying a large majority of the color cells (say 90-95%) during the first phase, then the barcode graph produced in the second phase may be small enough to be handled efficiently by the force-directed algorithm. The trade-off between the first and the second phase is a significant tuning parameter that will be addressed in more details later. We call this algorithm `Hybrid Force-Directed`: we next describe how to implement its two phases in more detail.

4.1 First Phase

At the beginning of the first phase, we try to evaluate the quality of the barcode image under processing. Intuitively speaking, if we realize that we are processing an "easy" barcode image, we could try to decode large portions of its color cells

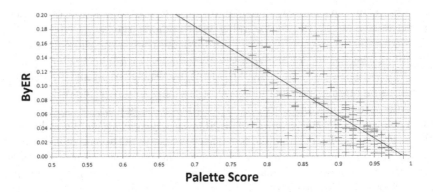

Fig. 3. Palette Score and Euclidean ByER. (Viewed better in color).

with simple methods, such as `Euclidean`. Conversely, "hard" barcode images might require the use of more sophisticated methods for many of its color cells. How can we decide whether the barcode image under consideration is easy or hard to decode? For which color cells do we decide to stop using simple methods and start using more sophisticated algorithms? This is a difficult task, since color cells may reveal to be easy or hard to be classified depending on several factors (e.g., chromatic distortions or light variations) which may occur under different variations in barcode images and even within the very same barcode image.

We do this as follows. At the beginning, we let all color palettes in the barcode (for which we know the true colors) classify each other according to their chromatic distances. Next, we assign a score to each color cell of a palette, reflecting how many color cells in the other palettes were classified correctly in this process. We can then compute a *palette score* for the entire barcode image, which takes into account the single scores obtained by the color cells in all the palettes. This score is normalized and ranges from 0 to 1, with higher scores reflecting better reciprocal classifications among the color cells in the palettes. In our experiments, the following simple definition of score revealed to be effective. Let c be a color cell of a palette and $T(c) \cup F(c)$ a partition of the set of color cells in all palettes, where $T(c)$ contains all color cells with the same color as c and $F(c)$ contains all color cells having color different from c. The *score* of c is the fraction of color cells $c_t \in T(c)$ such that the Euclidean distance $d(c_t, c) \leq \min_{c_f \in F(c)} \{d(c_f, c)\}$. The palette score for the entire barcode image is then the average of scores obtained by the color cells in all the palettes.

Since color palettes are replicated in different parts of the barcode area, they are supposedly distorted in the same way as the other color cells in the barcode: thus, we expect "easy" barcode images to have higher palette scores, and "hard" barcode images to have lower palette scores. This intuition was confirmed by our experiments. In particular, we report in Figure 3 the results of an experiment which compares the distribution of palette scores to the distribution of byte error rates (ByER) of the `Euclidean` method for the barcodes in our data set. As it can be seen from the figure, there is strong inverse correlation between the two distributions: the higher is the palette score of a barcode, and the lower is the byte error rate achieved by `Euclidean`. A Spearman correlation of -0.79 between the two distributions confirms that the palette score provides a good indication on whether the simple `Euclidean` method is able to decode successfully a given barcode image. We thus use the palette score to decide when to stop the first phase: all the color cells that have not been classified by `Euclidean` (and are expected to be harder to be classified) are passed to the second phase.

4.2   Second Phase

In the second phase, we analyze the color cells that were not classified before. In our experiments, few color cells survived to the second phase: the resulting barcode graph had typically much less than a thousand nodes, and it was substantially sparse, which made it possible to apply a force-directed method effectively. With this approach, cells which are very close in the color space will attract each other, and cells that are very far in the color space will repulse each other. To take into account also the effects of color cells that were classified before, we add 4 super-nodes: each super-node represents all the color cells that were already classified with a given color during the first phase. We connect the super-nodes to the color cells that are left in this phase, and set their edge costs suitably by taking into account both chromatic distances and sizes of super-nodes (total number of cells that were already classified with that color).

In particular, in this phase we use as a fast force-directed method OpenOrd [6], which is based on the Fruchterman-Reingold algorithm. OpenOrd allows the ability to control node clustering by ignoring, or cutting, the long edges. This can be done by varying the *edge-cut* parameter from 0 (no cutting) to 1. The iterations of OpenOrd are controlled via a simulated annealing type schedule consisting of five different phases (liquid, expansion, cool-down, crunch, and simmer). In our experiments we found that the best results were achieved with edge-cut set to 0.8, and when the schedule spent 25% of its time in each of the first three phases, 10% in the crunch phase, and 15% in the simmer phase.

Figure 4 depicts the layout of a barcode graph with 1,056 nodes, obtained after 300 iterations of OpenOrd. Note that in this particular case, one super-node (cyan) is of small degree, since most of the cyan color cells were already classified during the first phase. Finally, we use the layout produced by OpenOrd to produce four different clusters corresponding to the four colors. To do that, we try to select four different centers of mass in the layout graph and use them to identify the clusters. Among all the methods we experimented for the selection of the centers of mass, one of the most effective was based on PageRank: we picked nodes with the highest PageRanks in the layout graph, and computed the four centers of mass according to those nodes.

5   Experimental Results

In this section we report the results of our experimental study on the algorithm `Hybrid Force-Directed` described in Section 4. In particular, we are

Fig. 4. The result of the OpenOrd Layout algorithm on a color barcode graph with 1,056 nodes. The 4 super-nodes are highlighted in the layout. (Viewed better in color).

interested in comparing the new algorithm to the best methods resulting from our experiments: i.e., `Euclidean` for efficiency and `LMT` for effectiveness. This is reported in Figure 5, which shows the Box-and-Whisker plots for the byte error rates and the running times of the three algorithms. As it can be seen from this figure, `Hybrid Force-Directed` seems a good trade-off between the simpler (`Euclidean`) and the more complicated (`LMT`) methods. In particular, Figure 5(a) shows that `Hybrid Force-Directed` has accuracy in detection (measured in byte error rate) closer to `LMT` than to `Euclidean`. As far as the running times of the three different algorithms are concerned, as shown in Figure 5(b), `Euclidean` is the simplest and fastest method, and it is capable of classifying each color barcode in our data set in a few milliseconds. `Hybrid Force-Directed` was also fast (of the order of 1 second), while `LMT` classifier had a higher computational overhead, as it required almost 4 seconds on average. In summary, one can argue that `Hybrid Force-Directed` can be as effective for our problem as a machine learning method (such as `LMT`), but is simpler and more efficient in practice, so that it can be successfully deployed on low-end computing architectures.

To give more details on the accuracies obtainable with the different methods, and on the implication of this aspect on the achievable data density, we illustrate in Figure 6 the percentiles of ByER distributions (i.e., values below which specific

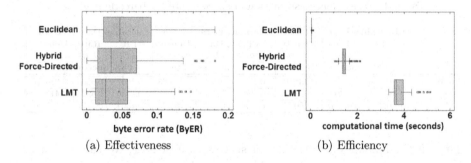

(a) Effectiveness   (b) Efficiency

Fig. 5. Box-and-Whisker plot for (a) ByER data and (b) computational time related to Euclidean, our Hybrid Force-Directed and LMT. (Viewed better in color).

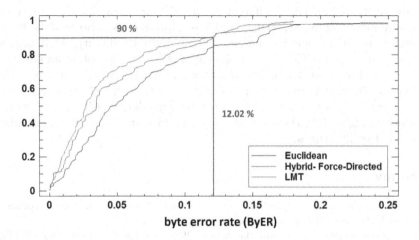

Fig. 6. Percentiles for ByER data related to the Euclidean classifier, our Hybrid Force-Directed algorithm and the LMT method. (Viewed better in color).

percentages of the data are found). In particular, we can read the percentile plot as follows. Consider the 90-th percentile (the value below which 90% of the cases fall) for ByER of Hybrid Force-Directed: this is highlighted in the figure, and has value 0.1202. This means that 90 barcodes out of 100 would be decoded if the symbols were robust to ByER up to 12.02%. If Hybrid Force-Directed is used for color classification, we may state that $Prob(ByER < 12.02\%) \approx 90\%$, even if this is just an estimation of the probability on the basis of ByER data collected in our experiments. If we wish to correct those errors, then we need twice as many redundant symbols. In other words, in order to achieve a success rate of 90%, the redundancy rate (RR) should be approximately 24.04%, which will reduce the data rate (DR) to about 75.96% of the overall capacity. Because color barcodes of our experiments are capable of storing 4,930 $bytes/inch^2$

Table 1. Performance as function of success rate for barcode reading

	Tolerable ER ($ByER$)	Data Rate (DR)	Data Density $B/inch^2$	Tolerable ER ($ByER$)	Data Rate (DR)	Data Density $B/inch^2$	Tolerable ER ($ByER$)	Data Rate (DR)	Data Density $B/inch^2$
	85% Success Rate			90% Success Rate			95% Success Rate		
Euclidean	0.1210	0.7580	3,736.94	0.1545	0.6910	3,406.63	0.1667	0.6666	3,286.34
Hybrid FD	0.1050	0.7900	3,894.70	0.1202	0.7596	3,744.83	0.1536	0.6928	3,415.50
LMT	0.0841	0.8318	4,100.77	0.1195	0.7610	3,751.73	0.1364	0.7272	3,585.10

(data plus redundancy), to achieve a success rate of 90%, we can obtain an effective data density of at most 3,744 $bytes/inch^2$.

Table 1 shows the trade-off between data density and reliability (in terms of success rate) for the tree methods that can be inferred from Figure 6. In this table, the error rates that can be tolerated to achieve the target success rates are illustrated for three different levels (85%, 90% and 95%), along with the corresponding data rate (DR), which is expressed as ratio of data bytes to overall bytes (data plus redundancy bytes), and data density. As it can be seen from Table 1, to achieve a success rate of 90%, we can obtain an effective data density of at most 3,406 $bytes/inch^2$ with Euclidean, 3,744 $bytes/inch^2$ with Hybrid Force-Directed, and 3,751 $bytes/inch^2$ with LMT. In other terms, Hybrid Force-Directed is able to achieve the same data density as LMT, which improves by roughly 10% the data density of Euclidean, while still maintaining reasonable decoding times.

References

1. Blondel, V., Guillaume, J., Lambiotte, R., Lefebvre, E.: Fast unfolding of communities in large networks. J. Stat. Mech. (10), P10008 (2008)
2. Bulan, O., Blasinski, H., Sharma, G.: Color QR codes: Increased capacity via per-channel data encoding and interference cancellation. In: Color and Imaging Conference, pp. 156–159. Society for Imaging Science and Technology, Springfield (2011)
3. Bulan, O., Sharma, G.: High capacity color barcodes: Per channel data encoding via orientation modulation in elliptical dot arrays. IEEE Trans. Image Process. 20(5), 1337–1350 (2011)
4. Fruchterman, T., Reingold, E.: Graph drawing by force-directed placement. Softw. Pract. Exp. 21(11), 1129–1164 (1991)
5. Grillo, A., Lentini, A., Querini, M., Italiano, G.: High capacity colored two dimensional codes. In: International Multiconference on Computer Science and Information Technology, pp. 709–716. IEEE, New York (2010)
6. Martin, S., Brown, W., Klavans, R., Boyack, K.: OpenOrd: an open-source toolbox for large graph layout. In: Visualization and Data Analysis. SPIE, Bellingham (2011)
7. Parikh, D., Jancke, G.: Localization and segmentation of a 2D high capacity color barcode. In: Workshop on Applications of Computer Vision. IEEE, New York (2008)
8. Querini, M., Italiano, G.: Color classifiers for 2D color barcodes. In: Federated Conference on Computer Science and Information Systems, pp. 611–618. IEEE (2013); Full version submitted to the Special Issue of the Conference in the Computer Science and Information Systems (ComSIS) Journal

Computing Consensus Curves

Livio De La Cruz[1], Stephen Kobourov[1,*], Sergey Pupyrev[1,2,*],
Paul S. Shen[1], and Sankar Veeramoni[1,*]

[1] Department of Computer Science, University of Arizona, Tucson, AZ, USA
[2] Institute of Mathematics and Computer Science,
Ural Federal University, Ekaterinburg, Russia

Abstract. We study the problem of extracting accurate average ant trajectories from many (inaccurate) input trajectories contributed by citizen scientists. Although there are many generic software tools for motion tracking and specific ones for insect tracking, even untrained humans are better at this task. We consider several local (one ant at a time) and global (all ants together) methods. Our best performing algorithm uses a novel global method, based on finding edge-disjoint paths in a graph constructed from the input trajectories. The underlying optimization problem is a new and interesting network flow variant. Even though the problem is NP-complete, two heuristics work well in practice, outperforming all other approaches, including the best automated system.

1 Introduction

Tracking moving objects in video is a difficult task to automate. Despite advances in machine learning and computer vision, the best way to accomplish such a task still is by hand. At the same time, people spend millions of hours each day playing games like *Solitaire*, *Angry Birds*, and *Farmville* on phones and computers. This presents an opportunity to harness some of the time people spend on online games for more productive but still enjoyable work. In the last few years it has been shown that citizen scientists can contribute to image processing tasks, such as the *Galaxy Zoo* project in which thousands of citizen scientists labeled millions of images of galaxies from the Hubble Deep Sky Survey and *FoldIt* in which online gamers helped to decode the structure of an AIDS protein — a problem which stumped researchers for 15 years.

AngryAnts is our online game available at http://angryants.cs.arizona.edu, which plays videos and allows citizen scientists to build the trajectory of a specified ant via mouse clicks. When we have enough data, we compute a most realistic *consensus* trajectory for each ant. Our motivation comes from biologists who wish to discover longitudinal behavioral patterns in ant colonies. The trajectories of individual ants in a colony extracted from videos are needed to answer questions such as how often do ants communicate, what different roles do ants play in a colony, and how does interaction and communication affect the success or failure of a colony? Existing automated solutions are not good enough, and there is only so much data that even motivated students can annotate in the research lab.

Related Work. The problem of computing the most likely trajectory from a set of given trajectories has been studied in many different contexts and here we mention a few

---

* Supported in part by NSF grants CCF-1115971 and DEB 1053573.

J. Gudmundsson and J. Katajainen (Eds.): SEA 2014, LNCS 8504, pp. 223–234, 2014.
© Springer International Publishing Switzerland 2014

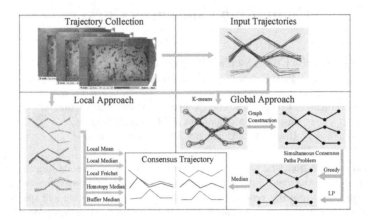

Fig. 1. Overview of consensus trajectory computation via citizen science. Local consensus may route different ants along the same trajectory (red and blue), while the global consensus ensures disjoint trajectories.

examples, which are similar to our approach. Buchin *et al.* [4] look for a representative trajectory for a given set of trajectories and compute a median representative rather than the mean. The Fréchet distance as similarity measure for trajectories is studied by Buchin *et al.* [3], who show how to incorporate time-correspondence and directional constraints. Trajcevski *et al.* [15] use the maximum distance at corresponding times as a measure of similarity between pairs of trajectories, and describe algorithms for matching under rotations and translations.

Yilmaz *et al.* [17] survey the state-of-the-art in object tracking methods. Some of the most recent methods include general approaches for tracking cells undergoing collisions by Nguyen *et al.* [13] and specific approaches for tracking insects by Fletcher *et al.* [7]. Also related are the automatic tracking method for tracking bees by Veeraraghavan *et al.* [16] and cluster-based, data-association approaches for tracking bats in infrared video by Betke *et al.* [2]. Tracking the motion and interaction of ants has also been studied by Khan *et al.* [8], who describe probabilistic methods, and by Maitra *et al.* [11] using computer vision techniques.

Our Contributions. We describe a citizen science approach for extracting consensus trajectories in an online game setting; see Fig. 1 for an overview. Combining human-generated trajectories with a new global approach for computing consensus curves outperforms even the most sophisticated and computationally expensive tracking algorithms [14], even in the most advantageous setting for automated solutions (e.g., high resolution video, sparse ant colony, individually painted ants).

Consider a *trajectory* as a sequence of T pairs $(p_i, t_i), 1 \leq i \leq T$, where $p_i = (x_i, y_i)$ is a point in the plane representing the position of an ant at timestamp t_i. We assume that between timestamps an ant keeps a constant velocity, and therefore, its trajectory is a polyline in 3D (or a possibly self-intersecting polyline in 2D). The input of our problem is a collection $\tau_1, \ldots, \tau_m$ of trajectories. Each trajectory corresponds to one of k ants, and we assume that there exists at least one trajectory for each ant, that is, $m \geq k$. Since the input data comes from the *AngryAnts* game, all trajectories have the same length (number of points). We also assume that the initial position of each ant is provided by the game (from a different game level called "Count the Ants" where users click on all the ants in the first frame of the video to identify their starting positions). Therefore, the first points of trajectories corresponding to the same ant are identical. Our goal is to compute a *consensus trajectory*, that is, our best guess for the actual route taken, for each of the k ants. While intuitively we are looking for the most probable ant trajectories, it is far from obvious how to measure the quality of a solution.

We designed, implemented, and evaluated several methods for computing accurate consensus trajectories from many (possibly inaccurate) trajectories submitted by citizen scientists. We consider several local strategies (clustering, median trajectories, and Fréchet matching) in one-ant-at-a-time setting. We also designed and implemented a novel global method in the all-ants-together setting. This approach is based on finding edge-disjoint paths in a graph constructed from all input trajectories. The underlying optimization problem is a new and interesting variant of network flow. Even though the problem is NP-complete, our two heuristics work well in practice, outperforming all other approaches including the automated system.

2   The Local Approach

Local Mean. Intuitively, a mean trajectory averages the locations for input trajectories. To compute the mean, we identify all input trajectories $\tau_1, \ldots, \tau_{m_c}$ for a particular ant c. For each timestamp t_i, we query points $p_1 = (x_1, y_1), \ldots, p_{m_c} = (x_{m_c}, y_{m_c})$ corresponding to t_i. The average point is (x_{avg}, y_{avg}), where $x_{avg} = (x_1 + \cdots + x_{m_c})/m_c$ and $y_{avg} = (y_1 + \cdots + y_{m_c})/m_c$ give the location of the mean trajectory at timestamp t_i. The sequence of these average points over time defines the local mean consensus trajectory, which is good when the number of input trajectories is very large or when all input trajectories are very accurate. In reality, however, this is often not the case; a single inaccurate input trajectory may greatly influence the result.

Local Median. The median point is more robust to outliers than the mean. For a set of points $p_1, p_2, \ldots, p_{m_c}$, we choose as median the point (x_{med}, y_{med}), where x_{med} is the median of the array $x_1, \ldots, x_{m_c}$ and y_{med} is the median of the array $y_1, \ldots, y_{m_c}$. The sequence of these median points over time defines the local median consensus trajectory. Note that the median of a set of points is not necessarily a point of the set: it could place an ant at a point that matches none of the input trajectories.

Local Fréchet. Informally, the Fréchet distance between two trajectories can be illustrated as the minimum dog-leash distance that allows a man to walk along one trajectory and his dog along the other, while connected at all times by the leash [1]. Computing the Fréchet distance produces an alignment of the trajectories: at each step, the position of the man is mapped to the position of the dog. Given the Fréchet alignment of two trajectories, we compute their consensus by taking the midpoint of the leash over time. To find the consensus, we repeatedly compute pairwise consensuses until only one trajectory remains. Since the results of the algorithm depends on the order in which the trajectories are merged, we try several different random orders choosing the best result. Note that, by definition, the Fréchet alignment of trajectories uses only the order of points and ignores the timestamps that are an essential feature of our input.

Homotopy Median and Buffer Median. In the above methods, the average of two locations could be an invalid location. In the median trajectory approach, the computed trajectory always lies on segments of input curves. Note that this is more restrictive than the point-based median method described above. We use two algorithms suggested by Buchin *et al.* [4] and Wiratma *et al.* [9]. In the *Homotopy median* algorithm, obstacles are placed in large faces bounded by segments of input trajectories to ensure that the median trajectory is homotopic to the set of trajectories that go around the obstacles. Hence, the median does not miss segments if the input trajectories are self-intersecting. The *Buffer median* (also known as Majority Median) algorithm is a combination of the buffer concept and Dijkstra's shortest path algorithm. A buffer is defined around a segment so that if the segment is a part of the median trajectory, then its buffer intersects all input trajectories. Thus, segments located near the middle of the trajectory have smaller buffer size and are good candidates for the median trajectory. Note again that both the Homotopy median and the Buffer median approaches ignore the timestamps.

3   The Global Approach

In the global approach we consider all input trajectories for all k ants together. The main motivation is that a trajectory corresponding to an ant may contain valid pieces of trajectories for other ants: a citizen scientist may mistakenly switch from tracking an ant x to tracking a different ant y at an intersection point where x and y cross paths. However, even when such mistakes occur, the trace after the intersection point is still useful as it contains a part of the trajectory of ant y. The global approach allows us to retain this possibly useful data as shown by an example in Fig. 1. Given a set of input trajectories, we compute the consensus trajectories in three steps: (1) create a graph G; (2) compute edge-disjoint paths in G; (3) extract consensus trajectories from the paths.

Step 1. We begin by creating a graph that models the interactions between ants in the video. For every timestamp, the graph has at most k vertices, which correspond to the positions of the k ants. If several ants are located close to each other, then we consider them to be at the same vertex. Intuitively, each such vertex is a possible point for a citizen scientist to switch to a wrong ant.

A weighted directed graph $G = (V, E)$ is constructed as follows. For every timestamp t_i, we extract points $p_1, \ldots, p_m$ from the given trajectories, where p_j is the position of trajectory τ_j at t_i. Using a modification of the k-means clustering algorithm [10], we partition the points into $\leq k$ clusters. Our clustering algorithm differs from the classical k-means in that it always merges the points into one cluster located closer than 50 pixels from each other. The vertices of G are the clusters for all timestamps; thus, G has at most kT vertices. We then add edges between vertices in consecutive timestamps. Let V_i represent a set of vertices at timestamp t_i. We add an edge (u, v) between two vertices $u \in V_i$ and $v \in V_{i+1}$ if there is an input trajectory with point p_i belonging to cluster u and point p_{i+1} belonging to cluster v. For each edge, we create k non-negative weights. For each ant $1 \leq x \leq k$ and for each edge $(u, v) \in E$, $u \in V_i, v \in V_{i+1}$, there is weight $w_{uv}^x \in \mathbb{Z}_{\geq 0}$, which equals to the number of trajectories between timestamps i and $i + 1$ associated with the ant x passing through the clusters u and v.

Note that by construction graph G is acyclic. Let $d^{in}(v)$ and $d^{out}(v)$ denote the indegree and the outdegree of the vertex v. Clearly, the only vertices with $d^{in}(v) = 0$ are in V_1, and the only vertices with $d^{out}(v) = 0$ are in V_T; we call them *source* and *destination* vertices and denote them by s_i and t_i, respectively. We assume that for all *intermediate* vertices $v \in V_i, 1 < i < T$, we have $d^{in}(v) = d^{out}(v)$; this is a realistic assumption because for each intersection point of trajectories, the number of incoming ants equals the number of outgoing ants. We say that a directed graph satisfies the *ant-conservation condition* if (1) the number of outgoing edges from sources and the number of incoming edges to destinations is equal to k, that is, $\sum_i d^{out}(s_i) = \sum_i d^{in}(t_i) = k$, and (2) indegree and outdegree of all its intermediate vertices are the same, that is, $d^{in}(v) = d^{out}(v)$.

Step 2. We compute k edge-disjoint paths in G, corresponding to the most "realistic" trajectories of the ants. The paths connect the k distinct vertices in V_1 to the k distinct vertices in V_T. Since the initial position of each ant is a part of input, we know the starting vertex of each path. However, it is not obvious which of the destination vertices in V_T correspond to each ant. To measure the quality of the resulting ant trajectories, that is, how well the edge-disjoint paths match the input trajectories, we introduce the following optimization problem.

SIMULTANEOUS CONSENSUS PATHS (SCP)

Input: A directed acyclic graph $G = (V, E)$ with k sources $s_1, \ldots, s_k$ and k destinations $t_1, \ldots, t_k$ satisfying the ant-conservation condition. The weight of an edge $e \in E$ for $1 \leq i \leq k$ equals $w_e^i \in \mathbb{Z}_{\geq 0}$.

Task: Find k edge-disjoint paths $P_1, \ldots, P_k$ so that path P_i starts at s_i and ends at t_j for some $1 \leq j \leq k$, and the total $cost = \sum_{i=1}^k \sum_{e \in P_i} w_e^i$ is maximized. Note that the objective is to simultaneously optimize all k disjoint paths. The decision problem is to find k edge-disjoint paths with total $cost \geq c$ for some constant $c \geq 0$.

The problem is related to the integer multi-commodity flow problem, which is known to be NP-hard [6]. In our setting, the weights on edges for different "commodities" are different, and the source-destination pairs are not known in advance. We study the SCP problem in the next section.

Step 3. We construct consensus trajectories corresponding to a solution of the SCP problem. Let P be a path for an ant x computed in the previous step. For each timestamp t_i, we consider the edge $(u, v) \in P, u \in V_i, v \in V_{i+1}$, and find a set S_{uv} of all input trajectories passing through both clusters u and v. We emphasize here that S_{uv} may contain (and often does contain) trajectories corresponding to more ants than just ant x. Next we compute the median of points of S_{uv} at timestamp t_i; the median is used as the position of ant x at timestamp t_i. The resulting trajectory of x is a polyline connecting its positions for all $t_i, 1 \leq i \leq T$.

4   The SIMULTANEOUS CONSENSUS PATHS Problem

The SCP problem is NP-complete even when restricted to planar graphs by a reduction from a variant of the edge-disjoint path problem [12]. On the other hand, the problem is fixed-parameter tractable in the number of paths. For a fixed constant k, it can be solved

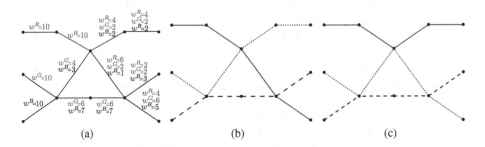

Fig. 2. The greedy algorithm may produce non-optimal solution. (a) The input graph with 3 paths: R, G, and B. (b) The paths computed by the greedy algorithm with $cost = 75$ (R is shown solid, G – dotted, B – dashed). (c) The optimal solution with $cost = 77$.

optimally via dynamic programming in time $O(|E| + k!|V|)$. The proofs of both claims are in the full version [5]. Next we suggest a greedy heuristic and provide an integer linear programming (ILP) formulation for SCP.

The Greedy Algorithm. The algorithm finds the longest path among all source-destination pairs s_i, t_j. The length of a path starting at s_i is the sum of w^i_{uv} for all edges (u, v) of the path. Since G is an acyclic directed graph, the longest path for the specified pair s_i, t_j can be computed in time $O(|E| + |V|)$ via dynamic programming. Once the longest path is found, we remove all edges of the path from G and proceed with the next longest path. The algorithm finds at most k paths on each iteration and the number of iterations is k; hence, the overall running time is $O(k^2(|V| + |E|))$.

The algorithm always yields a solution with k disjoint paths. Initially, G satisfies the ant-conservation condition: $d^{in}(v) = d^{out}(v)$ for all intermediate vertices v. Since G is connected, there exists a source-destination path. After removing the longest path, G may be disconnected, but the ant-conservation condition still holds for every connected component. The number of outgoing edges from sources and the number of incoming edges to destinations are equal for every connected component. Thus, the greedy algorithm produces a feasible solution but not necessarily the optimal one; see Fig. 2.

Linear Programming Formulation. Let P_i denote the path in G from s_i to t_j for some j (the path of the i-th ant). For each P_i and each edge e, we introduce a binary variable x^i_e, which indicates whether path P_i passes through the edge e. The SCP problem can be formulated as the following ILP:

$$\text{maximize } \sum_e \sum_i w^i_e x^i_e$$

$$\begin{array}{llll}
\text{subject to} & \sum_i x^i_e = 1 & \forall e \in E & (1) \\
& \sum_{uv} x^i_{uv} = \sum_{vw} x^i_{vw} & \forall v \in V \setminus \{s_1, t_1, \ldots, s_k, t_k\}, 1 \leq i \leq k & (2) \\
& \sum_v x^i_{s_i v} = 1 & \forall 1 \leq i \leq k & (3) \\
& x^i_e \in \{0, 1\} & \forall e \in E, 1 \leq i \leq k & (4)
\end{array}$$

Here constraint (1) guarantees that the paths are disjoint: there is exactly one path passing through every edge. Constraint (2) enforces consistency of the paths at every intermediate vertex: if a vertex v is contained in a path, then the path passes through

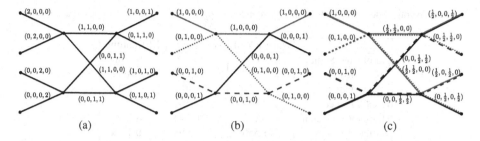

Fig. 3. Graph in which a fractional solution has cost greater than the cost of any integer solution. (a) The input graph with 4 paths. The vector on the edge e corresponds to the weights $(w_e^1, w_e^2, w_e^3, w_e^4)$ for the paths on e. (b) The optimum integer solution with $cost = 15$. The vector on the edge e corresponds to the solution $(x_e^1, x_e^2, x_e^3, x_e^4)$ on e. (c) A fractional solution with $cost = 16$. The vector on the edge e corresponds to the solution $(x_e^1, x_e^2, x_e^3, x_e^4)$ on e.

an edge (u, v) and an edge (v, w) for some $u, w \in V$. In constraint (3) we sum over vertices v with $(s_i, v) \in E$; it implies that the i-th path starts at source s_i. If we relax the integrality constraint (4) by $0 \le x_e^i \le 1$, we have a fractional LP formulation for the SCP problem, which can be solved in polynomial-time. However, the solution does not have a natural interpretation in the context of ants (fractional ants do not make sense in the biological problem). Further, we found an example for which the best fractional solution has cost strictly greater than the cost of any integer solution; see Fig. 3.

We can convert an optimal fractional LP solution x^* into a feasible integer solution as follows. Randomly pick an ant $1 \le i \le k$, with probability of choosing the i-th ant proportional to its weight $\sum_e w_e^i x^{*i}_e$ in the fractional solution. We then consider the graph with modified edge weights in which the weight of an edge $e \in E$ is $w_e^i x^{*i}_e$. We look for the longest path starting at source s_i in this graph, assign the path to the i-th ant, and remove this path from the graph. We then rerun the LP to find a fractional solution on the smaller instance. Our experiments suggest that this rounding scheme yields an integer solution that is close to optimal.

5   Experimental Results

We use a machine with Intel i5 3.2GHz, 8GB RAM and CPLEX Optimizer.

Real-World Dataset. Here we consider a real-world scenario and compare ground truth data to seven different consensus algorithms described in this paper, along with an automated solution. To evaluate our various algorithms, we work with a video of a *Temnothorax rugatulus* ant colony containing $10,000$ frames, recorded at 30 frames per second. This particular video contains ants that are individually painted and has been analyzed with the state-of-the-art automated multi-target tracking system of Poff *et al.* [14]. The method required about 160 minutes to analyze the video. To evaluate the automated system, they create a *ground truth* trajectory for each ant, by manually examining *every ant in every 100th frame* of the automated output and correcting when necessary. We use this ground truth data to evaluate the efficiency of the algorithms considered and

Table 1. Average and worst root-mean-square error (in pixels) computed for proposed algorithms

Algorithm	Worst RMSE	Average RMSE	Runtime
Automated Solution	95.308	9.593	160 min
Local Mean	105.271	12.531	< 100 ms
Local Median	112.741	9.801	< 100 ms
Local Fréchet	127.104	15.562	1.2 sec
Homotopy Median	146.267	20.244	8.2 sec
Buffer Median	171.556	23.998	9.7 sec
Global ILP	20.588	8.716	34 sec
Global Greedy	24.820	8.900	0.2 sec

the automated system. Note that just by the way the ground truth is generated, results are inherently biased in favor of the automated solution.

Our dataset consists of 252 citizen scientist generated trajectories for 50 ants, with between 2 and 8 trajectories per ant. To compute the ant trajectories, we construct the interaction graph G, as described in Section 3. For our dataset, the graph contains $4,246$ vertices and $10,494$ edges. We apply the five local methods and two of the global methods (greedy and integer linear programming) to build consensus trajectories.

We computed seven different consensus trajectories for each ant: five based on the local algorithms, and two based on the global algorithms. An overview of results is in Table 1 and in Fig. 4(a). We compare all seven, as well as the trajectories computed by the automated system, by measuring average and worst root-mean-square error of the Euclidean distance between pairs of points of computed and ground truth trajectories. We notice that the approximate dimensions of an ant in our video are 60×15 pixels.

Among the local approaches, the local median is best. The local mean is negatively impacted by the outliers in the data. The Fréchet approach, Homotopy median, and Buffer median perform poorly. This could be due to the very self-intersecting trajectories making these algorithms miss entire pieces. We used the default values for all the parameters in these algorithms; a careful tuning will likely improve accuracy. In the Fréchet approach we tried 50 different random orders of merging trajectories.

The two global approaches perform similarly with respect to quality. There are only few segments of trajectories in which the results differ. It is important to emphasize that the global approaches outperform the automated solution for the cases where many citizen scientists make the same mistake; see Fig. 4(a). Such cases do happen in practice; see the full version [5] for an example. In this case, none of the local algorithms have a chance to recover a correct trajectory. Only the global approaches allow us to identify the correct ant and produce accurate results. On the other hand, the automated system is more accurate for "simpler" ants. We stress again that the ground truth data is inherently biased towards the automated solution because it was obtained by modifying the trajectories obtained from the automated solution.

A big challenge for the existing automated systems is tracking ants in long videos. For long videos (e.g., hundreds or even thousands of hours), automated tracking methods are not reliable. Whenever such algorithms loose tracking, the error quickly accumulates and a trajectory often is not recovered. Our global approaches naturally

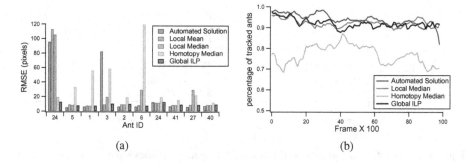

Fig. 4. (a) Root-mean-square error for the 10 "most movable" ants (according to the ground truth). (b) Comparison of tracking accuracy between automated solution and proposed algorithms.

resolve this problem. We evaluate tracking accuracy as the percentage of ants correctly "tracked" at a given timestamp; here, we consider the ant correctly tracked if the distance between the ground truth and our trajectory is less than 15 pixels (typical width of ant head). The accuracy of the automated solution decreases over time, and by the end of the 5-minute video it is below 87% accuracy; see Fig. 4(b). Our algorithms are steadily over 90% accuracy over the entire video.

Synthetic Dataset. In order to validate our global approach on a larger dataset, we generate a collection of synthetic graphs. Here we test our algorithms for the SCP problem, that is, the algorithms for computing disjoint paths on graphs, rather than for extracting optimal trajectories. We construct a set of directed acyclic graphs having approximately the same characteristics as the interaction graph computed for the real-world dataset. The graph construction follows the same pipeline as described in Section 3; every graph is generated for $k \leq 50$ ants and $T \leq 100$ timestamps. The ants form k vertices for the initial timestamp; on every subsequent timestamp pairs of ants meet with probability 0.4 (the constant estimated for the real-world graph). Thus, for every timestamp, we have from $k/2$ to k vertices. The pairs of ants meeting at a timestamp are chosen randomly with the restriction that indegrees and outdegrees of every vertex in a graph are equal and at most 2. By construction of the graphs, we naturally get "ground truth" paths for all ants. Next we generate citizen scientist trajectories in two scenarios: $2 - 8$ trajectories (as in the real-world dataset) and $15 - 20$ trajectories per ant. A citizen scientist tracking an ant is modeled by a path starting at the source of the ant. At every junction vertex (with outdegree 2) there is fixed probability $P(error)$ of making a mistake by switching from tracking the current ant to the other one. If $P(error) = 0$, then user trajectories always coincide with the ground truth paths; if $P(error) = 0.5$, the trajectories may be considered as random walks on the graph.

We evaluate the greedy heuristic (GREEDY) and the linear program with rounding (LP+ROUNDING). As the latter heuristic is a randomized algorithm, we report the best result over 5 runs for a given input. For small instances, we also compute an exact solution using the integer linear program (ILP). We analyze the precision of the algorithms under various parameters. For every edge of the graph G, we say that it is correctly identified if both the algorithm and the ground truth assign the edge to the same path.

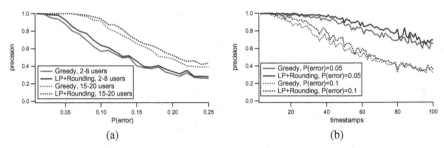

Fig. 5. Precision of the algorithms on the synthetic dataset. Solid lines represent average values over 5 runs of the algorithms for a given error/timestamp. (a) Results for $k = 50$ ants, $T = 100$ timestamps (5-minute video segment), and various values of $P(error)$. (b) Results for $k = 50$ ants, $2 - 8$ user trajectories per ant, and various number of timestamps.

The precision is measured as the fraction of correctly identified edges in G: a value of 1 means that all paths are correct. As in the real-world dataset, we consider a scenario with $k = 50$ ants. As expected, increasing the probability of making a mistake decreases the quality of the solution; see Fig. 5(a). However, increasing the number of user trajectories does help. Both algorithms recover all paths correctly if the number of trajectories for each ant is more than 15, even in a case with $P(error) = 0.1$. On the other hand, with $P(error) > 0.2$ the precision drops to under 50%. Although we cannot definitively measure the accuracy of all citizen scientists, empirical evidence from our experiment indicates that the error rate was very low: $P(error) = 0.02$. This is an order of magnitude lower than the upper limit on errors that our algorithms can handle. We also consider the impact of video length on the precision of the algorithms; see Fig. 5(b). Not surprisingly, precision is higher for short videos: for 2-minute segments (40 timestamps) and $P(error) = 0.05$, LP+ROUNDING produces the correct paths, while for 5-minute segments (100 timestamps) only 60% of paths are correct.

We also analyze the effectiveness of the GREEDY and LP+ROUNDING algorithms for the SCP problem with $P(error) = 0.2$ and $2 - 8$ trajectories per ant; see Fig. 6. To normalize the results, values are given as a percentage of the *cost* of an optimal fractional solution (FLP) for the SCP problem. Note that the ILP results are in the range $[0.98, 1.0]$, which means that an optimal integer solution is always very close to the optimal fractional solution. Both GREEDY and LP+ROUNDING perform very well, achieving ≈ 0.85 of the optimal solution. These two algorithms produce similar results, with LP+ROUNDING usually outperforming GREEDY.

Running times are shown in Fig. 7. As expected, the greedy algorithm, with complexity dependent linearly on the size of the graph, finishes in under few milliseconds. The ILP is also relatively quick on the real-world dataset, computing the optimal solution within a minute. On synthetic data and more erroneous real-world data the ILP approach is applicable only when the number of ants is small, e.g., $k < 25$. For larger values of k, the computation of optimal disjoint paths takes hours. On the other hand, LP+ROUNDING is fast: the real-world instances with $k = 50$ ants and $T = 100$ timestamps are processed within a minute. GREEDY takes only $2 - 3$ seconds on the largest instances. Hence, the running times of our algorithms (except ILP) are practical.

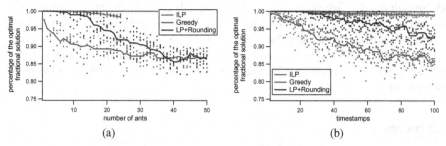

Fig. 6. Quality of the algorithms: ratio between the *cost* of obtained solution and the optimal (fractional) *cost*. Results for single instances are depicted by dots and solid curves show the average values over 5 runs for a given number of ants/timestamps. Note that the y-axis starts at value 0.75. (a) Results for $T = 100$ timestamps, $P(error) = 0.2$, and $2 - 8$ user trajectories per ant. (b) Results for $k = 20$ ants, $P(error) = 0.2$, and $2 - 8$ user trajectories per ant.

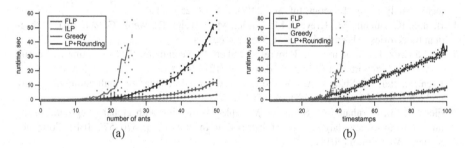

Fig. 7. Running time of the algorithms with $P(error) = 0.2$ and $2 - 8$ trajectories per ant. The results for single instances are depicted by dots, while solid lines represent average values over 5 runs for a given number of ants/timestamps. (a) Results for $T = 100$ timestamps and various number of ants. (b) Results for $k = 50$ ants and various length of a video.

6 Conclusions and Future Work

We described a system for computing consensus trajectories from a large number of input trajectories, contributed by untrained citizen scientists. We proposed a new global approach for computing consensus trajectories and experimentally demonstrated its effectiveness. In particular, the global approach outperforms the state-of-the-art in computer vision tools, even in their most advantageous setting (high resolution video, sparse ant colony, individually painted ants). In reality, there are hundreds of thousands of hours of video in settings that are much more difficult for the computer vision tools and where we expect our citizen science approach to compare even more favorably.

A great deal of challenging problems remain. Arguably, the best method would be to track "easy ants" and/or "easy trajectory segments" automatically, while asking citizen scientists to solve the hard ants and hard ant trajectory segments. In such a scenario, every input trajectory will describe a part of the complete ant trajectory, which would require stitching together many short pieces of overlapping trajectories.

Acknowledgments. We thank the A. Dornhaus lab for introducing us to the problem, and M. Shin and T. Fasciano for automated solutions and ground truth. We thank A. Das, A. Efrat, F. Brandenburg, K. Buchin, M. Buchin, J. Gudmundsson, K. Mehlhorn, C. Scheideler, M. van Kreveld, and C. Wenk for discussions. Finally, we thank J. Chen, R. Compton, Y. Huang, Z. Shi, and Y. Xu for help with the game development.

References

1. Alt, H., Godau, M.: Computing the Fréchet distance between two polygonal curves. Internat. J. Comput. Geom. Appl. 5(1), 75–91 (1995)
2. Betke, M., Hirsh, D., Bagchi, A., Hristov, N., Makris, N., Kunz, T.: Tracking large variable numbers of objects in clutter. In: CVPR, pp. 1–8. IEEE Computer Society, Washington (2007)
3. Buchin, K., Buchin, M., Gudmundsson, J.: Constrained free space diagrams: a tool for trajectory analysis. Int. J. Geogr. Inf. Sci. 24(7), 1101–1125 (2010)
4. Buchin, K., Buchin, M., Kreveld, M., Löffler, M., Silveira, R., Wenk, C., Wiratma, L.: Median trajectories. Algorithmica 66(3), 595–614 (2013)
5. De La Cruz, L., Kobourov, S., Pupyrev, S., Shen, P., Veeramoni, S.: Computing consensus curves. Arxiv report (2013), http://arxiv.org/abs/1212.0935
6. Even, S., Itai, A., Shamir, A.: On the complexity of timetable and multicommodity flow problems. SIAM J. Comput. 5(4), 691–703 (1976)
7. Fletcher, M., Dornhaus, A., Shin, M.: Multiple ant tracking with global foreground maximization and variable target proposal distribution. In: WACV, pp. 570–576. IEEE Computer Society, Washington (2011)
8. Khan, Z., Balch, T., Dellaert, F.: MCMC-based particle filtering for tracking a variable number of interacting targets. IEEE TPAMI 27(11), 1805–1819 (2005)
9. van Kreveld, M., Wiratma, L.: Median trajectories using well-visited regions and shortest paths. In: GIS, pp. 241–250. ACM, New York (2011)
10. Lloyd, S.: Least squares quantization in PCM. IEEE Trans. Inf. Theory 28(2), 129–137 (1982)
11. Maitra, P., Schneider, S., Shin, M.: Robust bee tracking with adaptive appearance template and geometry-constrained resampling. In: WACV, pp. 1–6. IEEE Computer Society, Washington (2009)
12. Marx, D.: Eulerian disjoint paths problem in grid graphs is NP-complete. Discrete Appl. Math. 143(1-3), 336–341 (2004)
13. Nguyen, N., Keller, S., Norris, E., Huynh, T., Clemens, M., Shin, M.: Tracking colliding cells in vivo microscopy. IEEE Trans. Biomed. Eng. 58(8), 2391–2400 (2011)
14. Poff, C., Hoan, N., Kang, T., Shin, M.: Efficient tracking of ants in long video with GPU and interaction. In: WACV, pp. 57–62. IEEE Computer Society, Washington (2012)
15. Trajcevski, G., Ding, H., Scheuermann, P., Tamassia, R., Vaccaro, D.: Dynamics-aware similarity of moving objects trajectories. In: GIS, pp. 1–8. ACM, New York (2007)
16. Veeraraghavan, A., Chellappa, R., Srinivasan, M.: Shape-and-behavior encoded tracking of bee dances. IEEE TPAMI 30(3), 463–476 (2008)
17. Yilmaz, A., Javed, O., Shah, M.: Object tracking: a survey. ACM Comput. Surv. 38, 1–45 (2006)

Evaluation of Labeling Strategies for Rotating Maps

Andreas Gemsa, Martin Nöllenburg, and Ignaz Rutter

Institute of Theoretical Informatics, Karlsruhe Institute of Technology,
Am Fasanengarten 5, 76131 Karlsruhe, Germany

Abstract. We consider the following problem of labeling points in a dynamic map that allows rotation. We are given a set of points in the plane labeled by a set of mutually disjoint labels, where each label is an axis-aligned rectangle attached with one corner to its respective point. We require that each label remains horizontally aligned during the map rotation and our goal is to find a set of mutually non-overlapping *active* labels for every rotation angle $\alpha \in [0, 2\pi)$ so that the number of active labels over a full map rotation of 2π is maximized.

We discuss and experimentally evaluate several labeling models that define additional consistency constraints on label activities in order to reduce flickering effects during monotone map rotation. We introduce three heuristic algorithms and compare them experimentally to an existing approximation algorithm and exact solutions obtained from an integer linear program. Our results show that on the one hand low flickering can be achieved at the expense of only a small reduction in the objective value, and that on the other hand the proposed heuristics achieve a high labeling quality significantly faster than the other methods.

1 Introduction

Dynamic digital maps, in which users can navigate by continuously zooming, panning, or rotating their personal map view, opened up a new era in cartography and geographic information science (GIS) from professional applications to personal mapping services on mobile devices. The continuously animated map view adds a temporal dimension to the map layout and thus many traditional algorithms for static maps do not extend easily to dynamic maps. Despite the popularity and widespread use of dynamic maps, relatively little attention has been paid to provably good or experimentally evaluated algorithms for dynamic maps.

In this paper we consider *dynamic map labeling* for points, i.e., the problem of deciding when and where to show labels for a set of point features on a map in such a way that visually distracting effects during map animation are kept to a minimum. In particular, we study rotating maps, where the mode of interaction is restricted to changing the map orientation, e.g., to be aligned with the travel direction in a car navigation system.

Been et al. [2, 3] defined a set of *consistency desiderata* for labeling zoomable dynamic maps, which include that (i) labels do not *pop* or *flicker* during monotone zooming, (ii) labels do not *jump* during the animation, and (iii) the labeling only depends on the current view and not its history. In our previous paper [8], we adapted the consistency model of Been et al. to rotating maps, showed NP-hardness and other properties of consistent labelings in this model, and provided efficient approximation algorithms.

J. Gudmundsson and J. Katajainen (Eds.): SEA 2014, LNCS 8504, pp. 235–246, 2014.

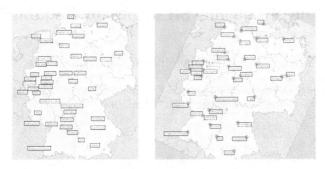

Fig. 1. Instance with 43 labeled cities in Germany. Input labeling (left), rotated by $\sim 25°$ (right). Background picture is in public domain. Retrieved from Wikipedia [Link]

Similar to the (NP-hard) label number maximization problem in static map labeling [6], the goal in dynamic map labeling is to maximize the number of visible or *active* labels integrated over one full rotation of 2π. The value of this integral is denoted as the *total activity* and defines our objective function. Figure 1 shows an example seen from two different angles. Without any consistency restrictions, we can select the active labels for every rotation angle $\alpha \in [0, 2\pi)$ independently of any other rotation angles. Clearly, this may produce an arbitrarily high number of flickering effects that occur whenever a label changes from active to inactive or vice versa. Depending on the actual consistency model, the number of flickering events per label is usually restricted to a very small number. Our goal in this paper is to evaluate several possible labeling strategies, where a labeling strategy combines both a consistency model and a labeling algorithm. First, we want to evaluate the loss in total activity caused by using a specific consistent labeling model rather than an unrestricted one. Second, we are interested in evaluating how close to the optimum total activity our proposed algorithms get for real-world instances in a given consistency model.

Related Work. Most previous work on dynamic map labeling covers maps that allow panning and zooming, e.g., [2, 3, 11, 12, 14]; there is also some work on labeling dynamic points in a static map [4, 5]. As mentioned above, the dynamic map labeling problem for rotating maps has first been considered in our previous paper [8]. We introduced a consistency model, and proved NP-completeness even for unit-square labels. For unit-height labels we described an efficient ¼-approximation algorithm as well as a PTAS. Yokosuka and Imai [15] considered the label size maximization problem for rotating maps, where the goal is to find the maximum font size for which all labels can be constantly active during rotation. Finally Gemsa et al. [7] studied a trajectory-based labeling model, in which a locally consistent labeling for a viewport moving along a given smooth trajectory needs be computed. Their model combines panning and rotation of the map view.

Our Contribution. In this paper we take a practical point of view on the dynamic map labeling problem for rotating maps. In Section 2 we formally introduce the algorithmic problem and discuss our original rather strict consistency model [8], as well as two possible relaxations that are interesting in practice. Section 3 introduces three greedy heuristics (one of which is a ⅛-approximation for unit square labels) and presents a formulation as an integer linear program (ILP), which provides us with optimal solutions against which to compare the algorithms. Our main contribution is the experimental evaluation in Section 4. We extracted several real-world labeling instances from OpenStreetMap data and make them available as a benchmark set. Based on these data, we evaluate both the trade-off between the consistency and the total activity, and the performance of the proposed labeling algorithms. The experimental results indicate that a high degree of labeling consistency can be obtained at a very small loss in activity. Moreover, our greedy algorithms achieve a high labeling quality and outperform the running times of the other methods by several orders of magnitude. We conclude with a suggestion of the most promising labeling strategies for typical use cases.

Due to space constraints we omitted all proofs in this paper. They can be found, together with a more detailed experimental analysis, in the full version [9].

2 Preliminaries

In this section we describe a general labeling model for rotating maps with axis-aligned rectangular labels. This model extends our earlier model [8].

Let M be an (abstract) map, consisting of a set $P = \{p_1, \ldots, p_n\}$ of points in the plane together with a set $L = \{\ell_1, \ldots, \ell_n\}$ of pairwise disjoint, closed, and axis-aligned rectangular labels in the plane. Each point p_i must coincide with a corner of its corresponding label ℓ_i; we denote that corner (and the point p_i) as the *anchor* of label ℓ_i. Since each label has four possible positions with respect to p_i this widely used model is known in the literature as the 4-position model (4P) [6].

As M rotates, each label ℓ_i in L must remain horizontally aligned and anchored at p_i. Thus, new label intersections form and existing ones disappear during the rotation of M. We take the following alternative perspective on the rotation of M. Rather than rotating the points, say clockwise, and keeping the labels horizontally aligned we may instead rotate each label counterclockwise around its anchor point and keep the set of points fixed. Both rotations are equivalent in the sense that they yield exactly the same intersections of labels and occlusions of points.

We consider all rotation angles modulo 2π. For convenience we introduce the interval notation $[a, b]$ for any two angles $a, b \in [0, 2\pi]$. If $a \le b$, this corresponds to the standard meaning of an interval, otherwise, if $a > b$, we define $[a, b] := [a, 2\pi] \cup [0, b]$. For simplicity, we refer to any set of the form $[a, b]$ as an interval. We define the length of an interval $I = [a, b]$ as $|I| = b - a$ if $a \le b$ and $|I| = 2\pi - a + b$ if $a > b$.

A *rotation* of L is defined by a rotation angle $\alpha \in [0, 2\pi)$. We define $L(\alpha)$ as the set of all labels, each rotated by an angle of α around its anchor point. A *rotation labeling* of M is a function $\phi \colon L \times [0, 2\pi) \to \{0, 1\}$ such that $\phi(\ell, \alpha) = 1$ if label ℓ is visible or *active* in the rotation of L by α, and $\phi(\ell, \alpha) = 0$ otherwise. We call a labeling ϕ *valid* if, for any rotation α, the set of labels $L_\phi(\alpha) = \{\ell \in L(\alpha) \mid \phi(\ell, \alpha) = 1\}$ consists of pairwise disjoint labels. If two labels ℓ and ℓ' in $L(\alpha)$ intersect, we say that they have a (soft) *conflict* at α, i.e., in a valid labeling at most one of them can be active at α. We define the set $C(\ell, \ell') = \{\alpha \in [0, 2\pi) \mid \ell \text{ and } \ell' \text{ are in conflict at } \alpha\}$ as the *conflict set* of ℓ and ℓ'. Further, we call a contiguous range in $C(\ell, \ell')$ a *conflict range*. The begin and end of a maximal conflict range are called *conflict events*.

For a label ℓ we call each maximal interval $I \subseteq [0, 2\pi)$ with $\phi(\ell, \alpha) = 1$ for all $\alpha \in I$ an *active range* of label ℓ and define the set $A_\phi(\ell)$ as the set of all active ranges of ℓ in ϕ. We call an active range where both boundaries are conflict events a *regular* active range. Our optimization goal is to find a valid labeling ϕ that shows a maximum number of labels integrated over one full rotation from 0 to 2π. The value of this integral is called the *total activity* $t(\phi)$ and can be computed as $t(\phi) = \sum_{\ell \in L} \sum_{I \in A_\phi(\ell)} |I|$. The problem of optimizing $t(\phi)$ is called *total activity maximization problem* (MAXTOTAL).

A valid labeling is not yet consistent in terms of the definition of Been et al. [2, 3]: while labels clearly do not jump and the labeling is independent of the rotation history, labels may still *flicker* multiple times during a full rotation from 0 to 2π, depending on how many active ranges they have in ϕ. In the most restrictive consistency model, which avoids flickering entirely, each label is either active for the full rotation $[0, 2\pi)$ or never at all. We denote this model as *0/1-model*. In our previous paper [8] we defined a rotation labeling as consistent if each label has only a single active range, which we denote here as the *1R-model*. This immediately generalizes to the *kR-model* that allows at most k active ranges for each label. Analogously, the unrestricted model, i. e., the model without restrictions on the number of active ranges per label, is denoted as the *∞R-model*.

We may apply another restriction to our consistency models, which is based on the occlusion of anchors. Among the conflicts in set $C(\ell, \ell')$ we further distinguish *hard conflicts*, i.e., conflicts where label ℓ intersects the anchor point of label ℓ'. If a labeling ϕ sets ℓ active during a hard conflict with ℓ', the anchor of ℓ' is occluded. This may be undesirable in some situation in practice, e.g., if every point in P carries useful information in the map, even if it is unlabeled. Thus we may optionally require that $\phi(\ell, \alpha) = 0$ during any hard conflict of a label ℓ with another label ℓ' at angle α. Note that we can include other obstacles (e. g., important landmarks on a map) which must not be occluded by a label in the form of hard conflicts. Note that a soft conflict is always a label-label conflict, while a hard conflict is always a label-point conflict (in our definition every label-point conflict induces also a label-label conflict). We showed earlier [8] that MAXTOTAL is NP-hard in the 1R-model avoiding hard conflicts and presented approximation algorithms.

3   Algorithmic Approaches

In this section we describe four algorithmic approaches for computing consistent active ranges that we evaluate in our experiments. We also evaluate our previous ¼-approximation algorithm [8] for MAXTOTAL, but omit its description due to space constraints; the full version [9] contains a sketch. Section 3.1 describes three simple greedy heuristics and Section 3.2 formulates an exact ILP model that we use primarily for evaluating the quality of the other solutions.

3.1   Greedy Heuristics

In this section we describe three new greedy algorithms to construct valid and consistent labelings with high total activity. These algorithms are conceptually simple and easy to implement, but in general we cannot give quality guarantees for the solutions computed by these algorithms.

All three greedy algorithms follow the same principle of iteratively assigning active ranges to all labels. The algorithm first initializes a set L' with all labels in L. Then it computes for each label ℓ its *maximum active range* $I_{\max}(\ell)$, which is the active range of maximum length $|I_{\max}(\ell)|$ such that (i) ℓ is not active while in conflict with another active label that was already considered by the algorithm, and (optionally) such that (ii) ℓ is not active while it has a hard conflict with another label. Initially the maximum active range of each label is either the full interval $[0, 2\pi]$ or the largest range that avoids hard conflicts. Then the algorithm repeats the following steps. It selects and removes a label ℓ from L', assigns it the active range $I_{\max}(\ell)$, and updates those labels in L' whose maximum active range is affected by the assignment of ℓ's active range. If we consider the kR-model with $k > 1$, we keep a counter for the number of selected active ranges and add another copy of ℓ with the next largest active range to L' if the counter value is less than k. The three algorithms differ only in the criterion that determines which label is selected from L' in each iteration.

The first algorithm we propose is called GreedyMax. In each step the algorithm selects the label with the largest maximum active range among all labels in L'. Ties are broken arbitrarily. The second algorithm, GreedyLowCost, determines for the maximum active range of each label the cost of adding it to the solution. This means that for each label $\ell \in L'$ with maximum active range $I_{\max}(\ell)$ the algorithm determines for all labels $\ell' \in L'$ that are in conflict with ℓ during $I_{\max}(\ell)$ by how much their maximum active range would shrink. The sum of this is the *cost* $c(\ell)$ of assigning the active range $I_{\max}(\ell)$ to ℓ. Among all labels in L' GreedyLowCost chooses the one with lowest cost. Finally, the last algorithm, GreedyBestRatio is a combination of the two preceding ones. In each step the algorithm chooses the label ℓ whose ratio $|I_{\max}(\ell)|/c(\ell)$ is maximum among all labels in L'. We conclude with a brief performance analysis of our algorithms.

Theorem 1. *In the kR-model with constant k the algorithm* GreedyMax *can be implemented to run in time $O(cn \cdot (c + \log n))$ and the algorithms* GreedyLowCost *and* GreedyBestRatio *can be implemented to run in time $O(cn \cdot (c^2 + \log n))$, where n is the number of labels and c is the maximum number of conflicts per label in the input instance. The space consumption of all algorithms is in $O(cn)$.*

The running time of GreedyMax can be further improved to $O(cn \log n)$. Moreover, if all labels are unit squares, GreedyMax is a $\frac{1}{8}$-approximation algorithm.

3.2 Integer Linear Program

In this section we present an ILP-based approach to find optimal solutions for MAXTO-TAL. This is justified since MAXTOTAL is NP-hard and we cannot hope for an efficient algorithm unless $P = NP$. We note that the same ILP formulation can also be used in the $\frac{1}{4}$-approximation algorithm to compute optimal solutions in the subinstances it considers.

The key idea of the ILP presented here is to determine regular active ranges induced by the ordered set of all conflict events. Our model contains for each label ℓ and each interval I a binary decision variable, which indicates whether or not ℓ is active during I. We add constraints to ensure that (i) no two conflicting labels are active at the same time within their conflict range and (ii) at most k disjoint contiguous active ranges can be selected for each label as required in the kR-model.

Model. For simplicity we assume in this section that the length of each conflict range is strictly larger than 0. This assumption is not essential for our ILP formulation, but makes the description easier.

Let E be the ordered set of conflict events that also contains 0 and 2π, and let $E[j]$ be the interval between the j-th and the $(j+1)$-th element in E. We call such an interval $E[j]$ an *atomic interval* and always consider its index j modulo $|E| - 1$. For each label $\ell_i \in L$ and for each atomic interval $E[j]$ we introduce two binary variables x_i^j and b_i^j to our model. We refer to the variables of the form x_i^j as *activity variables*. The intended meaning of x_i^j is that its value is 1 if and only if the label ℓ_i is active during the j-th atomic interval; otherwise x_i^j has value 0. We use the binary variables b_i^j to indicate the start of a new active range and to restrict their total number to k. This is achieved by adding the following constraints to our model.

$$x_i^j - b_i^j \leq x_i^{j-1} \qquad \forall \ell_i \in L \quad \forall j \in \{0, \ldots, |E| - 2\} \tag{1}$$

$$\sum_{0 \leq j \leq |E|-2} b_i^j \leq k \qquad \forall \ell_i \in L \tag{2}$$

The effect of constraint (1) is that it is only possible to start a new active range for label ℓ_i with atomic interval $E[j]$ (i.e., $x_i^{j-1} = 0$ and $x_i^j = 1$) if we account for that range by setting $b_i^j = 1$. Due to constraint (2) this can happen at most k times per label. We can also allow arbitrarily many active ranges per label as in the ∞R-model by completely omitting the variables b_i^j and the above constraints.

It remains to guarantee that no two labels can be active when they are in conflict. This can be done straightforwardly since we can compute for which atomic intervals two labels are in conflict and we ensure that not both activity variables can be set to 1. More specifically, for every pair of labels ℓ_i, ℓ_k and for every atomic interval j during which they are in conflict, we add the constraint

$$x_i^j + x_k^j \leq 1. \tag{3}$$

Optionally, incorporating hard conflicts can also be done easily as a hard conflict simply excludes certain atomic intervals from being part of an active range. We determine for each label all such atomic intervals in a preprocessing step and set the corresponding activity variables to 0.

Among all feasible solutions that satisfy the above constraints, we maximize the objective function $\sum_{\ell_i \in L} \sum_{0 \leq j \leq |E|-2} x_i^j \cdot |E[j]|$, which is equivalent to the total activity $t(\phi)$ of the induced labeling ϕ.

This ILP considers only regular active ranges, since label activities change states only at conflict events. However, there always exists an optimal solution that is regular [8, Lemma 4], and hence we are guaranteed to find a globally optimal solution. Let e be the number of conflict events and let c be the maximum number of conflict events per label in a MAXTOTAL instance, respectively. In the worst case the number of constraints that ensure that the solution is conflict-free (i. e., constraint (3)) is $O(c \cdot e)$ per label, whereas we require only $O(e)$ constraints of the other two types of constraints per label.

Theorem 2. *The ILP (1)–(3) solves* MAXTOTAL *and has at most $O(e \cdot n)$ variables and $O(c \cdot e \cdot n)$ constraints, where n is the number of labels, e the number of conflict events, and c the maximum number of conflicts per label.*

4   Experimental Evaluation

In this section we present the experimental evaluation of different labeling strategies based on the consistency models and algorithms introduced in Sections 2 and 3. We implemented our algorithms in C++ and compiled with GCC 4.7.1 using optimization level -O3. As ILP solver we used Gurobi 5.6. The running time experiments were performed on a single core of an AMD Opteron 2218 processor running Linux 2.6.34.10. The machine is clocked at 2.6 GHz, has 16 GiB of RAM and 2×1 MiB of L2 cache. All reported running times are wall-clock times.

4.1   Benchmark Instances

Since our labeling problem is immediately motivated by dynamic mapping applications, we focus on gathering real-world data for the evaluation. As data source we used the publicly available data provided by the OpenStreetMap project (www.osm.org). We extracted the latitudes, longitudes and names of *all* cities with a population of at least 50 000 for six countries (France, Germany, Italy, Japan, United Kingdom, and the United States of America) and created maps at three different scales.

To obtain a valid labeling instance several additional steps are necessary. First, the width and height of each label need to be chosen. Second, we need to map latitude and longitude to the two-dimensional plane. Third, recall that the input is a statically labeled map, and hence we need to compute such a static input labeling. For the first issue we used the same font that is used in Google Maps, i.e., Roboto Thin. The dimensions of each label were obtained by rendering the label's corresponding city name in Roboto Thin with font size 13, computing its bounding box, and adding a small additional buffer. For obtaining two-dimensional coordinates from the latitude and longitude of each point, we used a Mercator projection (where we approximate the ellipsoid with a sphere of radius $r = 6371$km). For the map scales we again wanted to be close to Google Maps. Hence, we derived instances in three different scales (65 pixel $\hat{=}$ 20km, 50km, 100km) for each country. For simplicity we refer to the scale of 65 pixel $\hat{=}$ 20km only by 20km (and likewise for the remaining scales). The last remaining step was to compute a valid input labeling. For this we used the 4P fixed-position model [6] and solved a simple ILP model to obtain a weighted maximum independent set in the label conflict graph, in which any two conflicting label positions are linked by an edge and weights are proportional to the population. Table 1 shows the characteristics of our benchmark data, which can be downloaded from i11www.iti.kit.edu/projects/dynamiclabeling/.

Table 1. Number of labels in each benchmark instance, the number of labels in the largest connected component (lcc) and the number of connected components (cc) in the conflict graph

	countries					
	FR	DE	GB	IT	JP	US
scales	#labels (#labels in lcc / #cc)					
20km	86 (12/51)	52 (20/26)	99 (73/19)	131 (28/48)	99 (12/34)	403 (26/203)
50km	80 (39/9)	43 (39/4)	68 (66/2)	111 (87/5)	80 (69/7)	359 (88/89)
100km	69 (69/1)	33 (33/1)	37 (37/1)	68 (68/1)	49 (44/3)	288 (213/16)

4.2 Evaluation of the Consistency Models

Here, we evaluate the different consistency models introduced in Section 2. The models differ by the admissible number of active ranges per label and the handling of hard conflicts. We begin by analyzing the effect of limiting the number of active ranges and consider the five models 0/1, 1R, 2R, 3R, and ∞R, all taking hard conflicts into account. As discussed in Section 2, the 0/1-model is flicker-free but expected to have a low total activity, especially in dense instances. On the other hand, the ∞R-model achieves the maximum possible total activity in any valid labeling, but is likely to produce a large number of flickering effects. Still, it serves as an upper bound on the total activities of the other models. The two most important quality criteria in our evaluation are (i) the total activity of the solution, and (ii) the average length of the active ranges.

In Table 2 we report the total activity of the optimal solution for the tested models relative to the solution in the ∞R-model. The results of the instances are aggregated by scale. We observe that the total activity of the optimal solutions in the 0/1-model drops significantly, namely to less than 55% compared to the optimal solution in the ∞R-model even for the least dense instance at scale 20km and to only 6% for a scale of 100km. Hence this model is of very little interest in practice.

We see a strong increase in the average total activity values already for the 1R-model compared to the optimal solution in the 0/1-model. For the large-scale instance 20km 1R reaches almost 95% of the ∞R-model, which has more than 19 times the number of flickering effects and active ranges of average length shorter by a factor of $1/9$. For map scales of 50km and 100km, the total activities drop to 88% and 81%, respectively, but at the same time the number of flickering effects and the average active range lengths in the ∞R model are extremely poor. Thus the 1R-model achieves generally a very good labeling quality by using only one active range per label.

Finally, we take a look at the middle ground between the 1R- and the ∞R-models. It turns out that total activity of the 2R-model is off from the ∞R-model by less than 1% at scale 20km and less than 5% at scale 100km, but this increase in activity over the 1R model comes at the cost of producing twice as many flickering effects and decreasing the average active range length by 30–40%. If we allow three active ranges per label, the total activity increases to more than 99% of the upper bound in the ∞R-model at all three scales, while having significantly fewer flickering effects and longer average active ranges. The activity gain by considering the kR-model for $k > 3$ is negligible and the disadvantage of increasing the number of flickering effects dominates.

Table 2. Average total activity of the optimal solutions with respect to the maximum possible objective value. Instances grouped by scale. Additionally we report the average interval length normalized to one full rotation, and for the ∞R-model the average number of intervals per label.

model	0/1	1R		2R		3R		∞R	
scale	total act.	total act.	$\sim$len	total act.	$\sim$len	total act.	$\sim$len	$\sim$len	$\sim$intervals
20km	54.04%	94.56%	0.76	99.36%	0.56	99.92%	0.47	0.08	19.13
50km	22.42%	87.79%	0.58	97.69%	0.35	99.54%	0.26	0.01	79.22
100km	6.19%	81.01%	0.44	95.83%	0.27	99.24%	0.19	< 0.01	128.4

We conclude that the 1R-model achieves the best compromise between total activity value and low flickering, at least for maps at larger scales with lower feature density. For dense maps the 2R- or even the 3R-models yield near-optimal activity values while still keeping the flickering relatively low. Going beyond three active ranges per label only creates more flickering but does not provide noticeable additional value.

It remains to investigate the impact of hard conflicts. For this we apply the 1R-model and compare the variant where all conflicts are treated equally (soft-conflict model) with the variant where hard conflicts are disallowed (hard-conflict model). We consider for each map scale the average relative increase in activity value of the soft-conflict model over the stricter hard-conflict model. For 20km instances the increase is 8.51%, for the intermediate scale 50km it is 19.25%, and for the small-scale map 100km the increase reaches 31.9%. These results indicate that, unsurprisingly, the soft-conflict model improves the total activity at all scales, and in particular for dense configurations of point features, where labels usually have several hard conflicts with nearby features. As discussed before, this improvement comes at the cost of temporarily occluding unlabeled but possibly important points. It is an interesting open usability question to determine user preferences for the two models and the actual effect of temporary point occlusions on the readability of dynamic maps, but such a user study is out of scope of this evaluation and left as an interesting direction for future work.

4.3   Evaluation of the Algorithms

In this section we evaluate the quality (total activity) and running time of the ¼-approximation algorithm and the three greedy heuristics GreedyMax, GreedyLowCost, and GreedyBestRatio (Section 3.1), which we abbreviate as QAPX, GM, GLC, and GBR, respectively. Additionally, we include the ILP (Section 3.2) as the only exact method in the evaluation. The ILP is also applied to optimally solve the independent subinstances considered by QAPX. In our implementation we heuristically improve the running time of the ILP by partitioning the conflict graph of the labels into its connected components and solving each connected component individually; see Table 1 for the number of labels in the largest connected component and the number of connected components in the conflict graph of each instance. For the ILP we set a time limit of 1 hour and restrict the ILP solver to a single thread. The same restrictions are applied to the ILP when solving the small subinstances in algorithm QAPX. By the design of the algorithm, a solution obtained by **QAPX** will consist of many labels that have no active range, although they could be assigned one (all labels that are discarded to obtain independent cells have active range set to length 0). To overcome this drawback, we propose a combination of **QAPX** with the greedy algorithms. More specifically, we apply one of our greedy algorithms to each of the four solutions computed by the ¼-approximation and determine among the four resulting solutions the best one. In the following we refer to the combination of the ¼-approximation with a greedy algorithm by adding a Q in front of the greedy algorithm's name (e. g., **QGLC**). We report the results of the 1R-soft-conflict model, which turned out as a reasonable compromise between low flickering and high total activity in Section 4.2.

We give a general overview of the performance of all evaluated algorithms as a scatter plot (Fig. 2). In this scatter plot each disk represents the result of an algorithm (indicated by color) applied to a single country instance. The size of the disk indicates the scale of the instance (the smaller the disk, the smaller the scale). We omitted the algorithms **QGM** and **QGLC** in this plot to increase readability, because the difference in running time and quality of the solutions

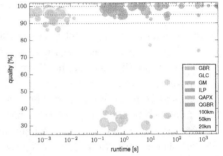

Fig. 2. Running time (log scale) and solution quality of the algorithms in the 1R soft-conflict model

between **QGM**, **QGLC**, and **QGBR** is negligible and creates extra overplotting.

We observe that the performance of the greedy algorithms is very good with respect to running time as well as quality of the solutions. As expected, the total activity of **QAPX** is always better than 25%, but generally much worse than for the remaining algorithms. It never gets close to the solutions produced by the greedy algorithms while being considerably slower. However, combining **QAPX** with a greedy algorithm achieves better solutions than greedy algorithms and **QAPX** alone, while the increase in running time over **QAPX** is negligible. Finally, we observe that the **ILP** solves the tested instances in a reasonable time frame. To obtain the optimal solution, the ILP required on average 758s, with a median of only 30.65s. However, we concede that larger instances may require significantly more time to solve, and may even be infeasible.

We now turn to a more detailed analysis of the two most promising approaches (i) using the greedy algorithms, and (ii) combining **QAPX** with the greedy algorithms. For a detailed depiction of the performance of the algorithms with respect to the quality of the solution see the diagrams in Fig. 3. We observe that among the three greedy algorithms **GBR** performs best with respect to quality with an average of 93.7%, but the difference to the other greedy algorithms is small. Even the greedy algorithm **GM** with the lowest total activity produces solution with an average of 91.8% of the optimal solution. Each of the combinations of **QAPX** with subsequent execution of a greedy algorithm outperforms each of the greedy algorithms alone in terms of quality. However, since the solutions produced by the greedy algorithms are already very close to the optimal solution, we observe only a slight increase in total activity for **QGM**, **QGLC**, and

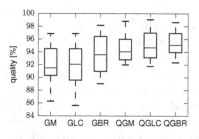

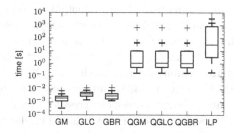

(a) Quality of the solutions as a percentage of the (optimal) ILP solution

(b) Running time (log scale) of the algorithms

Fig. 3. Performance of the greedy algorithms and **QAPX** with greedy postprocessing

QGBR over the greedy algorithms. The difference between both approaches becomes much more visible when considering the running time. While the average running time for the three greedy algorithms is between 2.5ms and 3.9ms, the average running time for the ¼-approximation algorithms is roughly 46s. However, we note that this large difference is mostly caused by one instance, which required over 664s to find the solution. The median running time for the enriched ¼-approximation algorithms is about 1.08s.

Our observations in this section are strengthened by the additional experiments (hard-conflict model and larger instances) reported in the full version of the paper [9], in which the performance of the greedy algorithms is even better. In order to give a final recommendation for an algorithm, it is necessary to make a choice on the time–quality trade-off that is acceptable in a particular situation. If running time is not the primary concern, e.g., for offline applications with high computing power available, we can recommend the ILP, which ran reasonably fast in our experiments. On the other hand, if computing power is limited and real-time labeling is necessary, e.g., on a mobile device, all three greedy heuristics can be recommended as the methods of choice; a slight advantage of **GBR** was observed in our experiments. All three algorithms run very fast (a few milliseconds) and empirically produce high activities of more than 90% of the optimum solution. If one wants to invest a few seconds of running time, the combination of **QAPX** with a greedy algorithm may be of interest as it produces slightly better solutions than the stand-alone greedy algorithms.

5   Conclusion

In this work, we evaluated different strategies for labeling dynamic maps that allow continuous rotation, where a labeling strategy consists of a consistency model and a labeling algorithm. In the first part of the evaluation, we considered the quality of optimal solutions in different consistency models. It turned out that the restriction to one or two active ranges per label (1R- and 2R-models) yields the best compromise in terms of low flickering and high total activity value of more than 95% of the upper bound obtained from the unrestricted model (∞R). Additionally, treating all pairwise label conflicts as soft conflicts increased the total activity values between 8% and 32% at the cost of occasional occlusion of unlabeled point features.

In the second part of the evaluation, we investigated the performance of three new greedy heuristics and our previous ¼-approximation algorithm [8] in terms of labeling quality and running time. It turned out that the greedy heuristics performed very well in both total activity (well above 90%) and running time (a few ms). The unmodified ¼-approximation performs much worse, but the combination of ¼-approximation and greedy heuristics yields slightly higher total activity than the greedy heuristics alone; the running time, however, can grow to several seconds. In conclusion, we believe that the 1R model in combination with any of the three greedy algorithms is, in most cases, the best labeling strategy for labeling dynamic rotating maps. Whether the soft-conflict or the hard-conflict model is more appropriate depends on requirements of the application.

Acknowledgment. This work was partially supported by a Google Research Award.

References

1. Agarwal, P.K., van Kreveld, M., Suri, S.: Label placement by maximum independent set in rectangles. Comput. Geom. Theory Appl. 11(3-4), 209–218 (1998)
2. Been, K., Daiches, E., Yap, C.: Dynamic map labeling. IEEE Trans. Vis. Comput. Graph. 12(5), 773–780 (2006)
3. Been, K., Nöllenburg, M., Poon, S.-H., Wolff, A.: Optimizing active ranges for consistent dynamic map labeling. Comput. Geom. Theory Appl. 43(3), 312–328 (2010)
4. de Berg, M., Gerrits, D.H.P.: Approximation algorithms for free-label maximization. Comput. Geom. Theory Appl. 45(4), 153–168 (2011)
5. de Berg, M., Gerrits, D.H.P.: Labeling moving points with a trade-off between label speed and label overlap. In: Bodlaender, H.L., Italiano, G.F. (eds.) ESA 2013. LNCS, vol. 8125, pp. 373–384. Springer, Heidelberg (2013)
6. Formann, M., Wagner, F.: A packing problem with applications to lettering of maps. In: Proc. 7th Ann. ACM Symp. Comput. Geom., pp. 281–288. ACM, New York (1991)
7. Gemsa, A., Niedermann, B., Nöllenburg, M.: Trajectory-based dynamic map labeling. In: Cai, L., Cheng, S.-W., Lam, T.-W. (eds.) ISAAC 2013. LNCS, vol. 8283, pp. 413–423. Springer, Heidelberg (2013)
8. Gemsa, A., Nöllenburg, M., Rutter, I.: Consistent labeling of rotating maps. In: Dehne, F., Iacono, J., Sack, J.-R. (eds.) WADS 2011. LNCS, vol. 6844, pp. 451–462. Springer, Heidelberg (2011)
9. Gemsa, A., Nöllenburg, M., Rutter, I.: Evaluation of Labeling Strategies for Rotating Maps. CoRR, abs/1404.1849 (2014)
10. Hochbaum, D.S., Maass, W.: Approximation schemes for covering and packing problems in image processing and VLSI. J. ACM 32(1), 130–136 (1985)
11. Nöllenburg, M., Polishchuk, V., Sysikaski, M.: Dynamic one-sided boundary labeling. In: Proc. 18th ACM SIGSPATIAL GIS, pp. 310–319. ACM, New York (2010)
12. Ooms, K., Kellens, W., Fack, V.: Dynamic map labeling for users. In: Proc. 24th Internat. Cartographic Conf., pp. 1–12. Military Geographic Institute, Santiago (2009)
13. van Kreveld, M., Strijk, T., Wolff, A.: Point labeling with sliding labels. Comput. Geom. Theory Appl. 13(1), 21–47 (1999)
14. Vaaraniemi, M., Treib, M., Westermann, R.: Temporally coherent real-time labeling of dynamic scenes. In: Proc. 3rd Internat. Conf. Computing for Geospatial Research and Applications, pp. 17:1–17:10. ACM, New York (2012)
15. Yokosuka, Y., Imai, K.: Polynomial time algorithms for label size maximization on rotating maps. In: Proc. 25th Canadian Conf. Comput. Geom., University of Waterloo, Waterloo, pp. 187–192 (2013)

Experimental Comparison of Semantic Word Clouds

Lukas Barth[1], Stephen G. Kobourov[2,*], and Sergey Pupyrev[2,3,*]

[1] Institute of Theoretical Informatics, Karlsruhe Institute of Technology, Karlsruhe, Germany
[2] Department of Computer Science, University of Arizona, Tucson, AZ, USA
[3] Institute of Mathematics and Computer Science,
Ural Federal University, Ekaterinburg, Russia

Abstract. We study the problem of computing semantics-preserving word clouds in which semantically related words are close to each other. We implement three earlier algorithms for creating word clouds and three new ones. We define several metrics for quantitative evaluation of the resulting layouts. Then the algorithms are compared according to these metrics, using two data sets of documents from Wikipedia and research papers. We show that two of our new algorithms outperform all the others by placing many more pairs of related words so that their bounding boxes are adjacent. Moreover, this improvement is not achieved at the expense of significantly worsened measurements for the other metrics.

1 Introduction

In the last few years, word clouds have become a standard tool for abstracting, visualizing, and comparing text documents. For example, word clouds were used in 2008 to contrast the speeches of the US presidential candidates Obama and McCain. Word clouds, or their close relatives tag clouds, are often used to represent importance among items (e.g., bands popularity on Last.fm) or serve as a navigation tool (e.g., Google search results).

A practical tool, Wordle [12], with its high quality design, graphics, style and functionality popularized word cloud visualizations as an appealing way to summarize the text. While tools like this are popular and widely used, most of them, including Wordle itself, have a potential shortcoming: they do not capture the relationships between the words in any way, as word placement is independent of context. But humans, as natural pattern-seekers, cannot help but perceive two words that are placed next to each other in a word cloud as being related in some way. In linguistics and in natural language processing if a pair of words often appears together in a sentence, then this is seen as evidence that this pair of words is linked semantically [7]. Thus, when using a word cloud, it makes sense to place such related words close to each other; see Fig. 1. In fact, recent empirical studies show that semantically clustered word clouds provide improvements over random layouts in specific tasks such as searching and recognition, and have higher user satisfaction compared to other layouts [11,3].

Nearly all recent word cloud visualization tools aim to incorporate semantics in the layout [2,5,13]. However, none provide any guarantees about the quality of the layout in terms of semantics. The existing algorithms are usually based on force-directed

---

* Supported in part by NSF grants CCF-1115971 and DEB 1053573.

J. Gudmundsson and J. Katajainen (Eds.): SEA 2014, LNCS 8504, pp. 247–258, 2014.

(a) RANDOM   (b) STAR FOREST

Fig. 1. Word clouds for the Wikipedia page "Copenhagen" (top 50 words)

graph layout heuristics to add such functionality. In contrast, we propose several new algorithms with such performance guarantees. Consider the following natural formal model of semantics-preserving word cloud visualization, based on a vertex-weighted and edge-weighted graph [1]. The vertices in the graph are the words in the document, with weights corresponding to some measure of importance (e.g., word frequency). The edges capture the semantic relatedness between pairs of words (e.g., co-occurrence), with weights corresponding to the strength of the relation. Each vertex must be drawn as an axis-aligned rectangle (box) with fixed dimensions determined by its weight. A contact between two boxes is a common boundary, and if two boxes are in contact, these boxes are called *touching*. The *realized adjacencies* are the sum of the edge weights for all pairs of touching boxes. The goal is a representation of the given boxes, maximizing the realized adjacencies.

In this paper we first define metrics that quantitatively measure the more abstract goal of "semantic preservation": realized adjacencies, distortion, compactness, and uniform area utilization. Then we implement and extensively test six algorithms for generating word clouds: three earlier methods and three new ones. Two of the new algorithms outperform the rest in terms of realized adjacencies, while not negatively impacting any of the remaining metrics. The online system implementing the algorithms, and which also provides source code and data sets is available at http://wordcloud.cs.arizona.edu.

2   Experimental Setup

All the algorithms for producing word clouds take as input an edge-weighted graph and rectangles of fixed dimensions associated with every vertex. There are several parameters to consider in a preprocessing step, needed to extract this information from input texts. Our preprocessing pipeline is illustrated in Fig. 2.

Term Extraction. We first split the input text into sentences, which are then tokenized into a collection of words using the open-source toolkit Apache OpenNLP. Common stop-words such as "a", "the", "is" are removed from the collection. The remaining words are grouped by their stems using the Porter Stemming Algorithm [10], so that related words such as "dance", "dancer", and "dancing" are reduced to their root, "danc". The most common variation of the word is used in the final word cloud.

Ranking. In the next step we rank the words in order of relative importance. We have three different ranking functions, depending on word usage in the input text.

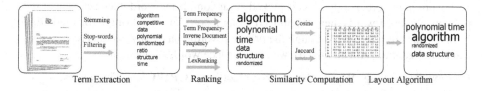

Fig. 2. Overview of creating a semantic word cloud visualization

Each ranking function orders words by their assigned weight (rank), and the top n of them are selected, where n is the number of words shown in the word cloud. *Term Frequency* (TF) is the most basic ranking function and one used in most traditional word cloud visualizations. Even after removing common stop-words, however, TF tends to rank highly many semantically meaningless words. *Term Frequency-Inverse Document Frequency* (TF-IDF) addresses this problem by normalizing the frequency of a word by its frequency in a larger text collection. In our case TF-IDF$(w, d, D) = $ TF$(w, d) \times$ IDF(w, D), where w is the word, d is the current input text and D is the collection of all our text documents. Our third ranking function uses the *LexRank* algorithm [4], a graph-based method for computing relative importance of words, and already used for semantic-preserving word clouds [13]. In the graph $G = (V, E)$, vertices represent words and edges represent co-occurrence of the words in the same sentences. A weight w_{ij} of an edge (i, j) is the number of times word i appears in the same sentence as word j. The rank values are computed using eigenvector centrality in G.

Similarity Computation. Given the ranked list of words, we calculate an $n \times n$ matrix of pairwise similarities so that related words receive high similarity values. We use two similarity functions depending on the input text. The *Cosine Similarity* between words i and j can be computed as $sim_{ij} = \frac{w_i \cdot w_j}{|w_i| \cdot |w_j|}$, where $w_i = (w_{i1}, \ldots, w_{in})$ and $w_j = (w_{j1}, \ldots, w_{jn})$ are the vectors representing co-occurrence of the words with other words in the input text. The *Jaccard Similarity* coefficient is the number of sentences two words appeared together in divided by the number of sentences either word appeared in: $sim_{ij} = \frac{|S_i \cap S_j|}{|S_i \cup S_j|}$, where S_i is the set of sentences containing word i. In both cases the similarity function produces a value between 0, indicating that a pair of words is not related, and 1, indicating that words are very similar.

3   Word Cloud Layout Algorithms

Here we briefly describe three early and three new word cloud layout algorithms, all of which we implemented and tested extensively. The input for all algorithms is a collection of n rectangles, each with a fixed width and height proportional to the rank of the word, together with $n \times n$ matrix with entries $0 \leq sim_{ij} \leq 1$. The output is a set of non-overlapping positions for the rectangles.

Wordle (RANDOM). The algorithm places one word at a time in a greedy fashion, aiming to use space as efficiently as possible [12]. First the words are sorted by weight (proportional to the height of the corresponding rectangle) in decreasing order. Then for each word in the order, a position is picked at random. If the word intersects one of the previously placed words then it is moved along a spiral of increasing radius radiating from its starting position. Although the original Wordle has many variants (e.g., words can be horizontal, vertical, mixed), we always place words horizontally.

Context-Preserving Word Cloud Visualization (CPWCV). The algorithm of Cui *et al.* [2] aims to capture semantics in two steps. First a dissimilarity matrix Δ is computed, where $\Delta_{ij} = 1 - sim_{ij}$ represents ideal distances between words i and j in n-dimensional space. Multidimensional scaling (MDS) is performed to obtain two-dimensional positions for the words so that the given distances are (approximately) preserved. Since the step usually creates a very sparse layout, the second step compacts the layout via a force-directed algorithm. Attractive forces between pairs of words reduce empty space, while repulsive forces ensure that words do not overlap. An additional force attempts to preserve semantic relations between words. To this end, a triangular mesh (Delaunay triangulation in the implementation) is computed from the initial word positions, and the additional force attempts to keep the mesh planar.

Seam Carving. The algorithm of Wu *et al.* [13] uses seam carving, a content-aware image resizing technique, to capture semantics. Here a preliminary overlap-free word layout is computed, using a force-directed algorithm adapted from CPWCV [2]. Then the screen space is divided into regions, and for each region an energy function is computed. A connected left-to-right or top-to-bottom path of low energy regions is called a seam. The major step of the algorithm is to iteratively carve out seams of low energy to remove empty spaces between words. Since the order of seam removal greatly affects the final result, a dynamic programming approach is used to find an optimal order. The final result is a word cloud in which no further seam can be removed.

Inflate-and-Push (INFLATE). We designed and implemented an alternative simple heuristic method for word layout, which aims to preserve semantic relations between pairs of words. The heuristic starts by scaling down all word rectangles by some constant $S > 0$ (in our implementation $S = 100$) and computing MDS on dissimilarity matrix Δ in which $\Delta_{ij} = \frac{1 - sim_{ij}}{S}$. At this point, the positions of the words respect their semantic relations; that is, semantically similar words are located close to each other. We then iteratively increase the dimensions of all rectangles by 5% ("inflate"). After each iteration some words may overlap. We resolve the overlaps using the repulsive forces from the force-directed model of the CPWCV algorithm ("push"). Since the dimensions of each rectangle grows by only 5%, the forces generally preserve relative positions of the words. In practice, 50 iterations of the "inflate-push" procedure suffice.

Star Forest. A star is a tree of depth at most 1, and in a star forest all connected components are stars. Our algorithm has three steps. First we partition the given graph, obtained from the dissimilarity matrix Δ, into disjoint stars (extracting a star forest). Then we build a word cloud for every star (realizing a star). Finally, the individual solutions are packed together to produce the result. The steps are described next.

We extract stars from the given graph greedily. We find a vertex v with the maximum adjacent weight, that is, the one for which $\sum_{u \in V} sim(v, u)$ is maximized. We then treat the vertex as a star center and the vertices $V \setminus \{v\}$ as leaves. A set of words that will

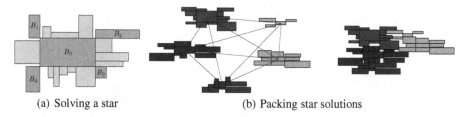

(a) Solving a star   (b) Packing star solutions

Fig. 3. Star Forest algorithm

be adjacent to star center v is computed, and these words are removed from the graph. This processes is repeated with the smaller graph, until the graph is empty.

Selecting the best set of words to be adjacent to a star center v is related to the Knapsack problem, where given a set of items, each with a size and a value, we want to pack a maximum-valued subset of items into a knapsack of given capacity. Let B_0 denote the box corresponding to the center of the star; see Fig. 3(a). In any optimal solution there are four boxes B_1, B_2, B_3, B_4 whose sides contain one corner of the center box B_0 each. Given B_1, B_2, B_3, B_4, the problem reduces to assigning each remaining box B_i to at most one of the four sides of B_0, which completely contains the contact between B_i and B_0. The task of assigning boxes to a side of B_0 is naturally converted to an instance of Knapsack: The size of a box B_i is its width for the horizontal sides and its height for the vertical sides of B_0, the value is the edge weight between B_i and B_0. Now we run the algorithm for the Knapsack problem for the top side of B_0, remove the realized boxes, and proceed with the bottom, left, and then right sides of the central box. To solve the Knapsack instances, we use the polynomial-time approximation scheme described in [6].

Finally, the computed solutions for individual stars are packed together in a compact drawing, which preserves the semantic relations between words in different stars. We begin by placing the stars on the plane so that no pair of words overlap; see Fig. 3(b). For every pair of stars s_1, s_2, we compute its similarity as the average similarity between the words comprising s_1 and s_2, that is, $sim(s_1, s_2) = \frac{\sum_{v \in s_1} \sum_{u \in s_2} sim(u,v)}{|s_1||s_2|}$, where $|s_1|, |s_2|$ are the number of words in the stars. MDS is utilized to find the initial layout with $k(1 - sim(s_1, s_2))$ being the ideal distance between the pair of stars. In our implementation, we set the scaling factor $k = 10$ to ensure an overlap-free result. Then a force-directed algorithm is employed to obtain a compact layout. Note that the algorithm adjusts the positions of whole stars rather than individual words; thus, the already realized adjacencies between words are preserved. The algorithm utilizes attractive forces aiming at removing empty space and placing semantically similar words close to each other. The force between the stars is defined as $f_a(s_1, s_2) = k_a(1 - sim(s_1, s_2))\Delta l$, where Δl represents the minimum distance between two central rectangles of the stars. The repulsive force is used to prevent overlaps between words. The force only exists if two words occlude each other. It is defined as $f_r(s_1, s_2) = k_r min(\Delta x, \Delta y)$, where Δx (Δy) is the width (height) of the overlapping region. We found that the priorities of the forces $k_a = 15, k_r = 500$ work well. As in a classical force-directed scheme, the algorithm iteratively adjust positions of the stars. We impose a limit of 500 iterations, although the process converges very quickly.

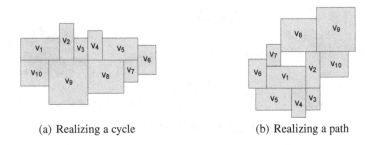

(a) Realizing a cycle   (b) Realizing a path

Fig. 4. The CYCLE COVER algorithm

Cycle Cover. This algorithm is also based on extracting a heavy planar subgraph from the graph $G = (V, E)$ defined by the similarity matrix. In particular, finding a cycle cover (vertex-disjoint set of cycles) with maximum weight, and realizing all cycles in the cover by touching boxes is a $\frac{2}{d_{max}+1}$ approximation algorithm for total realized adjacencies, for G with maximum degree d_{max}.

Although the heaviest cycle cover can be found in polynomial time in the size of the graph (see Chapter 10 in [8]), the algorithm is too slow in practice as it requires computation of a maximum weighted matching in a not necessarily bipartite graph. We use the following simpler algorithm instead, which transforms G into a related bipartite graph H. We create $V(H)$ by first copying all vertices of G. Then for each vertex $v \in V(G)$, we add a new vertex $v' \in V(H)$, and for each edge $(u, v) \in E(G)$, we create two edges $(u', v) \in E(H)$ and $(u, v') \in E(H)$ with weights $sim(u, v)$. The resulting graph H is bipartite by construction, thus making it easy to compute a maximum weight matching. The matching induces a set of vertex-disjoint paths and cycles in G as every u is matched with one v' and every u' is matched with one v.

Once cycles and paths are extracted, we place the corresponding boxes so that all edges are realized as follows. For a given cycle $(v_1, v_2, \ldots, v_n)$, let t be the maximum index such that $\sum_{i \leq t} w_i < \sum_{i \leq n} w_i / 2$, where w_i is the width of the i-th word. Vertices $v_1, v_2, \ldots, v_t$ are placed side by side in order from left to right with their bottom sides aligned on a shared horizontal line, while vertices $v_n, v_{n-1}, \ldots, v_{t+2}$ are placed from left to right with their top sides aligned on the same line; see Fig. 4(a). It remains to place v_{t+1} in contact with v_t and v_{t+2}, which can be done by adding v_{t+1} to the side of minimum width, or straddling the line in case of a tie. It is possible that the resulting layout has poor aspect ratio (as cycles can be long), so we convert cycles with more than 10 vertices into paths by removing the lightest edge. When realizing the edges of a path, we start with words v_1 and v_2 placed next to each other. The i-th word is added to the layout so that it touches v_{i-1} using its first available side in clockwise order radiating from the side of the contact between v_{i-2} and v_{i-1}. This strategy tends to create more compact, spiral-like layouts; see Fig. 4(b).

In the final step, the computed solutions for individual cycles and paths are packed together, aiming for a compact drawing which preserves the semantic relations between words in different groups. We use the same force-directed algorithm (described in the previous section for STAR FOREST) for that task.

4   Metrics for Evaluating Word Cloud Layouts

While a great deal of the world cloud appeal is *qualitative* and depends on good graphic design, visual appeal, etc., we concentrate of *quantitative* metrics that capture several desirable properties. We use these metrics to evaluate the quality of the six algorithms under consideration. The metrics are defined so that the measurement is a real number in the range $[0, 1]$ with 1 indicating the best value and 0 indicating the worst one.

Realized Adjacencies. Our primary quality criterion for semantics-preserving algorithms is the total realized adjacency weight. In practice, proper contacts are not strictly necessary; even if two words do not share a non-trivial boundary, they can be considered as "adjacent" if they are located very close to each other. We assume that two rectangles touch each other if the distance between their boundaries is less than 1% of the width of the smaller rectangle. For each pair of touching rectangles, the weight of the edge is added to the sum. We normalize the sum by dividing it by the sum of all edge weights, thus measuring the fraction of the adjacency weight realized. Hence, the metric is defined by $\frac{\sum_{(u,v)\in E_{realized}} sim(u,v)}{\sum_{(u,v)\in E} sim(u,v)}$. Note that this value is always less than 1 as the input graph (as described in Section 2) is a complete graph and thus highly non-planar. On the other hand, the contact graph of rectangles is always planar, which means that in most cases it is impossible to realize contacts between all pairs of words.

Distortion. This metric is used to measure another aspect of how well the desired similarities are realized, by comparing the distances between all pairs of rectangles to the desired dissimilarities of the words. In order to compute the metric for a given layout, we construct a matrix of ideal distances between the words with entry $\Delta_{uv} = 1 - sim(u, v)$ for words u and v, and a matrix of actual distances between words in which an entry d_{uv} denotes the distance between the centers of u and v. We modify the definition of distance to reflect the use of non-zero area boxes for vertices (instead of points), by measuring the distance between two words as the minimal distance between any pair of points on the corresponding two rectangles. We then consider the matrices as two random variables and compute the correlation coefficient between them:

$$r = 1 - \frac{\sum_{(u,v)\in E}(\Delta_{uv} - \overline{\Delta})(d_{uv} - \overline{d})}{\sqrt{\sum_{(u,v)\in E}(\Delta_{uv} - \overline{\Delta})^2 \sum_{(u,v)\in E}(d_{uv} - \overline{d})^2}},$$

where $\overline{\Delta}$ and $\overline{d}$ are the average values of the corresponding distances. Since the correlation coefficient takes values between -1 and 1, *Distortion* is defined by $(r + 1)/2$. The value of 1 indicates a perfect correspondence between dissimilarities and distances, while 0.5 means that the values are independent.

Compactness. Efficient use of drawing area is a natural goal, captured by this metric. We first compute the *used area* as the area covered by words, by adding up the areas of all rectangles. Clearly, any layout needs at least that much space as overlaps are not allowed in our setting. We then compute the *total area* of the layout using the bounding box containing all rectangles, or the area of the convex hull of all rectangles. The metric is $\frac{used\ area}{total\ area}$, with value 1 corresponding to the best possible packing.

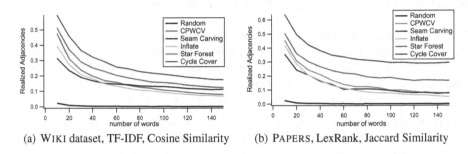

(a) WIKI dataset, TF-IDF, Cosine Similarity   (b) PAPERS, LexRank, Jaccard Similarity

Fig. 5. Realized adjacencies for word clouds of various size

Uniform Area Utilization. Arguably, a desired property of a word cloud is the randomness of the distribution of the words in the layout. In highly non-random distributions some parts of the cloud are densely populated while others are sparse. The metric captures the uniformity of the word distribution. We first divide the drawing of a word cloud with n words into $\sqrt{n} \times \sqrt{n}$ cells. Then for each rectangle, find indices (x, y) of the cell containing the center of the rectangle. We may now consider the words as a discrete random variable, and measure its information entropy, or the amount of uncertainty. To this end, we compute the relative entropy of the variable (induced by a given layout of words) with respect to the uniform distribution: $H = \sum_{i,j} p(i,j) \log \frac{p(i,j)}{q(i,j)}$, where the sum is taken over the created cells, $p(i,j)$ is the actual number of words in the cell (i,j), and $q(i,j) = 1$ is the number of words in the cell in the uniformly distributed word cloud. Since the entropy is maximized as $\log n$, the metric is defined by $1 - \frac{H}{\log n}$, where the value of 1 corresponds to an "ideal" word cloud.

Aspect Ratio and Running Time. While not formal evaluation metric for word cloud visualization, we also measure the aspect ratio of the final layouts and the running times of the algorithms. We use a standard measurement for the *Aspect Ratio* of the word clouds: W/H, where W and H are the width and height of the bounding box. The running time of word cloud algorithms is an important parameter as many such tools are expected to work in real-time and delays of more than a few seconds would negatively impact their utility. We measure *Running Time* for the execution of the layout phase of the evaluated algorithms, excluding the time needed to parse text, compute similarities between words, and draw the result on the display.

5   Results and Discussion

We evaluate all algorithms on two datasets: (1) WIKI, a set of 112 plain-text articles extracted from the English Wikipedia, each consisting of at least 200 distinct words, and (2) PAPERS, a set of 56 research papers published in conferences on experimental algorithms (SEA and ALENEX) in 2011-2012. The texts are preprocessed using the pipeline described in Section 2; that is, for every trial, the words are first ranked and then pairwise similarities between the top $10 \leq n \leq 150$ of them are computed. Every algorithm is executed 5 times on each text; hence, the results present average

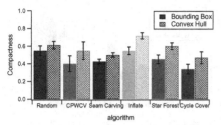 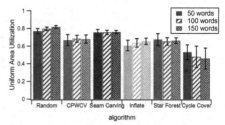

(a) WIKI, TF-IDF Ranking, Jaccard Similarity   (b) PAPERS, Lex Ranking, Cosine Similarity

Fig. 6. (a) Mean and standard deviation of *Compactness* for texts with 100 words. (b) Mean and standard deviation of *Uniform Area Utilization* for word clouds of various size.

values of metrics over the runs. Our implementation is in Java, and the experiments were performed on a machine with an Intel i5 3.2GHz processor with 8GB RAM.

Realized Adjacencies. The CYCLE COVER algorithm outperforms all other algorithms on all tested instances; see Fig. 5. Depending on the method for extracting pairwise word similarities, the algorithm realizes 30-50% of the total edge weights for small graphs (up to 50 words) and $15 - 30\%$ for larger graphs. It is worth noting here that CYCLE COVER improves upon the best existing heuristic by a factor of nearly 2. Furthermore, CYCLE COVER performs better than all the existing heuristics for almost all inputs. STAR FOREST also realizes more adjacencies than the existing semantics-aware methods. However, for large word clouds (with over 100 words) the difference between STAR FOREST and the existing algorithms drops to $2 - 5\%$.

It is noticeable that the CPWCV algorithm and our INFLATE algorithm perform similarly in all settings. This is the expected behavior, as both methods are based on similar force-directed models. For instances with up to $80 - 100$ words, these techniques are better than the SEAM CARVING algorithm. On the other hand, the more sophisticated SEAM CARVING algorithm is preferable for larger word clouds, which confirms earlier results [13]. Not surprisingly, the RANDOM algorithm realizes only a small fraction of the total weight as it is not designed for preserving semantics.

Compactness. We measure compactness using the bounding box and convex hull of the layout; see Fig. 6(a). We first observe that the results in general do not depend on the used dataset, the ranking algorithm, and the way similarities are computed. The word clouds constructed by the INFLATE algorithm are the most compact, while the remaining algorithms have similar performance. In practice, such "overcompacted" drawings are not very desirable since adjacent words are difficult to read. Hence, we increase the dimensions of each rectangle by $10 - 20\%$ and run the algorithm for the new instance.

Uniform Area Utilization. As expected, the RANDOM algorithm generates word clouds with the most uniform word placement; see Fig. 6(b). SEAM CARVING also utilizes area uniformly. The remaining methods are all mostly comparable, with CYCLE COVER being the worst. The standard deviation for these 4 algorithms is relatively high, which means that some of the created word clouds may be rather non-uniform. Similar to the *Compactness* metric, we do not observe any significant difference of area utilization for different setting (dataset, ranking and similarity algorithms).

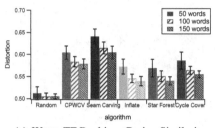 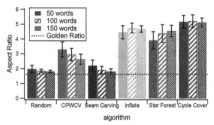

(a) WIKI, TF Ranking, Cosine Similarity   (b) PAPERS, TF Ranking, Cosine Similarity

Fig. 7. (a) Measurements of *Distortion*. Note that the y-axis starts at value 0.5. (b) *Aspect Ratio*.

Distortion. The SEAM CARVING algorithm steadily produces the most undistorted layouts (note as usual, although a bit counterintuitive here, high values are good); Fig. 7(a). Not surprisingly, the correlation coefficient between the dissimilarity matrix and the actual distance matrix produced by RANDOM is very close to 0 (hence, the value of *Distortion* is 0.5). For the remaining algorithms the results are slightly worse, but comparable. Note that all considered algorithms (except RANDOM) have a common feature: they start with an initial layout for the words (or for clusters of words) constructed by multidimensional scaling. At that point, the *Distortion* measurements are generally very good, but the layout is sparse. Next the algorithms try to compact the drawing, and the compaction step worsens *Distortion* significantly. Hence, there is an inherent tradeoff between compactness and distortion.

Aspect Ratio and Running Time. RANDOM and SEAM CARVING produce word clouds with aspect ratio close to the golden ratio ($\frac{1+\sqrt{5}}{2}$), which is commonly believed to be aesthetically pleasing; see Fig. 7(b). INFLATE, STAR FOREST, and CYCLE COVER generate drawings with the aspect ratio from $7:2$ to $9:2$, and the measurements are mostly independent of the number of words in a word cloud. We emphasize here that none of the considered algorithms is designed to preserve a specific aspect ratio. If this turns out to be a major aesthetic parameter, optimizing the layout algorithms, while maintaining the desired aspect ratio might be a meaningful direction for future work.

The running times of all but one algorithm are reasonable, dealing with $100 - 150$ words within 2 seconds; see Fig. 8. The exception is SEAM CARVING, which requires $15 - 25$ seconds per graph on the PAPERS dataset and $30 - 50$ on the WIKI dataset.

Discussion. The CYCLE COVER and the STAR FOREST algorithms are better at realizing adjacencies, and they are comparable to the other algorithms in compactness, area utilization, and distortion. A likely explanation is that the distribution of edge weights is highly non-uniform for real-world texts and articles; see Fig. 9(a). The weights follow a Zipf distribution with many light edges and few heavy ones. Further, a small fraction of the edges (about 8% for WIKI and 5% for PAPERS) carry half of the total edge weight. It is known that words in text documents follow this distribution (Chapter 5 in [9]) and that might explain why pairs of co-related words behave similarly. Both CYCLE COVER and STAR FOREST are based on the same principle: first extract "heavy" subgraphs and then realize contacts within each subgraph independently. On the other hand, the existing semantics-preserving approaches try to optimize all contacts simultaneously by

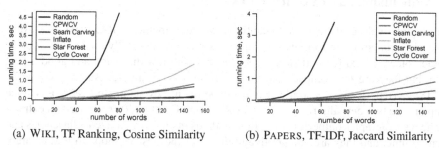

(a) WIKI, TF Ranking, Cosine Similarity   (b) PAPERS, TF-IDF, Jaccard Similarity

Fig. 8. Running time of the algorithms for word clouds of various size

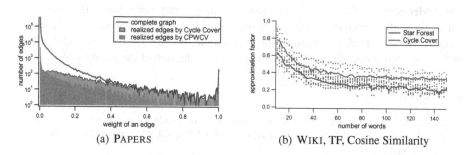

(a) PAPERS   (b) WIKI, TF, Cosine Similarity

Fig. 9. (a) Distribution of similarities between pairs of words among edges (red line), and the fraction of realized edges with a given weight (closed regions) for a new and an existing algorithms. (b) Comparing the realized weight to the upper bound for maximum realized edge weights. Dots represent single instances, solid lines represent average values over 5 runs.

considering the complete graph. Our new strategy is more effective as it realizes most of the heavy edges; see Fig. 9(a). It is possible that in a different scenario (where edge weights are not similarities between words) the distribution may be close to uniform; even in such a scenario CYCLE COVER and STAR FOREST outperform existing methods, but the fraction of the realized edge weights is smaller.

None of the presented semantics-preserving algorithms make any guarantees about the optimality of realized edge weights when the input graph is complete (as in real-world examples). However, it is still interesting to analyze how well these two algorithms realize adjacencies. Note that the sum of all edge weights in a graph is not a good upper bound for an optimal solution since the realized subgraph is necessarily planar and thus contains at most $3n - 6$ edges. Instead, we compute a maximum weight spanning tree of the graph. Since the weight w of the tree is a $1/3$-approximation for the maximum planar subgraph of G, the value of $3w$ is also an upper bound for the maximum realized edge weights. On average CYCLE COVER produces results which are at most 3 times less than the optimum for graphs with 150 vertices; see Fig. 9(b).

6   Conclusions and Future Work

We quantitatively evaluated six semantic word cloud methods. RANDOM uses available area in the most uniform way, but does not place semantically related words close to each other. SEAM CARVING produces layouts with low distortion and good aspect ratio, but they are not compact and the algorithm is very time-consuming. CPWCV and INFLATE perform very similarly in our experiments, even though INFLATE is much simpler. The two new algorithms STAR FOREST and CYCLE COVER, based on extracting heavy subgraphs, outperform all other methods in terms of realized adjacencies and running time, and they are competitive in the other metrics. We hope to find an algorithm with guaranteed approximation factor for extracting heavy planar subgraphs from general graphs, which would lead to a guaranteed fraction of realized adjacencies.

Acknowledgements. The problem of semantic-preserving word cloud algorithms began at Dagstuhl Seminar 12261. We thank organizers and participants, in particular T. Biedl. We also thank P. Gopalakrishnan, S. Pande, B. Yee, J. Li, J. Lu, and Z. Zhou for productive discussions and help with the implementation of the online tool.

References

1. Barth, L., et al.: Semantic word cloud representations: Hardness and approximation algorithms. In: Pardo, A., Viola, A. (eds.) LATIN 2014. LNCS, vol. 8392, pp. 514–525. Springer, Heidelberg (2014)
2. Cui, W., Wu, Y., Liu, S., Wei, F., Zhou, M., Qu, H.: Context-preserving dynamic word cloud visualization. IEEE Comput. Graph. Appl. 30(6), 42–53 (2010)
3. Deutsch, S., Schrammel, J., Tscheligi, M.: Comparing different layouts of tag clouds: Findings on visual perception. In: Ebert, A., Dix, A., Gershon, N.D., Pohl, M. (eds.) HCIV (INTERACT) 2009. LNCS, vol. 6431, pp. 23–37. Springer, Heidelberg (2011)
4. Erkan, G., Radev, D.R.: LexRank: Graph-based lexical centrality as salience in text summarization. J. Artificial Intelligence Res. 22(1), 457–479 (2004)
5. Koh, K., Lee, B., Kim, B.H., Seo, J.: ManiWordle: Providing flexible control over Wordle. IEEE Trans. Vis. Comput. Graphics 16(6), 1190–1197 (2010)
6. Lawler, E.L.: Fast approximation algorithms for knapsack problems. Math. Oper. Res. 4(4), 339–356 (1979)
7. Li, H., Abe, N.: Word clustering and disambiguation based on co-occurrence data. In: Int. Conf. Comput. Linguistics, vol. 2, pp. 749–755. Association for Computational Linguistics, Stroudsburg (1998)
8. Lovász, L., Plummer, M.: Matching Theory. Akadémiai Kiadó, Budapest (1986)
9. Manning, C.D., Raghavan, P., Schütze, H.: Introduction to Information Retrieval. Cambridge University Press, Cambridge (2008)
10. Porter, M.F.: An algorithm for suffix stripping. Program: Electron. Lib. 14(3), 130–137 (1980)
11. Schrammel, J., Leitner, M., Tscheligi, M.: Semantically structured tag clouds: An empirical evaluation of clustered presentation approaches. In: SIGCHI Conference on Human Factors in Computing Systems, pp. 2037–2040. ACM, New York (2009)
12. Viégas, F.B., Wattenberg, M., Feinberg, J.: Participatory visualization with Wordle. IEEE Trans. Vis. Comput. Graphics 15(6), 1137–1144 (2009)
13. Wu, Y., Provan, T., Wei, F., Liu, S., Ma, K.L.: Semantic-preserving word clouds by seam carving. Comput. Graph. Forum 30(3), 741–750 (2011)

Hub Labels: Theory and Practice[*]

Daniel Delling[1], Andrew V. Goldberg[1], Ruslan Savchenko[2],
and Renato F. Werneck[1]

[1] Microsoft Research, 1065 La Avenida, Mountain View, CA 94043, USA
{dadellin,goldberg,renatow}@microsoft.com
[2] Department of Mech. and Math., Moscow State University
1 Leninskiye Gory, Moscow 119991, Russia
ruslan.savchenko@gmail.com

Abstract. The Hub Labeling algorithm (HL) is an exact shortest path algorithm with excellent query performance on some classes of problems. It precomputes some auxiliary information (stored as a *label*) for each vertex, and its query performance depends *only* on the label size. While there are polynomial-time approximation algorithms to find labels of approximately optimal size, practical solutions use hierarchical hub labels (HHL), which are faster to compute but offer no guarantee on the label size. We improve the theoretical and practical performance of the HL approximation algorithms, enabling us to compute such labels for moderately large problems. Our comparison shows that HHL algorithms scale much better and find labels that usually are not much bigger than the theoretically justified HL labels.

1 Introduction

In this paper we study the *hub labeling* (HL) algorithm [12], a powerful technique to answer exact point-to-point shortest path queries with excellent performance on some real-world networks. Although HL has been studied both theoretically and experimentally [2,3,4,5,6,8,21], in practice one uses fast heuristics instead of the algorithms with theoretical solution quality guarantee. Since the heuristics are used on real-life networks [2,3,4,5], it is important to gauge their solution quality and their potential for improvement.

The theoretically justified algorithms have escaped extensive experimental studies in the past because, although polynomial, they do not scale well. Our goal is to speed up the algorithms without losing the theoretical guarantees. Even if the algorithms do not scale as well as the heuristics, experiments on moderate-size problems give a useful measure of solution quality. A small gap would justify the heuristics, while a large gap would motivate their improvement.

During preprocessing, labeling algorithms [19] compute a *label* for every vertex of the graph and answer *s–t* shortest path queries using only the labels of *s* and *t* (and not the graph itself). HL is a labeling algorithm where the label $L(v)$ of v is a collection of vertices (*hubs* of v) with distances between v and the hubs. The

---

[*] Part of this work done while the author was at Microsoft Research.

J. Gudmundsson and J. Katajainen (Eds.): SEA 2014, LNCS 8504, pp. 259–270, 2014.
© Springer International Publishing Switzerland 2014

label $L(v)$ consists of two parts, the *forward label* $L_f(v)$ and the *backward label* $L_b(v)$. The labels must obey the *cover property*: for any two vertices s and t, the set $L_f(s) \cap L_b(t)$ must contain at least one hub v that is on the shortest s–t path. Given the labels, HL queries are straightforward: to find dist(s,t), simply find the hub $v \in L_f(s) \cap L_b(t)$ that minimizes dist(s,v) + dist(v,t). The *size* of the label is its number of hubs. The memory footprint of the algorithm is dominated by the sum of all label sizes, while query times are determined by the maximum label size. To measure solution quality, we mostly use the average label size, which is equivalent to the sum of all forward and backward label sizes; in practice, the maximum label size is not much higher than the average.

Finding the smallest HL for general graphs is NP-hard. Cohen et al. [8] give an $O(n^5)$-time $O(\log n)$-approximation algorithm for minimizing the size of the labeling. Babenko et al. [6] generalize this result to an $O(\log n)$-approximation algorithm for general p-norms (as defined in Section 2); we refer to their algorithm as GHL^p.

A special case of HL is *hierarchical hub labeling* (HHL), where vertices are globally ranked by "importance" and the label for a vertex can only have more important hubs than itself. HHL labels can be polynomially bigger than HL labels for some graph classes [14], but small HHL labels exist for other classes, such as trees. For general graphs, finding the smallest HHL is NP-hard [15], and no polylog-approximation algorithm is known. Practical heuristics for computing HHL have been studied in [3,5,9].

Known approximation algorithms for HL [1,6,8], although polynomial, have high time bounds. Cohen et al. [8] describe an implementation of their algorithm with a speedup based on lazy evaluation, but it is still slow and can only handle small problems. For reachability queries (a special case with zero arc lengths), Schenkel et al. [21] implement this algorithm with a different variant of lazy evaluation, and use divide-and-conquer to handle larger problems. Other implementations focus on HHL and have no theoretical guarantees. The implementations of Abraham et al. [2,3] work well on large road networks and some other networks of moderate size. The implementation of Akiba et al. [5] scales to large complex networks. Delling et al. [9] produce small labels for a wider range of inputs than either method.

Our Contributions. In this paper, we improve the GHL^p algorithm and study its practical performance. First, we propose a refinement of GHL^p that achieves an $O(n^3 \min(p, \log n) \log n)$ time bound and is more efficient in practice; for $p = 1$, this improves the bound of Cohen et al. from $O(n^5)$ to $O(n^3 \log n)$. Second, we investigate the tradeoff between average and maximum sizes for the labels computed by GHL^p for different values of p. Finally, our detailed experimental analysis confirms that GHL^1 produces smaller labels than HHL.

The preprocessing algorithms we study require $\Omega(n^2)$ memory and compute shortest path distances between all pairs of vertices. In principle, one could instead compute and store a distance table and answer queries by table lookup. For graphs for which HL queries work well, however, the labels are small. Thus one can run preprocessing on a large server, but run queries on a less powerful device. Furthermore, studying more sophisticated algorithms with large memory footprint allows us to judge the quality of more practical heuristics and potentially improve them.

2   Preliminaries

The input to the point-to-point shortest path problem is a directed graph $G = (V, A)$, a length function $\ell : A \to \mathcal{R}$, and a pair s, t of vertices. Let $n = |V|$ and $m = |A|$. The goal is to find $\mathrm{dist}(s, t)$, the length of the shortest s–t path in G, where the length of a path is the sum of the lengths of its arcs. We assume that the length function is non-negative and that there are no zero-length cycles.

The size of a forward (backward) label, $|L_f(v)|$ ($|L_b(v)|$), is the number of hubs it contains. We define the size of the full label $L(v) = (L_f(v), L_b(v))$ as $|L(v)| = (|L_f(v)| + |L_b(v)|)/2$. We generalize this definition as follows. Suppose vertex IDs are $1, 2, \ldots, n$. Define a $(2n)$-dimensional vector $\mathcal{L}$ by $\mathcal{L}_{2i-1} = |L_f(i)|$ and $\mathcal{L}_{2i} = |L_b(i)|$. We consider the ℓ_p norm of $\mathcal{L}$, defined as $\|\mathcal{L}\|_p = (\sum_{i=0}^{2n-1} \mathcal{L}_i^p)^{1/p}$, where p is a natural number or ∞; $\|\mathcal{L}\|_\infty = \max_i \mathcal{L}_i$. The hub labeling algorithm uses $O(\|\mathcal{L}\|_1)$ memory and has worst-case query time $O(\|\mathcal{L}\|_\infty)$.

3   Approximation Algorithms

We now discuss existing $O(\log n)$-approximation algorithms for $\|\mathcal{L}\|_p$. We start with Cohen et al.'s algorithm [8] for $p = 1$; Section 3.1 deals with arbitrary p.

Starting with an empty labeling, the algorithm in each iteration adds a vertex to some labels, until the labeling satisfies the cover property. The algorithm also maintains the set U of *uncovered* vertex pairs: $(u, w) \in U$ if $L_f(u) \cap L_b(w)$ does not contain a vertex on a shortest u–w path. Initially U contains all vertex pairs u, w such that w is reachable from u. The algorithm terminates when U becomes empty. In each iteration, the algorithm adds a vertex v to forward labels of $u \in S' \subseteq V$ and to backward labels of $w \in S'' \subseteq V$ such that ratio of the number of newly-covered paths over the total increase in the size of the labeling is (approximately) maximized. Formally, let $U(v, S', S'')$ be the set of pairs in U which become covered if v is added to $L_f(u) : u \in S'$ and $L_b(w) : w \in S''$. The algorithm maximizes $|U(v, S', S'')|/(|S'| + |S''|)$ over all $v \in V$ and $S', S'' \subseteq V$.

A *center graph* of v, $G_v = (X, Y, A_v)$, is a bipartite graph with $X = V, Y = V$, and an arc $(u, w) \in A_v$ if some shortest path from u to w goes through v. To do the maximization, the algorithm finds densest subgraphs of center graphs. The *density* of a graph $G = (V, A)$ is $|A|/|V|$. The *maximum density subgraph (MDS)* problem can be solved in polynomial time using maximum flows (e.g., [11]). For a vertex v, arcs of a subgraph of G_v induced by $S' \subseteq X$ and $S'' \subseteq Y$ correspond to the pairs of vertices in U that become covered if v is added to $L_f(u) : u \in S'$ and $L_b(w) : w \in S''$. Therefore, the MDS of G_v maximizes $|U(v, S', S'')|/(|S'| + |S''|)$ over all S', S''.

Cohen et al. [8] show that the use of a linear-time 2-approximation algorithm [18] instead of an exact MDS algorithm results in a faster algorithm while preserving the $O(\log n)$ approximation ratio. We refer to the *approximate* MDS problem as AMDS. The 2-AMDS algorithm works by iteratively deleting the minimum-degree vertex from the current graph, starting from the input graph and ending in a single-vertex graph. Kortsarz and Peleg [18] show that the subgraph of maximum density in the resulting sequence of subgraphs is a 2-AMDS. To find the desired triple (v, S', S''), Cohen et al. solve the AMDS problem for all $G_v : v \in V$, and take v that gives the approximately densest subgraph.

Zero-Weight Vertex Heuristic. Cohen et al. [8] note that G_v can have arcs (u, w) such that the label of one endpoint contains v but the other does not. In this case, if u is included in a subgraph, it contributes to the denominator of its density, even though we do not need to add v to $L_f(u)$. They propose an intuitive modification of the algorithm that maintains the $O(\log n)$ approximation guarantee but performs better in practice. They assign zero weight to a vertex $u \in X$ if $v \in L_f(u)$ and to a vertex $w \in Y$ if $v \in L_b(w)$ and unit weights to all other vertices. They also modify the 2-AMDS algorithm to repeatedly delete vertices minimizing the ratio of degree over weight, where $x/0 = \infty$. Zero-weight vertices will be removed last; in fact, the algorithm can stop as soon as it reaches such a vertex.

3.1   Optimizing Arbitrary Norms

The $\|\mathcal{L}\|_p$ algorithm for $p > 1$ by Babenko et al. [6] is similar to the case $p = 1$, but uses weighted MDS, with vertex weights determined by the current labeling. The *maximum weighted densest subgraph* problem, takes as input a graph with vertex weights $c(v) \geq 0$. The goal is to find a subgraph of maximum weighted density, defined as the number of arcs divided by the total weight of the vertices. The 2-AMDS algorithm generalizes to this case: instead of choosing a vertex with minimum degree, each iteration chooses a vertex v that minimizes the ratio between the degree and the weight $c(v)$.

Consider an iteration of GHL^p with labels L_f and L_b. For the center graphs, we define the weight of a vertex $u \in X$ by $c_p(u) := (|L_f(u)| + 1)^p - |L_f(u)|^p$ and for $w \in Y$, $c_p(w) := (|L_b(w)| + 1)^p - |L_b(w)|^p$. For $p = 1$ we get the standard (unweighted) MDS problem. The $O(\log n)$-approximation algorithm for $\|\mathcal{L}\|_p$ is similar to the algorithm for $\|\mathcal{L}\|_1$, except that each iteration finds the triple (v, S', S'') that maximizes $|U(v, S', S'')|/(c_p(S') + c_p(S''))$. A 2-approximation for such triple (v, S', S'') is sufficient, and can be found by running a weighted 2-AMDS algorithm for all center graphs G_v.

The observation that for $p \geq \log n$, the ℓ_∞ norm is within a constant factor of the ℓ_p norm yields a $\log(n)$-approximation algorithm for ℓ_∞.

4   Improved Time Bound

The time bound for the Cohen et al. [8] algorithm (GHL^1) is $O(n^5)$: each iteration computes n AMDSes on graphs with $O(n^2)$ arcs, and the number of iterations is bounded by $O(n^2)$, the worst-case size of the labeling. A naive implementation of GHL[1] maintains all center graphs and uses $O(n^3)$ space, but this can be reduced to $O(n^2)$. We propose an *eager-lazy* variant of GHL[1] that runs in $O(n^3 \log n)$ time and $O(n^2)$ space, and still achieves an $O(\log n)$ approximation ratio. It also extends to an $O(n^3 \min(p, \log n) \log n)$ implementation of GHL[p].

First we describe our data structures. We precompute, in $O(n\mathrm{Dij}(n, m))$ time, a table of all pairs of shortest path distances. ($\mathrm{Dij}(n, m)$ denotes the running time of Dijkstra's algorithm [10] on a network with n vertices and m arcs.) We maintain an $n \times n$ Boolean array, with bit (i, j) indicating whether the current labeling covers the pair i, j. These data structures use $O(n^2)$ space.

Eager evaluation is a variant of the 2-AMDS algorithm that guarantees that deleting the edges of an approximate densest subgraph reduces the MDS bound for the remaining graph. We use this fact to bound the number of times the algorithm selects the center graph of a vertex. Consider a graph G, let μ be an upper bound on its MDS value, and fix $\alpha \geq 1$. The α-*eager evaluation* repeatedly deletes the minimum degree vertex of G while the density of G is less than $\mu/(2\alpha)$. If we use the zero-weight heuristic, we do not remove zero-weight vertices. Let G' be the subgraph that remains at the end of this procedure. Let G_α be the subgraph of G induced by the vertices deleted during this process. Let $\tilde{G}$ be the graph G with all edges from G' deleted; if we use the zero-weight heuristic, we also change the weight of vertices of G' from one to zero.

Lemma 1. *During the algorithm, (1) the density of G' is at least $\mu/(2\alpha)$ and (2) the MDS value of $\tilde{G}$ is at most μ/α.*

Proof. By construction, we stop when the density of G is at least $\mu/(2\alpha)$, which implies (1). For a vertex v in G_α, let m_v be the number of edges adjacent to v when v was deleted from G. We have $m_v < \mu/\alpha$, since v is the smallest-degree vertex before the deletion and $m_v \geq \mu/\alpha$ would imply a $\mu/(2\alpha)$ lower bound on the graph density. Consider a subgraph H of $\tilde{G}$ and let H_α be H with G' deleted. Let m_h and n_h be the number of edges in H and vertices in H_α respectively. If we use zero-weight heuristic, m_h/n_h is the density of H; otherwise, it is an upper bound. We have $m_h \leq \sum_{v \in H_\alpha} m_v < n_h \cdot \mu/\alpha$. Therefore $m_h/n_h < \mu/\alpha$. □

Note that G' may be empty, in which case (μ/α) is an improved bound on the MDS value of $G = G_\alpha$.

Next we discuss *lazy evaluation*. Cohen et al. [8] observed that the MDS value of a center graph does not increase when edges are removed. They thus propose keeping a priority queue of AMDS values and processing the center graph corresponding to the maximum value. In addition, they maintain a marker for each center graph indicating whether it has changed since its last AMDS computation. Each iteration processes the subgraph with maximum value. If the AMDS value is invalid (outdated), they recompute it. Otherwise they augment the labeling according to this subgraph, delete the AMDS of the center graph used in this iteration, and mark the affected AMDS values as invalid.

Schenkel et al. [21] use a variant of lazy evaluation. Instead of maintaining the AMDSes, they maintain their density values. An iteration considers the maximum value, d_v, and recomputes an AMDS of G_v. If the density of this AMDS is at least d_v,[1] this subgraph is used to update the labels and the center graphs. Otherwise d_v is updated.

Combining the second variant of lazy evaluation with eager evaluation, we get our *eager-lazy labeling* version of GHL[1]. Let $\alpha > 1$ be a constant. During initialization, the algorithm computes upper bounds μ_v on the MDS values of the center graphs using the 2-AMDS algorithm and keeps the values in a max-heap Q. At each iteration, the algorithm extracts the maximum μ_v from the heap and applies α-eager evaluation to G_v. If eager evaluation finds a non-empty subgraph G', it adds v to the labels of the vertices of G'. Then it updates U (the set of uncovered pairs) by iterating over all uncovered pairs of vertices (u, w) and, if $v \in L_f(u)$, $v \in L_b(w)$ and $\text{dist}(u, v) + \text{dist}(v, w) = \text{dist}(u, w)$, it marks (u, w) as covered. By Lemma 1 we know that after the update the MDS value of G_v is at most μ_v/α. The same bound holds if G' is empty. Finally, we set $\mu_v = \mu_v/\alpha$ and add μ_v back to Q. (One can use the bound found by the 2-AMDS algorithm, if it is smaller.)

Theorem 1. *The eager-lazy variant of GHL[1] is an $O(\log n)$ approximation algorithm running in $O(n^3 \log n)$ time and $O(n^2)$ space.*

Proof. The results of [8] imply that, for an $O(\log n)$ approximation, one can use constant factor approximations of the maximum density subgraph instead of the exact solutions. When the algorithm adds v to the labels, G' is a 2α approximation. The space bound is $O(n^2)$, since it is dominated by the distance table and the coverage matrix U. For the time bound, note that each iteration performs eager evaluation. In addition, if we find a non-empty G', we iterate over all vertex pairs to mark the newly covered ones, taking $O(n^2)$ time overall. To bound the number of iterations, note that each iteration that considers G_v decreases μ_v by a factor of α. For a graph with at least one edge, the maximum subgraph density is between $1/2$ and n. Therefore each G_v can be selected $O(\log n)$ times, yielding an $O(n \log n)$ bound on the number of iterations.   □

Now consider GHL[p] for $p > 1$. The MDSes of the center graphs are monotonically non-increasing, so we can use lazy evaluation. To use eager evaluation, we generalize the α-eager evaluation algorithm to the weighted case. Let μ be an upper bound on the weighted density and fix $\alpha > 1$. Define the *score* of a vertex to be its degree divided by its weight. The algorithm repeatedly deletes the minimum-score vertex from the graph while the graph density if less than μ/α. As before, let G' be the graph left when the α-eager evaluation terminates, and let $\tilde{G}$ be G with the edges from G' deleted and the weights of vertices from G' adjusted if we use zero-weight heuristic. We generalize Lemma 1 as follows:

Lemma 2. *During the algorithm, (1) the weighed density of G' is at least $\mu/(2\alpha)$ and (2) the weighted MDS value of $\tilde{G}$ is at most μ/α.*

---

[1] Although the MDS value of G_v is monotonically decreasing, an approximate computation may find a subgraph with higher value than the previous one.

We defined the weight of a vertex $u \in X$ by $c_p(u) = (|L_f(u)| + 1)^p - |L_f(u)|^p$ and for $w \in Y$, $c_p(w) = (|L_b(w)| + 1)^p - |L_b(w)|^p$. The maximum weighted subgraph density is $O(n)$ and the minimum non-zero density is $\Omega(1/n^p)$, so the density range is $O(n^{p+1})$. If an iteration of GHL^p chooses G_v, the density of G_v is reduced by a factor of α, yielding an $O(p \log n)$ bound on the number of times G_v is chosen. For $p > \log n$, $\|L\|_p = O(\|L\|_{\log n})$, and we get an $O(\log^2 n)$ bound.

Theorem 2. *The eager-lazy variant of GHL^p is an $O(\log n)$ approximation algorithm runs in $O(n^3 \min(p, \log n) \log n)$ time and $O(n^2)$ space.*

The implementation of GHL^p does not assume shortest path uniqueness. We could apply an a-priori length function perturbation to make the paths unique and center graphs less dense. We do not know how to improve theoretical bounds in this case; experimental results appear in Section 5.

4.1 Practical Improvement

The main bottlenecks of an iteration of our algorithm are updating the set U after adding a vertex to the labels and determining the set of center graph arcs adjacent to a vertex in the eager evaluation and AMDS subroutines. We propose a *shortest path graph* heuristic to speed up these operations in practice.

Let SPO_v be the graph induced by the arcs on shortest paths out of v. Similarly, SPI_v is induced by the arcs on shortest paths into v. Both SPO_v and SPI_v are acyclic (we assume no zero cycles). If shortest paths are unique, SPO_v and SPI_v are trees. We maintain these graphs implicitly: arc (u, w) is in SPO_v if $\text{dist}(v, u) + \ell(u, w) = \text{dist}(v, w)$, and in SPI_v if $\ell(u, w) + \text{dist}(w, v) = \text{dist}(u, v)$.

Suppose we add v to $L_f(u) : u \in S'$ and to $L_b(w) : w \in S''$. Then a pair $(u, w) : u \in S', w \in S''$ is covered if v is on a u-w shortest path. So we can iterate over all $u \in S'$ and perform a DFS of SPO_u starting at v. For each vertex w visited by the DFS, if $w \in S''$ we mark the pair (u, w) as covered.

We also use the SPO graphs to find the set of outgoing arcs from a vertex u in G_v (the case of incoming arcs is similar, using the SPI graphs). Again, we perform a DFS on SPO_u starting at v. When visiting a vertex w, we know that (u, w) is an arc of G_v if the pair u, w is not covered.

Although a DFS takes $O(m)$ time in the worst case, it tends to visit a small fraction of the graph, leading to a speedup in practice.

5 Experiments

We implemented all algorithms in this paper in C++ and compiled them with Microsoft Visual C++ 2012. We conducted our experiments on a machine running Windows Server 2008 R2. It has 96 GiB of DDR3-1333 RAM and two 6-core Intel Xeon X5680 3.33 GHz CPUs, each with 6×64 KB of L1, 6×256 KB of L2, and 12 MB of shared L3 cache. All our executions are single-threaded.

We test our algorithm on a wide range of realistic and synthetic instances. We test road networks from the 9th DIMACS Implementation Challenge [13]:

Table 1. Running times and label sizes for GHL[1] with $\alpha = 1.0, 1.1, 1.5$ and HHL

instance	$|V|$	$|A|$	time (s)				average label			
			G1.0	G1.1	G1.5	HHL	G1.0	G1.1	G1.5	HHL
PGPgiant	10680	48632	19113.8	3338.9	1721.2	969.4	19.1	19.4	20.3	20.5
alue5067	3524	11120	2970.8	2486.1	1796.0	158.6	23.4	24.5	25.4	24.1
beethoven	2521	15090	607.5	336.4	219.8	39.3	25.4	26.0	27.9	26.4
berlin10k	10370	24789	16026.8	8648.8	6168.2	1102.7	20.5	21.3	23.4	25.7
berlin5k	5307	12640	2475.1	1494.6	1082.7	191.5	18.0	18.6	20.2	21.6
email	1133	10902	108.8	46.9	22.6	4.4	30.0	30.4	31.4	36.8
grid10	961	3720	77.5	68.2	46.7	3.9	18.6	19.2	20.0	18.2
grid12	3969	15624	7191.2	5861.7	4038.8	276.3	27.9	28.7	29.6	27.6
hep-th	5835	27630	6374.6	1479.4	716.1	344.3	38.7	39.2	41.2	47.3
ksw-32-1	1024	6118	47.9	26.1	14.8	2.8	42.8	43.5	45.3	58.6
ksw-45-1	2025	12090	320.3	154.7	87.9	17.7	59.1	59.6	62.0	84.9
ksw-64-1	4096	24482	2319.4	900.9	489.3	124.2	81.4	82.3	84.9	126.0
polblogs	1222	33428	375.5	144.9	74.4	8.2	25.2	25.5	26.4	29.1
power	4941	13188	696.2	387.8	318.3	69.8	13.7	13.9	14.9	14.0
rgg10u	993	6162	42.5	32.4	24.6	4.0	14.0	14.4	15.0	14.6
rgg10w	993	6162	27.6	18.6	14.1	3.5	15.4	15.9	17.4	17.2
rgg12u	4088	31746	3494.6	2599.5	1853.5	300.2	24.3	25.2	27.2	26.3
rgg12w	4088	31746	1959.9	1182.2	795.8	103.1	29.9	31.0	34.5	36.7
rome99	3353	8859	587.6	399.6	269.2	41.2	23.8	24.9	27.3	28.4
venus	2838	17016	977.9	557.7	365.8	57.9	27.3	28.0	29.9	28.1

rome99, berlin5k, and berlin10k (the latter two are subgraphs from the Western Europe network). From the 10th DIMACS Implementation Challenge [7], we consider various complex networks: PGPgiant (communication), email (social), hep-th (collaboration), polblogs (links), and power (power grid). From the SteinLib [17], we take a grid graph with holes from VLSI applications (alue5067). We also consider triangulations from Computer Graphics applications [20] (beethoven and venus). Finally, we generated synthetic instances representing random geometric graphs with unit (rgg-u) and Euclidean (rgg-w) edge lengths, square grids (grid), and Kleinberg's small world graphs [16] (ksw). A ksw graph with parameter N consists of an $N \times N$ toroidal grid with additional long-distance edges: for each vertex, we add an edge to a random vertex at Manhattan distance d in the grid, where the probability of picking a particular value of d is proportional to d^{-2}.

Road networks are weighted and directed; rgg-w are weighted and undirected; all others are unweighted and undirected. All instances are (strongly) connected, although all algorithms still work otherwise.

Table 1 gives preprocessing times and average label sizes for our GHL[1] algorithm for $\alpha = 1.0, 1.1, 1.5$ (denoted by Gα). For comparison, we also include the HHL implementation of Delling et al. [9] that produces the best label quality; it uses on-line tie-breaking and runs in $O(n^3)$ time. Although there are much faster variants of HHL [9,3], they produce slightly worse labels. For example, the

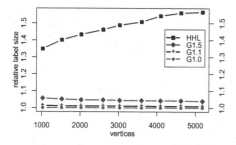 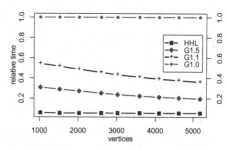

Fig. 1. Label size and running time (relative to G1.0) on small-world (ksw) problems as a function of input size

default algorithm of Delling et al. [9] takes less than 4 seconds to find labels for a road network with 32 413 vertices representing a bigger region of Berlin.

The table shows that increasing α from 1.0 (which produces the same results of Cohen et al.'s algorithm) to 1.1 barely degrades label quality and results in speedups ranging from less than two (for most graphs tested) to more than six (for complex networks such as hep-th). Increasing α to 1.5 leads to further speedups with more noticeable degradation of solution quality. In fact, G1.5 labels are sometimes bigger than HHL labels. Better approximations of the maximum density subgraph lead to smaller labels, as the theory predicts.

HHL is much faster than GHL[1], but usually produces worse labels. The difference is negligible for some graph classes (such as grids and triangulations), but HHL labels are larger by 20% for road networks and up to 50% for larger small-world instances (ksw).

Figure 1 analyzes the ksw family in more detail. It reports the average label sizes and running times of HHL, G1.1, and G1.5 relative to G1.0. Each data point represents five instances with different random seeds. The figure shows that HHL produces asymptotically worse labels than GHL[1]. Similarly, G1.1 and G1.5 have slight asymptotic advantage over G1.0 in terms of running times. Although the version of HHL we test has very similar asymptotic complexity to G1.0, recall that much faster (often asymptotically so in practice) versions of HHL [9] (with slightly worse labels) exist.

Figure 2 examines the tradeoff between preprocessing time and quality for GHL[1] in more detail. Each instance is represented by 6 points, each corresponding (from left to right) to a distinct value of α: 1.0, 1.05, 1.1, 1.2, 1.5, and 2.0. For each instance, all results are relative to $\alpha = 1.0$: increasing α leads to higher speedups but larger labels. For complex networks, such as email, hep-th, and PGPgiant, the tradeoff is favorable: speedups of an order of magnitude lead to labels that are bigger by less than 15%. For other classes, the tradeoff is not as good, with even small speedups leading to non-trivial losses in quality. Given that, we use $\alpha = 1.1$ (a relatively small value) for the remainder of the experiments.

We now compare the quality of the labels computed by the GHL^p algorithm for different values of p: we use $p = 1, 2, 4, \lceil \log n \rceil$. Recall that GHL^p approximates

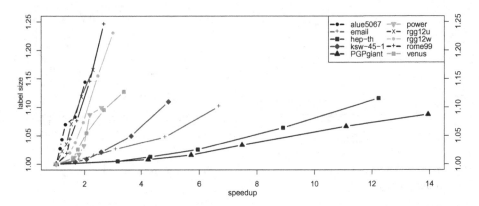

Fig. 2. Tradeoff between label size and running time. Each curve considers six values of α (1.0, 1.05, 1.1, 1.2, 1.5, and 2.0, from left to right); all values are relative to $\alpha = 1.0$.

the ℓ_p norm, and for $p = \lceil \log n \rceil$ it also approximates the ℓ_∞-norm. For select instances, Table 2 shows the average time as well as the average and maximum label sizes for each norm. The $L_1 L_\infty$ problems were designed by Babenko et al. [6] to show that there can be a large gap between the best labelings according to ℓ_1 and ℓ_∞ norms, and indeed we see a big difference between GHL^1 and $GHL^{\lceil \log n \rceil}$ labels. For other problems, ℓ_1 and ℓ_2 labels have similar average and maximum sizes, with no clear winner. As p increases further, the average label size increases and the maximum size decreases. Intuitively, this is because large labels contribute more to the p-norm when p increases. Overall, however, the difference in either measure is not overwhelming. GHL^p is slower for larger p values, often by a large amount. These results suggest that, except in pathological cases, using $p > 1$ will increase running time but will not result in substantially different label norms.

Table 2. GHL^p performance for different p values ($\alpha = 1.1$)

instance	time (s)				average label				maximum label			
	L_1	L_2	L_4	L_∞	L_1	L_2	L_4	L_∞	L_1	L_2	L_4	L_∞
$L_1 L_\infty$-13	8.8	21.6	30.7	28.6	3.8	7.5	13.7	14.1	171	171	33	21
$L_1 L_\infty$-16	30.9	88.3	122.9	123.5	3.8	9.0	16.6	17.2	258	258	42	25
alue5067	2486.1	3930.6	6398.1	9415.8	24.5	24.4	24.4	25.9	41	39	37	37
berlin5k	1494.6	2661.6	4597.0	6697.7	18.6	18.6	19.0	20.1	40	37	36	34
email	46.9	93.3	159.3	222.5	30.4	30.5	31.6	32.9	59	52	49	46
grid10	68.2	147.9	191.1	298.9	19.2	19.3	19.7	20.2	29	27	29	27
hep-th	1479.4	2607.9	5939.2	7609.9	39.2	39.4	40.7	43.6	88	85	72	67
ksw-45-1	154.7	337.2	716.2	1110.4	59.6	59.4	60.1	61.0	82	76	74	70
polblogs	144.9	281.4	442.7	614.5	25.5	25.7	27.0	28.0	86	62	54	52
power	387.8	602.2	977.8	1491.9	13.9	14.0	14.3	15.1	28	30	29	29
rgg12u	2599.5	4440.9	7876.7	11976.3	25.2	25.1	25.7	26.9	45	45	42	42
rgg12w	1182.2	2234.8	3965.4	6226.2	31.0	30.8	31.0	32.2	56	50	51	48
rome99	399.6	769.5	1620.2	2534.8	24.9	24.7	25.4	26.8	50	47	43	44
venus	557.7	945.7	1725.4	2602.4	28.0	27.8	28.3	29.3	46	44	40	41

Table 3. Results on perturbed instances; all values are relative to the corresponding unperturbed instance

Instance	speedup				relative label size			
	G1.0	G1.1	G1.5	HHL	G1.0	G1.1	G1.5	HHL
alue5067	4.63	5.27	5.51	3.17	1.30	1.29	1.34	1.38
berlin5k	0.99	1.01	1.00	1.00	1.00	1.00	1.00	1.00
email	3.12	2.96	2.84	1.77	1.61	1.60	1.61	1.61
grid10	4.93	5.61	5.50	2.21	1.25	1.24	1.28	1.35
hep-th	1.94	1.86	1.83	1.99	1.40	1.40	1.40	1.44
ksw-45-1	1.70	1.84	1.88	1.88	1.33	1.34	1.33	1.29
polblogs	4.98	5.00	5.41	1.94	1.89	1.89	1.88	2.06
power	1.03	1.16	1.29	1.13	1.08	1.09	1.06	1.09
rgg12u	1.91	2.36	2.51	2.35	1.19	1.19	1.22	1.32
rgg12w	1.01	1.02	1.01	1.01	1.00	1.00	1.00	1.00
rome99	1.01	1.00	1.00	0.99	1.00	1.00	1.00	1.00
venus	1.64	2.04	2.27	2.21	1.34	1.34	1.35	1.41

The algorithms we have tested so far perform on-line tie-breaking: they consider that a vertex v covers a pair s, t if there exists at least one shortest s–t path that contains v. To confirm that this leads to better labels than committing to one specific s–t path in advance, we ran the same algorithms on *perturbed* versions of some instances, with arc lengths increased by up to 1% at random, independently and uniformly. This mostly preserves the shortest path structure, but makes ties much less likely.

For each perturbed instance and algorithm, Table 3 shows the speedup (in terms of label generation time) and the average label size *relative to the original instance*. For graphs that were originally weighted (berlin5k, rgg12w, and rome99), all algorithms behave almost exactly as before. For the remaining inputs (in which ties are much more common), all algorithms become consistently faster, but produce worse labels. For example, all algorithms become about twice as fast on hep-th, but labels become 40% larger; for email, labels increase by as much as 50%. This confirms that, for best label quality, HL should be used with on-line tie-breaking.

6   Concluding Remarks

Our improvements to the running time and our compact data structures for the theoretically justified HL algorithm allow us to solve bigger problems than previously possible: instances with about 10 000 vertices can be solved in an hour. Our results provide some justification for using hierarchical labels (HHL) in practice: on real-world instances, HHL labels are not much bigger than those found by the theoretically justified (non-hierarchical) algorithms, and HHL is usually much faster. That said, the difference in quality (sometimes higher than 20%) is not negligible; faster approximation algorithms would still be quite useful.

References

1. Abraham, I., Delling, D., Fiat, A., Goldberg, A.V., Werneck, R.F.: VC-dimension and shortest path algorithms. In: Aceto, L., Henzinger, M., Sgall, J. (eds.) ICALP 2011, Part I. LNCS, vol. 6755, pp. 690–699. Springer, Heidelberg (2011)
2. Abraham, I., Delling, D., Goldberg, A.V., Werneck, R.F.: A hub-based labeling algorithm for shortest paths on road networks. In: Pardalos, P.M., Rebennack, S. (eds.) SEA 2011. LNCS, vol. 6630, pp. 230–241. Springer, Heidelberg (2011)
3. Abraham, I., Delling, D., Goldberg, A.V., Werneck, R.F.: Hierarchical hub labelings for shortest paths. In: Epstein, L., Ferragina, P. (eds.) ESA 2012. LNCS, vol. 7501, pp. 24–35. Springer, Heidelberg (2012)
4. Akiba, T., Iwata, Y., Kawarabayashi, K.I., Kawata, Y.: Fast shortest path distance queries on road networks by pruned highway labeling. In: ALENEX 2014, pp. 147–154. SIAM, Philadelphia (2014)
5. Akiba, T., Iwata, Y., Yoshida, Y.: Fast exact shortest-path distance queries on large networks by pruned landmark labeling. In: SIGMOD 2013, pp. 349–360. ACM, New York (2013)
6. Babenko, M., Goldberg, A.V., Gupta, A., Nagarajan, V.: Algorithms for hub label optimization. In: Fomin, F.V., Freivalds, R., Kwiatkowska, M., Peleg, D. (eds.) ICALP 2013, Part I. LNCS, vol. 7965, pp. 69–80. Springer, Heidelberg (2013)
7. Bader, D., Meyerhenke, H., Sanders, P., Wagner, D. (eds.): Graph Partitioning and Graph Clustering. AMS, Boston (2013)
8. Cohen, E., Halperin, E., Kaplan, H., Zwick, U.: Reachability and distance queries via 2-hop labels. SIAM J. Comput. 32(5), 1338–1355 (2003)
9. Delling, D., Goldberg, A.V., Pajor, T., Werneck, R.F.: Robust exact distance queries on massive networks. TR MSR-TR-2014-12, Microsoft Research (2014)
10. Dijkstra, E.W.: A note on two problems in connexion with graphs. Numer. Math. 1, 269–271 (1959)
11. Gallo, G., Grigoriadis, M.D., Tarjan, R.E.: A fast parametric maximum flow algorithm and applications. SIAM J. Comput. 18, 30–55 (1989)
12. Gavoille, C., Peleg, D., Pérennes, S., Raz, R.: Distance labeling in graphs. J. Algorithms 53(1), 85–112 (2004)
13. Goldberg, A.V., Kaplan, H., Werneck, R.F.: Reach for A*: shortest path algorithms with preprocessing. In: The Shortest Path Problem: Ninth DIMACS Implementation Challenge, DIMACS Book, vol. 74, pp. 93–139. AMS, Boston (2009)
14. Goldberg, A.V., Razenshteyn, I., Savchenko, R.: Separating hierarchical and general hub labelings. In: Chatterjee, K., Sgall, J. (eds.) MFCS 2013. LNCS, vol. 8087, pp. 469–479. Springer, Heidelberg (2013)
15. Kaplan, H.: Personal communication (2013)
16. Kleinberg, J.M.: The small-world phenomenon: An algorithmic perspective. In: STOC 2000, pp. 163–170. ACM, New York (2000)
17. Koch, T., Martin, A., Voß, S.: SteinLib: An updated library on Steiner tree problems in graphs. Tech. Rep. ZIB-Report 00-37, Konrad-Zuse-Zentrum für Informationstechnik, Heidelberg (2000), http://elib.zib.de/steinlib
18. Kortsarz, G., Peleg, D.: Generating sparse 2-spanners. J. Alg. 17, 222–236 (1994)
19. Peleg, D.: Proximity-preserving labeling schemes. J. Gr. Th. 33(3), 167–176 (2000)
20. Sander, P.V., Nehab, D., Chlamtac, E., Hoppe, H.: Efficient traversal of mesh edges using adjacency primitives. ACM Trans. Graphics 27(5), 144:1–144:9 (2008)
21. Schenkel, R., Theobald, A., Weikum, G.: HOPI: An efficient connection index for complex XML document collections. In: Bertino, E., Christodoulakis, S., Plexousakis, D., Christophides, V., Koubarakis, M., Böhm, K., Ferrari, E. (eds.) EDBT 2004. LNCS, vol. 2992, pp. 237–255. Springer, Heidelberg (2004)

Customizable Contraction Hierarchies*

Julian Dibbelt, Ben Strasser, and Dorothea Wagner

Karlsruhe Institute of Technology (KIT), Kaiserstruhe 12, 76131 Karlsruhe, Germany
{dibbelt,strasser,dorothea.wagner}@kit.edu

Abstract. We consider the problem of quickly computing shortest paths in weighted graphs given auxiliary data derived in an expensive preprocessing phase. By adding a fast weight-customization phase, we extend Contraction Hierarchies [12] to support the three-phase workflow introduced by Delling et al. [6]. Our Customizable Contraction Hierarchies use nested dissection orders as suggested in [3]. We provide an in-depth experimental analysis on large road and game maps that clearly shows that Customizable Contraction Hierarchies are a very practicable solution in scenarios where edge weights often change.

1 Introduction

Computing optimal routes in road networks has many applications such as navigation devices, logistics, traffic simulation or web-based route planning. For practical performance on large road networks, preprocessing techniques that augment the network with auxiliary data in an (expensive) offline phase have proven useful. See [1] for an overview. Among the most successful techniques are Contraction Hierarchies (CH) [12], which have been utilized in many scenarios. However, their preprocessing is in general metric-dependent, e.g., edge weights are needed a-priori. Substantial changes to the metric, e.g., due to user preferences, may require expensive recomputation. For this reason a Customizable Route Planning (CRP) approach was proposed in [6], extending the multi-level-overlay MLD techniques of [20,15]. It works in three phases: In a first expensive phase auxiliary data is computed that solely exploits the topological structure of the network, disregarding its metric. In a second much less expensive phase this auxiliary data is *customized* to the specific metric, enabling fast queries in the third phase. In this work we extend CH to support such a three-phase approach.

Besides road networks, game maps are an interesting application where fast shortest path computations are important (c.f. [21]). In real-time strategy games the basic topology of the map often is fixed. However, since buildings are constructed or destroyed, fields are rendered impassable or freed up. Yet, because the *fog of war*, every player has his own knowledge of the map: A unit must not route around a building that the player has not yet seen. Furthermore, units such as hovercrafts may traverse water and land, while other units are bound

---

* Partial support by DFG grant WA654/16-2, EU grant 288094 (eCOMPASS), and Google Focused Research Award.

to land. This results in vastly different, evolving metrics for different unit types per player, making metric-dependent preprocessing difficult to apply.

One of the central building blocks of this paper is to use metric-independent nested dissection orders (ND-orders) for CH precomputation instead of the metric-dependent order of [12]. This approach was proposed by [3], and a preliminary case study can be found in [23]. A similar idea was followed by [9], where the authors employ partial CHs to engineer subroutines of their customization phase (they also had preliminary experiments on full CH). Worth mentioning are also the works of [18]. They consider small graphs of low treewidth and leverage this property to compute good orders and CHs (without explicitly using the term CH). Interestingly, our experiments show that also large road networks have relatively low treewidth. Furthermore, note that while customizable speedup techniques for shortest path queries may be a very recent development, the idea to use ND-orders to compute CH-like structures is far older and widely used in the *sparse matrix solving* community. We refer the interested reader to [13,17].

Our Contribution. The main contribution of our work is to show that Customizable Contraction Hierarchies (CCH) solely based on the ND-principle are feasible and practical. Compared to CRP [6] we achieve a similar preprocessing–query tradeoff, albeit with slightly better query performance at slightly slower customization speed (and somewhat more space). Interestingly, for less well-behaved metrics such as travel distance, we achieve query times below the original metric-dependent CH of [12].

An *extended version* of this paper with more detailed experiments is available at http://arxiv.org/abs/1402.0402. It contains concepts that are not strictly necessary for the considered workflow but might be useful for extensions of it.

2   Basics

We denote by $G = (V, E)$ an *undirected n-vertex graph* where V is the set of *vertices* and E the set of *edges*. Furthermore, $G = (V, A)$ denotes a *directed graph* where A is the set of *arcs*. We consider *simple* graphs that have no loops or multi-edges. We denote by $N(v)$ the set of adjacent vertices of v. A *vertex separator* is a vertex subset $S \subseteq V$ whose removal separates G into two disconnected subgraphs induced by the vertex sets A and B. The separator S is *balanced* if $|A|, |B| \leq 2n/3$. A *vertex order* $\pi : \{1 \ldots n\} \to V$ is a bijection. Its inverse π^{-1} assigns each vertex a *rank*. *Undirected edge weights* are denoted using $w : E \to \mathbb{R}^+$. With respect to a vertex order π we define an *upward weight* $w_u : E \to \mathbb{R}^+$ and a *downward weight* $w_d : E \to \mathbb{R}^+$. The *vertex contraction* of v in G consists of removing v and all incident edges and inserting edges between all neighbors if not already present. We iteratively contract all vertices according to their rank resulting in a set Q of additional edges. We direct the edges in E and in Q upward from lower ranks to higher ranks resulting in the upward directed search graph $G_\pi^\wedge$. The search space SS(v) of a vertex v is the subgraph of $G_\pi^\wedge$ reachable from v. For every vertex pair s and t, it has been shown that a shortest up-down path must exist. This up-down path can be found by running a bidirectional search from from s restricted to $SS(s)$ and from t restricted to $SS(t)$ [12]. The elimination tree $T_{G,\pi}$ is a tree directed towards its root $\pi(n)$. The parent of vertex $\pi(i)$ is its upward neighbor v of minimal rank $\pi^{-1}(v)$. As shown in [3] the set of vertices on the path from v to $\pi(n)$ is the set of vertices in $SS(v)$. Note that the elimination tree is only defined for undirected unweighted CHs. A *lower triangle* of an arc (x, y) in $G_\pi^\wedge$ is a triple (x, y, z) such that arcs (z, x) and (z, y) exist. Similarly, an *intermediate triangle* is a triple such that (x, z) and (z, y) exist, and an *upper triangle* a triple such that (x, z) and (y, z) exist. Weights on $G_\pi^\wedge$ are called *metrics*. A weight on G is extended to an *initial metric* on $G_\pi^\wedge$ by assigning ∞ to all non-original arcs. A metric m is called *customized* if for all lower triangles (x, y, z), it holds that $m(x, y) \leq m(z, x) + m(z, y)$.

Note that our CHs build on undirected graphs. To model one-way streets we use two different up- and downward metrics, setting either m_u or m_d to ∞.

3   Phase 1: Preprocessing the Graph Topology

Metric-Independent Order. To support metric-independence, we use *nested dissection* orders (ND-orders) as suggested in [3]. An order π for G is computed recursively by determining a minimum balanced separator S that splits G into parts induced by the vertex sets A and B. The lowest ranks are assigned to the vertices in A, the next ranks to the vertices in B, and finally the highest ranks are assigned to S. Computing ND-orders requires good graph partitioning, and recent years have seen heuristics that solve the problem very well even for continental road graphs [19,8,7]. Note that these partitioners compute edge cuts. On our instances, a good vertex separator can be derived by arbitrarily picking one incident vertex for each edge.

Theorem 1. *Let G be a graph with recursive balanced separators with $O(n^\alpha)$ vertices. If G has a minimum balanced separator with $\Theta(n^\alpha)$ vertices then a ND-order gives an $O(1)$-approximation of the average and maximum search spaces of an optimal metric-independent CH in terms of vertices and arcs.*

Proof. (sketch, see extended version) In a lemma the authors of [17] show that a clique exists in the CH of the size of the minimum balanced separator. We observe that this is the top level separator and that it dominates all lower separators.

Constructing the Contraction Hierarchy. While the CH can be constructed given the order and the input graph as described in [12], we improve construction performance by (temporarily) considering all inputs as undirected and unweighted graphs. By designing a specialized *Contraction Graph* datastructure, detailed in the extended version, we significantly accelerate construction time while limiting space use during construction. Denote by n the number of vertices, m the number of edges in G, by m' the number of edges in $G_\pi^\wedge$, and by $\alpha(n)$ the inverse $A(n, n)$ Ackermann function. Our algorithm needs $O(m'\alpha(n))$ time and $O(m)$ space. The core idea, introduced in [14], consists of reducing vertex contraction to edge contraction, maintaining an independent set of virtually contracted vertices. Also, we build the elimination tree and derive for every vertex its level $\ell(x)$ in $G_\pi^\wedge$.

Memory Order. After having obtained the nested dissection order we reorder the in-memory vertex IDs of the input graph accordingly, i.e., the contraction order of the reordered graph is the identity. This greatly improves cache locality.

4   Phase 2: Customizing the Metric

In this section, we describe how to customize a metric and parallelize the process. Furthermore, we show how to update the weight of a single arc. A base operation for our algorithms is efficiently enumerating all lower triangles of an arc. It can be implemented using adjacency arrays or accelerated using extra preprocessing.

Basic Triangle Enumeration. Construct an upward and a downward adjacency array for $G_\pi^\wedge$, where incident arcs are ordered by their head vertex ID. Unlike common practice, we also assign and store arc IDs. (By lexicographically assigning arc IDs we eliminate the need for arc IDs in the upward adjacency array.) Denote by $N_u(v)$ the upward neighborhood of v and by $N_d(v)$ the downward neighborhood. All lower triangles of an arc (x, y) are enumerated by simultaneously scanning $N_d(x)$ and $N_d(y)$ by increasing vertex ID to determine their intersection $N_d(x) \cap N_d(y) = \{z_1 \ldots z_k\}$. The lower triangles are all triples (x, y, z_i). The corresponding arc IDs are stored in the adjacency arrays. This approach requires space proportional to the number of arcs in $G_\pi^\wedge$. All upper and intermediate triangles are found by merging $N_u(x)$ and $N_u(y)$ (respectively $N_d(y)$).

Triangle Preprocessing. Instead of merging the neighborhoods, we propose to create an adjacency-array-like structure that maps the arc ID of (x, y) onto the pair of arc ids of (z, x) and (z, y) for every lower triangle (x, y, z). This requires space proportional to the number of triangles in $G_\pi^\wedge$ but allows for faster access.

Metric Customization. Our algorithm iterates over all levels from the bottom to the top. On level i, it iterates (using multiple threads) over all arcs (x, y) with level $\ell(x) = i$. For each such arc (x, y), the algorithm enumerates all lower triangles (x, y, z) and performs $m(x, y) \leftarrow \min\{m(x, y), m(z, x) + m(z, y)\}$. The operation maintains the shortest path structure. Furthermore, the resulting metric is customized by definition. Note that we synchronize the threads between levels. Then, since we only consider lower triangles, no read/write conflicts occurs. Hence, no locks or atomic operations are needed.

Vectorization. A metric can be replaced by an interleaved set of k metrics by replacing every $m(x, y)$ by a vector of k elements. All k metrics are customized in one go, amortizing triangle enumeration time. To customize directed graphs, recall that we first extract upward and downward weights w_u and w_d. These are independently transformed into initial upward and downward metrics m_u and m_d.

However, they must not be customized independently: For every lower triangle (x, y, z) we set $m_u(x, y) \leftarrow \min\{m_u(x, y), m_d(z, x) + m_u(z, y)\}$ and $m_d(x, y) \leftarrow \min\{m_d(x, y), m_d(z, y) + m_u(z, x)\}$. When using interleaved metrics it is straightforward to use SSE (avoid addition overflow by setting $\infty = \text{int}_{\max}/2$).

Partial Updates. Denote by $U = \{((x_i, y_i), n_i)\}$ the set of arcs whose weights should be updated where (x_i, y_i) is the arc ID and n_i the new weight. Observe that modifying the weight of one arc can trigger new changes, but only to higher arcs. We therefore organize U as a priority queue ordered by the level of x_i. We iteratively remove arcs from the queue and apply the change. If new changes are triggered, we insert these into the queue. Let (x, y) be the arc removed from the queue, n its new weight, and o its old weight. We first check if (x, y) can be bypassed via a lower triangle, improving n: We iterate over all lower triangles (x, y, z) and perform $n \leftarrow \min\{n, m(z, x) + m(z, y)\}$. Finally, if $\{x, y\}$ is an edge in the original graph G, we ensure that n is not larger than the original weight. If after both checks $n = m(x, y)$ holds, no change is necessary and no further changes are triggered. Otherwise, we iterate over all upper triangles (x, y, z) and test whether $m(x, z) + o = m(y, z)$ holds. (Note that (y, z, x) is a lower triangle of (y, z).) If so, we add the change $((y, z), m(x, z) + n)$ to the queue. Intermediate triangles are handled analogously.

5   Phase 3: At Query Time

In this section we describe how to compute a shortest up-down path in $G_\pi^\wedge$, given a source and target vertex s and t and a customized metric.

Basic and Stalling. The basic query alternates two instances of Dijkstra's algorithm on $G_\pi^\wedge$ from s and t, maintaining a tentative distance of the shortest path discovered so far (initially ∞). If G is undirected then both searches use the same metric. Otherwise, the search from s uses the upward metric m_u, and the search from t the downward metric m_d. In either case, in contrast to [12], the searches operate on the same upward search graph $G_\pi^\wedge$. Each search is stopped once its radius is larger than the tentative distance. Additionally, we evaluate a basic version of the stall-on-demand optimization presented in [12].

Elimination Tree. Inspired by [3], we us the precomputed elimination tree to efficiently enumerate all vertices in $\text{SS}(s)$ and $\text{SS}(t)$ by increasing rank at query time without using a priority queue. We store two tentative distance arrays $d_f(v)$ and $d_b(v)$. Initially these are all set to ∞. First, we compute the lowest common ancestor (LCA) x of s and t in the elimination tree: We simultaneously enumerate all ancestors of s and t by increasing rank until a common ancestor is found. Second, we iterate over all vertices y on the branch from s to x, relaxing all forward arcs of such y. Third, we do the same for all vertices y from t to x in the backward search. Fourth, we iterate over all vertices y' from x to the root (the top-level vertex), simultaneously relaxing the respective outgoing arcs of each y'

for the forward and backward searches. We further determine the vertex z that minimizes $d_f(z) + d_b(z)$, deriving the path distance and preparing unpacking. We finish by iterating over all vertices from s and t to the root, resetting all distances d_f and d_b to ∞ (which proved cheaper than timestamping).

Path Unpacking. The original CH of [12] unpacks an up-down path by storing for every arc (x, y) the vertex z of the lower triangle (x, y, z) that caused the weight at $m(x, y)$. This information depends on the metric and we want to avoid storing additional metric-dependent information. We therefore resort to a different strategy. Denote by $p_1 \ldots p_k$ the up-down path found by the query. As long as a lower triangle (p_i, p_{i+1}, x) exists with $m(p_i, p_{i+1}) = m(x, p_i) + m(x, p_{i+1})$ insert the vertex x between p_i and p_{i+1}.

6   Comparison with CRP

Both CRP and CCH are multi-level overlay techniques in spirit of [20], but CRP typically utilizes less levels. Yet, regarding customization, both approaches clearly converge, since CRP uses further guidance levels and contraction as a subroutine [9]. Note, however, that CRP does not contract the top-level separator. Queries differ more, with CRP running plain Dijkstra on the lowest cells before employing preprocessed data. Note that by exploiting partition information at query time, CRP offers unidirectional queries.

Our triangle preprocessing has similarities with micro and macro code [9]. While their approach does not allow for a—in our context crucial—random access, we can atleast compare space consumption: In that regard, micro code is clearly the most expensive. Macro code, on the other hand, is compactest as the dominating substructure only stores one ID per triangle (instead of our two IDs). However, we enumerate undirected triangles and thus, depending on the instance, may have only half as many triangles.

One major advantage of CRP over other techniques is that it works well with turn costs. Since our benchmark instances lack realistic turn cost data (while synthetic data tends to be very simplistic), we deemed it unproper to experimentally evaluate CCH performance on turn costs. However, using a theoretical argument we predict that turn costs have no major impact: They can be incorporated by adding turn cliques to the graph. Small edge cuts in the original graph correspond to small cuts in the turn-aware graph. Analyzing the exact growth of cuts (in the extended version), we conclude that the impact on search space size is at most a factor of 2 to 4. Practical performance might be better. Note, that the game scenario does not need turn costs.

7   Experiments

The experiments were run on a dual 8-core Intel Xeon E5-2670 processor clocked at 2.6 GHz, with 64 GiB of DDR3-1600 RAM, 20 MiB of L3 and 256 KiB of L2 cache (internally called compute11). We use g++ 4.7.1 with -O3.

Table 1. The number of vertices and of directed arcs of the benchmark graphs. We further present the number of edges in the induced undirected graph.

	#Vertices	#Arcs	#Edges	symmetric?
TheFrozenSea	754 195	5 815 688	2 907 844	yes
Europe	18 010 173	42 188 664	22 211 721	no

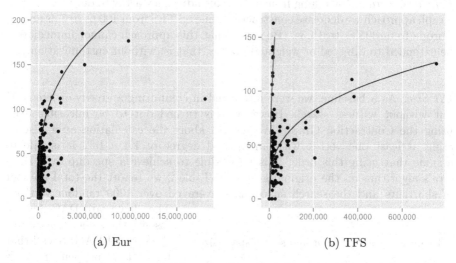

(a) Eur   (b) TFS

Fig. 1. Vertices in the separator (vertical) vs vertices in the subgraph being partitioned (horizontal). The red function is a cubic root ($y = \sqrt[3]{x}$ for Eur and $y = 1.4 \cdot \sqrt[3]{x}$ for TFS) and the blue function is a square root ($y = \sqrt{x}$ for TFS).

We consider two large instances of practical relevance (c.f. Table 1): First, the road network of Europe (Eur in short) as made available by PTV AG for the DIMACS challenge [10]. It is the defacto standard for benchmarking road route planning research. Second, the game map TheFrozenSea (TFS in short), one of the largest of Star Craft maps available at [22]. The map is composed of square tiles, each having at most eight neighbors. Some tiles are non-walkable. Walkable tiles form pockets of free space connected through *choke points* of very limited space. The corresponding graph contains for every walkable tile a vertex and for every pair of adjacent walkable tiles an edge. Diagonal edges are weighted by $\sqrt{2}$, others by 1. The graph is symmetric and contains large grid subgraphs.

ND-Order. To compute a ND-order, we use KaHIP in its "strong" configuration, with imbalance set to 20%. We recursively bisect the graph using edge cuts from which vertex separators are derived. We repeat each cut ten times with different random seeds and pick the smallest one. The resulting separator sizes are illustrated in Figure 1 (in the extended version we also consider Metis [16]). The top-level separators of Europe are curiously small due to the special topology of the continent (Great Britain, Spain, and France are easily separated). Once all such features are exploited, the cut sizes seem to follow a $\Theta(\sqrt[3]{n})$-law, making the approximation result from Theorem 1 applicable. On TFS, the top-level separators that cut through choke points also follow a $\Theta(\sqrt[3]{n})$-law. The residual subgraphs are grids with $\Theta(\sqrt{n})$ cuts. This explains the two peaks in the plot.

CH Construction. Switching from a dynamic adjacency array to our Contraction Graph approach we decrease construction time on TFS from 490.6s to 3.8s and on Europe from 305.8s to 15.5s. (Be aware that this approach cannot immediately be extended to directed or weighted graphs, that is, without customization.)

CH Size. As a baseline, we were interested in computing a greedy-ordered CH but without witness search. However, this turned out to be infeasible even using the Contraction Graph. We had to abort the calculation after discovering at least one vertex with an upward degree of 1.4×10^6. It is safe to assume that using this order it is impossible to achieve a speedup over Dijkstra's algorithm on the original graph. In Table 2 we report the total number of shortcuts and the search space sizes averaged over 1000 random vertices.

The downward-DAG has in essence the same structure as the upward-DAG. Recall that for metric independent CHs both DAGs are the same and therefore it only has to be stored once in memory. For a CH with witness search two separate upward and downward DAGs must be stored. For a ND-order we can perform a perfect witness search (i.e. we remove every arc not part of a shortest path). For a greedy order we only know of the heuristic approach of [12], but on a smaller road instance we were able to show that the difference between perfect and heuristic is very small. The details are in the extended version. Interestingly, the number of vertices in the ND search spaces are smaller than for the greedy order. However, the number of arcs in the ND search spaces is significantly higher, dominating query running time. The witness search decreases the number of arcs for the ND order by a factor of 2 to 4. On game maps the greedy order decreases the number of arcs by another factor of 2 and on Europe even by another factor of 10. Additionally, we counted the number of triangles which is a performance relevant figure for customization: TFS has 864M and Europe has 578M triangles. Precomputing all triangles as suggested requires 6.6GB space on TFS and 4.6GB on Europe. (Using Metis reduces space on TFS to 4.6GB.)

Table 2. Shortcut count and search space sizes

	Order	Witness search	#Arcs upward	Avg. up SS size vertices	arcs
TFS	Greedy	heuristic	6 399 080	1 281	13 330
TFS	ND	none	25 099 646	674	89 567
TFS	ND	perfect	10 161 889	645	24 782
Europe	Greedy	heuristic	33 911 692	709	4 808
Europe	ND	none	73 920 453	652	117 406
Europe	ND	perfect	55 657 315	616	44 677

Treewidth. As detailed in [4], the treewidth of a graph is a measure deeply coupled with chordal super graphs. The authors show in their Theorem 6 that the maximum upward degree over all vertices in a metric-independent CH of a graph G is an upper bound to the treewidth of G. Using this, we are able to upper bound the treewidth of TFS by 287 and of Europe by 479.

Customization. In Table 3 we report the times needed to customize a metric. Using all optimizations presented we customize Europe in below one second. When amortized[1], we even achieve 0.415ms which is only slightly above the (non-amortized) 0.347ms reported in [9]. (Note that their experiments were run on a different machine with a faster clock but 2×6 instead of 2×8 cores. Also, their implementation is turn-aware, touching more memory, making an exact comparison difficult.) However, their overlays are more space efficient than our metric-independent CHs. The running time of 0.415ms is fast but comes at a high price as many cores are needed. We thus evaluate the time needed for partial updates on a single core. We averaged over 10 000 runs in which we set a random arc in

Table 3. The running times needed to compute or update a metric. The $\times 4$ indicates that 4 interleaved metric pairs are customized in one go.

SSE	precompute triangles?	thread count	TFS	Eur time [ms]
no×1	none	1	10.08	10.88
yes×1	none	1	9.34	9.55
yes×1	all	1	3.75	3.22
yes×1	all	16	0.61	0.74
yes×2	all	16	0.76	1.05
yes×4	all	16	1.50	1.66

	Queue removals			time [ms]		
	med.	avg.	max.	med.	avg.	max.
TFS	8	382.4	12035	0.017	2.0	99.2
Eur	1	38.8	10666	0.003	0.2	87.2

the CH to a random value. The median, average and maximum running times significantly differ. This is because there are a few arcs on highways or choke points that trigger a lot of subsequent changes whereas for most arcs a weight change has nearly no effect.

Distance Query. In Table 4 we report the performance of our query algorithms, comparing them to the original CH of [12]. We manage to come very close to their query times, but as we also support a fast customization even achieving the same order of magnitude is a major result. We ran 10^6 shortest path distance queries with the source and target vertices picked uniformly at random. The presented times are averaged single core running times without any SSE. The evaluated metric-dependent (greedy order) CH is comparable to [12]. (Because of a slower machine their reported running times are all slightly higher. The only exception is the Europe graph with the distance metric. Here, our measured running time of only 0.540ms is disproportionally faster. We assume that our greedy order is better since we do not use lazy updates, spending more preprocessing time.) As in [12] the stall-on-demand heuristic improves running times by a factor 2 to 5.

---

[1] We refer to a server scenario of multiple active users that require simultaneous customization, e.g., due to traffic updates.

Table 4. Query running times and explored forward search space sizes

	Metric	Order	Query Type	Settled Vertices	Relaxed Arcs	Time [ms]
TFS	Map-Dist.	Greedy	Basic	1 199	12 692	0.539
		Greedy	Stalling	319	3 459	0.286
		ND	Basic	603	82 824	0.644
		ND	Stalling	560	74 244	0.774
		ND	Tree	674	89 567	0.316
Europe	Travel-Time	Greedy	Basic	546	3 623	0.283
		Greedy	Stalling	113	668	0.107
		ND	Basic	581	107 297	0.810
		ND	Stalling	418	75 694	0.857
		ND	Tree	652	117 406	0.413
	Distance	Greedy	Basic	3 653	104 548	2.662
		Greedy	Stalling	286	7 124	0.540
		ND	Basic	581	107 080	0.867
		ND	Stalling	465	84 718	0.992
		ND	Tree	652	117 406	0.414

Recall that in addition to the obvious modifications we also reordered the vertices in memory according to the nested dissection order for the metric-independent CH. Preliminary experiments have shown that this has a measurable effect on cache performance and results in 2 to 3 times faster query times. We do not reorder the vertices for the metric-dependent CHs to remain comparable to [12] and because it is a lot less clear what a good cache-friendly order is. For the basic CCH query (ND order) on Europe with travel time metric, the number of settled vertices only increases by a small factor compared to metric-dependent case. On TFS or Europe with distance metric this difference is even smaller. However, the number of relaxed arcs is significantly larger. As this increases the costs of the stalling test, stall-on-demand does not pay off anymore. We observe that the basic query algorithm (despite the use of a stopping criterion) still visits large portions of the possible search space. In contrast, the elimination tree based query visits slightly more vertices. However, it does not use a priority queue and needs therefore less time per vertex. Furthermore, for the elimination tree based algorithm the code paths do not depend on the metric. Hence, query times are completely independent of the metric as can be seen by comparing the tree ND query times for travel time metric to distance metric. In contrast, for the basic ND query algorithms the metric still has a slight influence on the performance. A possible downside of the elimination tree query currently is that local queries are not faster than global queries. However, one might be able to exploit LCA information as a locality filter, switching to the basic query if sufficiently local.

A stalling query on the metric-dependent CH with travel time is on Europe about a factor of 5 faster than the elimination tree query. However for the distance metric the order is inversed and our CHs are even faster by a factor of about 20%. We see three factors contributing here: 1) The distance metric is a comparatively hard metric, 2) the in-memory ID reordering, 3) the lack of a priority queue.

CRP without turn costs on Europe needs 0.72ms with travel time and 0.79ms with distance [6]. However, they parallelize by running the forward and the backward searches in different threads. This parallelization can be added to our query but even without it our query outperforms their query times. Table 5 summarizes the situation.

Finally we measured the time needed to fully unpack the paths. For travel time on Europe the time is 0.25ms (and 0.27ms for TFS), which is below a query time. For Europe with distance the time is 0.52ms, which is slightly higher than query time, as paths tend to contain more arcs because the input graph is modeled a lot less fine grained on the highways used by the travel metric. In [12] they report full path extraction costs of 0.323ms for travel time on Europe but with precomputed middle nodes. This is slightly faster when accounting for the processor differences. However, we are positive that we could achieve similar performance exploiting preprocessed triangle information for path unpacking.

Table 5. Comparison with CRP/MLD. CCH hits a Pareto-point.

[ms]	Cust.	Query
CRP	0.347	0.72
CCH	0.415	0.41

8 Conclusions

We have extended CHs to a three phase customization approach and demonstrated that the approach is practicable and efficient not only on real world road graphs but also on game maps.

Future Work. While a graph topology with small cuts is one of the main driving force behind the running time performance of CHs it is clear from Table 2 that better metric-dependent orders can be constructed by exploiting additional travel time specific properties. We would like to further investigate this gap. Better ND-orders directly result in better CHs and thus further partitioning research is useful. Further investigation into algorithms explicitly exploiting treewidth [5,18] might be promising. Also, determining the precise treewidth could prove useful. Revisiting all of the existing CH extensions to see which can profit from an ND-order is worthwhile. An interesting candidate are Time-Dependent CHs [2] where computing a good metric-dependent order has proven relatively expensive.

Acknowledgements We would like to thank Ignaz Rutter and Tim Zeitz for very inspiring conversations on the topic.

References

1. Bast, H., Delling, D., Goldberg, A.V., Müller–Hannemann, M., Pajor, T., Sanders, P., Wagner, D., Werneck, R.F.: Route planning in transportation networks. In: Technical Report MSR-TR-2014-4. Microsoft Research, Mountain View (2014)
2. Batz, G.V., Geisberger, R., Sanders, P., Vetter, C.: Minimum time-dependent travel times with contraction hierarchies. ACM J. Exp. Algorithmics 18, 1–43 (2013)

3. Bauer, R., Columbus, T., Rutter, I., Wagner, D.: Search-space size in contraction hierarchies. In: Fomin, F.V., Freivalds, R., Kwiatkowska, M., Peleg, D. (eds.) ICALP 2013, Part I. LNCS, vol. 7965, pp. 93–104. Springer, Heidelberg (2013)
4. Bodlaender, H.L., Koster, A.M.C.A.: Treewidth computations I. Upper bounds. Inform. and Comput. 208, 259–275 (2010)
5. Chaudhuri, S., Zaroliagis, C.: Shortest paths in digraphs of small treewidth. Part I: Sequential algorithms. Algorithmica 27, 212–226 (2000)
6. Delling, D., Goldberg, A.V., Pajor, T., Werneck, R.F.: Customizable route planning. In: Pardalos, P.M., Rebennack, S. (eds.) SEA 2011. LNCS, vol. 6630, pp. 376–387. Springer, Heidelberg (2011)
7. Delling, D., Goldberg, A.V., Razenshteyn, I., Werneck, R.F.: Graph partitioning with natural cuts. In: 2011 IEEE International Parallel & Distributed Processing Symposium (IPDPS 2011), pp. 1135–1146. IEEE Computer Society, Los Alamitos (2011)
8. Delling, D., Goldberg, A.V., Razenshteyn, I., Werneck, R.F.: Exact combinatorial branch-and-bound for graph bisection. In: Bader, D.A., Mutzel, P. (eds.) ALENEX 2012, pp. 30–44. SIAM, Philadelphia (2012)
9. Delling, D., Werneck, R.F.: Faster customization of road networks. In: Bonifaci, V., Demetrescu, C., Marchetti-Spaccamela, A. (eds.) SEA 2013. LNCS, vol. 7933, pp. 30–42. Springer, Heidelberg (2013)
10. Demetrescu, C., Goldberg, A.V., Johnson, D.S. (eds.): The Shortest Path Problem: Ninth DIMACS Implementation Challenge. American Mathematical Society, Rhode Island (2009)
11. Fulkerson, D.R., Gross, O.A.: Incidence matrices and interval graphs. Pacific J. Math. 15, 835–855 (1965)
12. Geisberger, R., Sanders, P., Schultes, D., Vetter, C.: Exact routing in large road networks using contraction hierarchies. Transportation Science 46, 388–404 (2012)
13. George, A.: Nested dissection of a regular finite element mesh. SIAM J. Numer. Anal. 10, 345–363 (1973)
14. George, A., Liu, J.W.: A quotient graph model for symmetric factorization. In: Duff, I.S., Stewart, G.W. (eds.) Sparse Matrix Proceedings, pp. 154–175. SIAM, Philadelphia (1978)
15. Holzer, M., Schulz, F., Wagner, D.: Engineering multilevel overlay graphs for shortest-path queries. ACM J. Exp. Algorithmics 13, 1–26 (2008)
16. Karypis, G., Kumar, V.: A fast and high quality multilevel scheme for partitioning irregular graphs. SIAM J. Sci. Comput. 20, 359–392 (1999)
17. Lipton, R.J., Rose, D.J., Tarjan, R.: Generalized nested dissection. SIAM J. Numer. Anal. 16, 346–358 (1979)
18. Planken, L., de Weerdt, M., van Krogt, R.: Computing all-pairs shortest paths by leveraging low treewidth. J. Artificial Intelligence Res. 43, 353–388 (2012)
19. Sanders, P., Schulz, C.: Think locally, act globally: Highly balanced graph partitioning. In: Bonifaci, V., Demetrescu, C., Marchetti-Spaccamela, A. (eds.) SEA 2013. LNCS, vol. 7933, pp. 164–175. Springer, Heidelberg (2013)
20. Schulz, F., Wagner, D., Weihe, K.: Dijkstra's algorithm on-line: An empirical case study from public railroad transport. ACM J. Exp. Algorithmics 5, 1–23 (2000)
21. Storandt, S.: Contraction hierarchies on grid graphs. In: Timm, I.J., Thimm, M. (eds.) KI 2013. LNCS, vol. 8077, pp. 236–247. Springer, Heidelberg (2013)
22. Sturtevant, N.: Benchmarks for grid-based pathfinding. Transactions on Computational Intelligence and AI in Games (2012)
23. Zeitz, T.: Weak contraction hierarchies work! Bachelor thesis. KIT, Karlsruhe (2013)

Experimental Evaluation of Dynamic Shortest Path Tree Algorithms on Homogeneous Batches*

Annalisa D'Andrea, Mattia D'Emidio, Daniele Frigioni,
Stefano Leucci, and Guido Proietti

Department of Information Engineering, Computer Science and Mathematics,
University of L'Aquila, Via Vetoio, 67100 L'Aquila, Italy
{annalisa.dandrea,stefano.leucci}@graduate.univaq.it,
{mattia.demidio,daniele.frigioni,guido.proietti}@univaq.it

Abstract. In this paper we focus on *dynamic batch* algorithms for single-source shortest paths in graphs with positive real edge weights. A dynamic algorithm is called *batch* if it is able to handle graph changes that consist of multiple edge updates at a time, i.e., a batch. Unfortunately, most of the algorithmic solutions known in the literature for this problem are analyzed with respect to heterogeneous parameters, and this makes unfeasible an effective comparison on a theoretical basis. Thus, for a full comprehension of their actual performance, in the past these solutions have been assessed experimentally. In this paper, we move ahead along this direction, by focusing our attention on *homogeneous* batches, i.e., either *incremental* or *decremental* batches, which model realistic dynamic scenarios like node failures in communication networks and traffic jams in road networks. We provide an extensive experimental study including both the most effective previous batch algorithms and a recently developed one, which was explicitly designed (and was shown to be theoretically efficient) exactly for homogeneous batches. Our work complements previous studies and shows that the various solutions can be consistently ranked on the basis of the type of homogeneous batch and of the underlying network. As a result, we believe it can be helpful in selecting a proper solution depending on the specific application scenario.

1 Introduction

The problem of updating shortest paths in real networks whose topology dynamically changes over time is a core functionality of today's communication and transportation networks. In fact, it finds application in many real-world scenarios as Internet routing, routing in road networks, timetabling in railway networks. In these scenarios, shortest paths are stored and have to be updated whenever the underlying graph undergoes dynamic updates.

In general, the typical update operations that can occur on a network can be modeled as insertions and deletions of edges and edge weight changes (weight

---

* Research partially supported by the Research Grant 2010N5K7EB PRIN 2010 "ARS TechnoMedia" from the Italian Ministry of University and Research.

J. Gudmundsson and J. Katajainen (Eds.): SEA 2014, LNCS 8504, pp. 283–294, 2014.

decrease or weight increase) in the underlying graph. A dynamic algorithm is a *batch algorithm* if it is able to handle graph changes that consist of multiple edge updates at a time, i.e. a *batch*. If a batch consists of only delete and weight increase (resp., insert and weight decrease) operations, then it is called *decremental batch* (resp., *incremental batch*), otherwise it is called *full batch*.

We focus on the single-source shortest path problem in graphs with positive real edge weights, which can be solved by running Dijkstra's algorithm [8].

Related Work. The problem of updating single-source shortest paths in dynamic scenarios has been widely studied in the literature [4,5,6,9,10,11,13,16,17], but none of the proposed solutions is better than the re-computation from scratch in the worst case. Concerning the batch problem, very few solutions are known in the literature [7,13,15]. More formally, let $\beta = (\mu_1, \ldots, \mu_h)$ be a full batch of h edge update operations to be performed on a positively real-weighted input graph $G = (V, E, w)$ with n vertices and m edges. In the following, we denote by $\delta(G, \mu_i)$ the set of vertices of G which are *affected* by the edge update μ_i, i.e., the set of vertices that change either their parent towards the source or their distance from the source in G as a consequence of μ_i, while we denote by $\Delta(G, \beta)$ the set of vertices affected by the entire batch β. In other words, $\delta(G, \mu_i)$ (resp., $\Delta(G, \beta)$) represents the change in the *output* induced by μ_i (resp., β). Moreover, we denote by $\lambda(G, \mu_i)$ the set of components (i.e., edges and vertices) of G which are *modified* by the edge update μ_i, i.e., the two vertices and the set of edges that are adjacent to the edge updated by μ_i, plus the set of edges that are incident to vertices in $\delta(G, \mu_i)$. Similarly, $\Lambda(G, \beta)$ will denote the set of components modified by the entire batch β. Finally, we denote by $|\beta|$ the number of operations in β, and by $||\beta|| = |\Delta(G, \beta)| + |\Lambda(G, \beta)|$.

In [15] the authors propose an algorithm running in $O(||\beta|| \log ||\beta||)$ time. On the other hand, in [13] the authors propose a framework containing various algorithms, whose running time is given as function of many different parameters. In particular the best of these algorithms requires $O(|\delta(G, \mu)| \cdot \log(|\delta(G, \mu)|) + \gamma \cdot D_{\max} \cdot |\delta(G, \mu)|)$ time, where γ can be as large as the number of nodes in $\delta(G, \mu)$ that change *both* their parent and their distance, and $D_{\max}$ is the maximum vertex degree of the graph. Finally, in [7] two new algorithms have been proposed for separately managing decremental and incremental batches. These algorithms require the computation of a *k-bounded accounting function* (*k-baf* in the following) for the input graph G, i.e., a function that, for each edge (x, y) of G, determines either vertex x or vertex y as the *owner* of the edge, in such a way that the maximum number of edges assigned to a vertex is bounded by k. In general, a graph always admits an $O(\sqrt{m})$-baf, while determining the minimum k for which G admits a k-baf is known to be NP-hard and 2-approximable [11]. In particular, several classes of graphs admit a $O(1)$-baf, as for example planar graphs, graphs with bounded degree/treewidth, and others. Once that a k-baf for G has been found, then the first algorithm of [7] is able to process a decremental batch β in $O((|\beta| + |\hat{\delta}(G, \beta)| \cdot k) \cdot \log n)$ worst-case time, while the second one is able to process an incremental batch β in $O(|\beta| \log n + |\hat{\delta}(G, \beta)| \cdot \max\{k, k^*\} \cdot \log n)$ worst-case time, where k^* is the *minimum* integer such that a k^*-baf exists for the graph after β, and $\hat{\delta}(G, \beta)$ denotes the set of affected vertices caused by the *simultaneous* execution of all the operations in β. Notice that $|\hat{\delta}(G, \beta)|$ can be much smaller than $|\Delta(G, \beta)|$, as vertices that are affected multiple times w.r.t. the single operations in β are considered only once in $\hat{\delta}(G, \beta)$.

It is evident from the above discussion that the solutions in [7,13,15] are theoretically analyzed w.r.t. heterogeneous parameters, which makes unfeasible an effective comparison on a theoretical basis. Thus, a full comprehension of their actual performance should be supported by experimental evidence. In the past, this has been done for the solutions provided in [13] and [15]. More precisely, in [2], these two batch algorithms have been implemented (in several tuned variants) and compared, also against a repeated application of two classic single-edge update algorithms, namely those proposed by Ramalingam and Reps [15], and by Frigioni *et al.* [10]. The algorithms were tested on a wide set of input instances including: unit-disk graphs, railway networks, Internet graphs at the Autonomous System-level, Internet graphs at the router level provided by the CAIDA consortium [12], road networks provided by PTV [14], and synthetic graphs. These input instances were subject to various kinds of batch updates including: batches of multiple randomly chosen edges, node failure and recovery batches, and traffic jams. The first outcome of the experiments of [2] is the astonishing level of data dependency of the algorithms. In other words, it turned out that a proper assessment of the running time of an algorithm is not possible without a full knowledge of the actual running instance. The second outcome is that the experiments confirm the intuition that it is useful to process a set of updates as a batch only when the updated edges have a strong interference regarding their impact on the shortest path tree. While updates that are far away from each other usually do not interfere, and hence they can be handled iteratively. Notwithstanding the huge data dependency of the algorithms, in most of the cases, the algorithm given in [16], denoted from now on by RR (standing for *Ramalingan & Reps*), and the basic tuned implementation of that given in [15], denoted from now on by T-SWSF (standing for *Tuned-Strict Weakly Superior Function*), were the better performing ones among those analyzed in [2].

Contribution of the Paper. In this paper, we move ahead towards a full experimental characterization of the various known batch algorithms. More precisely, we provide the very first implementation of the algorithms in [7], which as we said are efficient for *homogeneous* (i.e., either incremental or decremental) batches, which happen to model realistic dynamic scenarios like node failures in communication networks and traffic jams in road networks. We denote these algorithms from now on as DDFLP-I and DDFLP-D, respectively (we will refer to them simply as DDFLP, when the meaning is clear from the context), and we compare them exactly against RR and T-SWSF. Our work complements previous studies by analyzing how the old algorithms and the new ones behave in handling various types of homogeneous batches in several classes of networks. In particular, once again we analyze node failures and traffic jams (both with recovery), which are a special type of decremental batch (followed by the corresponding incremental batch for the recovery). Moreover, we perform extensive experiments on random sequences of incremental/decremental batches of updates, and on a mix of them. The considered classes of networks were: road networks, an Internet graph at the router level provided by the CAIDA consortium [12], and synthetic graphs (either following a node-degree power law or a classic Erdős-Rényi model). Our experiments are consistent with those of [2], in the sense that also here the various solutions can be ranked on the basis of the type of batch and of the underlying network. Thus, our main contribution is a helpful instrument to select a proper solution depending on the specific application scenario. In particular, we newly learnt from the experiments that in the prominent node failure and recovery scenario, the combination of DDFLP-I and DDFLP-D is faster than RR and T-SWSF. Actually, this is not the only advantage brought by the use of the algorithms given in [7]: in basically every circumstance, they show to be very parsimonious in terms of number of edges scanned during the processing of a batch, and this feature can be of fundamental importance when the underlying graph is not directly available in main memory.

2   Implemented Algorithms

In this section we briefly describe RR, T-SWSF, and DDFLP. Throughout the paper, we use the following notation. Let $G = (V, E, w)$ be a weighted graph with n vertices and m edges, and let $s \in V$ be a fixed source. Each edge $(x, y) \in E$ has a positive real weight w_{xy} associated. Let $d : V \Rightarrow \mathbb{R}^+$ be a function giving, for each $x \in V$, the distance of x from s (the weight of a shortest path from s to x), let $T(s) = (V_T, E_T)$ be a shortest paths tree of G rooted at s, and, for any $x \in V$, let $T(x)$ be the subtree of $T(s)$, rooted at x. Every $x \in V$ has one parent, denoted as $parent(x)$, and a set of children, denoted as $children(x)$, in $T(s)$. An edge (x, y) is a *tree edge* if $(x, y) \in E_T$; otherwise it is a *non-tree edge*.

Description of RR. Recall that RR is a single-edge update algorithm. The algorithm maintains a subset of the edges of the graph G, denoted as SP, which contains exactly the edges of G that belong to at least one shortest path from s to the other vertices of the graph. The graph with vertex set V and edge set SP is a directed acyclic graph, denoted as $SP(G)$, and contains all the shortest paths from s to the other vertices of G.

In the case of an edge insertion (weight decrease), an adaptation of Dijkstra's algorithm is used that works on the vertices affected by that insertion. In particular, if (v, w) is the inserted edge, vertices are stored in a priority queue with priority equal to their distance from vertex w. When a vertex x with minimum priority is extracted from the heap, *all* edges (x, y) are traversed; vertex y is inserted in the heap, or its priority in the heap is updated, only if the shortest path from s to x, plus edge (x, y) yields a path shorter than the shortest path currently known for y. In this case the algorithm deletes all the edges incoming y in $SP(G)$ and adds edge (x, y) to $SP(G)$. On the other hand, if $d(x) + w_{xy} = d(y)$, then edge (x, y) is simply added to $SP(G)$. The algorithm halts when the distances of all the affected vertices, and the new $SP(G)$ have been computed.

In the case of an edge deletion (weight increase), RR works in two phases. Suppose (v, w) is the deleted edge. In the first phase the algorithm finds the vertices affected by that deletion, performing a computation similar to a topological sorting of $SP(G)$. A work set is maintained containing vertices that have been identified as affected, but have not yet been processed; a vertex z is placed in the work set exactly once, when the number of edges incoming z in $SP(G)$ drops to zero. Initially, vertex w is inserted in the work set, only if there are no further edges in $SP(G)$ incoming w after the deletion of (v, w). Vertices in the work set are processed one by one, and when vertex u is processed, all edges outgoing u are deleted from $SP(G)$. All vertices that are identified as affected during this process are inserted in the work set. In the second phase, the new distances of the affected vertices from the source are computed, using an approach similar to Dijkstra's one. In particular, if A is the set of affected vertices computed during the first phase, and B is the set of unaffected vertices, with $s \in B$, then the correct distance from the source is known for each vertex in B, and the new distance value has to be computed for each vertex in A. This problem is reduced to a Dijkstra's computation as follows: introduce a new source vertex s', by condensing into s' the subgraph of G induced by vertices in B; consider the subgraph induced by vertices in A and, for each edge (v, w) from a vertex v outside A to a vertex w inside A, add an edge from s' to w, with weight equal to the final distance of vertex v plus w_{vw}. Now Dijkstra's algorithm is performed on the graph built as above, starting from s'. During this second phase the algorithm also updates, for each vertex v, the set of edges incoming and outgoing v in $SP(G)$ in a way similar to that described in the case of an edge insertion.

Description of T-SWSF. Algorithm SWSF is able to manage *full* batches, and stores for each node v of the graph G, a label $D[v]$ such that, initially, $D[v]$ equals the distance $d(v)$ from a source vertex s. It is based on the use of the notion of *inconsistent node*. A node is *inconsistent* if, given a batch of edge modifications β, it is the target of an edge in β. The algorithm adjusts inconsistent nodes v as follows: the label $D[v]$ is set to the *consistent* value $con(v)$ and the node is inserted in a priority queue Q with priority $\min\{D[v], d[v]\}$, where $con(v) = \min_{(u,v) \in E}\{D[u] + w_{uv}\}$, if $v \neq s$, and it is equal to 0 when $v = s$. In case v is already in Q, SWSF only updates its priority.

When a node is removed from Q, it is said to be *adjusted*. Then, the algorithm executes a main phase where, while Q is not empty, the following operations are performed: the node w with minimum priority is extracted and deleted from Q. If $d(w) < D[w]$, then $D[w]$ is set to $d(w)$ and each neighbor of w is adjusted. If $d(w) > D[w]$, then $D[w]$ is set to infinity and each neighbor of w is adjusted.

T-SWSF works like SWSF, but the number of some time-consuming operations to be performed is reduced as follows. In particular, note that, SWSF needs to scan all adjacent edges of a node x in order to compute $con(x)$. T-SWSF relaxes fewer of such edges as follows: when a neighbor v of a node w is adjusted with $d[w] < D[w]$, the algorithm computes $con(v)$. The same strategy works in the initialization phase when we compute $con(x)$ for a node x that is the target node of an edge with decreased edge weight. When a neighbor v of w is adjusted with $d[w] > D[w]$, we set $D_{old} := D[w]$ and $D[w] := \infty$. We can skip v when $D_{old} + w_{wv}$ equals $d[v]$. The same strategy holds in the initialization phase for target nodes of edges with increased weight.

Description of DDFLP. In DDFLP the same technique proposed in [10] is used to partition the edges incident to each vertex of G and to store them. In particular, any edge (x, y) has an *owner*, denoted as $owner(x, y)$, that is either x or y. For each vertex x, $ownership(x)$ is the set of edges owned by x, and $non\text{-}ownership(x)$ is the set of edges with one endpoint in x, but not owned by x. If G has a k-baf then, for each $x \in V$, $ownership(x)$ contains at most k edges. Furthermore, given an edge (z, q) of G, the backward level (resp., forward level) of edge (z, q) and of vertex q, relative to vertex z, is the quantity $b\_level_z(q) = d(q) - w_{zq}$ (resp., $f\_level_z(q) = d(q) + w_{zq}$). The intuition is that the level of an edge (z, q) provides information about the shortest available path from s to q passing through z. To bound the number of edges scanned by DDFLP each time that a vertex is updated, the edges in $non\text{-}ownership(x)$ are stored as follows: in DDFLP-D, $non\text{-}ownership(x)$ is stored as a min-based priority queue named F_x; the priority of edge (x, y) (of vertex y) in F_x, denoted as $F_x(y)$, is the computed value of $f\_level_x(y)$; in DDFLP-I, $non\text{-}ownership(x)$ is stored as a max-based priority queue named B_x; the priority of edge (x, y) (of vertex y) in B_x, denoted as $B_x(y)$, is the computed value of $b\_level_x(y)$.

In what follows, for any $x \in V$, we say that edges in $ownership(x)$ are scanned by ownership, edges in $non\text{-}ownership(x)$ are scanned by priority.

The Decremental Algorithm. DDFLP-D uses the following notion of coloring. Given a vertex $q \in V$, $color(q)$ is: white if q changes neither the distance from s nor the parent in $T(s)$; red if q increases the distance from s; pink if q preserves its distance from s, but it must replace the parent in $T(s)$.

The algorithm works in three phases. In the first phase all the operations in the batch are considered, one at a time, and all the data structures are updated. If an operation in the batch affects a tree edge (z, q), and q is the target vertex, then q is inserted in a min-heap Q with priority equal to its distance from s.

In the second phase, the algorithm assigns a color to each vertex of the graph as follows. It first extracts from Q the min-priority vertex x, that is the affected vertex which was closer to s in G. Now, two cases can occur. If vertex x has a neighbor y such that $color(y) \neq$ red and $d(y) + w_{xy} = d(x)$, that is a vertex that allows x to keep its distance unchanged, then the color of x is set to pink and its parent $P(x)$ is set to y, otherwise the color of x is set to red and all the children v, in $T(s)$, of x are inserted in Q, if they have not been inserted yet, with priority equal to $d(v)$. In order to search for y, the algorithm performs both a *scan by ownership*, by examining the other endpoint v of every edge $(x, v) \in ownership(x)$, and a *scan by priority* by iteratively extracting an edge (x, v) from F_x until $f\_level_x(v)$ is greater than $d(x)$. The algorithm proceeds by extracting the next vertex from Q and by repeating the above strategy. When the heap becomes empty, the coloring phase terminates.

In the third phase the algorithm updates the distances of all the **red** vertices as follows. For each **red** vertex x, the algorithm tries to find the *best* **non-red** *neighbor*, that is a vertex y such that $d(y)+w_{xy} = \min_{\{k \in N(x):color(k) \neq \mathtt{red}\}}\{d(k)+w_{kx}\}$. As in the previous phase this is done by performing both a scan by ownership and a scan by priority. If such neighbor exists, it is inserted in a min-heap Q with priority equal to $d(y) + w_{xy}$, and the distance of x is set to the same value. Otherwise, x is inserted in Q with priority equal to infinity, the distance from s to x is set to infinity and the parent of x is set to *null*. Finally, a relaxing step is performed. In particular, the min-priority vertex x is extracted from Q and, for each neighbor v of x, the distance from s to v is relaxed. If $D(v) > D(x) + w_{xv}$, then $D(v)$ is set to $D(x) + w_{xv}$ and $P(v)$ is set to x. The priority of v in Q is also updated to $D(x) + w_{xv}$. Note that, once a vertex is extracted from Q, its distance does not change anymore. The algorithm proceeds by extracting the next vertex from Q and by iterating the above strategy. When the heap becomes empty both the shortest path tree and the distance function are updated.

The Incremental Algorithm. DDFLP-I uses the following notion of coloring. Given a vertex $q \in V$, $color(q)$ is **white** if q changes neither the distance from s nor the parent in $T(s)$, while it is **blue** if q changes its distance from s.

The algorithm works in two phases. In the first phase, all the operations in the incremental batch are considered one by one, and all the data structures are updated. If an operation in the batch, involving an edge (z, q), induces a decrease in the distance from s of one of the two endpoints $x \in \{z, q\}$, then the procedure sets $color(x)$ to **blue** and inserts it into a min-heap Q with priority equal to the new induced distance. Note that, if x is already in Q, the algorithm simply updates the priority.

In the second phase, the algorithm processes the vertices in Q: it extracts the min-priority vertex x from Q and, for each $y \in N(x)$, it performs a relaxing step: if a path from s to y passing through x in G' shorter than the shortest path from s to y in G is discovered, the color of y is set **blue**, its distance is updated, and it is inserted into the min-heap Q with priority equal to the new distance. As in the decremental version, the algorithm searches for x by scanning the edges incident to y both by ownership and by priority. The scan by ownership is exactly the same while the scan by priority iteratively extracts edges from the priority queue B_y (and thus can be stopped as soon as an edge (y, v) such that $b\_level_y(v)$ is less that the new distance of y is found).

3   Experimental Setup

Our experiments have been performed on a workstation equipped with a Quad-core 3.60 GHz Intel Xeon X5687 processor, with 24 GB of main memory.

Input Instances. As input to the algorithms we used instances belonging to the following graph categories.

1. **Road Graphs.** A road graph is a weighted directed graph G, used to model real road networks, where nodes represent points on the network, edges represent road segments between two points, and the weight function represents

an estimate of the travel time to traverse road segments. We considered two road graphs available from PTV [14]: the road graph of Belgium, denoted by BEL, having 458 598 nodes and 1 164 124 edges, and the road graph of Germany, denoted by GER, having 4 047 577 nodes and 9 814 894 edges.

2. **Random Graphs.** We considered random graphs generated by the *Erdős-Rényi* algorithm [3]. In particular, we generated two Erdős-Rényi graphs, denoted by ERD-D and ERD-S, with approximately 50 000 nodes and a density of 1% and 0.2% of that of a complete graph, respectively. The weight of each edge was uniformly chosen as an integer not larger than 1 000, as in [2].

3. **CAIDA Graphs.** A CAIDA graph is an Internet topology at the router level of the *CAIDA IPv4 topology dataset* [12]. We parsed the files provided by CAIDA to obtain a weighted undirected graph, denoted by CAI, with 20 000 nodes and 51 600 edges, where a node represents an IP address, edges represent links among hops, and weights are given by Round Trip Times.

4. **Power-law Graphs.** A power-law graph is a graph having a power-law node degree distribution. These graphs model many of the currently implemented communication infrastructures, like the Internet, the World Wide Web, and so on. Examples of power-law networks are the artificial instances generated by the *Barabási-Albert* algorithm [1]. A Barabási-Albert topology is generated by iteratively adding one node at a time, starting from a given connected graph with at least two nodes. A newly added node is connected to any other existing nodes with a probability proportional to the degree of the existing nodes. Accordingly to the approach used in [2], edge weights were set to uniform. Specifically, we considered a Barabási-Albert graph, denoted by BAR, with 631 912 nodes and 2 001 544 edges.

The details of such graphs are given in Table 1, where we report the number of nodes and edges, the average and the maximum node degree, the value k of a k-baf of the graph as computed by the linear-time 2-approximation algorithm proposed in [11], the average and maximum sizes of the *non-ownership* sets of the nodes of the graph (i.e., the average and the maximum sizes of the heaps F_x and B_x required by DDFLP, for each node x).

Table 1. Input graphs details

Graph	|V|	|E|	avgdeg	maxdeg	k	avgHeapSize	maxHeapSize
GER	4 047 577	9 814 894	4.85	7	3	1.21	7
BEL	458 598	1 164 124	5.08	8	3	1.27	7
ERD-D	50 000	12 505 622	500.22	642	215	125.07	306
ERD-S	49 672	250 112	10.07	16	3	2.52	14
CAI	20 000	51 630	5.16	203	7	1.29	202
BAR	631 912	2 001 544	6.33	3 214	3	1.58	3 212

Experiments. We performed four different types of experiments on the input instances of Table 1, in which each batch of updates is chosen as follows.

1. **Node Failure and Recovery.** First, a node v is chosen uniformly at random. The update consists of two steps. In the first step, v fails and the

weights of all edges adjacent to v are set to infinity. In the second step, v recovers and the weights of all edges adjacent to v are reset to their original values.

2. **Traffic Jam and Recovery.** This experiment models real-world traffic jams in road networks, which often occur along shortest paths. A shortest path containing b edges is chosen uniformly at random. The update consists of two steps: in the first step, the weights of edges in the path are multiplied by 10; in the second step, the edge weights are reset to their original values. We chose b equal to 10, as in [2], for all the graphs, except for BAR and ERD-D, where the shortest paths tree has very small height. For these instances we chose b equal to 5 and 4, respectively.

3. **Incremental.** 25 edges of the graph are chosen uniformly at random. The weight of each edge is decreased by a quantity between 1% and 99% of the actual weight of that edge.

4. **Decremental.** 25 edges of the graph are chosen uniformly at random. The weight of each edge is increased by a quantity between 1% and 200% of the actual weight of that edge.

For each experiment, 100 instances of the problem were generated. In details, a shortest path tree rooted at a randomly chosen node was computed, along with auxiliary preprocessed data, such as, for instance, the shortest path subgraph required by RR or forward/backward levels required by DDFLP. For each instance and for each type of experiment, we generated a sequence of 20 batches and executed the algorithms. In addition to the aforesaid experiments, we generated also mixed sequences of 20 incremental and decremental batches each containing 25 edge modifications. In these sequences, the type of each batch is chosen at runtime by a coin flip, and the weight changes are ranged as in the case of incremental and decremental batches.

To properly measure the performance of the implemented algorithms, a full Dijkstra run is performed directly after each update, and the corresponding *speed-up* (i.e., the time needed by Dijkstra's algorithm divided by the time needed by the update algorithm) is computed. The results of our experiments are shown in Tables 2–6, where we report the speed-up of each algorithm (columns 2, 3, and 4, resp.). In order to make our experimental analysis complete, we evaluated also some machine independent parameters, which give a different point of view on the performance of the algorithms implemented in this paper. In particular, we measured for each algorithm the number of edge scan operations. In the case of DDFLP, this number is made up by two components: the number of edges scanned by ownership, and the number of edges scanned by priority. We report the values of these parameters in the last 4 columns of each table.

4   Analysis

First of all, our experiments confirm the majority of the results of [2] regarding batch shortest-path algorithms. In particular, our experiments provide a clear

Table 2. Node failure and recovery batches

Graph	Speed-up			Edge Scans			
	RR	T-SWSF	DDFLP	RR	T-SWSF	DDFLP-OWNERSHIP	DDFLP-PRIORITY
GER	207 369	212 785	**247 947**	22 938	7 730	**2 554**	1 041
BEL	23 078	24 406	**26 781**	6 570	3 014	**1 030**	406
ERD-D	1 903	**4 399**	971	6 597	1 528	**604**	1
ERD-S	5 149	**6 954**	4 969	152	94	**36**	4
CAI	2 118	2 344	**2 413**	165	100	**34**	13
BAR	81 025	91 905	**95 206**	32	17	**7**	1

Table 3. Traffic jam and recovery batches

Graph	Speed-up			Edge Scans			
	RR	T-SWSF	DDFLP	RR	T-SWSF	DDFLP-OWNERSHIP	DDFLP-PRIORITY
GER	27 901	**42 662**	26 688	164 066	38 867	**12 520**	6 246
BEL	3 222	**4 454**	2 893	55 672	15 684	**5 268**	2 095
ERD-D	42	**235**	14	352 845	255 899	**100 480**	216
ERD-S	110	**167**	52	23 363	16 692	**5 928**	937
CAI	69	**109**	62	107 144	47 945	**11 476**	5 664
BAR	28 082	**36 054**	24 022	2 538	2 687	**751**	179

evidence of the huge data dependency of the algorithms (the performance of the three algorithms are quite different, depending on the graph and on the type of batches). They also show that it is practically useful to process a set of updates as a batch only when the updated edges have a strong interference regarding their impact on the shortest path tree, while updates that are far away from each other usually do not interfere, and hence these updates can be handled iteratively. This latter observation is strongly supported by the data of Table 2 and 3, where the speed-ups of the algorithms in case of node failure and traffic jam batches are reported. In these cases, it is highly likely that edge changes interfere with one another and, in fact, RR is basically outperformed by DDFLP and T-SWSF, respectively.

A second evidence of our experimental analysis is given by the machine independent parameters of Tables 2–6, which in fact show that the sum of the number of edges scanned by ownership and the number of edges scanned by priority by DDFLP is always smaller than the number of edge scans of RR and T-SWSF. This is clearly due to the ownership mechanism used by DDFLP, which determines a significant saving in the number of edges scanned by priority, thus reflecting the theoretical bounds of the algorithms presented in [7]. In the practical cases where scanning an edge requires some non-negligible amount of time, (e.g., the graph is too big to fit into the main memory, weights must be queried from an external database, etc.), edge scans are likely to become the predominant factor in the runtime of the algorithms. Thus, any improvement in these parameters, will immediately reflect on their performance.

In terms of computational time, our experiments also show that the performance of DDFLP is almost always comparable to that of RR and T-SWSF. This is clearly due to the fact that, although DDFLP scans less edges than the other algorithms, the scans by priority have logarithmic cost, which becomes

Table 4. Incremental batches

Graph	Speed-up			Edge Scans			
	RR	T-SWSF	DDFLP	RR	T-SWSF	DDFLP-OWNERSHIP	DDFLP-PRIORITY
GER	2 201	**2 904**	1 601	624 615	90 184	**45 096**	**27 840**
BEL	192	**264**	145	216 115	35 766	**17 886**	**10 277**
ERD-D	2 456	**7 050**	1 198	12 317	4 091	**2 036**	**7**
ERD-S	559	**854**	271	3 612	1 132	**570**	**159**
CAI	112	**129**	75	4 949	1 492	**768**	**588**
BAR	5 056	**7 317**	3 264	3 350	1 116	**567**	**320**

Table 5. Decremental batches

Graph	Speed-up			Edge Scans			
	RR	T-SWSF	DDFLP	RR	T-SWSF	DDFLP-OWNERSHIP	DDFLP-PRIORITY
GER	**3 068**	2 848	1 753	256 572	288 957	**96 084**	**31 854**
BEL	**347**	337	207	98 648	107 461	**36 568**	**11 998**
ERD-D	7 794	**17 150**	10 719	2 177	2 203	**812**	**1**
ERD-S	482	**1 127**	492	2 029	2 104	**768**	**54**
CAI	**191**	180	128	2 311	2 644	**894**	**224**
BAR	8 003	**13 537**	7 702	353	470	**165**	**3**

predominant when the graph is dense. With the exception of traffic jam and re-covery batches on the very dense ERD-D, the ratio between the speed-up provided by DDFLP and that provided by RR is always between 0.47 (ERD-S in Table 3) and 1.20 (GER in Table 2), while the ratio between the speed-up provided by DDFLP and that provided by T-SWSF is always between 0.31 (ERD-S in Table 3) and 1.17 (GER in Table 2).

The variability of the performance of DDFLP on different instances is surely due to the ownership mechanism whose performance tightly depends on the structure of the graph. In fact, when the ownership function is bounded by a small integer and the maximum degree is not a small constant w.r.t. the size of the graph (see Table 1), DDFLP is more likely to be the fastest algorithm. On the contrary, in very dense graphs, where the ownership function is not bounded by a small constant, DDFLP is more likely to be slower than both other algorithms, even though it is faster than RR in some cases (see Table 2 and 5). It is also worth noticing that the average speed-ups of RR and T-SWSF, reported in Tables 2–6, are almost always bigger than those reported in [2] for the very same instance GER. This improvement can be due to the use of either a more efficient framework or more optimized code and libraries.

In conclusion, notwithstanding the big level of data dependency, we feel that the following guidelines can be provided, once that a batch algorithm has to be selected. If one expects that either node failures or traffic jams have to be faced, then it should be definitely preferred either DDFLP or T-SWSF, respectively. On the other hand, on generic homogeneous batches, the better performing solutions are T-SWSF and RR, and the higher is the interference among the updates, the more the former is preferable to the latter, and vice versa.

Table 6. Mixed sequences of incremental and decremental batches

Graph	Speed-up			Edge Scans			
	RR	T-SWSF	DDFLP	RR	T-SWSF	DDFLP-OWNERSHIP	DDFLP-PRIORITY
GER	2 679	**2 905**	1 531	566 110	201 808	**73 400**	**30 661**
BEL	269	**297**	151	192 358	65 388	**25 579**	**10 925**
ERD-D	5 080	**11 120**	3 716	15 284	4 583	**2 271**	**13**
ERD-S	518	**959**	296	2 818	1 777	**719**	**103**
CAI	148	**151**	80	4 681	3 040	**1 070**	**523**
BAR	6 346	**10 163**	4 446	1 486	668	**305**	**125**

References

1. Albert, R., Barabási, A.L.: Emergence of scaling in random networks. Science 286, 509–512 (1999)
2. Bauer, R., Wagner, D.: Batch dynamic single-source shortest-path algorithms: An experimental study. In: Vahrenhold, J. (ed.) SEA 2009. LNCS, vol. 5526, pp. 51–62. Springer, Heidelberg (2009)
3. Bollobás, B.: Random Graphs. Cambridge University Press, Cambridge (2001)
4. Bruera, F., Cicerone, S., D'Angelo, G., Di Stefano, G., Frigioni, D.: Dynamic multi-level overlay graphs for shortest paths. Math. Comput. Sci. 1(4), 709–736 (2008)
5. Chan, E., Yang, Y.: Shortest path tree computation in dynamic graphs. IEEE Trans. Comput. 4(58), 541–557 (2009)
6. Cicerone, S., D'Angelo, G., Di Stefano, G., Frigioni, D.: Partially dynamic efficient algorithms for distributed shortest paths. Theoret. Comput. Sci. 411, 1013–1037 (2010)
7. D'Andrea, A., D'Emidio, M., Frigioni, D., Leucci, S., Proietti, G.: Dynamically maintaining shortest path trees under batches of updates. In: Moscibroda, T., Rescigno, A.A. (eds.) SIROCCO 2013. LNCS, vol. 8179, pp. 286–297. Springer, Heidelberg (2013)
8. Dijkstra, E.W.: A note on two problems in connexion with graphs. Numer. Math. 1, 269–271 (1959)
9. Frigioni, D., Marchetti-Spaccamela, A., Nanni, U.: Semidynamic algorithms for maintaining single source shortest paths trees. Algorithmica 22(3), 250–274 (1998)
10. Frigioni, D., Marchetti-Spaccamela, A., Nanni, U.: Fully dynamic algorithms for maintaining shortest paths trees. J. Algorithms 34(2), 251–281 (2000)
11. Frigioni, D., Marchetti-Spaccamela, A., Nanni, U.: Fully dynamic shortest paths in digraphs with arbitrary arc weights. J. Algorithms 49(1), 86–113 (2003)
12. Hyun, Y., Huffaker, B., Andersen, D., Aben, E., Shannon, C., Luckie, M., Claffy, K.: The CAIDA IPv4 routed/24 topology dataset, http://www.caida.org/data/active/ipv4_routed_24_topology_dataset.xml
13. Narváez, P., Siu, K.Y., Tzeng, H.Y.: New dynamic algorithms for shortest path tree computation. IEEE/ACM Trans. Netw. 8(6), 734–746 (2000)
14. PTV: (2008), http://www.ptv.de
15. Ramalingam, G., Reps, T.W.: An incremental algorithm for a generalization of the shortest paths problem. J. Algorithms 21, 267–305 (1996)
16. Ramalingam, G., Reps, T.: On the computational complexity of dynamic graph problems. Theoret. Comput. Sci. 158(1&2), 233–277 (1996)
17. Roditty, L., Zwick, U.: On dynamic shortest paths problems. Algorithmica 61(2), 389–401 (2011)

Exploiting GPS Data in Public Transport Journey Planners*

Luca Allulli[1], Giuseppe F. Italiano[2], and Federico Santaroni[2]

[1] Roma Servizi per la Mobilità s.r.l., Piazzale degli Archivi 40, 00144 Roma, Italy
luca.allulli@agenziamobilita.roma.it
[2] Univ. of Rome "Tor Vergata", Via del Politecnico 1, 00133 Roma, Italy
italiano@disp.uniroma2.it, santaroni@ing.uniroma2.it

Abstract. Current journey planners for public transport are mostly based on timetabling, i.e., they hinge on the assumption that all transit vehicles run on schedule. Unfortunately, this might not always be realistic, as unpredictable delays may occur quite often in practice. In this case, it seems quite natural to ask whether the availability of real-time updates on the geo-location of transit vehicles may help improving the quality of the solutions offered by routing algorithms. To address this question, we consider the public transport network of the metropolitan area of Rome, where delays are not rare events, and report the results of our experiments with two journey planners that are widely used for this city: one based on timetabling information only (Google Transit) and one which makes explicit use of GPS data on the geo-location of transit vehicles (Muovi Roma).

Keywords: Shortest Path Problems, Route Planning, Timetable-based Routing, Public Transport Networks.

1 Introduction

In the last decade, there have been many new advances on point-to-point shortest path algorithms, motivated by the widespread use of navigation software systems. Many new efficient algorithmic techniques have been proposed for this problem, including hierarchical approaches (e.g., Contraction Hierarchies) [13,16], reach-based algorithms [14,15], Transit Node Routing [5], and Hub-Based Labeling methods [3,2]. We refer the interested reader to [4] for an excellent survey of this area, and we only mention here that this research is not only relevant from a theoretical viewpoint, but it also had a big practical impact on commercial navigation systems. Indeed, some of the algorithmic approaches proposed in the literature are currently being deployed in journey planners offered by several companies, including Apple, Microsoft and Google.

Most of the point-to-point shortest path algorithms developed for road networks can be adapted to work on public transport networks, but unfortunately

---

* Work partially supported by the Italian Ministry of Education, University and Research, under Project AMANDA (Algorithmics for MAssive and Networked DAta).

J. Gudmundsson and J. Katajainen (Eds.): SEA 2014, LNCS 8504, pp. 295–306, 2014.

they fail to yield similar improvements [7,12]. As explained in [6], one of the reasons is that most public transportation networks, such as bus-only networks in big metropolitan areas, are far more complicated than road networks. Indeed, public transport networks are known to be less hierarchically structured and are inherently event-based, as transit vehicles have specific times at which they depart or arrive at given stops.

In the last years, several new algorithmic techniques, such as Transfer Patterns [7], RAPTOR (Round bAsed Public Transit Optimized Router) [9], and Connection Scan [10], have been specifically designed for public transport networks. Most of the routing algorithms for public transport networks use timetabling data, i.e., they compute shortest paths starting from a given timetable, and thus assume implicitly that all transit vehicles run on schedule. Unfortunately, this might not always be a realistic assumption: unavoidable delays may occur frequently and for many unplanned reasons, including traffic jams, accidents, road closures, bad weather, increased ridership, vehicle breakdowns and sometimes even unrealistic scheduling. In previous experimental work [11], we showed that in some practical cases (such as in the public transport network of the metropolitan area of Rome), there can be large discrepancies between the journey times that are predicted by timetabling and the actual travel times that may occur in practice. In fact, we believe that significant deviations from the transit schedule are not limited to the case of the transport network of Rome, but they happen often in many other urban areas worldwide. In all such scenarios, timetabled-based routing methods may suffer from many inaccuracies (such as incorrect estimates of the waiting/transfer times at transit stops), and thus, independently of their own merits, they might fail to deliver high quality solutions. In this scenario, it seems quite natural to ask whether the availability of real-time information about the actual geo-location of transit vehicles may be used to mitigate the possible oversights of timetabling-based methods.

To address this question, we consider the public transport network of the metropolitan area of Rome, where delays are not rare events, and augment our preliminary work [11] in several new and perhaps intriguing directions. First, we assess more precisely the difference between the original timetable and the actual movement of transit vehicles in the network. In order to accomplish this, we measure the actual arrival times of transit vehicles at their stops, and compare them against the corresponding times provided by the (theoretical) schedule. Even for lines of particular importance (such as buses joining top touristic hubs in the city), our experiments report significant fluctuations between the actual position of transit vehicles and the original timetable. As the main contribution of this paper, we provide the first attempt (to the best of our knowledge) to evaluate empirically the effectiveness of journey planners that use real-time information about the actual geo-location of transit vehicles. To do this, and differently from [11], we consider here both a journey planner based on timetabling information only (Google Transit), and a journey planner which makes explicit use of GPS data on the geo-location of transit vehicles, i.e., the new journey planner developed by the Transit Agency of Rome [1] (Muovi Roma). We remark that, like other complex software systems, both Google Transit and Muovi Roma include in their own implementations many different details which can influence the quality of their query results. Nevertheless, we believe that their different inputs (timetabling only for Google Transit, timetabling and GPS data for Muovi Roma) can still be used to explain some differences in their query results. In particular, as the main objective of this work we are not interested in comparing the two journey planners in general, but we are only trying to understand the behavior of the two different approaches (timetabling versus real-time GPS data) whenever there are large deviations from the original schedule.

To perform our experiments, we picked a certain day (precisely Wednesday January 8, 2014). This was a typical day, where the delays observed turn out to be quite standard (i.e., no major traffic jams, construction works or extreme weather conditions). On that day, we generated many random queries with origin and destination chosen uniformly at random among the stops of the transit network of the metropolitan area of Rome. We then submitted all the queries generated both to Google Transit and to Muovi Roma, and recorded all the journeys suggested by the two journey planners. On the very same day, we collected the GPS data on the geo-location of all transit vehicles, publicly available as open data, by querying every minute the information system of the Transit Agency of Rome [1]. With all these data, we built a simulator capable of following precisely the actual time taken by any journey on that given day, according to the GPS tracking of transit vehicles in the public transport network of Rome. Finally, we followed the journeys suggested by Google Transit and by Muovi Roma on this simulator, and analyzed all the data collected. One of the main findings of our experimental work is that the exploitation of GPS data can be actually an important asset for journey planners. Indeed, our experiments highlight that in the presence of significant fluctuations from the original timetable, Muovi Roma seemed to estimate with better accuracy the actual travel times in the network. Furthermore, GPS data seem to be helpful in obtaining better predictions for the transfer times, which is another important ingredient for journey planners.

2   Preliminaries

A public transport network consists of a set of *stops* interconnected by a set of *routes* and by a set of *footpaths*. In more detail, we use the following terminology. A *stop* corresponds to a location in the network where passengers may either enter into or exit from a transit vehicle (such as a bus stop or a subway station). A *hop* is a connection between two adjacent stops and models a vehicle departing from stop u and arriving at stop v without intermediate stops in between. A *trip* consists of a sequence of consecutive hops operated by the same transit vehicle. A *route* is a collection of trips which share the same sequence of stops. A *line* is a collection of routes sharing the same terminals. We usually, but not necessarily, have two routes per line: one going from a terminal to another and one going in the reverse direction. A *footpath* enables walking transfers between nearby stops. Each footpath consists of two stops and an associated (constant) walking time between the two stops. A *journey* connects a source stop s to a target stop t, and consists of a sequence of trips and footpaths in the order of travel. Each trip in the journey is associated with two stops, corresponding to the pick-up and drop-off points. Usually a journey describes also the times at which one should take the vehicles performing the specified trips, and the time estimations of the trips and of the footpaths.

3   Journey Planners Considered in Our Experiments

In our experiments, we considered two different journey planners which are available for the city of Rome: *Google Transit* and *Muovi Roma*. Google Transit is a well known journey planner for public transport networks, which is directly integrated in Google Maps. Since it is widely known and used, we only mention here that Google Transit is available in many cities worldwide and it uses the timetables provided by local Transport Authorities in the GTFS format (http://developers.google.com/transit/gtfs/).

Muovi Roma [1] is a journey planner developed by the Transit Agency of Rome. It works only in the metropolitan area of Rome, and makes use of data from GPS bus trackers whenever possible. The role of GPS data is twofold: on the one hand, Muovi Roma collects historical data with the aim of improving the accuracy of time estimations. On the other hand, real-time GPS data is directly fed to the shortest paths algorithm, in order to foresee events that will take place in the near future. Muovi Roma computes an optimal route using Dijkstra's single-source shortest paths algorithm on a time-dependent graph [6]. The computation begins from the source node, and keeps track of the time when each graph node is expected to be reached. In this way, the cost of each edge can be a function of the time τ when the edge traversal begins. Two kinds of edges are of particular interest in modeling a public transport network: *bus waiting edges* and *bus ride edges*. A bus waiting edge models the fact that the user is waiting at a particular bus stop. Edge cost depends on how much time the user is going to wait. If GPS data are available, the algorithm simply computes the waiting time $\tau' - \tau$, where τ' is the foreseen arrival time of the first bus that will reach the bus stop after time τ. In the absence of usable GPS data, the estimation is based on historical data: essentially, the average waiting time in a time slot σ is computed, where $\tau \in \sigma$. Time slots are chosen in such a way that it makes sense to use statistics to predict the expected waiting time in each slot. A bus ride edge models a hop, whose weight depends on the average bus speed on that edge, measured using GPS data. As before, when GPS data are available, they are used to measure the current edge speeds, which will remain valid for the near future. Otherwise, the edge function switches to using statistics based on historical data. Timetables are used by Muovi Roma for routes where GPS tracking is not available: in particular, for regional trains and the subway. Regional trains generally run on schedule, and have a low frequency: therefore, timetables are employed to determine the exact time when a train will leave every station, and hence the exact waiting times at stations. For the subway and other high-frequency lines, average waiting times are used instead. As any journey planner used in practice, Muovi Roma tries to suggest a journey that is both quick and comfortable, taking into account parameters that express user's journey preferences, e.g. how much the user is willing to walk. This is achieved by Muovi Roma by using generalized edge cost functions that model the users' perceived costs, taking into account both journey duration and preferences. The shortest paths algorithm actually minimizes this generalized cost.

Like other complex software systems, both Google Transit and Muovi Roma include, in their implementations, many different details which can influence the quality of the results obtained in response to a query. Nevertheless, we believe that their different inputs (timetabling only for Google Transit, timetabling and GPS data for Muovi Roma) could be a discriminating factor which can explain the main differences in their query results.

4   GPS Data Processing and Simulation System

In this section we describe briefly our simulation system, which is at the heart of our experiments. We first collect GPS data, as follows. We submit queries every minute to the information system of the Transit Agency of Rome [1], in order to obtain (from GPS data) the instantaneous geo-location of all buses in the network at a given time (the Transit Agency has no data about the subway network). As a response to a query, the following data are returned by the Transit Agency of Rome for each bus currently in the network: the bus identification number; the route in which the bus is operating; the relative position of the bus in this route; a timestamp identifying the time in which the bus last sent its GPS data to the information system. This GPS stream may not be always complete, i.e., we noticed that in some cases a bus could loose connection to the information system. In our experiments these limitations were not too problematic, as we were able to collect the vast majority of the vehicles' data, which revealed to be sufficient for our experiments.

After collecting this stream of data, we are able to reconstruct the actual times of the trips performed by all buses. In particular, our simulation system is able to simulate closely the experience of a user traveling in the network according to a given journey, and can measure the actual time required by that journey. Assume that we wish to simulate a user taking a given journey from an origin s to a destination t, at a given departure time τ. When the journey requires the user to take a bus of route r at time τ' from stop s_d to s_a, we use the GPS data to look for the first trip t_r of route r which passes through stop s_d at time $\tau_d > \tau'$; following the trip t_r we then retrieve the time τ_a at which the bus arrives at stop s_a. Note that we compute the actual required time by a journey as $\tau_a - \tau$, where τ_a is the time at which we reach the destination of our journey and τ is the departure time of the journey, as specified by the query. For subway trips and footpaths there are no GPS data available, and our simulator computes their times according to the subway timetables and to the information returned by Google Maps, respectively. We claim that this has a negligible impact in the

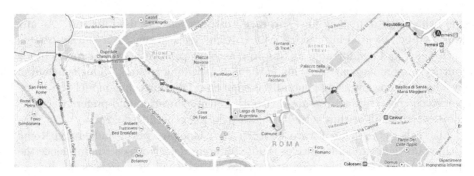

Fig. 1. Line 64, route from *S. Pietro Station* to *Termini Station* (better viewed in color)

overall reliability of our experiments, since subway and walking times are known to be extremely more reliable than bus trips.

5   Experiments and Discussion

As a case study, we considered the public transport networks of the metropolitan area of Rome, which consists of 309 buses and 2 subway lines, for a total of 7,089 stops (7,037 bus and 52 subway stops). We collected and processed (as described in Section 4) all the GPS data available on the geo-location of buses on a typical weekday. In particular we considered in our experiments Wednesday, January 8, 2014. As mentioned in the introduction, this was a typical day, and the delays observed turn out to be quite standard for Rome (i.e., no major traffic jams, construction works or extreme weather conditions). In the following, we report the results of our experiments.

5.1   Analysis of Bus Trips

In our first experiment, we assessed more precisely the difference between the official timetable and the actual movement of transit vehicles in the public transport network. In particular, we used the GPS data obtained from the Transit Agency of Rome to measure the actual arrival times of transit vehicles at their stops, and compared them against the (theoretical) arrival times of the same transit vehicles according to the official timetable. Our experiments highlighted significant discrepancies between the actual arrival times that could be measured with GPS data and the theoretical arrival times. This was clear even for bus lines of particular importance, such as buses joining top touristic hubs in the city, and outside of the typical rush hours (which are between 8am and 10am, and between 5pm and 7pm). In particular, we illustrate below the case of line 64, focusing on route going from *S. Pietro Station* to the main train station in Rome (*Termini*), which connects many important touristic attractions in the city center of Rome (see Figure 1).

Figure 2 illustrates the results of our experiment with line 64 in the time range from 1pm to 3pm. In particular, Figure 2(a) shows how the bus trips

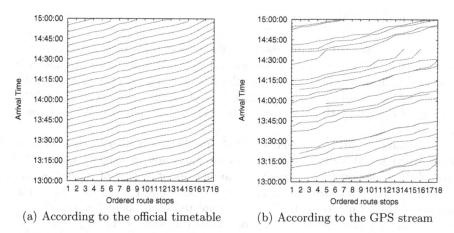

(a) According to the official timetable   (b) According to the GPS stream

Fig. 2. Line 64, route from San Pietro to Termini in the time range from 1pm to 3pm

should happen according to the official timetable, with a scheduled departure from the first stop every 5 minutes. Figure 2(b) reports the actual status of the bus trips in the same time range, this time according to GPS data. As it can be seen from a direct comparison between the two figures, there are several crucial discrepancies. According to the original timetabling, 30 buses are supposed to be running in that route between 1pm and 3pm, while according to GPS data only 25 buses appear to be running. Furthermore, there are relevant fluctuations between the actual positions of buses in the route and their theoretical positions according to the official timetable. In particular, there are several cases in which there is first a significant delay and then several bus trips merge into bursts, i.e., few buses travel very close to each other, as it is typical with bus networks. This can have a high impact in the travel times or the waiting times at a given bus stop: in theory (i.e., according to the schedule, as in Figure 2(a)), a user is supposed to wait at most 5 minutes at each bus stop; in practice (i.e, according to the GPS data, as in Figure 2(b)), this waiting time can be as high as 20 minutes. We remark that this is of particular relevance for journey planners. Indeed, a journey planner based on timetabling would estimate journey travel times by considering only the data in Figure 2(a): in some cases, it could thus estimate short (but unrealistic) transfer times for a possible connection with line 64, and would thus potentially include it in its suggested journeys. On the other side, a journey planner based on GPS data would be influenced by the actual location of buses (Figure 2(b)), and thus it is likely that it would get more realistic estimates on the waiting times for connections.

Note that in Figure 2(b) there are few bus trips that either start after the first stop or end prematurely before the last stop: this is due to the fact that some GPS transmitter may loose sporadically connection to the information system of the Transit Agency. We remark that this is not a problematic issue, since we were able to collect the vast majority of the vehicles' data, which revealed to be sufficient for our experiments.

5.2   Accuracy of Estimates

In this section, we report the results of the experiments performed with the two journey planners described in Section 3: Google Transit, based on timetable information only, and Muovi Roma, which makes explicit use of GPS data on the geo-location of transit vehicles. In our first set of experiments, we measured the accuracy of the time estimates provided by each journey planner. We performed this as follows.

On January 8, 2014, we generated 610 queries: each query q was defined by an origin s and a destination t, selected uniformly at random among the stops of the transportation network of the metropolitan area of Rome, and by a starting time τ from the origin s, chosen again uniformly at random in the time range from 7am to 8pm. We then submitted each query both to Google Transit and to Muovi Roma, exactly one minute before its starting time τ (so as to use "fresh" GPS data), and recorded all the journeys J_i, suggested by the two journey planners, together with the time estimates $t_e(J_i)$ given by the corresponding journey planner. On the very same day, we collected the GPS data on the geo-location of all transit vehicles and we used our simulation system to compute the actual time $t_a(J_i)$ required by journey J_i, according to the tracking of transit vehicles in the network. Since Muovi Roma returns only one journey per query, while Google Transit usually returns four alternative journeys, in this experiment we considered for each query only the "best" journey returned by Google Transit, i.e., the one having the smallest actual required time (according to GPS data).

We define the *error coefficient* of journey J_i to be $t_a(J_i)/t_e(J_i)$, i.e., its actual time (according to GPS data) divided by its estimated time (i.e., the time predicted by the corresponding journey planner). Note that a perfectly estimated journey will have error coefficient equal to 1, an optimistic estimate will have error coefficient larger than 1, while a pessimistic estimate will have error coefficient smaller than 1. The more the error coefficient will depart substantially from 1, the less accurate will be the time estimate of the corresponding journey planner.   We report the distribution of the error coefficients as a function of the journey times, as follows. For each journey J_i, we take as a reference the estimated journey time $t_e(J_i)$ provided by the corresponding journey planner. For the sake of representation, we group all the journeys into *time slots* with a 5-minute resolution. That is, we measure $t_e(J_i)$ in minutes and let the k-th time slot σ_k contain all journeys J_i such that $t_e(J_i)$ lies in the interval $[5k, 5(k+1))$. For each time slot σ_k, we compute the average, the 10th and the 90th percentile of the error coefficient of journeys falling in that time slot.

Figure 3 plots the distributions of the error coefficients (in log scale) for Google Transit (Figure 3(a)) and for Muovi Roma (Figure 3(b)), and illustrates the average, 10th and 90th percentile for each time slot, as defined above. We remark that very few points are not represented in those figures, as they fall outside of the plots (but they contribute to the computation of the average and of the percentiles). In both cases, for journeys that were estimated as short ones (i.e., journeys with estimated times less than 20 minutes), the error coefficient

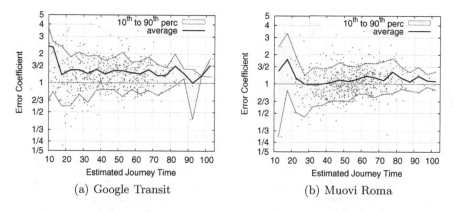

(a) Google Transit   (b) Muovi Roma

Fig. 3. Error Coefficient in log scale (better viewed in color)

seems to be quite large, and with a wide interval between its 10th and 90th percentile. Both journey planners seem to err on the optimistic side, as their estimates are smaller than the actual time required for the journey. This is not surprising, since journeys that tend to be estimated as short ones are likely to be affected more (in relative terms) by fluctuations on the schedule. A closer look at journeys that were estimated in a medium range (i.e., journeys with estimated times between 20 and 70 minutes) show marked differences between Google Transit and Muovi Roma. For Google Transit, the interval between the 10th and the 90th percentile of the error coefficient ranges from 0.58 to 2.07, while the average is contained in the interval [1.23, 1.38]. This implies that in 80% of those cases the actual time required for the journey ranges from 0.58 to 2.07 of the time predicted by Google Transit. On the other side, for journeys that were estimated in the medium range by Muovi Roma, the interval between the 10th and the 90th percentile of the error coefficient ranges from 0.46 to 1.92 (in 80% of the journeys the actual time required ranges from 0.46 to 1.92 of the time predicted by Muovi Roma), and the average of the error coefficient is now contained in a significantly smaller interval, namely [0.97, 1.16]. As the estimated journey times increase further (i.e., for journeys with estimated times larger than 70 minutes), then the average of the error coefficient seems to increase slightly, especially for Muovi Roma. This is mainly due to the fact the information about the geo-location of transit vehicles at a certain time might no longer be very useful more than one hour later. Note that, however, the range between the 10th and the 90th percentile becomes slightly bigger for Google Transit and slightly smaller for Muovi Roma. The overall behavior seems to suggest that, whenever there are significant fluctuations in the actual schedule, the use of GPS data provides a better accuracy in estimating the actual travel times, which is an important feature for a journey planner.

We remark that this experiment tries to measure how accurately a given journey planner is able to estimate the travel times in the network. However, it gives no information on the overall quality of the journeys returned in response

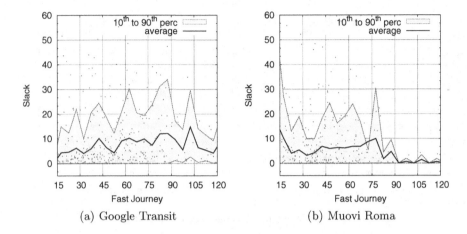

Fig. 4. Slacks and fast journeys (better viewed in color). Fast journey times and slacks are measured in minutes.

to queries. In other words, although the journey returned by a planner could have estimated travel times which are very close to their actual times, this journey might still not be among the best options available. This will be considered in the next section.

5.3   Quality of Solutions

We next assessed empirically the quality of the journeys suggested by the different planners. This is a very complicated issue, since journey planners try to optimize multiple cost criteria in public transport networks (such as travel times, number of transfers, total distance walked, etc.). For the sake of simplicity, here we do not consider multiple criteria and their tradeoffs, but restrict our attention only to the travel time of each journey.

In this framework, we performed the following analysis. For each query q, we considered the journeys returned by Google Transit and by Muovi Roma, and their actual travel times (according to GPS data). Among those journeys, we define the *fast journey* for query q to be the one with smallest travel time, and denote this travel time by $t_a^*(q)$. Given a journey $J_i(q)$ corresponding to query q, we define its *slack* as the difference between its actual travel time and the time required by the fast journey for the same query (i.e., as $t_a(J_i(q)) - t_a^*(q)$). Obviously, a fast journey will have slack equal to 0: the larger the slack, the more the corresponding journey will be slower than the fast journey.

Figure 4 plots the distributions of slacks for Google Transit (Figure 4(a)) and for Muovi Roma (Figure 4(b)), and illustrates the average, 10th and 90th percentile for each time slot. Once again, very few points are not represented in those figures, as they fall outside of the plots. Note that the x-axis is now different from Figure 3: in particular, Figure 3 has *estimated* journey times, while Figure 4 has *actual* journey times. As it can be seen from Figure 4, the effectiveness of

GPS data seems to increase with the length of the journey. Indeed, Muovi Roma was able to achieve small slacks and to return fast journeys, especially in the case of long range queries (i.e., for journeys with actual times larger than 70 minutes). This might seem in contradiction with our discussion in Section 5.2, as one would expect that the information about the geo-location of transit vehicles at a certain time might no longer be very useful more than one hour later. However, one should consider that long journeys are more likely to involve trips on suburban lines, which run on much lower frequencies: missing a transfer in this case will cause a significant delay. Thus, a long range journey based on an unreliable timetable is likely to incur in discrepancies which have a higher impact on the overall travel time. For such cases, indeed GPS data appear to be helpful in choosing more suitable transfers.

6   Conclusions and Future Work

In this paper, we have tried to evaluate empirically the effectiveness of journey planners that use real-time data on the actual geo-location of transit vehicles. In particular, we have considered as a case study the public transport network of the metropolitan area of Rome, where significant deviations from the original timetable seem quite common, and made several experiments with Google Transit (a journey planner based on timetabling information) and with Muovi Roma (a journey planner which makes explicit use of GPS data on the geo-location of transit vehicles, developed especially for the city of Rome). Although the main algorithmic ideas employed by Google Transit are known to be remarkably robust to fluctuations in the schedule [8], our experiments suggest that the exploitation of GPS data can actually be an important asset for journey planners.

Few words about the reproducibility of our experiments are in order. In particular, our experiments are based only on publicly accessible data: the API's for route planning offered by Google Transit and by Muovi Roma, and the real-time GPS bus tracks published as open data by the Transit Authority of Rome [1]. It is also possible to extend our experiments to any other city where both Google Transit and real-time GPS data are available, by adapting Muovi Roma (an open source project) to use the local GPS data. We underline that Muovi Roma expects as input: (1) a description of the local transit network (e.g., a GTFS file) and the local road network (e.g., an OpenStreetMap dataset), and (2) a periodical snapshot of the position of each vehicle on the transit network. From this data, Muovi Roma computes edge speeds and bus arrival predictions, that are fed to its shortest paths algorithm.

One of the limits of this work is that Google Transit and Muovi Roma are highly sophisticated systems, based on many different low-level implementation details, which might influence in several ways the results obtained in response to a query. In order to better highlight the differences between timetabling and GPS inputs in practical scenarios, we are planning to perform new experiments using exactly the same journey planner in the two different cases. In particular, we are currently experimenting with algorithms (such as RAPTOR [9]) which require no preprocessing, and thus could be adapted to work with GPS data.

References

1. Agenzia Roma servizi per la Mobilità. Muoversi a Roma (2014),
   http://www.agenziamobilita.roma.it/servizi/open-data/
   (Online; accessed January 2014)
2. Abraham, I., Delling, D., Fiat, A., Goldberg, A.V., Werneck, R.F.F.: VC-dimension
   and shortest path algorithms. In: Aceto, L., Henzinger, M., Sgall, J. (eds.) ICALP
   2011, Part I. LNCS, vol. 6755, pp. 690–699. Springer, Heidelberg (2011)
3. Abraham, I., Fiat, A., Goldberg, A.V., Werneck, R.F.F.: Highway dimension, shortest paths, and provably efficient algorithms. In: SODA 2010, pp. 782–793. SIAM,
   Philadelphia (2010)
4. Bast, H., Delling, D., Goldberg, A.V., Müller-Hannemann, M., Pajor, T., Sanders,
   P., Wagner, D., Werneck, R.: Route planning in transportation networks. Tech.
   Rep. MSR-TR-2014-4, Microsoft Research (2014)
5. Bast, H., Funke, S., Matijevic, D., Sanders, P., Schultes, D.: In transit to constant
   time shortest-path queries in road networks. In: ALENEX 2007. SIAM, Philadelphia (2007)
6. Bast, H.: Car or public transport - Two worlds. In: Albers, S., Alt, H., Näher,
   S. (eds.) Efficient Algorithms. LNCS, vol. 5760, pp. 355–367. Springer, Heidelberg
   (2009)
7. Bast, H., Carlsson, E., Eigenwillig, A., Geisberger, R., Harrelson, C., Raychev, V.,
   Viger, F.: Fast routing in very large public transportation networks using Transfer
   Patterns. In: de Berg, M., Meyer, U. (eds.) ESA 2010, Part I. LNCS, vol. 6346,
   pp. 290–301. Springer, Heidelberg (2010)
8. Bast, H., Sternisko, J., Storandt, S.: Delay-robustness of Transfer Patterns in public transportation route planning. In: ATMOS 2013. OASICS, vol. 33, pp. 42–54.
   Schloss Dagstuhl (2013)
9. Delling, D., Pajor, T., Werneck, R.F.F.: Round-based public transit routing. In:
   ALENEX 2012, pp. 130–140. SIAM, Philadelphia (2012)
10. Dibbelt, J., Pajor, T., Strasser, B., Wagner, D.: Intriguingly simple and fast transit
    routing. In: Bonifaci, V., Demetrescu, C., Marchetti-Spaccamela, A. (eds.) SEA
    2013. LNCS, vol. 7933, pp. 43–54. Springer, Heidelberg (2013)
11. Firmani, D., Italiano, G.F., Laura, L., Santaroni, F.: Is timetabling routing always
    reliable for public transport? In: ATMOS 2013. OASICS, vol. 33, Schloss Dagstuhl
    (2013)
12. Geisberger, R.: Contraction of timetable networks with realistic transfers. In: Festa,
    P. (ed.) SEA 2010. LNCS, vol. 6049, pp. 71–82. Springer, Heidelberg (2010)
13. Geisberger, R., Sanders, P., Schultes, D., Delling, D.: Contraction hierarchies:
    Faster and simpler hierarchical routing in road networks. In: McGeoch, C.C. (ed.)
    WEA 2008. LNCS, vol. 5038, pp. 319–333. Springer, Heidelberg (2008)
14. Goldberg, A.V., Kaplan, H., Werneck, R.F.: Reach for A*: Efficient point-to-point
    shortest path algorithms. In: ALENEX 2006, pp. 129–143. SIAM, Philadelphia
    (2006)
15. Gutman, R.J.: Reach-based routing: A new approach to shortest path algorithms
    optimized for road networks. In: ALENEX/ANALC 2004, pp. 100–111. SIAM,
    Philadelphia (2004)
16. Sanders, P., Schultes, D.: Highway hierarchies hasten exact shortest path queries.
    In: Brodal, G.S., Leonardi, S. (eds.) ESA 2005. LNCS, vol. 3669, pp. 568–579.
    Springer, Heidelberg (2005)

Order-Preserving Matching with Filtration[*]

Tamanna Chhabra and Jorma Tarhio

Department of Computer Science and Engineering
Aalto University
P.O. Box 15400, 00076 Aalto, Finland
{firstname.lastname}@aalto.fi

Abstract. The problem of order-preserving matching has gained attention lately. The text and the pattern consist of numbers. The task is to find all substrings in the text which have the same relative order as the pattern. The problem has applications in analysis of time series like stock market or weather data. Solutions based on the KMP and BMH algorithms have been presented earlier. We present a new sublinear solution based on filtration. Any algorithm for exact string matching can be used as a filtering method. If the filtration algorithm is sublinear, the total method is sublinear on average. We show by practical experiments that the new solution is more efficient than earlier algorithms.

1 Introduction

String matching [11] is a widely known problem in Computer Science. Given a text T of length n and a pattern P of length m, both being strings over a finite alphabet Σ, the task of string matching is to find all the occurrences of P in T.

The problem of order-preserving matching [2,3,8,9] has gained attention lately. It considers strings of numbers. The relative order of the positions in the pattern P is matched with the relative order of a substring u of T, where $|u| = |P|$. Suppose $P = (10, 22, 15, 30, 20, 18, 27)$ and $T = (22, 85, 79, 24, 42, 27, 62, 40, 32, 47, 69, 55, 25)$, then the relative order of P matches the substring $u = (24, 42, 27, 62, 40, 32, 47)$ of T, see Fig. 1. In the pattern P the relative order of the positions is: 1, 5, 2, 7, 4, 3, 6. This means 10 is the smallest number in the string, 15 is the second smallest, 18 the third smallest and so on. In order-preserving matching, the task is to find a substring in T which has the same relative order as P. In the given example, u has the same relative order as P. So u matches the pattern P.

At least three solutions [2,8,9] have been proposed for on-line order-preserving matching of strings. Kubica et al. [9] and Kim et al. [8] have presented solutions based on the Knuth–Morris–Pratt algorithm (KMP) [10]. Kim et al. doubted the applicability of the Boyer–Moore approach [1] to order-preserving matching. However, Cho et al. [2] demonstrated that the bad character heuristic works. We will present a new practical solution based on approximate string matching.

[*] Supported by the Academy of Finland (grant 134287).

J. Gudmundsson and J. Katajainen (Eds.): SEA 2014, LNCS 8504, pp. 307–314, 2014.
© Springer International Publishing Switzerland 2014

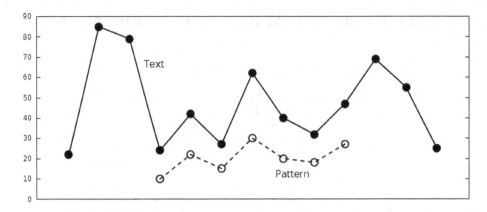

Fig. 1. Example of order-preserving matching

We form a modified pattern and use an algorithm for exact string matching as a filtration method. Our approach is simpler and in practice more efficient than earlier solutions. We transform the original pattern P into a binary string P' expressing increases (1), equalities (0), and decreases (0) between subsequent pattern positions. Then we search for P' in the similarly transformed text T'. For example, $P' = 101001$ corresponds to $P = (10, 22, 15, 30, 20, 18, 27)$ and $T' = 100101001100$ to T above. Each occurrence is a match candidate which is verified following the numerical order of the positions of the original pattern P. Note that in this approach any algorithm for exact string matching can be used as a filtration method. If the filtration algorithm is sublinear and the text is transformed on line, the total method is sublinear on average.

We made experiments with two sublinear string matching algorithms and two linear string matching algorithms as the filtering method. Our approach with sublinear filters was considerably faster than the algorithm by Cho et al. [2], which is the first sublinear solution of the problem and which was the previous winner.

The paper is organized as follows. Section 2 describes the previous solutions for the order-preserving matching, Section 3 presents our solution based on filtration, Section 4 analyses the new approach, Section 5 presents and discusses the results of practical experiments, and Section 6 concludes the article.

2   Previous solutions

There are three solutions presented so far, two algorithms are based on the KMP algorithm [10] and the third one is based on the Boyer–Moore–Horspool algorithm (BMH) [1,6].

In the first KMP approach presented by Kubica et al. [9], the fail function in the KMP algorithm is modified to compute the order-borders table. This can be achieved in linear time. The KMP algorithm is mutated such that it determines if the text contains substring with the same relative order as that of the pattern using the order-borders table B, which is defined for the pattern P as follows:

$$B[1] = 0$$
$$B[i] = \max\{j < i : P[1..j] \approx P[i-j+1..j]\} \text{ for } i \geq 2$$

This computation can also be done in linear time. Hence, the time complexity of the algorithm is linear.

The second KMP approach by Kim et al. [8] is based on the prefix representation and it is further optimized according to the nearest neighbor representation.

The prefix representation is based on finding the rank of each integer in the prefix. It can be computed easily by inserting each character to the dynamic order statistic tree and then computing the rank of each character in the prefix. The functions insert and rank both take $O(\log m)$ time and since there are $O(m)$ operations, the time complexity of computing prefix representation is $O(m \log m)$. The failure function is computed as in the KMP algorithm using the previous values in $O(m \log m)$ time. Finally the text is searched using the function KMP-order-matcher. It is different from the KMP algorithm as it is based on the ranks of the prefix. The functions involved in it are rank, insert and delete, and all of them take $O(\log m)$ time, and since the number of operations can be at most n, these functions take $O(n \log m)$ time. The failure function takes $O(m \log m)$ time and computing the prefix representation takes $O(m \log m)$ time. Thus the total time complexity is $O(n \log m)$.

This approach is optimized to overcome the overhead involved in computing the rank function using the nearest neighbor representation. It checks if the order of each character in the text matches the corresponding character in the pattern. In this, the characters itself are compared and there is no need to compute the rank. It takes $O(m \log m)$ time to compute the nearest neighbor representation of the pattern, $O(m)$ time for the computation of the failure function and $O(n)$ time for searching the text. So the total time complexity is $O(n + m \log m)$.

The BMH approach [2] is based on the bad character rule applied to q-grams, i.e. strings of q characters. A q-gram is treated as a single character to make shifts longer. In this way, a large amount of text can be skipped for long patterns, and the algorithm is obviously sublinear on the average. This algorithm is the first sublinear solution for order-preserving matching.

3 Our Solution

In Section 1 we gave an informal description of order-preserving matching. That definition causes problems in handling equal values which can appear in real data. The concept of order-isomorphism removes these problems.

Problem definition. Two strings u and v of the same length over Σ are called *order-isomorphic* [9], written $u \approx v$, if

$$u_i \leq u_j \Leftrightarrow v_i \leq v_j \text{ for } 1 \leq i, j \leq |u|.$$

In the *order-preserving pattern matching problem*, the task is to find all the substrings of T which are order-isomorphic with P.

Our solution for order-preserving matching consists of two parts: filtration and verification. First the text is filtered with some exact string matching algorithm and then the match candidates are verified using a checking routine.

Filtration. For filtration, the consecutive numbers in the pattern $P = p_1 p_2 ... p_m$ are compared pairwise in the preprocessing phase and the result is encoded as a string $P' = b_1 b_2 ... b_{m-1}$ of binary numbers: b_i is 1 if $p_i < p_{i+1}$ holds, otherwise b_i is 0. In the search phase, some algorithm for exact string matching (let us call it A) is applied to filter out the text. When Algorithm A reads an alignment window of the original text, the text is encoded on line in the same way as the pattern. Algorithm A is run, as if the whole text would have been encoded. Because Algorithm A may recognize an occurrence of P' which does not correspond to an actual match of P in T, each occurrence of P' is only a match candidate which should be verified. It is clear that this filtration method cannot skip any occurrence of P in T.

Verification. In the preprocessing phase, the positions of the pattern $P = p_1 p_2 ... p_m$ are sorted. The result is an auxiliary table r: $p_{r[i]} \leq p_{r[j]}$ holds for each pair $i < j$ and $p_{r[1]}$ is the smallest number in P. In addition, we need a binary vector E representing the equalities: $E[i] = 1$ denotes that $p_{r[i]} = p_{r[i+1]}$ holds. The match candidates found by Algorithm A are traversed in accordance with the table r. If the candidate starts from t_j in T, the first comparison is done between $t_{j-1+r[1]}$ and $t_{j-1+r[2]}$. There is a mismatch when

$$t_{j-1+r[i]} > t_{j-1+r[i+1]} \text{ or}$$
$$(t_{j-1+r[i]} = t_{j-1+r[i+1]} \text{ and } E[i] = 0) \text{ or}$$
$$(t_{j-1+r[i]} < t_{j-1+r[i+1]} \text{ and } E[i] = 1)$$

is satisfied. The candidate is discarded when a mismatch is encountered. Verification is efficient because sorting is done only once during preprocessing.

Remark. We use binary characters in encoding. We also tried encoding of three characters 0, 1, and 2 corresponding to '<', '=', and '>', but the binary approach was faster in practice, because testing of one condition is faster than testing of two conditions. Also the frequency of consecutive equalities is low in real data.

4   Analysis

We will prove that our approach is sublinear in the average case, if the filtration algorithm is sublinear. Sublinearity means that on average all the characters in the text are not examined.

Let us assume that the numbers in P and T are integers and they are statistically independent of each other and the distribution of numbers is discrete uniform. Let P' and T' be the transformed pattern and text. Let c be the count of the integer range (i.e. the alphabet size). The probability of one in a position of P' or T' (as a result of a comparison) is $p = (c^2/2 - c/2)/c^2 = (c-1)/2c$, because there are c^2 integer pairs and c equalities. So the probability of a character match q is

$$p^2 + (1-p)^2 = 2p(p-1) + 1 = 1 - \frac{c-1}{c} \cdot \frac{c+1}{2c} = 1 - \frac{c^2-1}{2c^2} = \frac{1}{2} + \frac{1}{2c^2}.$$

The probability of a match of P' (i.e. a match candidate of P) at a certain position of T' is q^{m-1}, which approaches to zero, when m grows. This is true even for $c = 2$. This means that the verification time approaches zero when m grows, and the filtration time dominates. If the filtration method is sublinear, the total algorithm is sublinear.

In the worst case, the total algorithm requires $O(nm)$ time, if for example P is 1^m and T is 1^n.

The preprocessing phase requires $O(m \log m)$ due to sorting of the pattern positions. The space requirement is $O(m)$.

5 Experiments

We tested four string matching algorithms as filtration methods for order-preserving matching. Two of them, SBNDM2 and SBNDM4 [4] are based on the Backward Nondeterministic DAWG Matching (BNDM) algorithm [11]. In BNDM, each alignment window is processed from right to left like in the Boyer–Moore algorithm [1]. BNDM simulates the nondeterministic automaton of the reversed pattern with bitparallelism. SBNDMq starts the processing of each alignment window by reading a q-gram. The third algorithm is Fast Shift-Or (FSO) [5]. We utilized a version of FSO coded by B. Ďurian [4]. FSO was selected because it is fast on short binary patterns [4]. The fourth algorithm is the KMP algorithm [10]; together with verification it was supposed to approximate the two earlier methods [8,9] based on KMP. Of the algorithms, SBNDM2 and SBNDM4 are sublinear, whereas FSO and KMP are linear.

The tests were run on Intel 2.70 GHz i7 processor with 16 GB of memory running Ubuntu 12.10. All the algorithms were implemented in C in the 64 bit mode and run in the testing framework of Hume and Sunday [7]. Our solution based on filtration was compared with the earlier BMH approach by Cho et al. [2] (the authors generously let us use their implementation).

For testing we used three texts: a random text and two real texts, which were time series of the Dow Jones index and Helsinki temperatures. The random data contains 1,000,000 random integers between 0 and 2^{30}. The Dow Jones data contains 15,248 integers pertaining to the daily values of the stock index in the years 1950–2011 and the Helsinki temperature data contains 6,818 integers referring to the daily mean temperatures in Fahrenheit (multiplied by ten) in Helsinki in the years 1995–2005. From each text we picked random patterns of length $5, 8, 10, 12, 15$, and 20. Each set contains 1,000 patterns for the random

Table 1. Execution time of algorithms (random: in seconds, others: in 10 milliseconds)

Data	Algorithm	5	8	10	15	20	30	50
DOW	KOPM	2.02	1.94	1.94	2.00	1.94	1.96	1.95
	BMOPM-3	1.64	1.06	0.91	0.81	0.79	0.76	0.78
	BMOPM-4	2.16	0.96	0.72	0.50	0.41	0.34	0.31
	BMOPM-5	4.34	1.13	0.78	0.47	0.35	0.27	0.22
	FSO-OPM	1.01	0.48	0.46	0.46	0.44	0.46	0.46
	S2OPM	1.05	0.58	0.41	0.24	0.16	0.09	0.06
	S4OPM	1.03	0.31	0.20	0.31	0.10	0.08	0.06
Hel temp	KOPM	0.85	0.81	0.75	0.77	0.76	0.76	0.76
	BMOPM-3	0.70	0.46	0.46	0.40	0.39	0.39	0.44
	BMOPM-4	0.91	0.40	0.42	0.28	0.24	0.21	0.21
	BMOPM-5	1.87	0.49	0.52	0.31	0.23	0.18	0.17
	FSO-OPM	0.34	0.21	0.22	0.22	0.21	0.21	0.21
	S2OPM	0.40	0.18	0.13	0.08	0.06	0.03	0.03
	S4OPM	0.43	0.12	0.09	0.06	0.04	0.04	0.03
Random	KOPM	6.90	4.91	6.66	6.06	4.94	6.72	6.70
	BMOPM-3	8.08	4.01	4.69	4.17	4.08	4.10	4.07
	BMOPM-4	11.48	3.89	3.47	1.77	1.39	1.47	1.18
	BMOPM-5	22.83	5.97	4.70	2.67	1.88	1.22	0.79
	FSO-OPM	5.36	1.93	1.64	1.68	1.65	1.69	1.68
	S2OPM	3.90	2.47	1.89	1.17	1.26	0.79	0.35
	S4OPM	3.90	2.01	1.76	1.15	0.85	0.57	0.35

text and 200 patterns for the real texts. Table 1 shows the average execution times of all the algorithms. In addition, a graph on the times for the Dow Jones data is shown in Fig. 2. The real texts were tested with 180 repeated runs and the random text was tested with 60 repeated runs. In Table 1, S2OPM represents the algorithm based on SBNDM2 filtration, S4OPM represents the algorithm based on SBNDM4 filtration, BMOPM-q represents the BMH approach for $q = 3, 4, 5$, KOPM represents the algorithm based on KMP filtration and FSO-OPM represents the algorithm based on Fast Shift-Or.

From Table 1, it can be clearly seen that in case of real data, S4OPM is faster than all the other algorithms for all tested values of m except for $m = 5$. For $m = 5$, FSO-OPM is the fastest. In all the three cases, S2OPM is slower than S4OPM but its execution time approaches the execution time of S4OPM as the value of m increases. And for $m = 50$, the execution times of S2OPM and S4OPM are almost equal. Relatively, S2OPM and S4OPM performed better on the real data than on the random data. In the case of Helsinki daily temperatures, S2OPM and S4OPM both give almost the same performance. In the case of random data, for $m = 5$ S2OPM and S4OPM are the best and for $m = 8$ and for $m = 10$, FSO-OPM is the best.

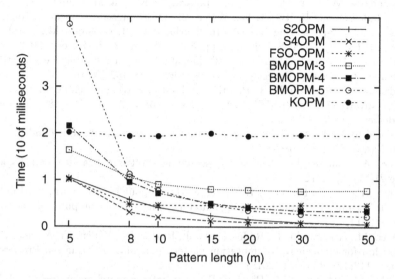

Fig. 2. Execution times of algorithms for the Dow Jones data

In case of real data like the Dow Jones stock index, it is also interesting to consider the number of matches. For $m = 5$, the number of matches is 56,048, for $m = 8$ number of matches is 2,133, and for $m = 10$ number of matches is 360.

6 Concluding Remarks

We introduced a new practical solution based on filtration for order-preserving matching. Any exact string matching algorithm can be used as the filtration algorithm. In this paper we have used SBNDM2, SBNDM4, FSO, and KMP as the filtration method of our solution. The results of our practical experiments prove that our SBNDM2 and SBNDM4 based solutions are much faster than the earlier BMOPM algorithm. Moreover, our FSO based solution is the fastest algorithm for certain short pattern lengths.

Acknowledgement. We thank Hannu Peltola for his help.

References

1. Boyer, R.S., Moore, J.S.: A fast string searching algorithm. Communications of the ACM 20(10), 762–772 (1977)
2. Cho, S., Na, J.C., Park, K., Sim, J.S.: Fast order-preserving pattern matching. In: Widmayer, P., Xu, Y., Zhu, B. (eds.) COCOA 2013. LNCS, vol. 8287, pp. 295–305. Springer, Heidelberg (2013)

3. Crochemore, M., Iliopoulos, C.S., Kociumaka, T., Kubica, M., Langiu, A., Pissis, S.P., Radoszewski, J., Rytter, W., Waleń, T.: Order-preserving incomplete suffix trees and order-preserving indexes. In: Kurland, O., Lewenstein, M., Porat, E. (eds.) SPIRE 2013. LNCS, vol. 8214, pp. 84–95. Springer, Heidelberg (2013)

4. Ďurian, B., Holub, J., Peltola, H., Tarhio, J.: Improving practical exact string matching. Information Processing Letters 110(4), 148–152 (2010)

5. Fredriksson, K., Grabowski, S.: Practical and optimal string matching. In: Consens, M.P., Navarro, G. (eds.) SPIRE 2005. LNCS, vol. 3772, pp. 376–387. Springer, Heidelberg (2005)

6. Horspool, R.N.: Practical fast searching in strings. Software–Practice and Experience 10(6), 501–506 (1980)

7. Hume, A., Sunday, D.: Fast string searching. Software–Practice and Experience 21(11), 1221–1248 (1991)

8. Kim, J., Eades, P., Fleischer, R., Hong, S.-H., Iliopoulos, C.S., Park, K., Puglisi, S.J., Tokuyama, T.: Order preserving matching. Theoretical Computer Science 525, 68–79 (2014)

9. Kubica, M., Kulczynski, T., Radoszewski, J., Rytter, W., Walen, T.: A linear time algorithm for consecutive permutation pattern matching. Information Processing Letters 113(12), 430–433 (2013)

10. Knuth, D.E., Morris, J.M., Pratt, V.R.: Fast pattern matching in strings. SIAM Journal on Computing 6(2), 323–350 (1977)

11. Navarro, G., Raffinot, M.: Flexible pattern matching in strings. Practical on-line search algorithms for texts and biological sequences. Cambridge University Press, New York (2002)

Approximate Online Matching
of Circular Strings*

Tommi Hirvola and Jorma Tarhio

Department of Computer Science and Engineering
Aalto University, P.O. Box 15400, 00076 Aalto, Finland
`firstname.surname@aalto.fi`

Abstract. Recently a fast algorithm based on the BNDM algorithm
has been presented for exact matching of circular strings. From this
algorithm, we derive several sublinear methods for approximate online
matching of circular strings. The applicability of the new algorithms is
demonstrated with practical experiments. In many cases, the new algorithms are faster than an earlier solution.

1 Introduction

The task of string matching is to find all occurrences of a *pattern* $P = p_1 \ldots p_m$
in a *text* $T = t_1 \ldots t_n$ where all characters in P and T are drawn from a finite
alphabet Σ of size σ. In *approximate string matching* the occurrences are allowed
to contain errors. We consider two variations of approximate string matching.
In the *k differences problem*, the total number of mismatches, deletions and
insertions between the pattern and its occurrence should be at most k. In the
k mismatches problem, only mismatches are allowed. Many algorithms [21] for
both variations of approximate string matching have been developed.

The *circular string* is a set of strings $C(S) = \{S^{(1)}, \ldots, S^{(m)}\}$ corresponding
to a given string $S = s_1 \ldots s_m$, where $S^{(i)} = s_i \ldots s_m s_1 \ldots s_{i-1}$ is a *rotation* (or
conjugate) of S. Given a text T and a circular pattern $C(P)$ for $P = p_1 \ldots p_m$,
the *circular pattern matching problem* (CPM) is to find all occurrences of $C(P)$
in T.

In the *exact CPM problem* (ECPM), the Z algorithm [12] finds all the exact
occurrences in time $O(n + m)$. Lin and Adjeroh [16] present an $O(n \log \sigma)$ algorithm based on suffix trees. Chen et al. [6] give a solution that is sublinear on
the average for ECPM. Susik et al. [26] describe a filtering method that solves
the ECPM problem in average sublinear time. By *sublinearity* we mean that a
part of text characters can be skipped. There are several solutions [11,16,18,23]
for the *approximate CPM problem* (ACPM), but none of them are sublinear. For
example, Lin and Adjeroh [16] give an $O(km^2n)$ algorithm.

In this paper, we develop sublinear online algorithms for the ACPM problem.
Our point of view is the practical efficiency of algorithms. There are several

---

* Work supported by the Academy of Finland (grant 134287).

J. Gudmundsson and J. Katajainen (Eds.): SEA 2014, LNCS 8504, pp. 315–325, 2014.

sublinear algorithms [21] for a single noncircular pattern, but a straightforward application of them does not guarantee sublinearity in ACPM, because a run is needed for each of the m rotations and the total time complexity is even in the best case at least $m \cdot O(n/m)$. However, Fredriksson and Navarro [9] give a sublinear multi-pattern algorithm (FN for short), which can be applied to the ACPM problem with k differences in $O((k + \log_\sigma m) \, n/m)$ average time at moderate error levels. As far as we know, the FN algorithm is the fastest one among earlier online algorithms for the ACPM problem. In this paper, we introduce three new algorithms ASB, ACB, and ACBq for the ACPM problem. The experimental results presented in Section 4 show that ASB and ACB are mostly faster than FN for English and random data. For DNA data, ACBq is mostly faster than FN.

Matching of circular strings has several applications. In computational geometry, one needs to find a polygon within a set of polygons. If the initial vertex is not provided, a polygon of k vertices has k representations and the problem can be reduced to the search of circular strings [4,13]. Music retrieval is another application area [16]. Circular strings are also found in bioinformatics. There are hundreds of protein pairs having sequences which are rotations of each other [17]. Most of the rotations retain their original three-dimensional protein structure and biological function [2], which makes their identification important. Traditional sequence alignment methods have been developed for linear sequences, and they are not well suited for circular strings, hence there is a need for new methods.

In the algorithms, we use C-like notations: '|', '&', '≪', and '≫' represent bitwise operations OR, AND, left shift, and right shift, respectively. The size of the computer word is denoted by w.

2   CBNDM

Our method for approximate matching of circular strings is based on the CBNDM algorithm introduced by Chen et al. [6]. Therefore we start by explaining it.

After the advent of the Shift-Or [1] algorithm, bit-parallel string matching methods have gained more and more interest. The BNDM (Backward Nondeterministic DAWG Matching) algorithm [20] is a nice example of an elegant, compact, and efficient piece of code for exact string matching. Superficially BNDM appears to be a cross of the Shift-Or and Boyer–Moore algorithms [3]. However, the Boyer–Moore algorithm searches for suffixes of the pattern while BNDM searches for factors of the pattern. BNDM simulates the nondeterministic finite factor automaton of the reverse pattern. The precomputed table B associates each character with an *instance vector* expressing its locations in the pattern. The inner loop of BNDM checks an alignment of the pattern (i.e. an alignment window in the text) in the right-to-left order. At the same time the loop recognizes prefixes of the pattern. The leftmost one of the found prefixes determines the next alignment window of the algorithm. BNDM can be modified for approximate matching [14].

SBNDM [22,24] (short for Simple BNDM) is a simplified version of BNDM. SBNDM does not explicitly take care of prefixes, but shifts the pattern simply over the text character which caused the *state vector* D to become zero or over the first character of the alignment window in case of a match. In practice, SBNDM is slightly faster than BNDM especially for short patterns even though it examines more text characters [24]. The pseudocode of SBNDM is given as Algorithm 1. This version works for patterns of at most w characters.

Algorithm 1 (SBNDM)
1: for $a \in \Sigma$ do $B[a] \leftarrow 0$
2: for $i \leftarrow 1$ to m do $B[p_i] \leftarrow B[p_i] \mid (1 \ll (m - i))$
3: $i \leftarrow 1$
4: while $i \leq n - m + 1$ do
5:   $j \leftarrow 1; D \leftarrow B[t_{i+m-j}]$
6:   while $D \neq 0^m$ and $j < m$ do
7:     $j \leftarrow j + 1$
8:     $D \leftarrow (D \ll 1) \ \& \ B[t_{i+m-j}]$
9:   if $j = m$ then output i
10: $i \leftarrow i + (m - j + 1)$

The Circular BNDM algorithm (CBNDM) is based on SBNDM. In CBNDM the left shift is replaced by the rotating left shift of m bits. The state vector on line 8 is updated as follows:

$$D \leftarrow (D \overleftrightarrow{\ll} 1) \ \& \ B[t_{i+m-j}]$$

For example, $10100 \overleftrightarrow{\ll} 1$ results in 01001 for $m = 5$.

Let $D[1], \ldots, D[m]$ be the last m bits of D. Now $D[r] = 1$ means that the substring $t_{i+m-j} \ldots t_{i+m-1}$ is the same as the prefix u of $P^{(r)}$, $|u| = j$. Especially, if $D[r] = 1$ holds when $j = m$, the alignment window contains an occurrence of $P^{(r)}$.

The only difference between the CBNDM and SBNDM algorithms is the updating of the state vector—other parts of the algorithms are identical. Actually the condition $j < m$ on line 6 is necessary only in CBNDM, not in SBNDM [7].

SBNDMq [7] is a variation of SBNDM applying q-grams. In each alignment window, SBNDMq first processes q text characters $t_i, \ldots, t_{i+q-1}$ before testing the state vector D. In practice, SBNDMq is considerably faster than SBNDM or BNDM. The running time of CBNDM can be improved in a similar way.

3   New Algorithms

A filtering method for approximate matching finds match candidates, which are then checked by another method. First, we introduce a filtering algorithm based on SBNDM for *noncircular* patterns with k allowed mismatches. Later this algorithm is modified for circular patterns and for k differences. We call this algorithm ASB for **A**pproximate **SB**NDM. Its idea is related to the Chang–Lawler algorithm (CL) [5] which applies the so-called *matching statistics*:

$$M(r, P) = \max\{j \mid j = -1 \text{ or } t_{r-j} \ldots t_r \text{ is a factor of } P\} + 1.$$

ASB searches up to $k + 1$ break points in M within an alignment window of m text characters. The search starts at the last character of the window: $M(i + m - 1, P) = v$. Now $t_{i+m-1-v}$ is the first break point, i.e., text character that causes state vector D to become zero. See an example in Table 1.

Table 1. Example of recognition of break points for $P = $ abbab

text	c	a	b	a	b	b	a	a
M	0	1	2	2	3	3	4	1
breaks	*		*				*	
D	00000	01001	00000	00001	00010	10110	00000	01001

Each break point corresponds to either a character not appearing in the pattern or a starting character of a new factor in a factor switch. When $k + 1$ break points have been found, the window is shifted forward. If k or less break points are found, the window contains a match candidate. This is because an approximate occurrence can contain at most $k + 1$ non-overlapping maximal factors of P [5].

In the CL algorithm, the distance of subsequent alignment windows is fixed, whereas ASB uses dynamic shifting. The matching statistics of CL is based on a suffix tree, which is considerably slower than the bit-parallel technique of ASB. There has been recent improvements in computation of matching statistics (e.g., [15,25]), but these methods require construction of heavy data structures which makes them slower than ASB for short patterns.

Algorithm 2 (ASB, k mismatches)
1: **for** $a \in \Sigma$ **do** $B[a] \leftarrow 0$
2: **for** $i \leftarrow 1$ **to** m **do** $B[p_i] \leftarrow B[p_i] \mid (1 \ll (i - 1))$
3: $E \leftarrow 1^{m+1}; i \leftarrow 1$
4: **while** $i \leq n - m + 1$ **do**
5:    $e \leftarrow 0; D \leftarrow E; j \leftarrow 0$
6:    **while** $e \leq k$ **and** $j < m$ **do**
7:       $j \leftarrow j + 1; D \leftarrow (D \gg 1) \ \& \ B[t_{i+m-j}]$
8:       **if** $D = 0^m$ **then** $e \leftarrow e + 1; D \leftarrow E$
9:    **if** $j = m$ **and** $e \leq k$ **then** check candidate $t_i \ldots t_{i+m-1}$
10:   $i \leftarrow i + (m - j + 1)$

The pseudocode of ASB is given as Algorithm 2. The algorithm itself does not use the matching statistics, but the break points are recognized as the state vector D becoming zero, see the example in Table 1. This is because instance vectors B form a factor automaton of the pattern. The state vector D stays nonzero while a factor is processed from right to left. If the factor ends at t_r, then D becomes zero at t_{r-1}. After that $B[t_{r-2}]$ is assigned to D, and the computation is resumed. Note that ASB applies the right shift instead of the left shift in SBNDM. Therefore the instance vectors B are reversed and we need $m+1$ bits for the constant vector E. For this reason, it is assumed that $m \le w-1$ holds. Alternatively, m bits could be used for E in order to be able to handle the case $m = w$ but then the code would be more complicated.

ASB can be applied directly to circular patterns by considering the discontinuation point of a circular occurrence as an extra error. So we search for $k+2$ break points, and if less break points are found in the alignment window, we have found a match candidate. ASB can also be modified to handle circular patterns by switching the bit shift to the bit rotation as done for SBNDM in Section 2. We call this algorithm ACB (short for **A**pproximate **CB**NDM). In ACB, the width of the constant vector E can be reduced to m to handle the case $m = w$.

As with most bit-parallel algorithms, ASB and ACB can be extended for patterns longer than w by using arrays for the state vector D and for each instance vector $B[a]$. ASB can also support long patterns by searching the prefix of length w of the pattern and checking for the full pattern on line 9 of Algorithm 2. In the case of ACB, one may consider both the prefix and the suffix of length w.

It is possible to speed up ASB and ACB by using q-grams as with SBNDMq. However, this variation needs that also the number of seen break points e is precomputed for each q-gram because q-grams can contain break points that would otherwise go undetected. Moreover, it is beneficial to fetch the precomputed values at each break point in addition to the right end of the alignment. We call this algorithm ACBq.

The k mismatches variation of ASB is easily extended to handle k differences. In the k differences problem, the width of the alignment window is $m - k$ to ensure that if an occurrence starts at the window position then any suffix of the window is a factor of the pattern with at most k differences so that the window is not abandoned. Moreover, in the verification step, we consider a text substring of length $m + k$ as the match candidate. The pseudocode of ASB with k differences is given as Algorithm 3. The pseudocode assumes that $m < w$ holds.

Algorithm 3 (ASB, k differences)
```
1:  for a ∈ Σ do B[a] ← 0
2:  for i ← 1 to m do B[pᵢ] ← B[pᵢ] | (1 ≪ (i − 1))
3:  E ← 1^{m+1}; i ← 1
4:  while i ≤ n − (m − k) + 1 do
5:      e ← 0; D ← E; j ← 0
6:      while e ≤ k and j < m − k do
7:          j ← j + 1; D ← (D ≫ 1) & B[t_{i+(m−k)−j}]
8:          if D = 0^m then e ← e + 1; D ← E
9:      if j = m − k and e ≤ k then check candidate tᵢ...t_{i+(m+k)−1}
10:     i ← i + ((m − k) − j + 1)
```

As done previously, circular patterns can be handled by switching the bit shift to the bit rotation or allowing one extra break point.

In the ACPM problem with k mismatches, the reported candidates can be verified by searching the match candidate S with k allowed mismatches in a text string PP. This can be done because the string PP contains all rotations of P. In ACPM with k differences, the lengths of the occurrences are not known in advance and the same method cannot be used. Instead, we search each rotation of P in S using an approximate matching algorithm that allows k differences, e.g., Myers' algorithm [19].

4   Analysis

In this section, we prove that the k differences variation of ACB is sublinear on the average. Sublinearity of our other algorithms can be proven with the same method by adjusting the window size and constants which do not change the resulting average time complexity. The proof is similar to that of the CL algorithm [5].

The goal is to first determine the expected number of character comparisons done in each alignment window. This number gives us the average shift length, which has to exceed half of the window width in order to skip characters in the text. Moreover, we need to consider the probability of a match candidate occurring in the window to show that the verification step does not worsen the average performance with reasonable values of m and k.

Assume that the characters of P and T are chosen independently and uniformly from Σ of size $\sigma \geq 2$. The inner loop of ACB finds up to $k + 1$ factors and break points of P in the alignment window. Let X_i be a random variable representing the length of the ith factor. Since the circular pattern P contains at most m distinct substrings of length $\log_\sigma m + d$, and there are $m\sigma^d$ different strings of that length, we get:

$$\forall \text{ integer } d \geq 0, \Pr[X_i = \log_\sigma m + d] < \sigma^{-d}$$

This gives us the following upper bound for the expected factor length:

$$E[X_i] < \log_\sigma m + \sum_{d=0}^{\infty} d\sigma^{-d} \leq \log_\sigma m + 2$$

Thus, ACB reads $(k + 1)(\log_\sigma m + 3)$ characters per window on the average. This has to be less than $(m - k)/2$ to achieve sublinearity. Equating $(k + 1)(\log_\sigma m + 3) < (m - k)/2$ yields us the following threshold for k:

$$k < \frac{m - 2\log_\sigma m - 6}{2\log_\sigma m + 7}$$

Now, we get the average time complexity of ACB by multiplying the expected number of examined windows with the average work done in each window:

$$O\left(\frac{n - m + 1}{m - k - (k + 1)(\log_\sigma m + 3)}\right) \cdot O((k + 1)(\log_\sigma m + 3))$$

This is equal to $O((n/m)k\log_\sigma m)$ given that $n \gg m$ and $k < m/(\log_\sigma m + O(1))$.

In conclusion, ACB is sublinear on the average when k is bounded appropriately and the match verification time is excluded. It is evident that match candidates are rare and do not affect the average time complexity if m is large and k is small. This can be formally proven by computing an upper bound for $\Pr[k + X_1 + X_2 + \cdots + X_{k+1} \geq m - k]$ with the help of Chernoff bounds. In [5], the probability is shown to be less than $1/m^3$ when k is upper bounded by the threshold $m/(\log_\sigma m + O(1))$.

5 Experiments

The test data consisted of three different types of text: DNA (4.5 MB), English (4.0 MB) and Rand256 (5.0 MB). The DNA text was the genome of E.coli. The English text was the King James Bible. The Rand256 text contained randomly generated data in the alphabet of 256 symbols. The texts were taken from the corpus of the SMART tool [8].

The patterns used in the tests were selected from the texts, and the resulting strings were randomly rotated and substituted in 0–5 characters positions to simulate mismatches. The pattern sets for the k differences algorithms allowed also insertions and deletions. In each test run, 1000 patterns were searched in the 2 MB long prefix of each text. Test runs were repeated 3 times for each algorithm and for each test set. Verification of match candidates was done as explained in Section 3.

As a reference method, we used the FN algorithm [9]. As far as we know, FN has been the best online algorithm for approximate matching of multiple patterns. We applied FN to the ACPM problem by forming all possible rotations of a pattern and searching them simultaneously. The implementation was obtained from the authors and ran with command-line options -D -t6 -B -Sb for the DNA text, and -A -t2 -B -Sb for the English and Rand256 texts. The option -s was used for searches allowing only mismatches. FN is also a filtering method and its implementation uses similar verification algorithms as our methods.

We also ran preliminary tests with an approximate multiple pattern matching algorithm proposed by Fulwider and Mukherjee [10], but that algorithm was not competitive.

The test computer has the Intel® Core™ i7 860 2.80 GHz processor with 16 GB DDR3 main memory. The operating system is Linux (kernel 3.2.0-48-generic). The test processes were run on a single core. The algorithms were implemented in C and compiled with gcc 4.6.3 using O2 or O3 optimization.

Table 2 shows test results in the case of k mismatches. ACB is faster than ASB on all the inputs and the values of k. This is because ASB gives more false positives than ACB. ACB is also significantly faster than FN for English and Rand256 data. When the preprocessing times of the patterns are included, our algorithms are much faster than FN in most tested cases because FN has a heavy preprocessing phase whereas ASB and ACB have negligible preprocessing

Table 2. Running times (in seconds) of algorithms for approximate circular matching with k mismatches. Preprocessing times of FN are in parentheses. Preprocessing times of ASB and ACB have been included in the running times.

Σ	m	$k = 1$			$k = 2$			$k = 5$		
		FN	ASB	ACB	FN	ASB	ACB	FN	ASB	ACB
DNA	20	2.16 (6.21)	8.64	4.41	4.75 (6.16)	16.01	8.23	442.68 (6.29)	1741.85	684.77
	40	1.05 (15.11)	3.55	2.13	1.97 (15.04)	5.14	3.49	4.14 (15.18)	14.47	10.49
	60	0.77 (28.70)	2.26	1.43	1.31 (28.61)	3.13	2.28	2.25 (28.73)	6.93	5.53
English	20	2.70 (56.81)	4.75	2.55	4.04 (56.95)	6.92	3.97	63.91 (78.91)	28.84	14.33
	40	1.68 (263.71)	2.44	1.41	2.38 (217.48)	3.40	2.09	5.62 (203.50)	7.05	4.83
	60	1.33 (443.65)	1.69	1.02	1.69 (397.06)	2.29	1.49	3.19 (382.84)	4.38	3.19
Rand256	20	1.69 (57.42)	1.12	0.82	2.23 (57.56)	1.53	1.13	5.99 (79.51)	3.04	2.32
	40	1.19 (266.26)	0.80	0.52	1.34 (219.86)	1.04	0.71	2.51 (205.70)	1.75	1.35
	60	0.94 (450.82)	0.67	0.43	0.94 (404.10)	0.85	0.57	1.74 (390.07)	1.37	1.04

times (at most a few milliseconds). However, excluding the preprocessing times, FN beats the other algorithms on DNA data and ASB on English data. The significance of preprocessing times depends much on user needs. The shorter texts are processed, the larger is the proportion of preprocessing in the total times. In addition, it must be noted that FN has a specific option for DNA data allowing tuned computation, whereas ASB and ACB make no assumptions about the input data. Also, decreasing m reduces the preprocessing times of FN accordingly, which makes FN more competitive in terms of total execution time for short patterns.

Table 3. Running times (in seconds) of algorithms for approximate circular matching with k differences. The preprocessing times of FN are shown in parentheses. Preprocessing times of ASB and ACB are included in the running times.

Σ	m	$k = 1$			$k = 2$			$k = 5$		
		FN	ASB	ACB	FN	ASB	ACB	FN	ASB	ACB
DNA	20	2.56 (5.60)	7.21	3.67	7.19 (5.56)	34.88	8.79	1932.83 (5.49)	6066.03	5540.40
	40	1.18 (8.92)	2.84	1.71	2.06 (8.81)	4.32	3.09	14.17 (8.80)	31.03	12.81
	60	0.98 (12.29)	1.83	1.16	1.19 (12.41)	2.71	2.01	2.91 (12.46)	7.26	5.84
English	20	2.99 (44.24)	3.65	2.39	5.16 (44.76)	6.83	4.35	661.07 (66.48)	840.25	254.65
	40	1.86 (169.99)	1.85	1.26	2.69 (123.98)	2.81	2.02	11.29 (109.40)	7.69	6.14
	60	1.59 (212.83)	1.33	0.91	1.86 (166.77)	1.96	1.43	4.01 (152.77)	4.33	3.81
Rand256	20	1.75 (44.27)	0.95	0.69	2.60 (44.25)	1.40	1.09	8.82 (66.65)	4.54	3.78
	40	1.18 (170.25)	0.61	0.44	1.45 (123.77)	0.84	0.66	3.00 (109.56)	1.99	1.80
	60	0.90 (213.54)	0.49	0.36	1.01 (167.20)	0.67	0.54	2.12 (152.99)	1.55	1.41

The results for approximate matching with k differences are presented in Table 3. Again ACB is the fastest algorithm in nearly all the test runs when the preprocessing times of FN are included. The only exception is DNA for $m = 20$ and $k = 5$ where FN beats both ASB and ACB. In this test run, the number of checked match candidates is high and FN handles the case better than our algorithms. In terms of scanning time only, ACB is the fastest algorithm on English and Rand256, but slightly slower than FN on DNA.

Note that in many cases ASB and ACB are faster when allowing differences instead of only mismatches. This is due to the faster occurrence verification

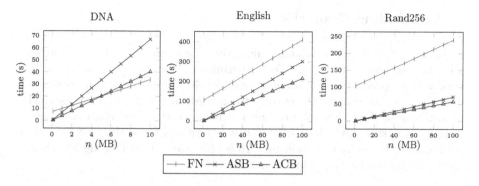

Fig. 1. Total running times of the k difference algorithms as a function of the input text size for $m = 30$ and $k = 3$. The range is 0–10 MB for DNA and 0–100 MB for English and Rand256.

algorithm applied in the k differences implementations of the algorithms. Also, the preprocessing of FN is faster for the k differences case than k mismatches.

Figure 1 shows that the running times of the algorithms grow linearly when the text length is increased. The long input texts were produced by concatenating the texts with themselves. We used 10 MB for DNA in order to visualize the intersection points clearly, while 100 MB was used for English and Rand256. On the DNA text, FN is slower than ASB and ACB for short texts due to the heavy preprocessing. However, FN becomes faster than the other algorithms when n increases. ASB and ACB are slower than FN on DNA when n exceeds 2 MB and 5 MB, respectively. On English and Rand256, our algorithms are faster than FN for all values of n.

Finally, we tuned our best algorithm, ACB, with q-grams for DNA data. The precomputed D and e values were stored into a table of σ^q ($\sigma = 4$) elements. The results are shown in Table 4.

Table 4. Running times (in seconds) of ACBq for the k differences problem. Preprocessing times are excluded. The times are comparable to those in Table 3.

Σ	m	$k = 1$ ACB4	ACB6	ACB8	$k = 2$ ACB4	ACB6	ACB8	$k = 5$ ACB4	ACB6	ACB8
DNA	20	2.62	1.97	1.36	6.19	7.10	6.63	5522.47	5524.56	5545.28
	40	1.28	1.10	0.75	2.07	2.40	2.59	9.51	9.42	11.42
	60	0.91	0.84	0.66	1.46	1.60	1.81	4.55	4.34	5.36

The q-gram variations are faster than the original ACB algorithm in all the test cases except for $q = 8$, $k = 5$ and $m = 20$. Interestingly, ACB8 was the fastest algorithm for $k = 1$, ACB4 for $k = 2$ and ACB6 for $k = 5$. ACBq also beats the running times of FN in several cases, especially for low k values.

6   Concluding Remarks

We have developed several sublinear algorithms for approximate online matching of circular strings with k errors. Our experiments show that the new algorithms work well in practice at reasonable error levels. According to our tests, our algorithms are faster than the current top algorithm FN [9] in many cases. The main advantages over FN are a compact implementation and significantly lower preprocessing times, which makes the new algorithms faster than FN by orders of magnitude for short texts and pattern lengths close to w. Our algorithms are also better suited for large alphabets than FN.

References

1. Baeza-Yates, R., Gonnet, G.H.: A new approach to text searching. Communications of the ACM 35(10), 74–82 (1992)
2. Bliven, S., Prlic, A.: Circular permutation in proteins. PLoS Computational Biology 8(3) (2012)
3. Boyer, R.S., Moore, J.S.: A fast string searching algorithm. Comm. ACM 20(10), 762–772 (1977)
4. Bunke, H., Bühler, U.: Applications of approximate string matching to 2D shape recognition. Pattern Recognition 26(12), 1797–1812 (1993)
5. Chang, W.I., Lawler, E.L.: Sublinear approximate string matching and biological applications. Algorithmica 12(4/5), 327–344 (1994)
6. Chen, K.H., Huang, G.S., Lee, R.C.T.: Exact circular pattern matching using the bit-parallelism and q-gram technique. In: Proc. of the 29th Workshop on Combinatorial Mathematics and Computation Theory, pp. 18–27. National Taipei College of Business (2012)
7. Ďurian, B., Holub, J., Peltola, H., Tarhio, J.: Improving practical exact string matching. Information Processing Letters 110(4), 148–152 (2010)
8. Faro, S., Lecroq, T.: Smart: a string matching algorithm research tool. University of Catania and Univeristy of Rouen (2011),
http://www.dmi.unict.it/~faro/smart/
9. Fredriksson, K., Navarro, N.: Average-optimal single and multiple approximate string matching. ACM Journal of Experimental Algorithmics 9, article 1.4 (2004)
10. Fulwider, S., Mukherjee, A.: Multiple Pattern Matching. In: The Second International Conferences on Pervasive Patterns and Applications, PATTERNS 2010, pp. 78–83 (2010)
11. Gregor, J., Thomason, M.G.: Dynamic programming alignment of sequences representing cyclic patterns. IEEE Trans. Pattern Anal. Mach. Intell. 15, 129–135 (1993)
12. Gusfield, D.: Algorithms on Strings, Trees and Sequences: Computer Science and Computational Biology. Cambridge University Press (1997)
13. Huh, Y., Yu, K., Heo, J.: Detecting conjugate-point pairs for map alignment between two polygon datasets. Computers, Environment and Urban Systems 35(3), 250–262 (2011)
14. Hyyrö, H., Navarro, G.: Bit-parallel witnesses and their applications to approximate string matching. Algorithmica 41(3), 203–231 (2005)

15. Kärkkäinen, J., Kempa, D., Puglisi, S.J.: Lightweight Lempel-Ziv parsing. In: Bonifaci, V., Demetrescu, C., Marchetti-Spaccamela, A. (eds.) SEA 2013. LNCS, vol. 7933, pp. 139–150. Springer, Heidelberg (2013)
16. Lin, J., Adjeroh, D.: All-against-all circular pattern matching. Computer Journal 55(7), 897–906 (2012)
17. Lo, W.C., Lee, C.C., Lee, C.Y., Lyu, P.C.: CPDB: A database of circular permutation in proteins. Nucleic Acids Research 37(suppl. 1), D328–D332 (2009)
18. Marzal, A., Barrachina, S.: Speeding up the computation of the edit distance for cyclic strings. In: Int'l Conference on Pattern Recognition, pp. 891–894. IEEE Computer Society Press (2000)
19. Myers, G.: A fast bit-vector algorithm for approximate string matching based on dynamic programming. Journal of the ACM 46(3), 395–415 (1999)
20. Navarro, G., Raffinot, M.: Fast and flexible string matching by combining bit-parallelism and suffix automata. Journal of Experimental Algorithmics 5 (2000)
21. Navarro, G.: A guided tour to approximate string matching. ACM Comput. Surv. 33(1), 31–88 (2001)
22. Navarro, G.: NR-grep: A fast and flexible pattern-matching tool. Softw. Pract. Exp. 31(13), 1265–1312 (2001)
23. Oncina, J.: The Cocke-Younger-Kasami algorithm for cyclic strings. In: ICPR 1996: Proc. 13th Int. Conf. Pattern Recognition, Vienna, Austria, pp. 413–416. IEEE Computer Society (1996)
24. Peltola, H., Tarhio, J.: Alternative algorithms for bit-parallel string matching. In: Nascimento, M.A., de Moura, E.S., Oliveira, A.L. (eds.) SPIRE 2003. LNCS, vol. 2857, pp. 80–93. Springer, Heidelberg (2003)
25. Schnattinger, T., Ohlebusch, E., Gog, S.: Bidirectional search in a string with wavelet trees and bidirectional matching statistics. Information and Computation 213, 13–22 (2012)
26. Susik, R., Grabowski, S., Deorowicz, S.: Fast and simple circular pattern matching. In: Gruca, A., Czachórski, T., Kozielski, S. (eds.) Man-Machine Interactions 3. AISC, vol. 242, pp. 541–548. Springer, Heidelberg (2014)

From Theory to Practice:
Plug and Play with Succinct Data Structures

Simon Gog[1], Timo Beller[2], Alistair Moffat[1], and Matthias Petri[1]

[1] Dept. Computing and Information Systems,
The University of Melbourne, Victoria 3010, Australia
[2] Inst. Theoretical Computer Science, Ulm University, 89069, Germany

Abstract. Engineering efficient implementations of compact and succinct structures is time-consuming and challenging, since there is no standard library of easy-to-use, highly optimized, and composable components. One consequence is that measuring the practical impact of new theoretical proposals is difficult, since older baseline implementations may not rely on the same basic components, and reimplementing from scratch can be time-consuming. In this paper we present a framework for experimentation with succinct data structures, providing a large set of configurable components, together with tests, benchmarks, and tools to analyze resource requirements. We demonstrate the functionality of the framework by recomposing two succinct solutions for top-k document retrieval which can operate on both character and integer alphabets.

1 Introduction

The field of succinct data structures (SDSs) has evolved rapidly in the last decade. New data structures such as the FM-Index, Wavelet Tree (WT), Range Minimum Query Structure (RMQ), Compressed Suffix Array (CSA), and Compressed Suffix Tree (CST) have been developed, and been shown to be remarkably versatile. These structures provide the same functionality as the corresponding uncompressed data structures, but do so using space which is asymptotically close to the information-theoretic lower bound needed to store the underlying data or objects. Using standard models of computation, the asymptotic runtime complexity of the operations performed by SDSs is also often identical to their classical counterparts. However, in practice, SDSs tend to be slower than the uncompressed structures, due to more complex memory access patterns on bitvectors, including non-sequential processing of unaligned bits. That is, they come in to their own only when the data scale means that an uncompressed structure would not fit in to main memory (or any particular level of the memory hierarchy), but a compressed structure would.

Accessing (ACCESS), counting (RANK), and selecting (SELECT) bits in bitvectors (BVs) are the fundamental operations from which more intricate operations are constructed. All three foundational operations can be supported in constant time adding only sublinear space. Wavelet trees build on BVs and generalize the three operations to alphabets of size $\sigma > 2$, with a corresponding increase

J. Gudmundsson and J. Katajainen (Eds.): SEA 2014, LNCS 8504, pp. 326–337, 2014.
© Springer International Publishing Switzerland 2014

in time to $\mathcal{O}(\log \sigma)$. A further layer up, some CSAs use WTs to realize their functionality; in turn, CSAs are the basis of yet more complex structures such as CSTs. Multiple alternatives exist at each level of this dependency hierarchy. For example, WTs differ in both shape (uniform versus Huffman-shaped) and in the choice made in the lower hierarchy levels (whether compressed or uncompressed BVs are used). The diversity of options allows structures to be composed that have a variety of time-space trade-offs. In practice, many of the complex structures that have been proposed are not yet fully implemented, since the implementations of underlying structures are missing. The use of non-optimized or non-composable substructures then prevents thorough empirical investigations, and makes it difficult to carry out impartial comparisons. The cost of implementing different approaches also creates a barrier that makes it difficult for new researchers, including graduate students, to enter the field.

As part of our investigation into SDSs, a library of flexible and efficient implementations has been assembled, providing a modular "plug and play, what you declare is what you get" approach to algorithm implementation. Instrumentation and visualization tools are also included in the library, allowing space costs to be accurately measured and depicted; as are efficient routines for constructing (in-memory as well as semi-external) and serializing all internal representations, allowing files containing succinct structures to be generated and re-read. We have also incorporated recent hardware developments, so that built-in popcount operations are used when available, and hardware hugepages can be enabled, to bypass address translation costs.

With this resource it is straightforward to *compose* complex structures; *measure* their behavior; and *explore* the range of alternatives. Having an established and robust code base also means that experimental comparisons of new data structures and algorithms can be made more resilient, providing greater consistency between alternative structures, and allowing better baseline systems and hence more reproducible evaluations. For example, different CST components might be appropriate to different types of input sequence (integer alphabet versus byte-based; highly-repetitive or not); and different sampling rates and access methods might be required. Using the library the right combination of components for any given application can be readily determined. The generality embedded in the library means that it is also useful in other fields such as information retrieval, natural language processing, bioinformatics, and image processing.

We illustrate the myriad virtues of the library – version 2 of SDSL – through two case studies: first, a detailed recomposition and re-evaluation of succinct document retrieval systems (Section 2); and second, a detailed examination of the construction processes used to create indexing structures (Section 3).

2   Document Retrieval Recomposed

The *top-k document retrieval problem* is fundamental in information retrieval (IR), and has become an active research topic in the succinct data structures community, see Navarro [13] for an excellent overview. For a collection of N

documents $\mathcal{C} = \{d_1, \ldots, d_N\}$ over an alphabet Σ of size σ, a query $\mathcal{Q}$ also a set of strings over Σ, and a ranking function $R : \mathcal{C} \times \mathcal{Q} \to \mathcal{R}$, the task is to return the k documents with highest values of $R(d_i, \mathcal{Q})$. The simple *frequency* version of the problem assumes that the query consists of a single sequence (term) q and that $R(d_i, q)$ is the frequency $tf(d_i, q)$ of q in d_i. The *tf-idf* version of the problem computes $R(d_i, \mathcal{Q}) = tf(d_i, q) \times \log(N/df(q))$, where $df(q)$ is the *document frequency* of q, the number of documents in which sequence q occurs. Note that the *tf-idf* formulation used here and in similar studies (for example, Sadakane [20]) is still a simplification of the measures used in modern retrieval systems, which incorporate factors like document length, static document components such as pagerank, and queries with multiple terms.

We focus here on the single term *frequency* and single term *tf-idf* versions of the problem. Sadakane [20] devised the first succinct structure for these problems, an approach we refer to as SADA. An alternative mechanism, GREEDY, was presented by Culpepper et al. [2]. Culpepper et al. [2] also describe implementations of SADA and GREEDY, based on components that were available in 2009. Other compressed index representations solving the top-k retrieval problem have been recently proposed. They provide different time and space trade-offs [8,15], but generally apply more complex storage, compression and query techniques to reduce storage costs or increase query performance. For the purpose of simplicity and highlighting the impact of state-of-the-art components to compose more complex structures, we focus on the two earlier, and comparatively simpler index structures SADA and GREEDY. Insights, techniques and performance gains however are transferable to more recent top-k retrieval techniques which are composed of the same basic succinct structures.

We first briefly explain both index types, and then recompose and reimplement them using the library, to study the impact of state-of-the-art components. For both solutions we use the conventions established by Sadakane [20] and Culpepper et al. [2]: the set of documents is concatenated to form a text $\mathcal{T}$ by appending a sentinel symbol $\#$ to each d_i; then joining them all in to one; then appending a further sentinel symbol $\$$ following d_N. Both sentinels are lexicographically smaller than any symbol in Σ, and $\$ < \#$. We use n to represent $|\mathcal{T}|$.

2.1 SADA and GREEDY Revisited

The SADA structure is composed of several components. First, a CSA (denoted `csa_full`) over $\mathcal{T}$ identifies, for any pattern p, the range $[sp..ep]$ of matching suffixes in $\mathcal{T}$, providing the functionality of a SA. Second, a BV is constructed (denoted `border[0..(n − 1)]`) with `border[i]` $= 1$ iff $\mathcal{T}[i] = \#$; supporting rank and select structures (`border_rank` and `border_select`) are also generated. A *document array* $\mathcal{D}[0..(n-1)]$ that maps the ith suffix in $\mathcal{T}$ to the document that contains it can then be emulated using `border` and `csa_full`, since $\mathcal{D}[i] = $ `border_rank`$(\mathcal{SA}[i])$. Third, to generate all distinct document numbers in a range $[sp..ep]$, an RMQ structure `rminq` is used. It is built over a conceptual array $C[0..(n-1)]$, defined as $C[i] = \max\{j \mid j < i \wedge D[j] = D[i]\}$; that is, $C[i]$ is the index in $\mathcal{D}$ of the last prior occurrence of $\mathcal{D}[i]$. The number of distinct values in $\mathcal{D}[sp..ep]$ corresponds to $df(q)$. Locations of the values are identified by computing $x = \mathtt{rminq}(sp, ep)$, the index of the minimum element in $C[sp..ep]$. A temporary BV of size N is used to check if $\mathcal{D}[x]$ was already retrieved; if it wasn't, the counting is continued by recursing into $[sp..(x-1)]$ and $[(x+1)..ep]$. Finally, a second RMQ structure ($\mathtt{rmaxq}$) and individual CSAs are built for each document d_i, in order to calculate $tf(d_i, q)$. The top-k items are then calculated by a partial sort on the $tf(d_i, q)$ values. In total, SADA uses $|\mathrm{CSA}(\mathcal{T})| + |\mathrm{BV}(\mathcal{T})| + 2|\mathrm{RMQ}| + \sum_{i=0}^{N-1} |\mathrm{CSA}(d_i)|$ bits. Choosing an H_k-compressed CSA for $\mathtt{csa\_full}$, a $2n + o(n)$-bit solution for $\mathtt{rminq}$ and $\mathtt{rmaxq}$, and a compressed BV for $\mathtt{border}$ [16], results in a total bit count bounded above by $nH_k(\mathcal{T}) + \sum_{i=0}^{N-1} |\mathrm{CSA}(d_i)| + 4n + o(n) + N(2 + \log(n/N))$. The space for the document CSAs was deliberately not substituted by the cost of a concrete solution, for two reasons: (1) each CSA has an alphabet-dependent overhead, which dominates the space for small documents; and (2) only the inverse SA functionality is used. Thus, in the recomposition, we opt for a *bit-compressed*[1] version of the inverse SA, with almost no overhead per document, and benefiting from constant access time.

The second solution – GREEDY – also uses $\mathtt{csa\_full}$ to translate patterns p to ranges $[sp..ep]$. In this solution the document array $\mathcal{D}$ is explicitly represented by a WT, denoted $\mathtt{wtd}$. The total size is then $|\mathrm{CSA}(\mathcal{T})| + |\mathrm{WT}(\mathcal{D})| = nH_k(\mathcal{T}) + nH_0(\mathcal{D})$, using compressed BVs for $\mathtt{wtd}$ [18]. Culpepper et al. [2] use these structures to solve the frequency variant of the top-k problem. In the first step the pattern is translated to a range in $\mathcal{D}$. A priority queue is then used to store pending nodes in the expansion in WT of the range $[sp..ep]$. At each step the largest node is extracted, and, if it is not a leaf node, its two children are inserted. This process is iterated until k leaves – corresponding to the top-k frequent documents – have emerged. The range size of a leaf corresponds to the term frequency $tf(d_i, q)$. In contrast to SADA, not all documents have to be listed in order to calculate the top-k. However, GREEDY does more work per document, since $\mathtt{wtd}$ has a height of $\log N$.

2.2 Experimental Setup

We adopt the experimental setup employed by Culpepper et al. [2], reusing the PROTEINS file and adding the ENWIKI file[2]. The ENWIKI datafiles come in four versions: parsed as either characters or words; and either a small prefix, or the whole collection, see Table 1. The character version was generated by removing markup from a wikipedia dump file; the word-based version by then applying the parser of the Stanford Natural Language Group.

We generated patterns of different lengths, again following the lead of Culpepper et al. [2], (see the commentary provided by Gog and Moffat [11] in regard

---

[1] That is, each item is encoded as a binary value in $\lceil \log |d_i| \rceil$ bits.

[2] We did not use the Wall Street Journal file, since licensing issues meant that we could not make it available for download. Our full setup is available at http://go.unimelb.edu.au/w68n

to random patterns), with 200 patterns of each length; reported query times are averages over sets of 200 patterns. All experiments were run on a server equipped with 144 GB of RAM and two Intel Xeon E5640 processors each with a 12 MB L3 cache. The experimental code is available within the benchmark suite of the library, and all used library classes (printed in fixed italic font) are linked to their definitions.

Table 1. Collection statistics: number of characters/words, number of documents, average document length, total collection size, and approximate H_k (determined using xz -best) in bits per character/word. The character based collections use one byte per symbol, while $\lceil \log \sigma \rceil$ bits are used the word based case.

Collection	n	N	n/N	σ	$|\mathcal{T}|$ in MB	$\approx H_k(\mathcal{T})$
character alphabet						
PROTEINS	58,959,815	143,244	412	40	56	0.90
ENWIKI-SML	68,210,334	4,390	15,538	206	65	2.01
ENWIKI-BIG	8,945,231,276	3,903,703	2,291	211	8,535	2.02
word alphabet						
ENWIKI-SML	12,741,343	4,390	2,902	281,577	29	5.03
ENWIKI-BIG	1,690,724,944	3,903,703	433	8,289,354	4,646	4.45

2.3 Plug and Play

We start by composing an instance of GREEDY. For csa_full, SDSL provides several CSA types. We opt for *csa_wt*, which is based on a WT. Choosing a Huffman-shaped WT [10] (*wt_huff*) and parameterizing it with a suitable BV (*rrr_vector*) results in an H_k-compressed CSA. The *rrr_vector* implements the on-the-fly decoding recently described by Navarro and Providel [14], which provides low redundancy. Finally, we minimize the space of the CSA by sampling (inverse) SA values only every millionth position. This does not affect the runtime of GREEDY, since it does not require SA access. For wtd we choose the alphabet-friendly WT class *wt_int* and parameterize it with a fast uncompressed BV (*bit_vector*) and small overhead rank structure (*rank_support_v5*). No select functionality is required in GREEDY.

We use the same toolbox to assemble a space- and time-efficient version of SADA. The full CSA in SADA has to provide fast element access; "plug-and-play" exploration with different CSAs showed that *csa_sada* [19] is the preferred choice in this situation. The suffix sampling rate is set to 32. Similar exploration revealed that a text-order sampling strategy yields a more attractive time-space trade-off than suffix-order sampling, provided the Elias-Fano compressed BV [16] (*sd_vector*) is used to mark the sampled suffixes. For components rminq and rmaxq we select the range min-max-tree based RMQ structure (*rmq_succinct_sct*). The inverse SAs of the documents are represented using bit-compressed vectors (*int_vector*), and the array of vectors is denoted by doc_isa.

Finally, we compose word alphabet versions of GREEDY and SADA by replacing the character alphabet strategy (`byte_alphabet`) in the definition of `csa_full` by the word alphabet strategy (`int_alphabet`).

Table 2. Sizes of indexes, in MB and as a multiple of the collection

Collection	character alphabet		word alphabet	
	GREEDY	SADA	GREEDY	SADA
PROTEINS	162 (2.87)	136 (2.42)	*– no word parsing –*	
ENWIKI-SML	130 (2.01)	204 (3.13)	38 (1.32)	50 (1.72)
ENWIKI-BIG	27,043 (3.17)	24,404 (2.86)	6,786 (1.46)	5,703 (1.23)

Memory Usage. Table 2 summarizes the space usage of the composed structures. They take considerably less space than those reported by Culpepper et al. [2]; for example, on PROTEINS, their SADA is 6.4 times larger than ours, and their GREEDY is 1.3 times larger. Also note that when the documents are short, our SADA is smaller than GREEDY. These space reductions result from the use of better-engineered components; the only algorithmic change made was the use of bit-compressed inverse SAs, instead of high-overhead CSAs.

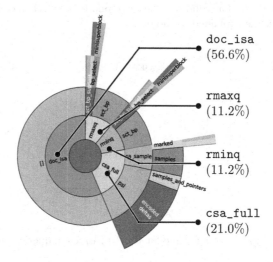

doc_isa
(56.6%)

rmaxq
(11.2%)

rminq
(11.2%)

csa_full
(21.0%)

Fig. 1. Sunburst visualization of the memory usage of the character-based SADA on file ENWIKI-BIG. A dynamic version is available at http://go.unimelb.edu.au/ba8n.

Figure 1 depicts a space visualization of the type that can be generated for any SDSL object. It reveals that `doc_isa` takes over half the space; that `rmaxq` and `rminq` are close to the optimal $2n$ bits ($2.6n$ bits); and that the CSA takes 5.3 bits per character. The latter differs from the optimal $H_k = 2.02$ bits reported in Table 2, as samples were added for fast SA access, which account for 2.4 of the 5.3 bits. The largest component of GREEDY (not shown here) is the document array $\mathcal{D}$, which requires $n \log N$ bits plus the overhead of the rank structure, around 92.2% of the total space for ENWIKI-BIG. Plugging in `rrr_vector` results in H_0-compression of `wtd` and reduces the overall size to 25,320 MB, while the query time is slowed down by a factor of between 2 and 4. Note that the word versions are smaller than the character-based ones, because n is smaller. SADA also benefits from smaller average document lengths. Overall the word indexes

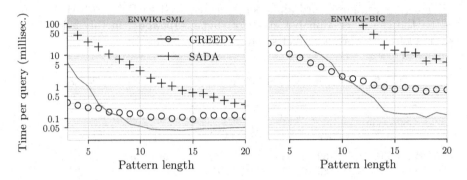

Fig. 2. Average time for top-10 *frequency* queries on the character-based collections. Results are omitted for pattern length ℓ if any single pattern of length ℓ took more than 5 seconds. The solid red line corresponds to SORT, a simple alternative implementation which is explained in the text.

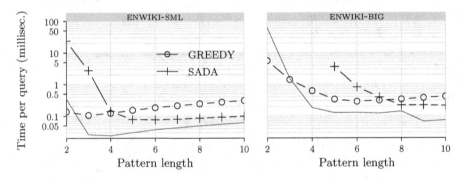

Fig. 3. Average time for top-10 *frequency* queries on the word-based collections

are smaller than the character-based indexes, and at around 2/3 of the original text size, are comparable to compressed positional inverted files.

Runtime. Query times are depicted in Figures 2 and 3. We make several observations. First, the runtime of both solutions depends on the collection size, since any given pattern occurs more often in a larger collection. This results in larger ranges, which are especially bad for SADA, since it processes all distinct documents in the range, even when computing top-10 queries. (This same behavior means that SADA can compute complex *tf-idf* queries in very similar times.) GREEDY is also dependent on size of the $[sp..ep]$ range, requiring two orders of magnitudes more time on ENWIKI-BIG than on ENWIKI-SML, matching the difference in their sizes. Second, for long patterns (>15 characters) SADA is now only one order of magnitude slower than GREEDY, in contrast to two orders reported by Culpepper et al. [2]. This is due to the faster extraction of inverse SA values and the use of a Ψ-based CSA instead of a WT-based one. For word indexes SADA now outperforms GREEDY for long queries, where the result range is very small. In this cases, the pattern matching becomes the dominating cost. In SADA, we used `csa_sada` which implements backward search by binary searches on Ψ using $\mathcal{O}(m \log n)$ time, while `csa_wt` in GREEDY performs $\mathcal{O}(m H_0(\mathcal{T}))$ non-local rank operations on the compressed BV. The GREEDY approach could be made significantly faster if the `rrr_vector` was replaced by an uncompressed BV, but the space would then become much greater than SADA.

We also compare our results to a simple baseline called SORT. Again, `csa_wt` is used as the CSA, but now the document array $\mathcal{D}$ is stored as a bit-compressed vector of $n \lceil \log N \rceil$ bits. After identifying the $[sp..ep]$ interval, the entries of $\mathcal{D}[sp..ep]$ are copied and sorted to generate (d_i, tf) pairs; the standard C++ partial sort is then used to retrieve the top-10. The red lines in Figures 2 and 3 show query times. The sequential or local processing of SORT is always superior to SADA, which, despite careful implementation, suffers from the non-locality of the range minimum/maximum queries. Moreover, GREEDY only dominates SORT in cases involving very wide intervals, emphasizing one of the key messages of this article – that it is only when careful implementations and large test instances are compared that the usefulness of advanced succinct data structures can be properly demonstrated..

We are aware that there are more recent proposals for top-k retrieval systems, such as the scheme of Hon et al. [7], which uses precomputed answers for selected ranges. The compressed suffix tree facility of the library enabled implementation of this solution in other experimentation [3,17]. The recent mechanism by Navarro and Nekrich [12] provides range-independent query performance; and the character-based implementation described by Konow and Navarro [8] would likely outperform SORT and GREEDY for small interval sizes. The presentation of these two schemas is beyond the scope of this article.

3 Efficient Construction of Complex Structures

Constructing SDSs over small data sets – up to hundreds of megabytes – is not a challenge from an engineering perspective, since commodity hardware supports memory-space many times larger than this. However, SDSs are explicitly intended to replace traditional data structures in resource-constrained environments; which means they are most applicable when the data is too large for uncompressed structures to be used, and hence that construction is also a critical issue. As well, complex structures composed of multiple sub-structures often contain dependencies between sub-structures which further complicate the construction process. For example, to construct a CST, a CSA is required; and to construct a CSA quickly, usually an uncompressed SA is needed. Under memory constraints, it is not possible to hold all of these structures in RAM concurrently. To alleviate this problem the library incorporates semi-external construction algorithms which stream data sequentially to and from disk. To facilitate this, the library provides serialization and save/load functionality for all substructures. Finally, as already noted, the library includes memory visualization techniques, which analyze space utilization during run-time and construction.

To demonstrate the complexities of the construction process in more detail, we examine the resource utilization during the construction of word-based SADA. Figure 4 shows the resource consumption for the 4.6 GB ENWIKI-BIG collection. In total, the construction process took 5,250 seconds and had a peak requirement of 13 GB. This corresponds to a throughput of 0.88 MB per second. In monetary terms, the SADA index for ENWIKI-BIG can be built for less than one dollar on the Amazon Cloud[3]. The majority of the time (65%) was spent constructing the CSA. First, the plain SA is constructed – phase 1 in Figure 4 – using the algorithm of Larsson and Sadakane [9]. The algorithm uses twice the memory space required by the resulting bit-compressed suffix array, accounting for the peak memory usage (13 GB) of the complete process. After construction, the SA is serialized to disk to construct the Burrows-Wheeler Transform (BWT) (phase 2). Only $\mathcal{T}$ is kept in memory, as it is the only place where random access is required, and the BWT sequence can be written to disk as it is formed. This semi-external construction process of the BWT requires 380 seconds, or 7% of the total time. The remainder of the CSA is constructed in 358 seconds. This includes constructing the Ψ mapping (shown as phase 3), and sampling the SA. The next major construction step, marked as phase 4 in the figure, constructs the N individual inverse SAs (`doc_isa`). Here for each document an SA is constructed, inverted, bit-compressed, and added to `doc_isa`. This process requires 785 seconds, or around 200 microseconds per document. Next, phase 5 constructs the document array $\mathcal{D}$ by streaming the full SA from disk and performing n RANK operations on `border`. Creating the complete array requires 271 seconds or 160 nanoseconds per $\mathcal{D}[i]$ value. This includes reading the SA from disk, performing the RANK on `border`, and storing the resulting $\mathcal{D}[i]$ value. Finally, `rminq` (phase 6) and `rmaxq` (phase 7) are created, including computing temporary C and C' arrays from $\mathcal{D}$. Creating `rminq` requires 360 seconds; computing `rmaxq` a further 396 seconds. Half of that time is spent creating C and C'; the balance on constructing the actual RMQ structures.

Semi-external construction is an important tool to minimize the resource consumption in each phase of the building process of SADA. Keeping all constructed components in memory would significantly increase the memory overhead during the later stages of the construction process. Thus, careful engineering of the construction phase of each individual component of a complex structure is important so that the overall resource requirements of the entire construction process is minimized. In our case, SADA is now so efficient that we can build the structure for the word tokenized GOV2 collection, used in the TREC Terabyte Track, on our experimental machine.

It is important to monitor construction costs, as the modularity of SDSs can lead to unintended resource consumption. For example, in an initial version of our SADA implementation, the CSA was not serialized to disk, but kept in memory. Visualization of the overall construction cost of SADA highlighted this inefficiency and allowed it to be rectified. Similarly, when reviewing the

---

[3] In November 2013 a "High-Memory Extra Large" instance with 17 GB RAM costs USD$0.41 per hour.

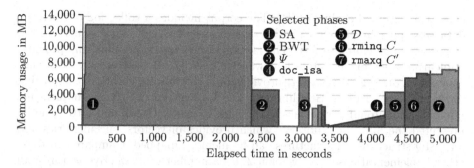

Fig. 4. A memory-time graph for construction of SADA over the word-based sequence ENWIKI-BIG (4.6 GB). The version generated by `memory_monitor` is available at http://go.unimelb.edu.au/7a8n.

character-based `csa_wt`, it became apparent that optimizing WT construction could not significantly improve the overall process, as that phase only accounts for around 4% of the total cost. An automatic resource tracker (`memory_monitor`) facilitates visualizing the space and time consumption of algorithms in order to provide such insights.

Operations on SDSs can be accelerated if 1 GB pages (hugepages) are used to address memory, rather than the standard 4 kB pages, since address translation becomes a bottleneck when pages are small [5]. The SDSL now includes memory management facilities that allow the use of hugepages during construction as well. We investigate the effect of hugepages on construction time by building `csa_wt`s of increasing size for prefixes of the character-based ENWIKI-BIG collection (Table 3). For small file sizes, hugepages have a modest effect on construc-

Table 3. Construction times (seconds) of `csa_wt` for prefixes of the character-based ENWIKI-BIG file

Prefix	Page size	
(MB)	4 kB	1 GB
10	2	2
100	29	25
1,000	379	307
5,000	2,524	1,877
8,535	5,282	4,482

tion time, and the 100 MB file is processed only 16% faster. As the file size increases the effect becomes more visible. The index of 1 GB prefix is constructed 24% faster, and the 5 GB index can be built 35% faster. The improvement in construction time then decreases for the full ENWIKI-BIG collection, as the TLB can only maintain a certain number of hugepage entries, after which address translation becomes more costly again.

4 Related Work

As part of their experimental work authors often make prototype implementations of their proposed structures available. Additionally, several experimental studies and publicly available libraries focusing on SDSs have emerged. The best-known is the PIZZA&CHILI corpus[4], which was released alongside an

---
[4] http://pizzachili.dcc.uchile.cl

extensive empirical evaluation [4]. The corpus includes a collection of reference data sets, plus implementations of several succinct text index structures accessed via a common interface. The LIBCDS library is also popular, and provides implementations of bitvectors and wavelet trees [1]. It has recently been subsumed by LIBCDS2[5]. Vigna [21] provides bitvector implementations supporting RANK and SELECT in the SUX library[6]. The Java version of SUX also implements minimal (monotone) perfect hash functions. Recently Grossi and Ottaviano [6] presented the SUCCINCT library[7] which provides bitvectors, succinct tries and an RMQ structure; all structures can be memory mapped. Compared to these other implementations, SDSL version 2 has a number of distinctive features: all indexes work on both character and *word inputs*; it is optimized for large-scale input (including *dynamic support for hugepages*); it provides coverage of a wide range of alternative structures (including several CSAs and CSTs), that can be composed and substituted in different ways; and it offers *dynamic visualizations* that allow detailed space evaluations to be undertaken. Finally, fully automated tests, *code coverage*, and *various benchmarks* are also included, and can be used by other researchers in the future to check the correctness and performance of further alternative implementations of the various modules. The italicized facets represent the enhancements in version 2 relative to the previous SDSL release [5].

5 Conclusion

We have explored the benefits that flow when modular and composable implementations of succinct data structure building blocks are available, and have showed that efficiency at all levels of the SDS hierarchy can be enhanced by careful attention to low-level detail, and to the provision of precisely-defined interfaces. As a part of that demonstration, we introduced the open source SDSL library, which contains efficient implementations of many SDSs. The library is structured to facilitate flexible prototyping of new high-level structures, and because it offers a range of options at each level of the data structure hierarchy, allows rapid exploration of implementation alternatives. The library is also robust in terms of scale, handling input sequences of arbitrary length over arbitrary alphabets. In addition, we have demonstrated that the use of hugepages can have a notable effect on construction times of large-scale SDSs; and shown that the advanced visualization features of the library provide important insights into the time and space requirements of SDSs.

Acknowledgment. The authors were supported by the Australian Research Council and by the Deutsche Forschungsgemeinschaft.

Software & Experiments. The library code, test suite, benchmarks, and a tutorial, are publicly available at https://github.com/simongog/sdsl-lite.

---

[5] https://github.com/fclaude/libcds2
[6] http://sux.di.unimi.it
[7] https://github.com/ot/succinct

References

1. Claude, F., Navarro, G.: Practical rank/select queries over arbitrary sequences. In: Amir, A., Turpin, A., Moffat, A. (eds.) SPIRE 2008. LNCS, vol. 5280, pp. 176–187. Springer, Heidelberg (2008)
2. Culpepper, J.S., Navarro, G., Puglisi, S.J., Turpin, A.: Top-k ranked document search in general text databases. In: de Berg, M., Meyer, U. (eds.) ESA 2010, Part II. LNCS, vol. 6347, pp. 194–205. Springer, Heidelberg (2010)
3. Culpepper, J.S., Petri, M., Scholer, F.: Efficient in-memory top-k document retrieval. In: Proc. SIGIR, pp. 225–234 (2012)
4. Ferragina, P., González, R., Navarro, G., Venturini, R.: Compressed text indexes: From theory to practice. J. Experimental Alg. 13 (2008)
5. Gog, S., Petri, M.: Optimized succinct data structures for massive data. In: Soft. Prac. & Exp. (2013) (to appear) , http://dx.doi.org/10.1002/spe.2198
6. Grossi, R., Ottaviano, G.: Design of practical succinct data structures for large data collections. In: Bonifaci, V., Demetrescu, C., Marchetti-Spaccamela, A. (eds.) SEA 2013. LNCS, vol. 7933, pp. 5–17. Springer, Heidelberg (2013)
7. Hon, W.-K., Shah, R., Vitter, J.S.: Space-efficient framework for top-k string retrieval problems. In: Proc. FOCS, pp. 713–722 (2009)
8. Konow, R., Navarro, G.: Faster compact top-k document retrieval. In: Proc. DCC, pp. 5–17 (2013)
9. Jesper Larsson, N., Sadakane, K.: Faster suffix sorting. Theor. Comp. Sc. 387(3), 258–272 (2007)
10. Mäkinen, V., Navarro, G.: Succinct suffix arrays based on run-length encoding. In: Apostolico, A., Crochemore, M., Park, K. (eds.) CPM 2005. LNCS, vol. 3537, pp. 45–56. Springer, Heidelberg (2005)
11. Moffat, A., Gog, S.: String search experimentation using massive data. Phil. Trans. Royal Soc. A (to appear, 2014)
12. Navarro, G., Nekrich, Y.: Top-k document retrieval in optimal time and linear space. In: Proc. SODA, pp. 1066–1078 (2012)
13. Navarro, G.: Spaces, trees and colors: The algorithmic landscape of document retrieval on sequences. ACM Comp. Surv. (to appear, 2014)
14. Navarro, G., Providel, E.: Fast, small, simple rank/select on bitmaps. In: Klasing, R. (ed.) SEA 2012. LNCS, vol. 7276, pp. 295–306. Springer, Heidelberg (2012)
15. Navarro, G., Puglisi, S.J., Valenzuela, D.: Practical compressed document retrieval. In: Pardalos, P.M., Rebennack, S. (eds.) SEA 2011. LNCS, vol. 6630, pp. 193–205. Springer, Heidelberg (2011)
16. Okanohara, D., Sadakane, K.: Practical entropy-compressed rank/select dictionary. In: Proc. ALENEX (2007)
17. Patil, M., Thankachan, S.V., Shah, R., Hon, W.-K., Vitter, J.S., Chandrasekaran, S.: Inverted indexes for phrases and strings. In: Proc. SIGIR, pp. 555–564 (2011)
18. Raman, R., Raman, V., Srinivasa Rao, S.: Succinct indexable dictionaries with applications to encoding k-ary trees and multisets. In: Proc. SODA, pp. 233–242 (2002)
19. Sadakane, K.: New text indexing functionalities of the compressed suffix arrays. J. Alg. 48(2), 294–313 (2003)
20. Sadakane, K.: Compressed suffix trees with full functionality. Theory Comp. Sys. 41(4), 589–607 (2007)
21. Vigna, S.: Broadword implementation of rank/select queries. In: McGeoch, C.C. (ed.) WEA 2008. LNCS, vol. 5038, pp. 154–168. Springer, Heidelberg (2008)

Improved ESP-index: A Practical Self-index for Highly Repetitive Texts*

Yoshimasa Takabatake[1], Yasuo Tabei[2], and Hiroshi Sakamoto[1]

[1] Kyushu Institute of Technology, Japan
{takabatake,hiroshi}@donald.ai.kyutech.ac.jp
[2] PRESTO, Japan Science and Technology Agency, Japan
tabei.y.aa@m.titech.ac.jp

Abstract. While several self-indexes for highly repetitive texts exist, developing a practical self-index applicable to real world repetitive texts remains a challenge. ESP-index is a grammar-based self-index on the notion of edit-sensitive parsing (ESP), an efficient parsing algorithm that guarantees upper bounds of parsing discrepancies between different appearances of the same subtexts in a text. Although ESP-index performs efficient top-down searches of query texts, it has a serious issue on binary searches for finding appearances of variables for a query text, which resulted in slowing down the query searches. We present an improved ESP-index (ESP-index-I) by leveraging the idea behind succinct data structures for large alphabets. While ESP-index-I keeps the same types of efficiencies as ESP-index about the top-down searches, it avoid the binary searches using fast rank/select operations. We experimentally test ESP-index-I on the ability to search query texts and extract subtexts from real world repetitive texts on a large-scale, and we show that ESP-index-I performs better that other possible approaches.

1   Introduction

Recently, highly repetitive text collections have become common. Examples are human genomes, version controlled documents and source codes in repositories. In particular, the current sequencing technology enables us to sequence individual genomes in a short time, resulting in generating a large amount of genomes. There is therefore a strong demand for developing powerful methods to store and process such repetitive texts on a large-scale.

Grammar compression is effective for compressing and processing repetitive texts, and it builds a *context free grammar* (CFG) that generates a single text. There are two types of problems: (i) building as small as possible of a CFG generating an input text and (ii) representing the obtained CFG as compactly as possible for various applications. Several methods have been presented for type (i). Representative methods are RePair [12] and LCA [17]. Methods for type (ii) have also been presented for processing repetitive texts, e.g., pattern matching [18], pattern mining [9] and edit distance computation [10].

* This work was supported by JSPS KAKENHI(24700140,23680016) and the JST PRESTO program.

J. Gudmundsson and J. Katajainen (Eds.): SEA 2014, LNCS 8504, pp. 338–350, 2014.

Table 1. Comparison with existing methods. Searching time and extraction time is presented in big O notation that is omitted for space limitations. u is text length, m is the length of a query text, n is the number of variables in a grammar, σ is alphabet size, z is the number of pharases in LZ77, d is the length of nesting in LZ77, occ is the number of occurrences of query text in a text, occ_c is the number of candidate apperances of queries, $\lg^*$ is the iterated logarithm and ϵ is a real value in $(0, 1)$. $\lg$ stands for $\log_2$.

	Space (bits)	Searching time	Extraction time
LZ-index [15]	$z \lg u + 5z \lg \sigma$ $-z \lg z + o(u) + O(Z)$	$m^2 d + (m + occ) \lg z$	md
Gagie et al. [6]	$2n \lg n + O(z \lg u$ $+z \lg z \lg \lg z)$	$m^2 + (m + occ) \lg \lg u$	$m + \lg \lg u$
SLP-index [2,3]	$n \lg u + O(n \lg n)$	$(m^2 + h(m + occ)) \lg n$	$(m + h) \lg n$
ESP-index [13]	$n \lg u + (1 + \epsilon)n \lg n$ $+4n + o(n)$	$(1/\epsilon)(m \lg n$ $+occ_c \lg m \lg u) \lg^* u$	$(1/\epsilon)(m + \lg u)$
ESP-index-I	$n \lg u + n \lg n$ $+2n + o(n \lg n)$	$(\lg \lg n)(m$ $+occ_c \lg m \lg u) \lg^* u$	$(\lg \lg n)(m + \lg u)$

Self-indexes aim at representing a text collection in a compressed format that supports extracting subtexts of arbitrary positions and provides query searches on the collection, and are fundamental in modern information retrieval. However, developing a grammar-based self-index remains a challenge, since a grammar-compressed text forms a tree structure named parse tree and variables attached to its nodes do not necessarily encode all portions of a text, which makes it difficult to search query texts from grammar compressed texts. Claude et al. [2,3] presented a grammar-based self-index named *SLP-index*. SLP-index uses two step approaches: (i) it finds variables called first occurrences that encode all prefixes and suffixes of a query text by binary searches on compactly encoded CFGs and (ii) then it discovers the remaining occurrences of the variables. However, finding first occurrences for moderately long queries is computationally demanding because the method needs to perform binary searches as many times as the query length, which resulted in reducing the practicality of their method.

Edit-sensitive parsing (ESP) [4] is an efficient parsing algorithm developed for approximately computing edit distances with moves between texts. ESP builds from a given text a parse tree that guarantees upper bounds of parsing discrepancies between different appearances of the same subtext. Maruyama et al. [13] presented another grammar-based self index called *ESP-index* on the notion of ESP. ESP-index represents a parse tree as a *directed acyclic graph* (DAG) and then encodes the DAG into succinct data structures for ordered trees and permutations. Unlike SLP-index, it performs top-down searches for finding candidates of appearances of a query text on the data structure by leveraging the upper bounds of parsing discrepancies in ESP. However, it has a serious issue on binary searches for finding appearances of variables.

In this paper, we present an *improved ESP-index* (ESP-index-I) for fast query searches. Our main contribution is to develop a novel data structure for encoding a parse tree built by ESP. Instead of encoding the DAG into two ordered trees using succinct data structures in ESP-index, ESP-index-I encodes it into a bit string and an integer array by leveraging the idea behind rank/select dictionaries for large alphabets [8]. Instead of performing binary searches for finding variables on data structures in SLP-index and ESP-index, ESP-index-I computes fast select queries in $O(1)$ time, resulting in faster query searches. Our results and those of existing algorithms are summarized in Table 1. We omit several details of ESP-index-I and proofs of theorems in the proceeding version of the paper for space limitaion, which is presented in the full version.

Experiments were performed on retrieving query texts from real-world large-scale texts. The performance comparison with other algorithms demonstrates ESP-index-I's superiority.

2   Preliminaries

The length of string S is denoted by $|S|$, and the cardinality of a set C is similarly denoted by $|C|$. The set of all strings over the alphabet Σ is denoted by Σ^*, and let $\Sigma^i = \{w \in \Sigma^* \mid |w| = i\}$. We assume a recursively enumerable set $calX$ of variables with $\Sigma \cup \mathcal{X} = \emptyset$. The expression a^+ ($a \in \Sigma$) denotes the set $\{a^k \mid k \geq 1\}$, and string a^k is called a *repetition* if $k \geq 2$. Strings x and z are said to be a prefix and suffix of $S = xyz$, respectively. Also, x, y, z are called substrings of S. $S[i]$ and $S[i, j]$ denote the i-th symbol of string S and the substring from $S[i]$ to $S[j]$, respectively. lg stands for $\log_2$. We let $\lg^{(1)} u = \lg u$, $\lg^{(i+1)} u = \lg \lg^{(i)} u$, and $\lg^* u = \min\{i \mid \lg^{(i)} u \leq 1\}$. In practice, we can consider $\lg^* u$ to be constant, since $\lg^* u \leq 5$ for $u \leq 2^{65536}$.

2.1   Grammar-Based Compression

A CFG is a quadruple $G = (\Sigma, V, D, X_s)$ where V is a finite subset of $\mathcal{X}$, D is a finite subset of $V \times (V \cup \Sigma)^*$ of production rules, and $X_s \in V$ represents the start symbol. Variables in V are called nonterminals. We assume a total order over $\Sigma \cup V$. The set of strings in Σ^* derived from X_s by G is denoted by $L(G)$. A CFG G is called *admissible* if for any $X \in \mathcal{X}$ there is exactly one production rule $X \to \gamma \in D$ and $|L(G)| = 1$. An admissible G deriving a text S is called a grammar compression of S. The size of G is the total of the lengths of strings on the right hand sides of all production rules; it is denoted by $|G|$. The problem of grammar compression is formalized as follows:

Definition 1 (Grammar Compression). *Given a string $w \in \Sigma^*$, compute a small, admissible G that derives only w.*

$S(D) \in \Sigma^*$ denotes the string derived by D from a string $S \in (\Sigma \cup V)^*$. For example, when $S = aYY$, $D = \{X \to bc, Y \to Xa\}$ and $\Sigma = \{a, b, c\}$, we obtain $S(D) = abcabca$. $|X|$, also denoted by $|X(D)|$, represents the length of the string derived by D from $X \in V$.

We assume any production rule $X \to \gamma$ satisfies $|\gamma| = 2$ because any grammar compression G can be transformed into G' satisfying $|G'| \leq 2|G|$.

The parse tree of G is represented by a rooted ordered binary tree such that internal nodes are labeled by variables, and the yields, i.e., the sequence of labels of leaves is equal to w. In a parse tree, any internal node $Z \in V$ corresponds to the production rule $Z \rightarrow XY$, and it has a left child labled by X and a right child labeled by Y. The height of a tree is the length of the longest one among paths from the root to leaves.

2.2   Phrase and Reverse Dictionaries

A phrase dictionary is a data structure for directly accessing a digram $X_i X_j$ from a given X_k if $X_k \rightarrow X_i X_j \in D$. It is typically implemented by an array requiring $2n \log n$ bits for storing n production rules. In this paper, D also represents its phrase dictionary. A reverse dictionary $D^{-1} : (\Sigma \cup \mathcal{X})^2 \rightarrow \mathcal{X}$ is a mapping from a given digram to a nonterminal symbol. D^{-1} returns a nonterminal Z associated with a digram XY if $Z \rightarrow XY \in D$; otherwise, it creates a new nonterminal symbol $Z' \notin V$ and returns Z'. For example, if we have a phrase $D = \{X_1 \rightarrow ab, X_2 \rightarrow cd\}$, then $D^{-1}(a, b)$ returns X_1, while $D^{-1}(b, c)$ creates a new nonterminal X_3 and returns it.

2.3   Rank/Select Dictionaries

Our method represents CFGs using a rank/select dictionary, a succinct data structure for a bit string B [11] supporting the following queries: $\text{rank}_c(B, i)$ returns the number of occurrences of $c \in \{0, 1\}$ in $B[0, i]$; $\text{select}_c(B, i)$ returns the position of the i-th occurrence of $c \in \{0, 1\}$ in B; $\text{access}(B, i)$ returns i-th bit in B. Data structures with only the $|B| + o(|B|)$ bit storage to achieve $O(1)$ time rank and select queries [16] have been presented.

GMR [8] is a rank/select dictionary for large alphabets and supports rank/ select/access queries for general alphabet strings $S \in \Sigma^*$. GMR uses $n \log n + o(n \log n)$ bits while computing both rank and access queries in $O(\log \log |\Sigma|)$ times and also computing select queries in $O(1)$ time. Space-efficient versions of GMR are also presented in [1].

3   ESP-index

3.1   Edit-Sensitive Parsing (ESP)

In this section, we review a grammar compression based on ESP [4], which is referred to as *GC-ESP*. The basic idea of GC-ESP is to (i) start from an input string $S \in \Sigma^*$, (ii) replace as many as possible of the same digrams in common substrings by the same variables, and (iii) iterate this process in a bottom-up manner until S is transformed to a single variable.

In each iteration, GC-ESP uniquely divides S into maximal non-overlapping substrings such that $S = S_1 S_2 \cdots S_\ell$ and each S_i is categorized into one of three types: (1) a repetition of a symbol; (2) a substring not including a type1 substring and of length at least $\lg^* |S|$; (3) a substring being neither type1 nor type2 substrings.

At one iteration of parsing S_i, GC-ESP builds two kinds of subtrees from strings XY and XYZ of length two and three, respectively. The first type is a

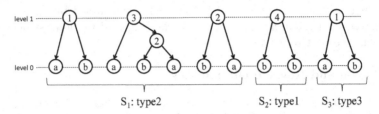

Fig. 1. An example for parsing $S = abababab bab$ by GC-ESP. There are three non-overlapping substrings $S_1 = ababa$ of type2, $S_2 = bb$ of type1, and $S_3 = ab$ of type3. They are individually parsed with a common reverse dictionary. The resulted string is 13241 that is parsed at one higher level.

2-tree corresponding to a production rule in the form of $A \rightarrow XY$. The second type is a 2-2-tree corresponding to production rules in the forms of $A \rightarrow XB$ and $B \rightarrow YZ$.

GC-ESP parses S_i according to its type. In case S_i is a type1 or type3 substring, GC-ESP performs the typical left aligned parsing where 2-trees are built from left to right in S_i and a 2-2-tree is built for the last three symbols if $|S_i|$ is odd, as follows:

- If $|S_i|$ is even, GC-ESP builds $A \rightarrow S_i[2j-1, 2j]$, $j = 1, ..., |S_i|/2$,
- Otherwise, it builds $A \rightarrow S_i[2j-1, 2j]$ for $j = 1, ..., (\lfloor |S_i|/2 \rfloor - 1)$, and builds $A \rightarrow BS_i[2j+1]$ and $B \rightarrow S_i[2j-1, 2j]$ for $j = \lfloor |S_i|/2 \rfloor$.

In case S_i is a type2 substring, GC-ESP further partitions S_i into several substrings such that $S_i = s_1 s_2 ... s_\ell$ ($2 \le |s_j| \le 3$) using *alphabet reduction* [4], which is detailed below. GC-ESP builds $A \rightarrow s_j$ if $|s_j| = 2$ or builds $A \rightarrow s_j[2, 3]$, $B \rightarrow s_j[1]A$ otherwise for $j = 1, ..., \ell$.

GC-ESP transforms S_i to S_i' and parses the concatenated string S_i' ($i = 1, ..., \ell$) at the next level of a parse tree (Figure 1). In addition, GC-ESP gradually builds a phrase dictionary D^k at kth level of a parse tree. The final dictionary D is the union of dictionaries built at each level of a parse tree, i.e., $D = D^1 \cup D^2 \cup ... \cup D^h$.

Alphabet Reduction: Alphabet reduction is a procedure for partitioning a string into substrings of length 2 and 3. Given a type2 substring S, consider $S[i]$ and $S[i-1]$ represented as binary integers. Let p be the position of the least significant bit in which $S[i]$ differs from $S[i-1]$, and let $bit(p, S[i]) \in \{0, 1\}$ be the value of $S[i]$ at the p-th position, where p starts at 0. Then, $L[i] = 2p + bit(p, S[i])$ is defined for any $i \ge 2$. Since S does not contain any repetitions as type2, the resulted string $L = L[2]L[3] ... L[|S|]$ does not also contain repetitions, i.e., L is type2. We note that if the number of different symbols in S is n which is denoted by $[S] = n$, clearly $[L] \le 2 \lg n$. Setting $S := L$, the next label string L is iteratively computed until $[L] \le \lg^* |S|$. At the final L^*, $S[i]$ of the original S is called *landmark* if $L^*[i] > \max\{L^*[i-1], L^*[i+1]\}$.

After deciding all landmarks, if $S[i]$ is a landmark, we replace $S[i-1, i]$ by a variable X and update the current dictionary with $X \rightarrow S[i-1, i]$. After replacing all landmarks, the remaining maximal substrings are replaced by the left aligned parsing.

Because L^* is type2 and $[L^*] \leq \lg^*|S|$, any substring of S longer than $2\lg^*|S|$ must contain at least one landmark. Thus, we have the following characteristic.

Lemma 1. (Cormode and Muthukrishnan [4]) Determining the closest landmark to $S[i]$ depends on only $\lg^*|S| + 5$ contiguous symbols to the left and 5 to the right.

This lemma tells us the following. Let S be type2 string containing α as $S = x\alpha y\alpha z$. Using Lemma 1, when α is sufficiently long (e.g., $|\alpha| \geq 2\lg^*|S|$), there is a partition $\alpha = \alpha_1\alpha_2$ such that $|\alpha_1| = O(\lg^*|S|)$ and whether $\alpha_2[i]$ is landmark or not is coincident in both occurrences of α.

Thus, we can construct a consistent parsing for all occurrences of α_2 in S, which almost covers whole α except a short prefix α_1. Such consistent parsing can be iteratively constructed for α_2 as the next S while it is sufficiently long.

Lemma 2. (Cormode and Muthukrishnan [4]) GC-ESP builds from a string S a parse tree of height $h = O(\lg|S|)$ in $O(|S|\lg^*|S|)$ time.

3.2 Algorithms

We present an algorithm for finding all the occurrences of pattern P in $S \in \Sigma^*$ parsed by ESP. Let T_S be the parsing tree for S by ESP and D be the resulted dictionary for T_S. We consider this problem of embedding a parsing tree T_P of P into T_S as follows. First, we construct T_P preserving the labeling in D and a new production rule is generated if its phrase is undefined. Second, T_P is divided into a sequence of maximal *adjacent* subtrees rooted by nodes $v_1, \ldots, v_k$ such that $yield(v_1 \cdots v_k) = P$, where $yield(v)$ denotes the string represented by the leaves of v and $yield(v_1 \cdots v_k)$ denotes the concatenation of strings $yield(v_1), yield(v_2), \ldots, yield(v_k)$.

If z is the lowest common ancestor of v_1 and v_k, which is denoted by $z = lca(v_1, v_k)$, the sequence $v_1, \ldots, v_k$ is said to be embedded into z, denoted by $(v_1 \cdots v_k) \prec z$. When $yield(v_1 \cdots v_k) = P$, z is called an *occurrence node* of P.

Definition 2. *Let $L(v)$ be the variable of v and let $L(v_1 \cdots v_k)$ be the concatenation. An* evidence *of P is defined as a string $Q \in (\Sigma \cup V)^*$ of length k satisfying the following condition: There is an occurrence node z of P iff there is a sequence $v_1 \cdots v_k$ such that $(v_1 \cdots v_k) \prec z$, $yield(v_1 \cdots v_k) = P$, and $L(v_1 \cdots v_k) = Q$.*

This is well defined because a trivial Q with $Q = P$ always exists. An evidence Q transforms the problem of finding an occurrence of P into that of embedding a shorter string Q into T_S. We present an algorithm for extracting evidences.

Evidence Extraction: The evidence Q of P is iteratively computed from the parsing of P as follows. Let $P = \alpha\beta$ for a maximal prefix α belonging to type1, 2 or 3. For i-th iteration of GC-ESP, α and β of P are transformed into α' and β', respectively. In case α is not type2, define $Q_i = \alpha$ and update $P := \beta'$. In this case, Q_i is an evidence of α and β' is an evidence of β. In case α is type2, define $Q_i = \alpha[1, j]$ with $j = \min\{p \mid p \geq \lg^*|S|, P[p]$ is landmark$\}$ and update $P := x\beta'$ where x is the suffix of α' deriving only $\alpha[j + 1, |\alpha|]$. In this case, by Lemma 1, Q_i is an evidence of $\alpha[1, j]$ and $x\beta'$ is an evidence of $\alpha[j + 1, |\alpha|]\beta$. Repeating this process until $|P| = 1$, we obtain the evidence of P as the concatenation of all Q_i. We obtain the upper bound of length Q.

Lemma 3. (Maruyama et al. [13]) There is an evidence Q of P such that $Q = Q_1 \cdots Q_k$ where $Q_i \in q_i^+$ ($q_i \in \Sigma \cup V$, $q_i \neq q_{i+1}$) and $k = O(\lg |P| \lg^* |S|)$.

Thus, we can obtain the time complexity of the pattern finding problem.

Counting, Locating, and Extracting: Given T_S and an evidence Q of P, a node z in T_S is an occurrence node of P iff there is a sequence $v_1, \ldots, v_k$ such that $(v_1, \ldots, v_k) \prec z$ and $L(v_1 \cdots v_k) = Q$. Thus, it is sufficient to adjacently embed all subtrees of $v_1, \ldots, v_k$ into T_S. We recall the fact that the subtree of v_1 is left adjacent to that of v_2 iff v_2 is a leftmost descendant of $right\_child(lra(v_1))$ where $lra(v)$ denotes the *lowest right ancestor of* v, i.e., v is the lowest ancestor of x such that the path from v to x contains at least one left edge. Because $z = lra(v_1)$ is unique and the height of T_S is $O(\lg |S|)$, we can check whether $(v_1, v_2) \prec z$ in $O(\lg |S|)$ time. Moreover, $(v_1, v_2, v_3) \prec z'$ iff $(z, v_3) \prec z'$ (possibly $z = z'$). Therefore, when $|Q_i| = 1$ for each i, we can execute the embedding of whole Q in $t = O(\lg |P| \lg |S| \lg^* |S|)$ time. For general case of $Q_i \in q_i^+$, the same time complexity t is obtained in Lemma 4.

Lemma 4. (Maruyama et al. [13]) The time complexity of embedding the evidence of P into T_S is $O(\lg |P| \lg |S| \lg^* |S|)$.

Thus, counting P in S is $O(|P| \lg^* |P| + occ_c \cdot t)$ where occ_c is the frequency of the largest embedded subtree that is called *core*. With a auxiliary data structure storing $|X|$, the length of string derived from $X \in V$, locating P can be computed in the same time complexity. Since T_S is balanced, the substring extraction of $S[i, i + m]$ can be computed in $O(m + \lg |S|)$ time.

ESP-index was implemented by LOUDS [5] and permutation [14] with the time-space trade-off parameter $\varepsilon \in (0, 1)$, and it supports queries of counting/locating patterns and extracting of substrings.

Theorem 1. (Maruyama et al. [13]) Let $|S| = u$, $|P| = m$, and $n = |V(G)|$ with the GC-ESP G of S. The time for counting and locating is $O(\frac{1}{\varepsilon}(m \lg n + occ_c \cdot \lg m \lg u) \lg^* u)$ and the time for extracting substring $S[i, i+m]$ is $O(\frac{1}{\varepsilon}(m + \lg u))$ with $(1 + \varepsilon)n \lg n + 4n + n \lg u + o(n)$ bits of space and any $\varepsilon \in (0, 1)$.

4   ESP-index-I

We present ESP-index-I for faster query searches than ESP-index. ESP-index-I encodes CFGs into a succinct representation by leveraging the idea behind GMR [8], a rank/select dictionary for large alphabets.

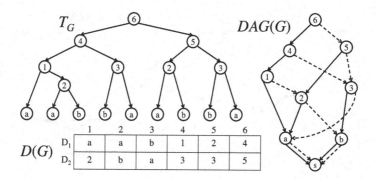

Fig. 2. Grammar compression G and its parsing tree T_G, DAG representation $DAG(G)$, and array representation $D(G)$, where $\Sigma = \{\mathrm{a}, \mathrm{b}\}$ and $V = \{1, 2, 3, 4, 5, 6\}$. In $DAG(G)$, the left edges are shown by solid lines. $D(G)$ itself is an implementation of the phrase dictionary.

DAG Representation: we represent a CFG G as a DAG where $Z \to XY \in P$ is considered as two directed left edge (Z, X) and right edge (Z, Y), i.e., G can be seen as a DAG with a single source and $|\Sigma|$ sinks. By introducing a super-sink s and drawing left and right edges from any sink to s, we can obtain the DAG with a single source/sink equivalent to G. We denote the DAG as $DAG(G)$ (Figure 2). $DAG(G)$ is decomposed into two spanning trees T_L and T_R consisting of the left edges and the right edges, respectively. ESP-index reconstructs G with a permutation $\pi : V(T_L) \to V(T_R)$ from (T_L, T_R, π). Instead, ESP-index-I reconstructs and traverses G by using GMR.

Succinct Encoding of Phrase Dictionary: For a grammar compression G with n variables, the set $D(G)$ of production rules is represented by a phrase dictionary $D[D_1[1, n], D_2[1, n]]$ such that $X_k \to X_i X_j \in D(G)$ iff $D_1[k] = i$ and $D_2[k] = j$. We consider a permutation $\pi : V \to V$ such that $\pi(D_1)$ is monotonic, i.e., $\pi(D_1[i]) \leq \pi(D_1[i + 1])$. Then D is transformed into an equivalent $\pi(D) = [\pi(D_1), \pi(D_2)]$ and let $D := \pi(D)$ (Figure 2). The monotonic sequence D_1 is encoded by the bit vector $B(D_1)$ as follows: $B(D_1) = 0^{D_1[1]} 1 0^{D_1[2]-D_1[1]} 1 \ldots 0^{D_1[n]-D_1[n-1]} 1$. We can get $D_1[k] = \mathrm{select}_1(B(D_1), k) - k$ in $O(1)$ time with $2n + o(n)$ bits of space.

GMR encodes the sequence D_2 into $A(D_2)$ with $n \lg n + o(n \lg n)$ bits of space. We can get $D_2[k] = \mathrm{access}(A(D_2), k)$ in $O(\lg \lg n)$ time. Thus, we can simulate the phrase dictionary D by $(B(D_1), A(D_2))$. The access/rank/select on $B(D_1)$ support to traverse T_L and the same operations on $A(D_2)$ support to traverse T_R. Thus, we can traverse the whole tree T_S equivalent to $DAG(G)$.

Simulation of Reverse Dictionary: ESP-index-I for string S is $(B(D_1), A(D_2))$. After indexing S, since the memory for D^{-1} is released, we must construct GC-ESP of pattern P simulating D^{-1} by $(B(D_1), A(D_2))$ for counting and locating P in S. To remember D^{-1}, the original ESP-index uses the binary search on T_L. On the other hand, we adopt $A(D_2)$ for simulating

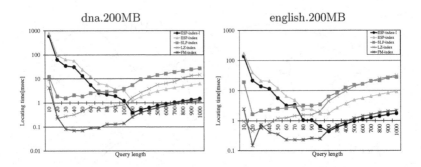

Fig. 3. Locating time of each method in milliseconds for dna.200MB (left) and english.200MB (right)

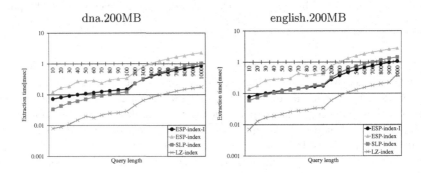

Fig. 4. Substring extraction time of each method in milliseconds for dna.200MB (left) and english.200MB (right)

D^{-1} by an advantage of response time. Indeed, we can improve the time $O(\lg n)$ to $O(\lg \lg n)$ for a query. To get $D^{-1}(X_i X_j) = X_k$, we can get the value of k in the following. let $B = B(D_1)$ and $A = A(D_2)$. (1)$p = \mathrm{select}_0(B, i) - i$ and $q = \mathrm{select}_0(B, i + 1) - (i + 1)$; (2)$r = \mathrm{select}_j(A, \mathrm{rank}_j(A, p) + 1)$; (3)$k = r$ if $r \leq q$, and no $X_k \rightarrow X_i X_j$ exists otherwise.

Since D_1 is monotonic, we restrict the range $k \in [p, q]$ by (1). By (2) and (3), we can check if $X_j \in D_2[p, q]$. If $X_j \in D_2[p, q]$, its position is the required k, and $X_j \notin D_2[p, q]$, there is no production rule of $X_k \rightarrow X_i X_j \in D$. The execution time of (1), (2), and (3) are $O(1)$, $O(\lg \lg n)$, and $O(1)$, respectively.

Theorem 2. *Counting time of ESP-index-I is $O((m + occ_c \lg m \lg u) \lg \lg n \lg^* u)$ with $2n + n \lg n + o(n \lg n)$ bits of space.*

Theorem 3. *With auxiliary $n \lg u + o(n)$ bits of space, ESP-index-I supports locating in the same time complexity and also supports extracting in $O((m + \lg u) \lg \lg n)$ time for any substring of length m.*

Table 2. Memory in megabytes for dna.200MB and english.200MB

	ESP-index-I	ESP-index	SLP-index	LZ-index	FM-index
dna.200MB	156	157	214	208	325
english.200MB	165	162	209	282	482

Table 3. Construction time in seconds for dna.200MB and english.200MB

	ESP-index-I	ESP-index	SLP-index	LZ-index	FM-index
dna.200MB	81.8	82.96	1,906.63	64.869	87.7
english.200MB	93.36	93.58	1,906.63	100.624	94.09

5 Experiments

5.1 Setups

We evaluated ESP-index-I in comparison to ESP-index, SLP-index, LZ-index on one core of an eight-core Intel Xeon CPU E7-8837 (2.67GHz) machine with 1024 GB memory. ESP-index is the previous version of ESP-index-I. SLP-index and LZ-index are state-of-the-arts of grammar-based and LZ-based indexes, respectively. SLP-index used RePair for building SLPs. We used LZ-index-1 as an implementation of LZ-index. We also compared ESP-index-I to FM-index, a self-index for general texts, which is downloadable from https://code.google.com/p/fmindex-plus-plus/. We implemented ESP-index-I in C++ and the other methods are also implemented in C++. We used benchmark texts named dna.200MB of DNA sequences and english.200MB of english texts which are downloadable from http://pizzachili.dcc.uchile.cl/texts.html. We also used four human genomes[1,2,3,4] of 12GB DNA sequences and wikipedia[5] of 7.8GB XML texts as large-scale repetitive texts.

5.2 Results on Benchmark Data

Figure 3 shows the locating time for query texts consisting of lengths from 10 to 1,000 in dna.200MB and english.200MB. Since the locating time of LZ-index depends quadratically on the query length, counting and locating query texts longer than 200 were slow on dna.200MB and english.200MB. SLP-index was also slow for locating query texts longer than 200, since SLP-index performs as many binary searches as query length for finding the first occurrences of variables. ESP-index-I and ESP-index were faster than LZ-index and SLP-index

---

[1] ftp://ftp.ncbi.nih.gov/genomes/H_sapiens/Assembled_chromosomes/
hs_ref_GRC37_chr*.fa.gz

[2] ftp://ftp.ncbi.nih.gov/genomes/H_sapiens/Assembled_chromosomes/
hs_alt_HuRef_chr*.fa.gz

[3] ftp://ftp.kobic.kr/pub/KOBIC-KoreanGenome/fasta/chromosome_*.fa.gz

[4] ftp://public.genomics.org.cn/BGI/yanhuang/fa/chr*.fa.gz

[5] http://dumps.wikimedia.org/enwikinews/20131015/enwikinews-20131015-
pages-meta-history.xml.bz2

Table 4. Counting, locating, compression and indexing times in seconds, and index and position size in megabytes for ESP-index-I on large-scale texts

	genome		wikipedia	
$|P|$	200	1, 000	200	1, 000
Counting time (msec)	1.06	2.29	139.56	13.04
Locating time (msec)	1.10	2.33	167.40	16.69
Compression time (sec)	4, 384		2, 347	
Indexing time (sec)	567		74	
Index size (MB)	3, 888		594	
Position size (MB)	1, 526		246	

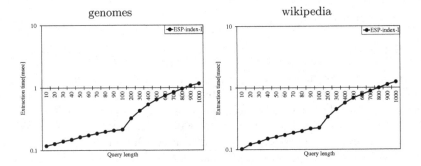

Fig. 5. Substring extraction time of four human genomes of 12GB DNA sequences (left) and wikipedia of 7.8GB english texts (right)

for locating query texts, which showed that top-down searches of ESP-index-I and ESP-index for finding occurrences of variables encoding query texts were effective. ESP-index-I was from 1.4 to 4.3 times faster than ESP-index with respect to locating time, which showed our encoding of CFGs were effective. The locating time of ESP-index-I was compatitive with that of FM-index but ESP-index-I showed a higher compressibility for repetitive texts than FM-index.

Figure 4 shows extraction time of subtexts for fixed positions and lengths. LZ-index based on LZ77 was fastest among all methods. On the otherhand, ESP-index-I was one of the fastest method among grammar-based self-indexes on dna.200MB and english.200MB.

Table 2 shows memory usage of each method in megabytes for dna.200MB and english.200MB. The methods except FM-index archived small memory, which demonstrated their high compressive abilities for texts. The memory usage of ESP-index-I was smallest among all methods, and it used 156MB and 165MB for dna.200MB and english.200MB, respectively. Since FM-index is a self-index for general texts, the memory usage of FM-index was largest among methods. Construction time is presented in Table 3.

5.3   Results on Large-Scale Repetitive Texts

We tried ESP-index-I on large-scale repetitive texts. The other methods except ESP-index-I did not work on these large texts, because they are applicable to only 32 bits inputs. Table 4 shows the results for counting and locating query texts of lengths of 200 and 1,000, construction time and encoding size. Figure 5 shows extraction time of substring of lengths from 10 to 1,000. These results showed an applicability of ESP-index-I to real world repetitive texts.

6 Conclusion

We have presented a practical self-index for highly repetitive texts. Our method is an improvement of ESP-index and performs fast query searches by traversing a parse tree encoded by rank/select dictionaries for large alphabets. Future work is to develop practical retrieval systems on our self-index. This would be beneficial to users for storing and processing large-scale repetitive texts.

Acknowledgments. We are thankful to Miguel A. Martínez-Prieto for providing us with a source code of SLP-index.

References

1. Barbay, J., Navarro, G.: On compressing permutations and adaptive sorting. Theor. Comp. Sci. 513, 109–123 (2013)
2. Claude, F., Navarro, G.: Self-indexed grammar-based compression. Fundam. Inform. 111, 313–337 (2010)
3. Claude, F., Navarro, G.: Improved grammar-based compressed indexes. In: Calderón-Benavides, L., González-Caro, C., Chávez, E., Ziviani, N. (eds.) SPIRE 2012. LNCS, vol. 7608, pp. 180–192. Springer, Heidelberg (2012)
4. Cormode, G., Muthukrishnan, S.: The string edit distance matching problem with moves. TALG 3, 2:1–2:19 (2007)
5. Delpratt, O., Rahman, N., Raman, R.: Engineering the louds succinct tree representation. In: Àlvarez, C., Serna, M. (eds.) WEA 2006. LNCS, vol. 4007, pp. 134–145. Springer, Heidelberg (2006)
6. Gagie, T., Gawrychowski, P., Kärkkäinen, J., Nekrich, Y., Puglisi, S.J.: A faster grammar-based self-index. In: Dediu, A.-H., Martín-Vide, C. (eds.) LATA 2012. LNCS, vol. 7183, pp. 240–251. Springer, Heidelberg (2012)
7. Gagie, T., Gawrychowski, P., Kärkkäinen, J., Nekrich, Y., Puglisi, S.J.: LZ77-based self-indexing with faster pattern matching. In: Pardo, A., Viola, A. (eds.) LATIN 2014. LNCS, vol. 8392, pp. 731–742. Springer, Heidelberg (2014)
8. Golynski, A., Munro, J.I., Rao, S.S.: Rank/select operations on large alphabets: a tool for text indexing. In: SODA, pp. 368–373 (2006)
9. Goto, K., Bannai, H., Inenaga, S., Takeda, M.: Fast q-gram mining on SLP compressed strings. JDA 18, 89–99 (2013)
10. Hermelin, D., Landau, G.M., Landau, S., Weimann, O.: A unified algorithm for accelerating edit-distance computation via text-compression. In: STACS, pp. 529–540 (2009)
11. Jacobson, G.: Space-efficient static trees and graphs. In: FOCS, pp. 549–554 (1989)
12. Larsson, N.J., Moffat, A.: Off-line dictionary-based compression. In: DCC, pp. 296–305 (1999)
13. Maruyama, S., Nakahara, M., Kishiue, N., Sakamoto, H.: ESP-Index: A compressed index based on edit-sensitive parsing. JDA 18, 100–112 (2013)

14. Munro, J.I., Raman, R., Raman, V., Rao, S.S.: Succinct representations of permutations. In: Baeten, J.C.M., Lenstra, J.K., Parrow, J., Woeginger, G.J. (eds.) ICALP 2003. LNCS, vol. 2719, pp. 345–356. Springer, Heidelberg (2003)
15. Navarro, G.: Indexing text using the ziv-lempel trie. JDA 2, 87–114 (2004)
16. Raman, R., Raman, V., Rao, S.S.: Succinct indexable dictionaries with applications to encoding k-ary trees, prefix sums and multisets. TALG 3 (2007)
17. Sakamoto, H., Maruyama, S., Kida, T., Shimozono, S.: A space-saving approximation algorithm for grammar-based compression. IEICE Trans. Inf. Syst. E92-D, 158–165 (2009)
18. Yamamoto, T., Bannai, H., Inenaga, S., Takeda, M.: Faster subsequence and don't-care pattern matching on compressed texts. In: Giancarlo, R., Manzini, G. (eds.) CPM 2011. LNCS, vol. 6661, pp. 309–322. Springer, Heidelberg (2011)

Partitioning Complex Networks
via Size-Constrained Clustering⋆,⋆⋆

Henning Meyerhenke, Peter Sanders, and Christian Schulz

Karlsruhe Institute of Technology (KIT), Karlsruhe, Germany
{meyerhenke,sanders,christian.schulz}@kit.edu

Abstract. The most commonly used method to tackle the graph partitioning problem in practice is the multilevel approach. During a coarsening phase, a multilevel graph partitioning algorithm reduces the graph size by iteratively contracting nodes and edges until the graph is small enough to be partitioned by some other algorithm. A partition of the input graph is then constructed by successively transferring the solution to the next finer graph and applying a local search algorithm to improve the current solution.

In this paper, we describe a novel approach to partition graphs effectively especially if the networks have a highly irregular structure. More precisely, our algorithm provides graph coarsening by iteratively contracting size-constrained clusterings that are computed using a label propagation algorithm. The *same* algorithm that provides the size-constrained clusterings can also be used during uncoarsening as a fast and simple local search algorithm.

Depending on the algorithm's configuration, we are able to compute partitions of very high quality outperforming all competitors, or partitions that are comparable to the best competitor in terms of quality, hMetis, while being nearly an order of magnitude faster on average. The fastest configuration partitions the largest graph available to us with 3.3 billion edges using a single machine in about ten minutes while cutting less than half of the edges than the fastest competitor, kMetis.

1   Introduction

Graph partitioning (GP) is very important for processing very large graphs, e.g. networks stemming from finite element methods, route planning, social networks or web graphs. Often the node set of such graphs needs to be partitioned (or clustered) such that there are few edges between the blocks (node subsets, parts). In particular, when you process a graph in parallel on k PEs (processing elements), you often want to partition the graph into k blocks of (about) equal size. Then each PE owns a roughly equally sized part of the graph. In this paper we focus on a version of the problem that constrains the maximum block size to $(1+\epsilon)$ times the average block size and tries to minimize the total cut size, i.e., the number of edges that run between blocks. Such edges are supposed to model the communication at block boundaries between the corresponding PEs. It

---

⋆ Partially supported by Deutsche Forschungsgemeinschaft DFG SA 933/10-1 and by the Ministry of Science, Research and the Arts Baden-Württemberg.
⋆⋆ This paper is a short version of the technical report [17].

J. Gudmundsson and J. Katajainen (Eds.): SEA 2014, LNCS 8504, pp. 351–363, 2014.
© Springer International Publishing Switzerland 2014

is well-known that there are more realistic (but more complicated) objective functions involving also the block that is worst and the number of its neighboring nodes [10], but the cut size has been the predominant optimization criterion. The GP problem is NP-complete [9] and there is no approximation algorithm with a constant ratio factor for general graphs [4]. Therefore heuristic algorithms are used in practice.

A successful heuristic for partitioning large graphs is the *multilevel graph partitioning* (MGP) approach depicted in Figure 1, where the graph is recursively *contracted* to obtain smaller graphs which should reflect the same basic structure as the input graph. After applying an *initial partitioning* algorithm to the smallest graph, the contraction is undone and, at each level, a *local search* method is used to improve the partitioning induced by the coarser level.

Recently the partitioning of *complex networks*, such as social networks or web graphs, has become a focus of investigation [7]. While partitioning meshes is a mature field, the structure of complex networks poses new challenges. Complex networks are often *scale-free* (many low-degree nodes, few high-degree nodes) and have the small-world property. Small world means that the network has a small diameter, so that the whole graph is discovered within a few hops from any source node. These two properties distinguish complex networks from traditional meshes and make finding small cuts difficult. Yet, to cope with massive network data sets in reasonable time, there is a need for parallel algorithms. Their efficient execution requires good graph partitions.

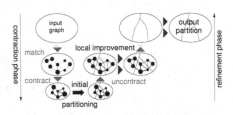

Fig. 1. Multilevel graph partitioning

The paper is organized as follows. We begin in Section 2 by introducing basic concepts and by summarizing related work. The main parts of the paper are Sections 3 and 4. The former introduces our rationale and the resulting size-constrained label propagation algorithm. We employ this algorithm during *both* coarsening *and* uncoarsening. During coarsening we iteratively compute a size-constrained graph clustering and contract it, and during uncoarsening the algorithm can be used as a fast local search algorithm. Section 4 augments the basic algorithm by several algorithmic components. The presented algorithms speed up computations and improve solution quality, in particular on graphs that have a irregular structure such as social networks or web graphs. Experiments in Section 5 indicate that our algorithms are able provide excellent partitioning quality in a short amount of time. For example, a web graph with 3.3 billion edges can be partitioned on a single machine in about ten minutes while cutting less than half of the edges than the partitions computed by kMetis. Finally, we conclude with Section 6.

2   Preliminaries

2.1   Basic Concepts

Consider an undirected graph $G = (V = \{0, \ldots, n-1\}, E, c, \omega)$ with edge weights $\omega : E \to \mathbb{R}_{>0}$, node weights $c : V \to \mathbb{R}_{\geq 0}$, $n = |V|$, and $m = |E|$. We extend c and ω to sets, i.e. $c(V') := \sum_{v \in V'} c(v)$ and $\omega(E') := \sum_{e \in E'} \omega(e)$. $\Gamma(v) := \{u : \{v, u\} \in E\}$

denotes the neighbors of v. We are looking for *blocks* of nodes $V_1,\ldots,V_k$ that partition V, i.e., $V_1 \cup \cdots \cup V_k = V$ and $V_i \cap V_j = \emptyset$ for $i \neq j$. The *balancing constraint* demands that $\forall i \in \{1..k\} : c(V_i) \leq L_{\max} := (1+\epsilon)c(V)/k + \max_{v \in V} c(v)$ for some parameter ϵ. The last term in this equation arises because each node is atomic and therefore a deviation of the heaviest node has to be allowed. Note that for unweighted graphs the balance constraint becomes $\forall i \in \{1..k\} : |V_i| \leq (1 + \epsilon)\lceil \frac{|V|}{k} \rceil$. The objective is to minimize the total *cut* $\sum_{i<j} w(E_{ij})$ where $E_{ij} := \{\{u,v\} \in E : u \in V_i, v \in V_j\}$. We say that a block V_i is *underloaded* if $|V_i| < L_{\max}$ and *overloaded* if $|V_i| > L_{\max}$. A clustering is also a partition, however, k is usually not given in advance and the balance constraint is removed. A size-constrained clustering constrains the size of the blocks of a clustering by a given upper bound U such that $c(V_i) \leq U$. Note that by adjusting the upper bound one can somehow control the number of blocks of a feasible clustering. For example, when using $U = 1$, the only feasible size-constrained clustering in an unweighted graphs is the clustering where each node forms a block of its own. A node $v \in V_i$ that has a neighbor $w \in V_j, i \neq j$, is a boundary node. A *matching* $M \subseteq E$ is a set of edges that do not share any common nodes, i.e. the graph (V, M) has maximum degree one. By default, our initial inputs will have unit edge and node weights. However, even those will be translated into weighted problems in the course of the multilevel algorithm. In order to avoid tedious notation, G will denote the current state of the graph before and after an (un)contraction in the multilevel scheme throughout this paper.

2.2 Related Work

There has been a *huge* amount of research on GP so that we refer the reader to [3,5] for most of the material. Here, we focus on issues closely related to our main contributions. All general-purpose methods that work well on large real-world graphs are based on the multilevel principle. The basic idea can be traced back to multigrid solvers for systems of linear equations. Recent practical methods are mostly based on graph theoretic aspects, in particular edge contraction and local search. There are different ways to create graph hierarchies such as matching-based schemes [30,12,8,20] or variations thereof [1] and techniques similar to algebraic multigrid [16,6,22]. Well-known MGP software packages include Jostle [30], Metis [12], and Scotch [20].

Graph clustering with the label propagation algorithm (LPA) has originally been described by Raghavan et al. [21]. In addition to its use for fast graph clustering, LPA has been used to partition networks, e.g. by Uganer and Backstrom [28]. The authors use partitions obtained by geographic initializations and improve the partition by combining LPA with linear programming.

KaHIP. KaHIP – Karlsruhe High Quality Partitioning – is a family of GP programs that tackle the balanced GP problem [23,25]. It includes KaFFPa (Karlsruhe Fast Flow Partitioner), which is a matching-based multilevel graph partitioning framework that uses for example flow-based methods and more-localized local searches to compute high quality partitions. We integrate our new techniques described in this paper into this multilevel algorithm. KaHIP also includes KaFFPaE (KaFFPa Evolutionary), which is a parallel evolutionary algorithm and KaBaPE (Karlsruhe Balanced Partitioner Evolutionary), which extends the evolutionary algorithm.

3   Cluster Contraction

We are now ready to explain the basic idea of our new approach for creating graph hierarchies, which is targeted at complex network such as social networks and web graphs. We start by introducing the size-constrained label propagation algorithm, which is used to compute clusterings of the graph. To compute a graph hierarchy, the clustering is contracted by replacing each cluster with a single node, and the process is repeated recursively until the graph is small. The hierarchy is then used by our partitioner. KaHIP uses its own initial partitioning and local search algorithms to partition the coarsest graph and to perform local improvement on each level, respectively. Due to the way the contraction is defined, it is ensured that a partition of a coarse graph corresponds to a partition of the input network with the same objective and balance. Note that cluster contraction is an aggressive coarsening strategy. In contrast to previous approaches, it enables us to drastically shrink the size of irregular networks. The intuition behind this technique is that a clustering of the graph (one hopes) contains many edges running inside the clusters and only a few edges running between clusters, which is favorable for the edge cut objective. Regarding complexity our experiments in Section 5 indicate that the number edges per node of the contracted graphs is smaller than the number of edges per node of the input network, and the clustering algorithm we use is fast.

3.1   Label Propagation with Size Constraints

The *label propagation algorithm* (LPA) was proposed by Raghavan et al. [21] for graph clustering. It is a fast, near-linear time algorithm that locally optimizes the number of edges cut. We outline the algorithm briefly. Initially, each node is in its own cluster/block, i.e. the initial block ID of a node is set to its node ID. The algorithm then works in rounds. In each round, the nodes of the graph are traversed in a random order. When a node v is visited, it is *moved* to the block that has the strongest connection

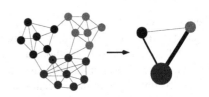

Fig. 2. Contraction of a clustering. The clustering of the graph on the left hand side is indicated by the colors. Each cluster of the graph on the left hand side corresponds to a node in the graph on the right hand side.

to v, i.e. it is moved to the cluster V_i that maximizes $\omega(\{(v, u) \mid u \in N(v) \cap V_i\})$. Ties are broken randomly. The process is repeated until the process has converged. Here, we perform at most ℓ iterations of the algorithm, where ℓ is a tuning parameter, and stop the algorithm if less then five percent of the nodes changed its cluster during one round. One LPA round can be implemented to run in $\mathcal{O}(n + m)$ time.

After we have computed a clustering, we *contract it* to obtain a coarser graph. Contracting the clustering works as follows: each block of the clustering is contracted into a single node. The weight of the node is set to the sum of the weight of all nodes in the original block. There is an edge between two nodes u and v in the contracted graph if the two corresponding blocks in the clustering are adjacent to each other in G, i.e. block u and block v are connected by at least one edge. The weight of an edge (A, B) is set to the sum of the weight of edges that run between block A and block B of the clustering. Note that due to the way the contraction is defined, a partition of the coarse graph corresponds to a partition of the finer graph with the same cut and balance. An example is shown in Figure 2.

In contrast to the original LPA [21], we have to ensure that each block of the cluster fulfills a size constraint. There are two reason for this. First, consider a clustering of the graph in which the weight of a block would exceed $(1 + \epsilon)\lceil \frac{|V|}{k} \rceil$. After contracting this clustering, it would be impossible to find a partition of the contracted graph that fulfills the balance constraint. Secondly, it has been shown that using more balanced graph hierarchies is beneficial when computing high quality graph partitions [11]. To ensure that blocks of the clustering do not become too large, we introduce an upper bound $U := \max(\max_v c(v), W)$ for the size of the blocks. Here, W is a parameter that will be chosen later. When the algorithm starts to compute a graph clustering on the input graph, the constraint is fulfilled since each of the blocks contains exactly one node. A neighboring block V_ℓ of a node v is called *eligible* if V_ℓ will not become overloaded once v is moved to V_ℓ. Now when we visit a node v, we move it to the *eligible block* that has the strongest connection to v. Hence, after moving a node, the size of each block is still smaller than or equal to U. Moreover, after contracting the clustering, the weight of each node is smaller or equal to U. One round of the modified version of the algorithm can still run in linear time by using an array of size $|V|$ to store the block sizes. We set the parameter W to $\frac{L_{\max}}{f}$, where f is a tuning parameter.

We repeat the process of computing a size-constrained clustering and contracting it, recursively. As soon as the graph is small enough, i.e. the number of remaining nodes is smaller than $\max(60k, n/(60k))$, it is initially partitioned by the initial partitioning algorithms provided in KaHIP. That means each node of the coarsest graph is assigned to a block. KaHIP uses a multilevel recursive bisection algorithm to create an initial partitioning [26]. Afterwards, the solution is transferred to the next finer level. We assign a node of the finer graph to the block of its coarse representative. Local improvement methods of KaHIP then try to improve the solution on the current level.

By using a different size-constraint – the constraint $W := L_{\max}$ of the original partitioning problem – the LPA can also be used as a simple and fast local search algorithm to improve a solution on the current level. However, one has to make small modifications to handle overloaded blocks. We modify the block selection rule when we use the algorithm as local search algorithm in case that the current node v under consideration is from an overloaded block V_ℓ. In this case it is *moved* to the eligible block that has the strongest connection to v without considering the block V_ℓ it is contained in. This way it is ensured that the move improves the balance of the partition (at the cost of the number of edges cut). Experiments in Section 5 show that the algorithm is a fast alternative to the local search algorithms provided by KaFFPa. Moreover, let us emphasize that the algorithm has a large potential to be efficiently parallelized.

4   Algorithmic Extensions

In this section we present numerous algorithmic extensions to the approach presented above. This includes using different orderings for size-constrained LPA, combining multiple clusterings into one clustering, advanced multilevel schemes, allowing additional amounts of imbalance on coarse levels of the multilevel hierarchy and a method to improve the speed of the algorithm.

Node Ordering for Label Propagation. The LPA traverses the nodes in a random order and moves a node to a cluster with the strongest connection in its neighborhood to compute a clustering. Instead of using a random order, one can use the ordering induced by the node degree (increasing). That means that in the first round of the label propagation algorithm, nodes with small node degree can change their cluster before nodes with a large node degree. Intuitively, this ensures that there is already a meaningful cluster structure when the LPA chooses the cluster of a high degree node. Hence, the algorithm is likely to compute better clusterings of the graph by using node orderings based on node degree. We also tried other node orderings such as weighted node degree. The overall solution quality and running time are comparable so that we omit more sophisticated orderings here.

Ensemble Clusterings. In machine learning, ensemble methods combine multiple weak classification (or clustering) algorithms to obtain a strong algorithm for classification (or clustering). Such an ensemble approach has been successfully applied to graph clustering by combining several base clusterings from different LPA runs. These base clusterings are used to decide whether pairs of nodes should belong to the same cluster [19,27]. We follow the idea to get *better* clusterings for the coarsening phase of our multilevel algorithm.

Given a number of clusterings, the *overlay clustering* is a clustering in which two nodes belong to the same cluster if and only if they belong to the same cluster in each of the input clusterings. Intuitively, if all of the input clusters agree that two nodes belong to the same block, then they are put into the same block in the overlay clustering. On the other hand, if there is one input clustering that puts the nodes into different blocks, then they are put into different blocks in the overlay clustering. More formally, given clusterings $\{C_1, \ldots, C_\ell\}$, we define the overlay clustering as the clustering where each block corresponds to a connected component of the graph $G_\mathcal{E} = (V, E \backslash \mathcal{E})$, where $\mathcal{E}$ is the union of the cut edges of each of the clusterings C_i, i.e. all edges that run between blocks in the clusterings C_i. Related definitions are possible, e.g. a cluster does not have to be a connected component. In our ensemble approach we use the clusterings obtained by size-constrained LPA as input to compute the overlay clustering. It is easy to see that the number of clusters in the overlay clustering cannot decrease compared to the number of clusters in each of the input clusterings. Moreover, the overlay clustering is feasible w.r.t. to the size constraint if each of the input clusterings is feasible.

Given ℓ clusterings $\{C_1, \ldots, C_\ell\}$, we use the following approach to *compute* the overlay clustering iteratively. Initially, the overlay clustering O is set to the clustering C_1. We then iterate through the remaining clusterings and incrementally update the current solution O. This is done by computing the overlay $\mathcal{O}$ with the current clustering C under consideration. More precisely, we use pairs of cluster IDs (i, j) as a key in a hash map $\mathcal{H}$, where i is a cluster ID of $\mathcal{O}$ and j is a cluster ID of the current clustering C. We then iterate through the nodes and initialize a counter c to zero. Let v be the current node. If the pair $(\mathcal{O}[v], C[v])$ is not contained in $\mathcal{H}$, we set $\mathcal{H}(\mathcal{O}[v], C[v])$ to c and increment c by one. Afterwards, we update the cluster ID of v in $\mathcal{O}$ to $\mathcal{H}(\mathcal{O}[v], C[v])$. Note that at the end of the algorithm, c is equal to the number of clusters contained in the overlay clustering. Moreover, it is possible to compute the overlay clustering directly by hashing ℓ-tuples [27]. However, we choose the simpler approach here since the computation of a clustering itself already takes near-linear time.

Iterated Multilevel Algorithms. A common approach to obtain high quality partitions is to use a multilevel algorithm multiple times using different random seeds and use the best partition that has been found. However, one can do better by transferring the solution of the previous multilevel iteration down the hierarchy. In the GP context, the notion of V-cycles has been introduced by Walshaw [29] and later has been augmented to more complex cycles [24]. These previous works use matching-based coarsening and cut edges are not eligible to be matched (and hence are not contracted). Thus, a given partition on the finest level can be used as initial partition of the coarsest graph (having the same balance and cut as the partition of the finest graph). For simplicity, we focus on iterated V-cycles. We *adopt this technique* also for our new coarsening scheme by ensuring that cut edges are not contracted after the first multilevel iteration. We present more details in the TR [17].

Allowing Larger Imbalances on Coarse Levels. It is well-known that temporarily allowing larger imbalance is useful to create good partitions [30,25]. Allowing an additional amount of imbalance $\hat{\epsilon}$ means that the balance constraint is relaxed to $(1 + \epsilon + \hat{\epsilon})\lceil\frac{|V|}{k}\rceil$, where ϵ is the original imbalance parameter and $\hat{\epsilon}$ is a parameter that has to be set appropriately. We adopt a simplified approach in this context and decrease the amount of additional allowed imbalance level-wise. In other words the largest amount of additional imbalance is allowed on the coarsest level and it is decreased level-wise until no additional amount of imbalance is allowed on the finest level. To be more precise, let the levels of the hierarchy be numbered in increasing order $G_1, \ldots, G_q$ where G_1 is the input graph G and G_q is the coarsest graph. The amount of allowed imbalance on a coarse level ℓ is set to $\hat{\epsilon}_\ell = \delta/(q - \ell + 1)$, where δ is a tuning parameter. No additional amount of imbalance is allowed on the finest level. Moreover, we only allow a larger amount of imbalance during the first V-cycle.

Active Nodes. The LPA looks at every node in each round of the algorithm. Assume for now that LPA is run without a size-constraint. After the first round of the algorithm, a node can only change its cluster if one or more of its neighbors changed its cluster in the previous round (for the sake of the argument we assume that ties are broken in exactly the same way as in the previous round). The active nodes approach keeps track of nodes that can potentially change their cluster. A node is called *active* if at least one of its neighbors changed its cluster in the *previous round*. In the first round all nodes are active. The original LPA is then modified so that only active nodes are considered for movement. This algorithm is always used when the label propagation algorithm is used as a local search algorithm during uncoarsening. A round of the modified algorithm can be implemented with running time linear in the amount of edges incident to the number of active nodes (for more details we refer the reader to the TR [17]).

5 Experiments

Methodology. We have implemented the algorithm described above using C++. We compiled it using g++ 4.8.2. The multilevel partitioning framework KaFFPa has

Table 1. Basic properties of the graphs test set

graph	n	m	Ref.	graph	n	m	Ref.
Large Graphs							
p2p-Gnutella04	6 405	29 215	[15]	citationCiteseer	268 495	≈1.2M	[2]
wordassociation-2011	10 617	63 788	[14]	coAuthorsDBLP	299 067	977 676	[2]
PGPgiantcompo	10 680	24 316	[15]	cnr-2000	325 557	≈2.7M	[2]
email-EuAll	16 805	60 260	[15]	web-Google	356 648	≈2.1M	[15]
as-22july06	22 963	48 436	[2]	coPapersCiteseer	434 102	≈16.0M	[2]
soc-Slashdot0902	28 550	379 445	[15]	coPapersDBLP	540 486	≈15.2M	[2]
loc-brightkite	56 739	212 945	[15]	as-skitter	554 930	≈5.8M	[15]
enron	69 244	254 449	[14]	amazon-2008	735 323	≈3.5M	[14]
loc-gowalla	196 591	950 327	[15]	eu-2005	862 664	≈16.1M	[2]
coAuthorsCiteseer	227 320	814 134	[2]	in-2004	≈1.3M	≈13.6M	[2]
wiki-Talk	232 314	≈1.5M	[15]				
Huge Graphs							
uk-2002	≈18.5M	≈262M	[14]	sk-2005	≈50.6M	≈1.8G	[14]
arabic-2005	≈22.7M	≈553M	[14]	uk-2007	≈106M	≈3.3G	[14]

different configurations. In this work, we look at the Strong and the Eco configuration of KaFFPa. The aim of KaFFPaEco is to be fairly fast and to compute partitions of high quality, whereas KaFFPaStrong targets very high solution quality. Unless otherwise mentioned, we perform ten repetitions for each configuration of the algorithm and report the arithmetic average of computed cut size, running time and the best cut found. When further averaging over multiple instances, we use the geometric mean in order to give every instance a comparable influence on the final score. For the number of partitions k, we choose the values used in [30]: 2, 4, 8, 16, 32, 64. Our default value for the allowed imbalance is 3% since this is one of the values used in [30] and the default value in kMetis. We performed a large amount of experiments to tune the algorithm's parameters. Their description is omitted due to space constraints. We use the following parameters of the algorithm which turned out to work well: the number of maximum label propagation iterations ℓ during coarsening and uncoarsening is set to 10, the factor f of the cluster size-constraint is set to 18, the number of V-cycles is set to three and the number of ensemble clusterings used to compute an ensemble clustering is set to 18 if k is smaller than 16, to 7 if k is 16 or 32 and to 3 if k is larger than 32.

Instances. We evaluate our algorithms on twenty-five graphs that have been collected from [2,15,14]. This includes a number of citation and social networks as well as web graphs. Table 1 summarizes the basic properties of these graphs. We use the large graph set to evaluate the performance of different algorithms in Section 5.1 and compare the performance of the fastest algorithms on the huge graphs in Section 5.2.

System. We use two machines for our experiments: *Machine A* is used for our experimental evaluation in Section 5.1. It is equipped with two Intel Xeon E5-2670 Octa-Core processors (Sandy Bridge) which run at a clock speed of 2.6 GHz. The machine has 64 GB main memory, 20 MB L3-Cache and 8x256 KB L2-Cache. *Machine B* is used for the experiments on the huge networks in Section 5.2. It is equipped with four Intel Xeon E5-4640 Octa-Core processors (Sandy Bridge) running at a clock speed of 2.4 GHz. The machine has 1 TB main memory, 20 MB L3-Cache and 8x256 KB L2-Cache.

5.1   Main Results and Comparison to Other Partitioning Packages

In this section we carefully compare our algorithms against other frequently used publicly available tools. We compare the average and minimum edge cut values produced by all of these tools on the large graphs from Table 1, as well as their average running time on these graphs. Experiments have been performed on machine A. For the comparison we used the k-way variant of hMetis 2.0 (p1) [13], kMetis 5.1 [12] and Scotch 6.0.0 [20] employing the quality option. In contrast to our algorithm, hMetis and Scotch often produce imbalanced partitions. Hence, these tools have a slight advantage in the following comparisons because we do not disqualify imbalanced solutions. In case of hMetis the partitions are imbalanced in 105 out of 1260 cases (up to 12% imbalance) and in case of Scotch the partitions are imbalanced in 218 out of 1260 cases (up to 226% imbalance). Note that the latest version of kMetis (5.1) improved the balancing on social networks by integrating a 2-hop matching algorithm that can match nodes if they share neighbors.

Table 2. Average cut, best cut and running time results for diverse algorithms on the large graphs. Configuration abbreviations: V-cycles (V), add. balance on coarse levels (B), ensemble clusterings (E), active nodes during coarsening (A), random node ordering (R).

Algorithm	avg. cut	best cut	t [s]
CEcoR	71 814.00	67 576.20	10.2
CEco	67 222.30	64 362.71	8.6
CEcoV	66 054.66	63 243.32	14.3
CEcoV/B	64 584.70	61 272.41	15.5
CEcoV/B/E	64 724.55	61 458.39	46.6
CEcoV/B/E/A	65 060.78	61 762.06	41.9
CFastR	74 414.33	69 877.59	4.7
CFast	68 839.33	65 909.11	3.9
CFastV	67 587.23	64 713.95	5.7
CFastV/B	70 514.36	66 783.70	5.8
CFastV/B/E	68 977.84	65 542.55	28.4
CFastV/B/E/A	68 942.89	65 616.23	24.4
UFast	69 169.78	65 965.31	1.5
UFastV	67 836.57	64 877.33	3.0
UEcoV/B	65 212.06	61 738.90	11.5
CStrong	60 178.71	58 440.74	422.1
UStrong	59 935.70	58 199.17	296.4
KaFFPaEco	85 920.07	80 577.80	36.2
KaFFPaStrong	63 141.24	60 623.96	640.8
Scotch	104 954.86	97 596.38	10.6
kMetis	71 977.70	68 434.58	0.4
hMetis	65 409.81	63 493.79	107.4

In addition to the default configurations KaFFPaEco and KaFFPaStrong, we have six base configurations, CEco, CFast, CStrong and UEco, UFast, UStrong. All of these configurations use the new clustering based coarsening scheme with the degree based node ordering. The configurations having an Eco in their name use the refinement techniques as used in KaFFPaEco, and the configurations having the word Fast in their name use the label propagation algorithm as local search algorithm instead. KaFFPa implements multilevel recursive bipartitioning as initial partitioning algorithm. The configurations starting with a C use the matching-based approach during initial partitioning and the configurations starting with a U use the clustering based coarsening scheme also during initial partitioning. We add additional letters to the base configuration name for each additional algorithmic component that is used. For example, CEcoV/B/E/A is based on CEco and uses V-cycles (V), additional imbalance on coarse levels (B), ensemble clusterings (E), and the active node approach (A). Moreover, if we use random node ordering (R) instead of degree based node ordering for the LP algorithm, we add the letter R to the base configuration. CStrong uses additional balance on coarse levels and ensemble clusterings for coarsening. It uses the refinement techniques of KaFFPaStrong. UStrong is the same as CStrong but uses the cluster based partitioning approach for initial partitioning.

Table 2 summarizes the results of our experiments. First of all, we observe large running time *and* quality improvements when switching from the matching-based coarsening scheme in KaFFPaEco to the new basic label propagation coarsening scheme (CEcoR vs. KaFFPaEco). In this case, running time is improved by a factor 3.5 and solution quality by roughly 20%. Experiments indicate that already the results obtained by the initial partitioning algorithm on the coarsest graphs are much better than before. Additionally enabling the node ordering heuristics yields an extra 8% improvement in solution quality and 20% improvement in running time (CEcoR vs CEco and CFastR vs. CFast). This is due to the fact that the node ordering heuristic improves the results of the size-constrained LPA by computing clusterings that have less edges between the clusters. As a consequence the contracted graphs have a smaller amount of total edge weight, which in turn yields better graph hierarchies for partitioning. Performing additional V-cycles and allowing additional imbalance on coarse levels improves solution quality but also increases running time (CEco vs. CEcoV vs. CEcoV/B). However, allowing additional imbalance does worsen solution quality if the size-constrained LPA is used as a refinement algorithm (CFastV vs. CFastV/B). This is caused by the poor ability of label propagation to balance imbalanced solutions. Using ensemble clusterings can enhance solution quality (CFastV/B vs. CFastV/B/E), but do not have to (CEcoV/B vs. CEcoV/B/E). Due to the size-constraints, the clusterings computed by the size-constrained LPA that uses the active nodes approach have more edges between the clusters than the LPA that does not use this technique. Hence, the active nodes approach improves running time but also degrades solution quality. Additional speedups are obtained when the label propagation coarsening is also used during initial partitioning, but sometimes solution quality is worsened slightly. For example, we achieve a 2.7 fold speedup when switching from CFast to UFast, and switching from CStrong to UStrong improves solution quality slightly while improving running time by 42%.

We now compare our algorithms against other partitioning packages. On average, the configuration UEcoV/B yields comparable quality to hMetis while being an order of magnitude faster. When repeating this configuration ten times and taking the best cut, we get an algorithm that has comparable running time to hMetis and improves quality by 6%. Our best configuration UStrong, cuts 9% less edges than hMetis and 20% less edges than kMetis. In this case, hMetis is a factor 3 faster than UStrong. However, when taking the best cut out of ten repetitions of hMetis and comparing it against the average results of UStrong, we still obtain 6% improvement. Overall, Scotch produces the worst partitioning quality among the competitors. It cuts 75% more edges than our best configuration UStrong. kMetis is about a factor 3.5 faster than our fastest configuration UFast, but also cuts more edges than this configuration.

5.2   Huge Web Graphs

In this section our experiments focus on the huge networks from Table 1 (which have up to 3.3 billion undirected edges). To save running time, we focus on the two fast configurations UFast and UFastV, and fix the number of blocks to $k = 16$.

Table 3. Avg. perf. on huge networks for $k = 16$

graph	arabic-2005			uk-2002		
algorithm	avg. cut	best cut	t [s]	avg. cut	best cut	avg. t [s]
UFast	1.91M	1.87M	111.2	1.47M	1.43M	71.7
UFastV	1.85M	1.79M	334.3	1.43M	1.39M	215.9
kMetis	3.58M	3.5M	99.6	2.46M	2.41M	63.7
	sk-2005			uk-2007		
UFast	23.01M	20.34M	387.1	4.34M	4.10M	626.5
UFastV	19.82M	18.18M	1166.4	4.19M	3.99M	1756.4
kMetis	19.43M	18.56M	405.3	11.44M	10.86M	827.6

We speed up UFast and UFastV even more, by only performing three label propagation iterations (instead of ten) during coarsening. Moreover, we also run kMetis and Scotch on these graphs and did not run hMetis due to the large running times on the set of large graphs. Scotch crashes when trying to partition sk-2005 and uk-2007-05, and did not finish after 24 hours of computations on the other two graphs. Table 3 summarizes the results (for detailed results see TR [17]). In three out of four cases our algorithms outperform kMetis in the number of edges cut by large factors while having a comparable running time. Moreover, in every instance the best cut produced by the algorithm UFastV is better than the best cut produced by kMetis. On average, we obtain 74% improvement over kMetis. The largest improvements are obtained obtained on the web graph uk-2007. Here, UFast cuts a factor 2.6 less edges than kMetis and is about 30% faster. The coarse graph after first contraction has already two orders of magnitude less nodes than the input graph and roughly three orders of magnitude less edges.

Note that related work employing label propagation techniques in ensemble graph clustering needs a few hours on a 50 nodes Hadoop cluster to cluster the graph uk-2007 [18], whereas for partitioning we currently need only one single machine for about ten minutes. Interestingly, on all graphs (except sk-2005) already the initial partitioning is much better than the final result of kMetis. For example, on average the initial partition of uk-2007 cuts 4.8 million edges, which improves by a factor of 2.4 on kMetis.

6   Conclusion and Future Work

Current state-of-the-art multilevel graph partitioners have difficulties when partitioning massive complex networks, at least partially due to ineffective coarsening. Thus, guided by techniques used in complex network clustering, we have devised a new scheme for coarsening based on the contraction of clusters derived from size-constrained label propagation. Additional algorithmic adaptations, also for local improvement, further improve running time and/or solution quality. The different configurations of our algorithm give users a gradual choice between the best quality currently available and very high speed. The quality of the best competitor, hMetis, is already reached with an order of magnitude less running time. The strengths of our techniques particularly unfold for huge web graphs, where we significantly improve on kMetis in terms of solution quality (up to $2.6\times$) with a mostly comparable running time.

Partitioning is a key prerequisite for efficient large-scale parallel graph algorithms. As huge networks become abundant, there is a need for their parallel analysis. Thus, as part of future work, we want to exploit the high degree of parallelism exhibited by label propagation and implement a scalable partitioner for distributed-memory parallelism.

References

1. Abou-Rjeili, A., Karypis, G.: Multilevel Algorithms for Partitioning Power-Law Graphs. In: Proc. of 20th Int. Parallel and Distributed Processing Symp. (2006)
2. Bader, D.A., Meyerhenke, H., Sanders, P., Schulz, C., Kappes, A., Wagner, D.: Benchmarking for Graph Clustering and Partitioning. In: Encyclopedia of Social Network Analysis and Mining (to appear)

3. Bichot, C., Siarry, P. (eds.): Graph Partitioning. Wiley (2011)
4. Bui, T.N., Jones, C.: Finding Good Approximate Vertex and Edge Partitions is NP-Hard. Information Processing Letters 42(3), 153–159 (1992)
5. Buluç, A., Meyerhenke, H., Safro, I., Sanders, P., Schulz, C.: Recent Advances in Graph Partitioning. In: Algorithm Engineering – Selected Topics, ArXiv:1311.3144 (to appear, 2014)
6. Chevalier, C., Safro, I.: Comparison of Coarsening Schemes for Multilevel Graph Partitioning. In: Stützle, T. (ed.) LION 3. LNCS, vol. 5851, pp. 191–205. Springer, Heidelberg (2009)
7. Costa, L.F., Oliveira Jr., O.N., Travieso, G., Rodrigues, F.A., Boas, P.R.V., Antiqueira, L., Viana, M.P., Rocha, L.E.C.: Analyzing and Modeling Real-World Phenomena with Complex Networks: A Survey of Applications. Adv. in Physics 60(3), 329–412 (2011)
8. Diekmann, R., Preis, R., Schlimbach, F., Walshaw, C.: Shape-optimized Mesh Partitioning and Load Balancing for Parallel Adaptive FEM. Par. Computing 26(12), 1555–1581 (2000)
9. Garey, M.R., Johnson, D.S., Stockmeyer, L.: Some Simplified NP-Complete Problems. In: Proc. of the 6th ACM Symp. on Theory of Computing, STOC 1974, pp. 47–63. ACM (1974)
10. Hendrickson, B., Kolda, T.G.: Graph Partitioning Models for Parallel Computing. Parallel Computing 26(12), 1519–1534 (2000)
11. Holtgrewe, M., Sanders, P., Schulz, C.: Engineering a Scalable High Quality Graph Partitioner. In: Proc. of the 24th Int. Parallal and Distributed Processing Symp., pp. 1–12 (2010)
12. Karypis, G., Kumar, V.: A Fast and High Quality Multilevel Scheme for Partitioning Irregular Graphs. SIAM J. on Scientific Computing 20(1), 359–392 (1998)
13. Karypis, G., Kumar, V.: Multilevel k-Way Hypergraph Partitioning. In: Proc. of the 36th ACM/IEEE Design Automation Conference, pp. 343–348. ACM (1999)
14. University of Milano Laboratory of Web Algorithms. Datasets, http://law.dsi.unimi.it/datasets.php
15. Leskovec, J.: Stanford Network Analysis Package (SNAP), http://snap.stanford.edu/index.html
16. Meyerhenke, H., Monien, B., Schamberger, S.: Accelerating Shape Optimizing Load Balancing for Parallel FEM Simulations by Algebraic Multigrid. In: Proc. of 20th Int. Parallel and Distributed Processing Symp. (2006)
17. Meyerhenke, H., Sanders, P., Schulz, C.: Partitioning Complex Networks via Size-Constrained Clustering. Technical Report arxiv:1402.3281 (2014)
18. Ovelgönne, M.: Distributed Community Detection in Web-Scale Networks. In: 2013 Int. Conf. on Advances in Social Networks Analysis and Mining, pp. 66–73 (2013)
19. Ovelgönne, M., Geyer-Schulz, A.: An Ensemble Learning Strategy for Graph Clustering. In: Graph Partitioning and Graph Clustering. Contemporary Mathematics. AMS and DIMACS, vol. (588) (2013)
20. Pellegrinim, F.: Scotch Home Page, http://wwwlabri.fr/pelegrin/scotch.
21. Raghavan, U.N., Albert, R., Kumara, S.: Near Linear Time Algorithm to Detect Community Structures in Large-Scale Networks. Physical Review E 76(3) (2007)
22. Safro, I., Sanders, P., Schulz, C.: Advanced Coarsening Schemes for Graph Partitioning. In: Klasing, R. (ed.) SEA 2012. LNCS, vol. 7276, pp. 369–380. Springer, Heidelberg (2012)
23. Sanders, P., Schulz, C.: KaHIP – Karlsruhe High Qualtity Partitioning Homepage, http://algo2.iti.kit.edu/documents/kahip/index.html
24. Sanders, P., Schulz, C.: Engineering Multilevel Graph Partitioning Algorithms. In: Demetrescu, C., Halldórsson, M.M. (eds.) ESA 2011. LNCS, vol. 6942, pp. 469–480. Springer, Heidelberg (2011)
25. Sanders, P., Schulz, C.: Think Locally, Act Globally: Highly Balanced Graph Partitioning. In: Bonifaci, V., Demetrescu, C., Marchetti-Spaccamela, A. (eds.) SEA 2013. LNCS, vol. 7933, pp. 164–175. Springer, Heidelberg (2013)
26. Schulz, C.: High Quality Graph Partititioning. PhD thesis, KIT (2013)

27. Staudt, C.L., Meyerhenke, H.: Engineering High-Performance Community Detection Heuristics for Massive Graphs. In: Proc. 42nd Conf. on Parallel Processing (ICPP 2013) (2013)
28. Ugander, J., Backstrom, L.: Balanced Label Propagation for Partitioning Massive Graphs. In: 6'th Int. Conf. on Web Search and Data Mining (WSDM 2013), pp. 507–516. ACM (2013)
29. Walshaw, C.: Multilevel Refinement for Combinatorial Optimisation Problems. Annals of Operations Research 131(1), 325–372 (2004)
30. Walshaw, C., Cross, M.: Mesh Partitioning: A Multilevel Balancing and Refinement Algorithm. SIAM J. on Scientific Computing 22(1), 63–80 (2000)

Tree-Based Coarsening and Partitioning of Complex Networks

Roland Glantz, Henning Meyerhenke, and Christian Schulz

Karlsruhe Institute of Technology (KIT), Karlsruhe, Germany

Abstract. Many applications produce massive complex networks whose analysis would benefit from parallel processing. Parallel algorithms, in turn, often require a suitable network partition. For solving optimization tasks such as graph partitioning on large networks, multilevel methods are preferred in practice. Yet, complex networks pose challenges to established multilevel algorithms, in particular to their coarsening phase.

One way to specify a (recursive) coarsening of a graph is to rate its edges and then contract the edges as prioritized by the rating. In this paper we (i) define weights for the edges of a network that express the edges' importance for connectivity, (ii) compute a minimum weight spanning tree T^m w.r.t. these weights, and (iii) rate the network edges based on the conductance values of T^m's fundamental cuts. To this end, we also (iv) develop the first optimal linear-time algorithm to compute the conductance values of *all* fundamental cuts of a given spanning tree.

We integrate the new edge rating into a leading multilevel graph partitioner and equip the latter with a new greedy postprocessing for optimizing the maximum communication volume (MCV). Bipartitioning experiments on established benchmark graphs show that both the postprocessing *and* the new edge rating improve upon the state of the art by more than 10%. In total, with a modest increase in running time, our new approach reduces the MCV of complex network partitions by 20.4%.

Keywords: Graph coarsening, multilevel graph partitioning, complex networks, fundamental cuts, spanning trees.

1 Introduction

Complex networks such as social networks or web graphs have become a focus of investigation recently [7]. Such networks are often scale-free, i.e. they have a power-law degree distribution with many low-degree vertices and few high-degree vertices. They also have a small diameter (small-world property), so that the whole network is discovered within a few hops from any vertex. Complex networks arise in a variety of applications; several of them generate massive data sets. As an example, the social network Facebook currently contains a billion active users (http://newsroom.fb.com/Key-Facts). On this scale many algorithmic tasks benefit from parallel processing. The efficiency of parallel algorithms on huge networks, in turn, is usually improved by *graph partitioning* (GP).

J. Gudmundsson and J. Katajainen (Eds.): SEA 2014, LNCS 8504, pp. 364–375, 2014.
© Springer International Publishing Switzerland 2014

Given a graph $G = (V, E)$ and a number of blocks $k > 0$, the GP problem asks for a division of V into k pairwise disjoint subsets $V_1, \ldots, V_k$ (*blocks*) such that no block is larger than $(1+\varepsilon) \cdot \left\lceil \frac{|V|}{k} \right\rceil$, where $\varepsilon \geq 0$ is the allowed imbalance. When GP is used for parallel processing, each processing element (PE) usually receives one block, and edges running between two blocks model communication between PEs. The most widely used objective function (whose minimization is $\mathcal{NP}$-hard) is the *edge cut*, the total weight of the edges between different blocks. However, it has been pointed out more than a decade ago [13] that the determining factor for modeling the communication cost of parallel iterative graph algorithms is the *maximum communication volume* (MCV), which has received growing attention recently, e.g. in a GP challenge [1]. MCV considers the worst communication volume taken over all blocks V_p $(1 \leq p \leq k)$ and thus penalizes imbalanced communication: $MCV(V_1, \ldots, V_k) := \max_p \sum_{v \in V_p} |\{V_i \mid \exists \{u, v\} \in E \text{ with } u \in V_i \neq V_p\}|$. While hypergraph (and some graph) partitioners minimize the *total* communication volume, we are not aware of other tools that *explicitly* target MCV. Note that parallel processing is only one of many applications for graph partitioning; more can be found in recent surveys [3,4].

All state-of-the-art tools for partitioning very large graphs in practice rely on the multilevel approach [3,4]. In the first phase a hierarchy of graphs $G_0, \ldots, G_l$ is built by recursive coarsening. G_l is supposed to be very small in size, but similar in structure to the input G_0. In the second phase a very good initial solution for G_l is computed. In the final phase, the solution is prolongated to the next-finer graph, where it is improved using a local improvement algorithm. This process of prolongation and local improvement is repeated up to G_0.

Partitioning static meshes and similar non-complex networks this way is fairly mature. Yet, the structure of complex networks (skewed degree distribution, small-world property) distinguishes complex networks from traditional inputs and makes finding small cuts challenging with current tools.

One reason for the difficulties of established multilevel graph partitioners is the coarsening phase. Most tools rely on edge contractions for coarsening. Traditionally, only edge weights have guided the selection of the edges to be contracted [17]. Holtgrewe *et al.* [14] recently presented a two-phase approach that makes contraction more systematic by separating two issues: An *edge rating* and a *matching algorithm*. The rating of an edge indicates how much sense it makes to contract the edge. The rating then forms the input to a matching algorithm, and the edges of the resulting matching are contracted.

Outline and Contribution. After the introduction we sketch the state of the art (Section 2) and settle necessary notation (Section 3). Our first technical contribution, described briefly in Section 4, results from our goal to minimize MCV rather than the edge cut: We equip a leading multilevel graph partitioner with greedy postprocessing that trades in small edge cuts for small MCVs.

Our main contributions follow in Sections 5 and 6. The first one is a new edge rating, designed for complex networks by combining local and non-local information. Its rationale is to find moderately balanced cuts of high quality quickly (by means of the clustering measure *conductance* [16] and its loose connection to MCV via isoperimetric graph partitioning [12]) and to use this information to indicate whether an edge is part of a small cut or not. Finding such cuts is done by evaluating conductance for all *fundamental cuts* (defined in Section 3) of a minimum spanning tree of the input graph with carefully chosen edge weights. The second main contribution facilitates an efficient computation of our new edge rating. We present the first optimal linear-time algorithm to compute the conductance values of all fundamental cuts of a spanning tree.

We have integrated both MCV postprocessing and our new edge rating ex_cond($\cdot$) into KaHIP [22,23], a state-of-the-art graph partitioner with a reference implementation of the edge rating ex_alg($\cdot$), which yielded the best quality for complex networks so far (see Section 2).

Experiments in Section 7 show that greedy MCV postprocessing improves the partitions of our complex network benchmark set in terms of MCV by 11.3% with a comparable running time. Additional extensive bipartitioning experiments (MCV postprocessing included) show that, compared to ex_alg($\cdot$), the fastest variant of our new edge rating further improves the MCVs by 10.3%, at the expense of an increase in running time by a factor of 1.79. Altogether, compared to previous work on partitioning complex networks with state-of-the-art methods [21], the total MCV reduction by our new techniques amounts to 20.4%.

Proofs, pseudocode and detailed experimental results can be found in the full version of this paper [11].

2   State of the Art

Multilevel graph partitioners such as METIS [17] and KaHIP [22,23] (more are described in recent surveys [3,4]) typically employ recursive coarsening by contracting edges, which are often computed as those of a matching. Edge ratings are important in guiding the matching algorithm; a successful edge rating is

$$expansion^{*2}(\{u,v\}) = \omega(\{u,v\})^2/(c(u)c(v)), \tag{2.1}$$

where the weights of the vertices $u, v \in V$ and of the edges $\{u, v\} \in E$ are given by $c(\cdot)$ and $\omega(\cdot)$, respectively [14]. In the experiments of this paper the vertex weights are all one on the finest level. Contraction of an edge $\{u, v\}$ yields a new node w with $c(w) = c(u) + c(v)$.

To broaden the view of the myopic rating above (it does not look beyond its incident vertices), Safro *et al.* [21] precompute the algebraic distance $\rho_{\{u,v\}}$ [5] for the end vertices of each edge $\{u, v\}$ and use the edge rating

$$ex\_alg(\{u,v\}) = (1/\rho_{\{u,v\}}) \cdot expansion^{*2}(u,v) \tag{2.2}$$

For graphs with power-law degree distributions, ex_alg($\cdot$) yields considerably higher partition quality than $expansion^{*2}(\cdot)$ [21]. This is due to the fact that algebraic distance expresses a semi-local connection strength of an edge $\{u, v\}$ [5]. Specifically, $\rho_{\{u,v\}}$ is computed from R randomly initialized vectors that are smoothed by a Jacobi-style over-relaxation for a few iterations. The idea is that the vector entries associated with well-connected vertices even out more quickly than those of poorly connected vertices. Thus, a high value of $\rho_{\{u,v\}}$ indicates that the edge $\{u, v\}$ constitutes a bottleneck and should not be contracted.

Another strategy for matching-based multilevel schemes in complex networks (e. g. for agglomerative clustering [8]) is to match unconnected vertices at 2-hop distance in order to eliminate star-like structures. Also, alternatives to matching-based coarsening exist, e. g. weighted aggregation schemes [6,19].

Pritchard and Thurimella [20] use a spanning tree to sample the *cycle space* of a graph in a uniform way and thus find small cuts (consisting of a single edge, two edges or a cut vertex) with high probability [20]. Our method uses a minimum weight spanning with carefully chosen edge weights. Moreover, we sample the *cut-space*. The aim of the sampling is to create a collection $\mathcal{C}$ of moderately balanced cuts which form the basis of our new edge rating.

We integrate our new algorithms into KAHIP [22,23]. KAHIP focuses on solution quality and has been shown recently to be a leading graph partitioner for a wide variety of graphs such as road networks, meshes, and complex networks [24]. It implements several advanced multilevel graph partitioning algorithms, meta-heuristics, and sophisticated local improvement schemes. KAHIP was the best tool in terms of edge cut and MCV in the 10th DIMACS Challenge [1].

3   Preliminaries

Let $G = (V, E, \omega)$ be a finite, undirected, connected, and simple graph. Its edge weights are given by $\omega : E \mapsto \mathbb{R}^+$. We write $\omega_{u,v}$ for $\omega(\{u, v\})$ and extend ω to subsets of E through $\omega(E') = \sum_{e \in E'} \omega(e)$.

For subsets V_1, V_2 of V with $V_1 \cap V_2 = \emptyset$, the set $S(V_1, V_2)$ consists of those edges in E that have one end vertex in V_1 and the other end vertex in V_2. If, in addition to $V_1 \cap V_2 = \emptyset$, it holds that (i) $V = V_1 \cup V_2$ and (ii) $V_1, V_2 \neq \emptyset$, then the pair (V_1, V_2) is called a *cut* of G, and $S(V_1, V_2)$ is called the *cut-set* of (V_1, V_2). The weight of a cut (V_1, V_2) is given by $\omega(S(V_1, V_2))$. The *volume* of any subset V' of V is the total weight of the edges incident on V' (which equals the sum over the weighted degrees of the vertices in V'):

$$vol(V') = \omega(\{e = \{v', v\} \in E \mid v' \in V', v \in V\}), \tag{3.1}$$

Definition 3.1 (Fundamental cut, cut-set $S_T(e_T)$, $\mathrm{cond}(e_T, T)$)
Let T be a spanning tree of G, and let $e_T \in E(T)$. If T_1 and T_2 are the connected components (trees) of the graph $(V, E(T) \setminus \{e_T\})$, then $(V(T_1), V(T_2))$ is the fundamental cut of G with respect to T and e_T, and

$$S_T(e_T) = S(V(T_1), V(T_2)). \tag{3.2}$$

is the fundamental cut-set of G with respect to T and e_T. Conductance is a common quality measure in graph clustering [16]. Its value for $(V(T_1), V(T_2))$ is

$$\mathrm{cond}(e_T, T) = \mathrm{cond}(V_1, V_2) = \frac{\omega(S(V_1, V_2))}{\min\{vol(V(T_1)), vol(V(T_2))\}} \tag{3.3}$$

4   Greedy MCV Optimization

The ultimate applications we target with our graph partitioning algorithm are iterative parallel algorithms executed on complex networks. As argued in Section 1, the maximum communication volume (MCV) is a more accurate optimization criterion than the edge cut. The graph partitioner KAHIP has so far solely focused on the edge cut, though. That is why, as a new feature, we equip KAHIP with a postprocessing that greedily optimizes MCV. This postprocessing is executed after local improvement on the finest level of the multilevel hierarchy and works in rounds. In each round, we iterate over all boundary vertices of the input partition in a random order and check whether moving the vertex from its own block to the opposite block reduces or keeps the MCV value. If this is the case, the current vertex will be moved to the opposite block. One round of the algorithm can be implemented in $\mathcal{O}(|E|)$ time (see [11] for more details). The total number of rounds of the algorithm is a tuning parameter. After preliminary experiments we have set it to 20.

5   A New Conductance-Based Edge Rating for Partitioning

An edge rating in a multilevel graph partitioner should yield a low rating for an edge e if e is likely to be contained in the cut-set of a "good" cut, e. g. if the cut-set consists of a bridge.

   In our approach a good cut is one that (i) has a low conductance and (ii) is at least moderately balanced. In complex networks (i) does not always imply (ii) (see below). A loose connection between conductance and MCV in bipartitions can be established via isoperimetric graph partitioning [12]. Our approach to define an edge rating and use it for partitioning is as follows.

1. Generate a collection $\mathcal{C}$ of moderately balanced bipartitions (cuts of G) that contain cuts with a low conductance value.
2. Define a measure $\text{Cond}(\cdot)$ such that $\text{Cond}(e)$ is low [high] if e is [not] contained in the cut-set of a cut in $\mathcal{C}$ with low conductance.
3. Instead of multiplying the edge rating $expansion^{*2}(\{u, v\})$ with the factor $(1/\rho_{\{u,v\}})$ as in [21], we replace one of the two (identical) myopic factors $\omega(\{u, v\})$ in $expansion^{*2}(\{u, v\})$ by the more far-sighted factor $\text{Cond}(\cdot)$. This yields the new edge rating

$$\text{ex\_cond}(\{u, v\}) = \omega(\{u, v\}) \, \text{Cond}(\{u, v\})/(c(u)c(v)) \qquad (5.1)$$

   The higher $\text{Cond}(e)$, the higher $\text{ex\_cond}(e)$, and thus the higher the chances for e to be contracted during coarsening.
4. Run a multilevel graph partitioner capable of handling edge ratings such as KAHIP with $\text{ex\_cond}(\cdot)$.

To specify $\text{ex\_cond}(\cdot)$, we need to define $\mathcal{C}$ and $\text{Cond}(\cdot)$.

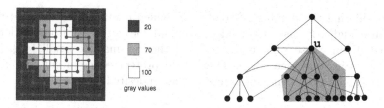

Fig. 1. (a) Example of MST (red) in GBIS. For the green arrow see the text. (b) Vertex attributes *intraWeight* and *interWeight*. Tree T is formed by the black edges, and the subtree with root u is contained in the shaded area. The weights of the blue and green edges contribute to *intraWeight*[u], and the weights of the red edges contribute to *interWeight*[u].

Specifics of $\mathcal{C}$. For the definition of $\mathcal{C}$, we resort to a basic concept of graph-based clustering, i. e. the use of minimum weight spanning trees (MSTs). We describe this concept in the context of graph-based image segmentation (GBIS) [9,26] for illustration purposes (see Figure 1a).

In GBIS one represents an image by a graph G whose vertices [edges] represent pixels [neighborhood relations of pixels]. The edges are equipped with weights that reflect the contrast between the gray values at the edges' end vertices. An MST T^m of G with respect to contrast has the following property (see [15, Thm. 4.3.3]). The contrast value associated with any $e \in E(T^m)$ is minimal compared to the contrast values of the edges in the fundamental cut-set $S_{T^m}(e)$ (see Eq. 3.2). Thus, for any $e \in E(T^m)$ with a high contrast value (see the green arrow in Figure 1a), the fundamental cut $S_{T^m}(e)$ yields a segmentation into two regions with a high contrast anywhere on the common border.

Here, we arrive at a collection $\mathcal{C}$ of $|V| - 1$ moderately balanced bipartitions (cuts) of G by (i) computing connectivity-based contrast values for the edges of G, (ii) computing an MST T^m of G w. r. t. these values, and (iii) letting $\mathcal{C}$ consist of G's fundamental cuts w. r. t. T^m. The contrast value of an edge $e = \{u, v\}$ should be low [high] if the edge is indispensable for "many" connections via shortest paths. Thus, the higher the contrast, the stronger the negative effect on G's connectivity if e is cut, and thus the more reasonable it is to cut e. To define the contrast values, we generate a random collection $\mathcal{T}$ of breadth-first-traversal (BFT) trees. A tree in $\mathcal{T}$ is formed by first choosing a root randomly. As usual, we let the trees grow out of the roots using a queue, but we process the edges incident on a vertex in a randomized order. Alternatively, SSSP trees may be used if edge weights are to be included.

Let $n_{\mathcal{T}}(u, v)$ denote the number of trees in $\mathcal{T}$ that contain e and in which u is closer to the tree's root than v (u and v cannot have the same distance to the root). We set the *contrast* value of an edge $\{u, v\}$ to

$$\gamma(\{u, v\}) = \min\{n_{\mathcal{T}}(u, v), n_{\mathcal{T}}(v, u)\}. \tag{5.2}$$

Just considering the number of trees in $\mathcal{T}$ which contain e, turned out to yield poorer partitions than using Eq. 5.2. We believe that this is due to small subgraphs which are connected to the graphs' "main bodies" via very few edges.

Just considering the number of trees in $\mathcal{T}$ which contain such an edge would result in a high contrast of the edge although the cut is far from moderately balanced. Even worse, the conductance of the cut may be small (e.g. if the cut-set contains only one edge). This would protect edges in cut-sets of very unbalanced cuts from being contracted — an undesired feature.

Specifics of Cond($\cdot$). Our plan is to define a measure Cond($\cdot$) such that Cond(e) is low [high] if e is [not] contained in the cut-set of a cut in $\mathcal{C}$ with low conductance. Hence, we set

$$\text{Cond}(e) = \min_{C \in \mathcal{C}, e \in S(C)} (\text{cond}(C)), \tag{5.3}$$

where $S(C)$ denotes the cut-set of the cut C. Let FC_e denote the set of edges in the (fundamental) cycle that arises if e is inserted into T^m. Then, the cuts $C \in \mathcal{C}$ with $e \in S(C)$ (see Eq. 5.3) are precisely the fundamental cuts $S_T(e_T)$ (see Eq. 3.2) with $e_T \in FC_e$ and $e_T \neq e$. Note that e is the only edge in FC_e that is not in $E(T^m)$. This suggests to first compute the Cond-values for all edges $e_T \in E(T^m)$ as specified in Section 6. For $e \notin E(T^m)$ the value of Cond(e) is then obtained by forming the minimum of the Cond-values of $FC_e \setminus \{e\}$. If $e = \{u, v\}$, then $FC_e \setminus \{e\}$ is the set of edges on the unique path in T^m that connects u to v.

6   An $\mathcal{O}(|E|)$-Algorithm for Computing All cond(e_T, T)

In this section we demonstrate how, for a given rooted spanning tree T of a graph $G(V, E)$, one can compute all conductance values cond(e_T, T), $e_T \in E(T)$, in time $\mathcal{O}(|E|)$ (the root can be chosen randomly). This algorithm facilitates an efficient computation of the edge rating introduced in the previous section. The key to achieving optimal running time is to aggregate information on fundamental cuts during a postorder traversal of T. The aggregated information is kept in the three vertex attributes *subtree Vol*, *intra Weight* and *inter Weight* defined in Definition 6.1 below. Technically, the three vertex attributes take the form of arrays, where indices represent vertices.

Definition 6.1 *Let $C_T(u)$ be the children of vertex u in T. Moreover, let $T(u)$ denote the subtree rooted at u, and let $D(u)$ (descendants of u) denote the set that contains the vertices of $T(u)$, i.e. $D(u) = V(T(u))$. We use the following three vertex attributes to aggregate information that we need to compute the conductance values:*

  – *subtree Vol$[u] = vol(D(u))$.*
  – *intra Weight$[u]$ equals twice the total weight of all edges $e = \{v, w\}$ with (i) $v, w \in D(u)$, (ii) $v, w \neq u$ and (iii) the lowest common ancestor of v and w in T is u (blue edges in Figure 1b) plus the total weight of all edges not in T with one end vertex being u and the other end vertex being contained in $D(u)$ (green edges in Figure 1b).*

- *interWeight*[u] *equals the total weight of all edges not in T with exactly one end vertex in $D(u)$ (red edges in Figure 1b).*

If u has a parent edge e_T, Eq. 3.3 takes the form

$$\text{cond}(e_T, T) = \frac{interWeight[u] + \omega(e_T)}{\min\{subtreeVol[u], vol(V) - subtreeVol[u]\}} \tag{6.1}$$

When computing *subtreeVol*, *intraWeight* and *interWeight*, we employ two vertex labellings (stored in arrays indexed by the vertices): *label*[u] indicates the preorder label of u in T, and *maxLabelDescendants*[u] indicates the maximum of *label*[t] over all $t \in T(u)$. We also need lowest common ancestors (LCAs). Queries LCA(T, u, v), i.e. the LCA of u and v on T, require constant time after an $\mathcal{O}(n)$-time preprocessing [2].

We start by initializing labels and vertex attributes to arrays of length $|V|$ with all entries set to 0 (for details see Algorithm 2 in [11]. Then we compute the entries of *label* and *maxLabelDescendants* in a single depth-first traversal of T and perform the preprocessing for LCA($\cdot, \cdot, \cdot$). Finally, we call a standard postorder traversal in T starting at the root of T. When visiting a vertex, either one of the subroutines LEAF($\cdot$) or NONLEAF($\cdot$) is called depending on the vertex type (see Algorithms 1 and 3 in [11].

If u is a leaf, Algorithm 3 in [11] sets entries *subtreeVol*[u] to *vol*($\{u\}$) and *interWeight*[u] to the total weight of all edges in $E \setminus E(T)$ that are incident on u. Likewise, the entry *intraWeight*[LCA(T, u, t)] is updated for any t with $\{u, t\} \notin E(T)$.

If u is not a leaf (see Algorithm 1), and if u has a parent edge in T, this edge is found in line 8 and the corresponding conductance value is computed in line 22 using *subtreeVol*[u] and *interWeight*[u]. The entry *intraWeight*[LCA(T, u, t)] is updated multiple times until the postorder traversal ascends from u towards the root of T (line 14). The update of *interWeight* is justified in the proof of Theorem 6.2. For this proof and the proof of Proposition 6.3 see [11]. Eq. 6.5 in Theorem 6.2 guarantees that the conductance values computed in line 22 are correct.

Theorem 6.2 *After having finished processing $u \in V$ in a traversal of T, the equalities given below hold (where in the last one we assume that u is not the root of T and that e_T is the parent edge of u in T).*

$$subtreeVol[u] = vol(D(u)), \tag{6.2}$$

$$intraWeight[u] = \sum_{c_i \neq c_j \in C(u)} \omega(S(D(c_i), D(c_j))) \tag{6.3}$$

$$+ \omega(S(D(u) \setminus \{u\}, \{u\}) \setminus E(T)) \text{ and} \tag{6.4}$$

$$interWeight[u] = \omega(S_T(e_T)) - \omega(e_T). \tag{6.5}$$

Proposition 6.3 *Given a rooted spanning tree T of $G = (V, E)$, the computation of all $\text{cond}(e_T, T)$, $e_T \in E(T)$, takes $\mathcal{O}(|E|)$ time.*

Algorithm 1. Procedure NONLEAF(T, u) called during postorder traversal of T

1: $parentEdge \leftarrow$ UNDEFINED_EDGE
2: **for all** $f = \{u, t\} \in E$ **do**
3:    **if** $f \in E(T)$ **then**
4:        **if** label[u] < label[t] **then**
5:            $subtreeVol[u] \leftarrow subtreeVol[u] + subtreeVol[t]$
6:            $interWeight[u] \leftarrow interWeight[u] + interWeight[t]$
7:        **else**
8:            $parentEdge \leftarrow f$
9:        **end if**
10:    **else**
11:        **if** $((label[t] < label[u]) \lor (label[t] > maxLabelDescendants[u]))$ **then**
12:                                    ▷ equivalent to test if $t \notin D(u)$
13:            $lca \leftarrow LCA(T, u, t)$
14:            $intraWeight[lca] \leftarrow intraWeight[lca] + \omega(f)$
15:            $interWeight[u] \leftarrow interWeight[u] + \omega(f)$
16:        **end if**
17:    **end if**
18: **end for**
19: $subtreeVol[u] \leftarrow subtreeVol[u] + vol(\{u\})$
20: $interWeight[u] \leftarrow interWeight[u] - intraWeight[u]$;
21: **if** $parentEdge \neq$ UNDEFINED_EDGE **then**
22:    $cond(parentEdge, T) \leftarrow \frac{interWeight[u] + \omega(parentEdge)}{\min\{subtreeVol[u], vol(V) - subtreeVol[u]\}}$
23: **end if**

7   Experimental Results

Approach and settings. The multilevel partitioner within the KAHIP package has three different algorithm configurations: strong, eco and fast. We use the eco configuration since this configuration was chosen in [21], too. The variable ε in the balance constraint is set to the common value 0.03.

We evaluate the postprocessing and compare ex_cond with ex_alg on the basis of the 15 complex networks listed in Table 1 (see also Table D.1 in [11]). The networks are from two popular archives [1,18]. The same networks have been used previously in [21] to evaluate ex_alg. All computations are done on a workstation with two 8-core Intel(R) Xeon(R) E5-2680 processors at 2.7 GHz. Our code is implemented in C/C++ and compiled with GCC 4.7.1. Note that we run sequential experiments only. First of all, we focus in this paper on solution quality, not on speed. Secondly, the standard of reference, ex_alg, is also implemented sequentially. Running times range from 0.4 seconds for PGPginatcompo to 26 minutes for wiki-Talk.

Since the results produced by KAHIP depend on many factors including random seeds, we perform 50 runs with different seeds for each network and compute the following three *performance indicators*:

- *minMCV* and *avgMCV*: minimal and average MCV found by KAHIP.
- *minCut* and *avgCut*: minimal and average cut found by KAHIP.
- *avgTime*: average time KAHIP needs for the complete partitioning process.

Postprocessing results. For ex_alg, the average reduction of avgMCV due to postprocessing amounts to 11.3% (see Table E.1 in [11]). Since the postprocessing trades in small edge cuts for small MCVs, values for *minMCV* and *avgMCV* [*minCut* and *avgCut*] are with [without] postprocessing. The increase in running time due to postprocessing is negligible.

Table 1. Complex networks used as benchmark set

Name	#vertices	#edges
p2p-Gnutella	6 405	29 215
PGPgiantcompo	10 680	24 316
email-EuAll	16 805	60 260
as-22july06	22 963	48 436
soc-Slashdot0902	28 550	379 445
loc-brightkite_edges	56 739	212 945
loc-gowalla_edges	196 591	950 327
coAuthorsCiteseer	227 320	814 134
wiki-Talk	232 314	1 458 806
citationCiteseer	268 495	1 156 647
coAuthorsDBLP	299 067	977 676
web-Google	356 648	2 093 324
coPapersCiteseer	434 102	16 036 720
coPapersDBLP	540 486	15 245 729
as-skitter	554 930	5 797 663

Edge rating results. Intriguingly, using an asymptotically optimal Range Minimum Query (RMQ) code (by Fischer and Heun [10]) within ex_cond for the algorithms in Section 6 does not decrease the running time. The straightforward asymptotically slower algorithm is slightly faster (1.1% in total) in our experiments. To investigate this effect further, we compare the results on a set of non-complex networks, Walshaw's graph partitioning archive [25]. Again, the implementation of the (in theory faster) RMQ algorithm does not play out, running time and quality remain comparable. Therefore, the running times in all tables refer to the implementation not using the Fischer/Heun RMQ code.

The edge rating ex_cond depends on the number of random spanning trees, i.e. $|\mathcal{T}|$. To make this clear we write ex_cond$_{|\mathcal{T}|}$ instead of ex_cond.

For a given network we measure the quality of the edge rating ex_cond$_{|\mathcal{T}|}$ through (three) quotients of the form (performance indicator using ex_cond$_{|\mathcal{T}|}$ divided by the same performance indicator using ex_alg). Tables E.2, E.3 and E.4 in [11] show the performance quotients of ex_cond$_{20}$, ex_cond$_{100}$ and ex_cond$_{200}$. The geometric means of the performance quotients over all networks are shown in Table 2.

As the main result we state that buying quality through increasing $|\mathcal{T}|$ is expensive in terms of running time. The rating ex_cond$_{20}$ already yields avgMCV that is 10.3% lower than avgMCV from ex_alg — at the expense of a relative increase in running time by only 1.79. The total reduction of average MCV from postprocessing *and* replacing ex_alg by ex_cond$_{20}$ amounts to 20.4% (see Tables E.1 and E.2 in [11]).

Table 2. Geometric means of minMCV, avgMCV and avgTime over all networks in Table 1. Number of trees: 20, 100 and 200. Reference is the edge rating ex_alg. A quotient < 1.0 means that ex_cond yields better results than ex_alg.

	minMCV	avgMCV	avgTime
Ratios ex_cond$_{20}$ / ex_alg	**0.892**	**0.897**	**1.793**
Ratios ex_cond$_{100}$ / ex_alg	**0.874**	**0.893**	**5.278**
Ratios ex_cond$_{200}$ / ex_alg	**0.865**	**0.890**	**9.411**

When we omit the postprocessing step and compare the average edge cut instead of MCV, ex_alg and ex_cond perform comparably well (see Table E.5 in [11]).

8   Conclusions and Future Work

Motivated by the deficits of coarsening complex networks during multilevel graph partitioning, we have devised a new edge rating for guiding edge contractions. Our linear-time algorithm for computing the conductance values of *all* fundamental cuts of a spanning tree facilitates an efficient computation of the rating. The evaluation shows a significant improvement over a previously leading code.

We would like to stress that good coarsening is not only of interest for graph partitioning, but can be employed in other applications that exploit hierarchies in networks. Future work should investigate the interaction of the contrast γ and the conductance values — possibly replacing γ. Our overall coarsening scheme is agnostic to such a replacement. Also, we would like to extend our methods to an arbitrary number of blocks. While the new edge rating should work out of the box, the greedy MCV minimization has to be adapted.

Acknowledgments. We thank Johannes Fischer for providing an implementation of the Range Minimum Query method presented in [10].

References

[1] Bader, D.A., Meyerhenke, H., Sanders, P., Wagner, D.: Graph Partitioning and Graph Clustering – 10th DIMACS Impl. Challenge. Contemporary Mathematics, vol. 588. AMS (2013)

[2] Bender, M.A., Farach-Colton, M.: The LCA problem revisited. In: Gonnet, G.H., Viola, A. (eds.) LATIN 2000. LNCS, vol. 1776, pp. 88–94. Springer, Heidelberg (2000)

[3] Bichot, C., Siarry, P. (eds.): Graph Partitioning. Wiley (2011)

[4] Buluç, A., Meyerhenke, H., Safro, I., Sanders, P., Schulz, C.: Recent Advances in Graph Partitioning. Technical Report ArXiv:1311.3144 (2014)

[5] Chen, J., Safro, I.: Algebraic distance on graphs. SIAM J. Comput. 6, 3468–3490 (2011)

[6] Chevalier, C., Safro, I.: Comparison of coarsening schemes for multi-level graph partitioning. In: Proc. Learning and Intelligent Optimization (2009)

[7] de Costa, L.F., Oliveira Jr., O.N., Travieso, G., Rodrigues, F.A., Boas, P.R.V., Antiqueira, L., Viana, M.P., Correa Rocha, L.E.: Analyzing and modeling real-world phenomena with complex networks: a survey of applications. Advances in Physics 60(3), 329–412 (2011)

[8] Fagginger Auer, B.O., Bisseling, R.H.: Graph coarsening and clustering on the GPU. In: Graph Partitioning and Graph Clustering. AMS and DIMACS (2013)

[9] Felzenszwalb, P.F., Huttenlocher, D.P.: Efficient graph-based image segmentation. Int. J. Comput. Vision 59(2), 167–181 (2004)

[10] Fischer, J., Heun, V.: Theoretical and Practical Improvements on the RMQ-Problem, with Applications to LCA and LCE. In: Lewenstein, M., Valiente, G. (eds.) CPM 2006. LNCS, vol. 4009, pp. 36–48. Springer, Heidelberg (2006)

[11] Glantz, R., Meyerhenke, H., Schulz, C.: Tree-based Coarsening and Partitioning of Complex Networks. Technical Report arXiv:1402.2782 (2014)

[12] Grady, L., Schwartz, E.L.: Isoperimetric graph partitioning for image segmentation. IEEE Trans. Pattern Anal. Mach. Intell. 28(3), 469–475 (2006)

[13] Hendrickson, B., Kolda, T.G.: Graph partitioning models for parallel computing. Parallel Computing 26(12), 1519–1534 (2000)

[14] Holtgrewe, M., Sanders, P., Schulz, C.: Engineering a scalable high quality graph partitioner. In: 24th Int. Parallel and Distributed Processing Symp, IPDPS (2010)

[15] Jungnickel, D.: Graphs, Networks and Algorithms, 2nd edn. Algorithms and Computation in Mathematics, vol. 5. Springer, Berlin (2005)

[16] Kannan, R., Vempala, S., Vetta, A.: On clusterings: Good, bad and spectral. J. of the ACM 51(3), 497–515 (2004)

[17] Karypis, G., Kumar, V.: A Fast and High Quality Multilevel Scheme for Partitioning Irregular Graphs. SIAM J. on Scientific Computing 20(1), 359–392 (1998)

[18] Leskovec, J.: Stanford Network Analysis Package (SNAP)

[19] Meyerhenke, H., Monien, B., Schamberger, S.: Graph partitioning and disturbed diffusion. Parallel Computing 35(10-11), 544–569 (2009)

[20] Pritchard, D., Thurimella, R.: Fast computation of small cuts via cycle space sampling. ACM Trans. Algorithms 46, 46:1–46:30 (2011)

[21] Safro, I., Sanders, P., Schulz, C.: Advanced coarsening schemes for graph partitioning. In: Klasing, R. (ed.) SEA 2012. LNCS, vol. 7276, pp. 369–380. Springer, Heidelberg (2012)

[22] Sanders, P., Schulz, C.: KaHIP – Karlsruhe High Qualtity Partitioning Homepage, http://algo2.iti.kit.edu/documents/kahip/index.html

[23] Sanders, P., Schulz, C.: Think Locally, Act Globally: Highly Balanced Graph Partitioning. In: Bonifaci, V., Demetrescu, C., Marchetti-Spaccamela, A. (eds.) SEA 2013. LNCS, vol. 7933, pp. 164–175. Springer, Heidelberg (2013)

[24] Schulz, C.: Hiqh Quality Graph Partititioning. PhD thesis, Karlsruhe Institute of Technology (2013)

[25] Soper, A.J., Walshaw, C., Cross, M.: A combined evolutionary search and multilevel optimisation approach to graph partitioning. Journal of Global Optimization 29(2), 225–241 (2004)

[26] Wassenberg, J., Middelmann, W., Sanders, P.: An efficient parallel algorithm for graph-based image segmentation. In: Jiang, X., Petkov, N. (eds.) CAIP 2009. LNCS, vol. 5702, pp. 1003–1010. Springer, Heidelberg (2009)

Improved Upper and Lower Bound Heuristics for Degree Anonymization in Social Networks

Sepp Hartung, Clemens Hoffmann, and André Nichterlein

Institut für Softwaretechnik und Theoretische Informatik, TU Berlin, Germany
{sepp.hartung,andre.nichterlein}@tu-berlin.de,
clemens.hoffmann@campus.tu-berlin.de

Abstract. Motivated by a strongly growing interest in anonymizing social network data, we investigate the NP-hard DEGREE ANONYMIZATION problem: given an undirected graph, the task is to add a minimum number of edges such that the graph becomes *k-anonymous*. That is, for each vertex there have to be at least $k-1$ other vertices of exactly the same degree. The model of degree anonymization has been introduced by Liu and Terzi [ACM SIGMOD'08], who also proposed and evaluated a two-phase heuristic. We present an enhancement of this heuristic, including new algorithms for each phase which significantly improve on the previously known theoretical and practical running times. Moreover, our algorithms are optimized for large-scale social networks and provide upper and lower bounds for the optimal solution. Notably, on about 26 % of the real-world data we provide (provably) optimal solutions; whereas in the other cases our upper bounds significantly improve on known heuristic solutions.

1 Introduction

In recent years, the analysis of (large-scale) social networks received a steadily growing attention and turned into a very active research field [6]. Its importance is mainly due the easy availability of social networks and due to the potential gains of an analysis revealing important subnetworks, statistical information, etc. However, as the analysis of networks may reveal sensitive data about the involved users, before publishing the networks it is necessary to preprocess them in order to respect privacy issues [8]. In a landmark paper [11] initiating a lot of follow-up work [4, 9, 12], Liu and Terzi transferred the so-called *k-anonymity* concept known for tabular data in databases [8, 13, 14, 15] to social networks modeled as undirected graphs. A graph is called *k-anonymous* if for each vertex there are at least $k - 1$ other vertices of the same degree. Therein, the larger k is, the better the expected level of anonymity is.

In this work we describe and evaluate a combination of heuristic algorithms which provide (for many tested instances matching) lower and upper bounds, for the following NP-hard graph anonymization problem:

DEGREE ANONYMIZATION [11]
Input: An undirected graph $G = (V, E)$ and an integer $k \in \mathbb{N}$.
Task: Find a minimum-size edge set E' over V such that adding E' to G results in a k-anonymous graph.

J. Gudmundsson and J. Katajainen (Eds.): SEA 2014, LNCS 8504, pp. 376–387, 2014.
© Springer International Publishing Switzerland 2014

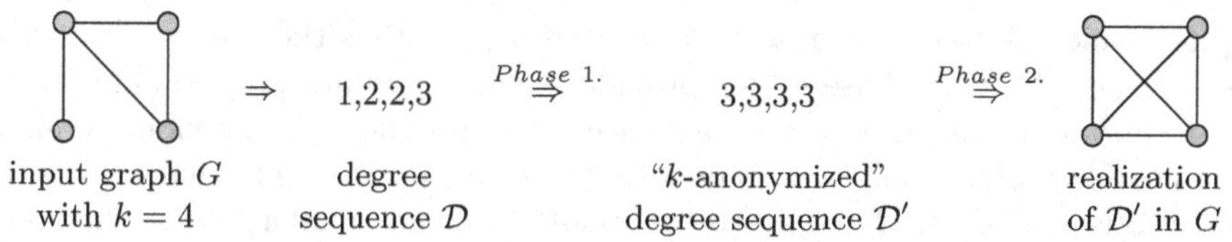

Fig. 1. A simple example for the two phases in the heuristic of Liu and Terzi [11]. Phase 1: Anonymize the degree sequence $\mathcal{D}$ of the input graph G by increasing the numbers in it such that each resulting number occurs at least k times. Phase 2: Realize the k-anonymized degree sequence $\mathcal{D}'$ as a super-graph of G.

As DEGREE ANONYMIZATION is NP-hard even for constant $k \geq 2$ [9], all known (experimentally evaluated) algorithms, are heuristics in nature [3, 11, 12, 16]. Liu and Terzi [11] proposed a heuristic which, in a nutshell, consists of the following two phases: i) Ignore the graph structure and solve a corresponding number problem and ii) try to transfer the solution from the number problem back to the graph instance. More formally (see Figure 1 for an example), given an instance (G, k), first compute the *degree sequence* $\mathcal{D}$ of G, that is, the multiset of positive integers corresponding to the vertex degrees in G. Then, Phase 1 consists of k-anonymizing the degree sequence $\mathcal{D}$ (each number occurs at least k times) by a minimum amount of increments to the numbers in $\mathcal{D}$ resulting in $\mathcal{D}'$. In Phase 2, try to realize the k-anonymous sequence $\mathcal{D}'$ as a super-graph of G, meaning that each vertex gets a *demand*, which is the difference of its degree in $\mathcal{D}'$ compared to $\mathcal{D}$, and then a "realization" algorithm adds edges to G such that for each vertex the amount of incident new edges equals its demand.

Note that, since the minimum "k-anonymization cost" of the degree sequence $\mathcal{D}$ (sum over all demands) is always a lower bound on the k-anonymization cost of G, the above described algorithm, if successful when trying to realize $\mathcal{D}'$ in G, optimally solves the given DEGREE ANONYMIZATION instance.

Related Work. We only discuss work on DEGREE ANONYMIZATION directly related to what we present here. Our algorithm framework is based on the two-phase algorithm due to Liu and Terzi [11] where also the model of graph (degree-)anonymization has been introduced. Other models of graph anonymization have been studied as well, see Zhou and Pei [18] (studying the neighborhood of vertices) and Chester et al. [4] (anonymizing vertex subsets). We refer to Zhou et al. [19] for a survey on anonymization techniques for social networks. DEGREE ANONYMIZATION is NP-hard for constant $k \geq 2$ and it is W[1]-hard (presumably not fixed-parameter tractable) with respect to the parameter size of a solution size [9]. On the positive side, there is polynomial-size kernel (efficient and effective preprocessing) with respect to the maximum degree of the input graph [9]. Lu et al. [12] and Casas-Roma et al. [3] designed and evaluated heuristic algorithms that are our reference points for comparing our results.

Our Contributions. Based on the two-phase approach of Liu and Terzi [11] we significantly improve the lower bound provided in Phase 1 and provide a simple heuristic for new upper bounds in Phase 2. Our algorithms are designed to deal with large-scale real world social networks (up to half a million vertices) and exploit some common features of social networks such as the power-law degree distribution [1]. For Phase 1, we provide a new dynamic programming algorithm of k-anonymizing a degree sequence $\mathcal{D}$ "improving" the previous running time $\mathcal{O}(nk)$ to $\mathcal{O}(\Delta k^2 s)$, where s denotes the solution size. Note that maximum degree Δ is in our considered instances about 500 times smaller than the number of vertices n. We also implemented a data reduction rule which leads to significant speedups of the dynamic program. We study two different cases to obtain upper bounds. If one of the degree sequences computed in Phase 1 is realizable, then this gives an optimal upper bound and otherwise we heuristically look for "near" realizable degree sequences. For Phase 2 we evaluate the already known "local exchange" heuristic [11] and provide some theoretical justification of its quality.

We implemented our algorithms and compare our upper bounds with a heuristic of Lu et al. [12], called *clustering-heuristic* in the following. Our empirical evaluation demonstrates that in about 26% of the real-world instances the lower bound matches the upper bound and in the remaining instances our heuristic upper bound is on average 40% smaller than the one provided by the clustering-heuristic. However, this comes at a cost of increased running time: the clustering-heuristic could solve all instances within 15 seconds whereas there are a few instances where our algorithms could not compute an upper bound within one hour.

Due to the space constraints, all proofs and some details are deferred to a full version. Most details and proofs are also given in an arXiv-version [10].

2   Preliminaries

We use standard graph-theoretic notation. All graphs studied in this paper are undirected and simple without self-loops and multi-edges. For a given graph $G = (V, E)$ with vertex set V and edge set E we set $n := |V|$ and $m := |E|$. Furthermore, by $\deg_G(v)$ we denote the degree of a vertex $v \in V$ in G and Δ_G denotes the maximum degree in G. For $0 \leq d \leq \Delta_G$ let $B_d^G := \{v \in V \mid \deg_G(v) = d\}$ be the *block* of degree d, that is, the set of all vertices with degree d in G. Thus, being k-anonymous is equivalent to each block being of size either zero or at least k. For a set S of edges with endpoints in a graph G, we denote by $G + S$ the graph that results from inserting all edges from S into G. We call S an *edge insertion set* for G and if $G + S$ is k-anonymous, then it is an *k-insertion set*.

A *degree sequence* $\mathcal{D}$ is a multiset of positive integers and $\Delta_{\mathcal{D}}$ denotes its maximum value. The degree sequence of a graph G with vertex set $V = \{v_1, \ldots, v_n\}$ is $\mathcal{D}_G := \{\deg_G(v_1), \ldots, \deg_G(v_n)\}$. For a degree sequence $\mathcal{D}$, we denote by b_d how often value d occurs in $\mathcal{D}$ and we set $\mathcal{B} = \{b_0, \ldots, b_{\Delta_{\mathcal{D}}}\}$ to be the *block sequence* of $\mathcal{D}$, that is, $\mathcal{B}$ is just the list of the block sizes of G. Clearly, the block sequence of a graph G is the block sequence of G's degree sequence. The block sequence can be viewed as a compact representation of a degree sequence (just storing the amount of vertices for each degree) and we use these two representations of vertex degrees interchangeably. Equivalently to graphs, a block sequence is k-anonymous if each value is either zero or at least k and a degree sequence is k-anonymous if its corresponding block sequence is k-anonymous.

Let $\mathcal{D} = \{d_1, \ldots, d_n\}$ and $\mathcal{D}' = \{d'_1, \ldots, d'_n\}$ be two degree sequences with corresponding block sequences $\mathcal{B}$ and $\mathcal{B}'$. We define $\|\mathcal{B}\| = |\mathcal{D}| = \sum_{i=1}^{n} d_i$. We write $\mathcal{D}' \geq \mathcal{D}$ and $\mathcal{B}' \ominus \mathcal{B}$ if for both degree sequences sorted in ascending order it holds that $d'_i \geq d_i$ for all i. Intuitively, this captures the interpretation "$\mathcal{D}'$ can be obtained from $\mathcal{D}$ by increasing some values". If $\mathcal{D}' \geq \mathcal{D}$, then (for sorted degree sequences) we define the degree sequence $\mathcal{D}' - \mathcal{D} = \{d'_1 - d_1, \ldots, d'_n - d_n\}$ and set $\mathcal{B}' \ominus \mathcal{B}$ to be its block sequence. We omit sub- and superscripts if the graph is clear from the context.

3   Description of the Algorithm Framework

In this section we present the details of our algorithm framework to solve DEGREE ANONYMIZATION. We first provide a general description how the problem is split into several subproblems (basically corresponding to the two-phase approach of Liu and Terzi [11]) and then describe the corresponding algorithms in detail.

3.1   General Framework Description

We first provide a more formal description of the two-phase approach due to Liu and Terzi [11] and then describe how we refine it: Let $(G = (V, E), k)$ be an input instance of DEGREE ANONYMIZATION.

Phase 1: For the degree sequence $\mathcal{D}$ of G, compute a k-anonymous degree sequence $\mathcal{D}'$ such that $\mathcal{D}' \geq \mathcal{D}$ and $|\mathcal{D} - \mathcal{D}'|$ is minimized.

Phase 2: Try to realize $\mathcal{D}'$ in G, that is, try to find an edge insertion set S such that the degree sequence of $G + S$ is $\mathcal{D}'$.

The minimum k-anonymization cost of $\mathcal{D}$, formally $|\mathcal{D}' - \mathcal{D}|/2$, is a lower bound on the number of edges in a k-insertion set for G. Hence, if succeeding in Phase 2 to realize $\mathcal{D}'$, then a minimum-size k-insertion set S for G has been found.

Liu and Terzi [11] gave a dynamic programming algorithm which exactly solves Phase 1 and they provided the so-called local exchange heuristic algorithm for Phase 2. If Phase 2 fails, then the heuristic of Liu and Terzi [11] relaxes the constraints and tries to find a k-insertion set yielding a graph "close" to $\mathcal{D}'$.

We started with a straightforward implementation of the dynamic programming algorithm and the local exchange heuristic. We encountered the problem that, even when iterating through all minimum k-anonymous degree sequences $\mathcal{D}'$, one often fails to realize $\mathcal{D}'$ in Phase 2. More importantly, we observed the difficulty that iterating through all minimum sequences is often to time consuming because the same sequence is recomputed multiple times. This is because the dynamic program iterates through all possibilities to choose "sections" of consecutive degrees in the (sorted) degree sequence $\mathcal{D}$ that end up in the same block in $\mathcal{D}'$. These sections have to be of length at least k (the final

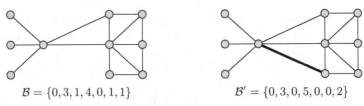

$$\mathcal{B} = \{0, 3, 1, 4, 0, 1, 1\} \qquad\qquad \mathcal{B}' = \{0, 3, 0, 5, 0, 0, 2\}$$

Fig. 2. A graph (left side) with block sequence $\mathcal{B}$ that can be 2-anonymized by adding one edge (right side) resulting in $\mathcal{B}'$. Another 2-anonymous block sequence (also of cost two) that will be found by the dynamic programming is $\mathcal{B}'' = \{0, 2, 2, 4, 0, 0, 2\}$. The realization of $\mathcal{B}''$ in G would require to add an edge between a degree-five vertex (there is only one) and a degree-one vertex, which is impossible.

block has to be full) but at most $2k - 1$ (longer sections can be split into two). However, if there is a huge block B (of size $\gg 2k$) in $\mathcal{D}$, then the algorithm goes through all possibilities to split B into sections, although it is not hard to show that at most $k - 1$ degrees from each block are increased. Thus, different ways to cut these degrees into sections result in the same degree sequence.

We thus redesigned the dynamic program for Phase 1. The main idea is to consider the block sequence of the input graph and exploiting the observation that at most $k-1$ degrees from a block are increased in a minimum-size solution. Therefore, we avoid to partition one block into multiple sections and the running time dependence on the number of vertices n can be replaced by the maximum degree Δ, yielding a significant performance increase.

We also improved the lower bound provided by $\mathcal{D}' - \mathcal{D}$ on the k-anonymization cost of G. To this end, the basic observation was that while trying to realize one of the minimum k-anonymous sequences $\mathcal{D}'$ in Phase 2 (failing in almost all cases), we encountered that by a simple criterion on the sequence $\mathcal{D}' - \mathcal{D}$ one can even prove that $\mathcal{D}'$ is not realizable in G. That is, a k-insertion set S for G corresponding to $\mathcal{D}'$ would induce a graph with degree sequence $\mathcal{D}' - \mathcal{D}$. Hence, the requirement that there is a graph with degree sequence $\mathcal{D}' - \mathcal{D}$ is a necessary condition to realize $\mathcal{D}'$ in G in Phase 2. Thus, for increasing cost c, by iterating through all k-anonymous sequences $\mathcal{D}'$ with $|\mathcal{D}' - \mathcal{D}| = c$ and excluding the possibility that $\mathcal{D}'$ is realizable in G by the criterion on $\mathcal{D}' - \mathcal{D}$, one can step-wisely improve the lower bound on the k-anonymization cost of G. We apply this strategy and thus our dynamic programming table allows to iterate through all k-anonymous sequences $\mathcal{D}'$ with $|\mathcal{D}' - \mathcal{D}| = c$. Unfortunately, even this criterion might not be sufficient because the already present edges in G might prevent the insertion of a k-insertion set which corresponds to $\mathcal{D}' - \mathcal{D}$ (see Figure 2 for an example). We thus designed a test which not only checks whether $\mathcal{D}' - \mathcal{D}$ is realizable but also takes already present edges in G into account while preserving that $|\mathcal{D}' - \mathcal{D}|$ is a lower bound on the k-anonymization cost of G. With this further requirement on the resulting sequences $\mathcal{D}'$ of Phase 1, in our experiments we observe that Phase 2 of realizing $\mathcal{D}'$ in G is in 26 % of the real-world instances successful. Hence, 26 % of the instances can be solved optimally. See Subsection 3.2 for a detailed description of our algorithm for Phase 1.

For Phase 2 the task is to decide whether a given k-anonymization $\mathcal{D}'$ can be realized in G. As we will show that this problem is NP-hard, we split the problem into two parts and try to solve each part separately by a heuristic. First, we find a degree-vertex mapping, that is, we assign each degree $d_i' \in \mathcal{D}'$ to a vertex v in G such that $d_i' \geq \deg_G(v)$. Then, the demand of vertex v is set to $d_i' - \deg_G(v)$. Second, given a degree-vertex mapping with the corresponding demands we try the find an edge insertion set such that the number of incident new edges for each vertex is equal to its demand. While the second part could in principle be done optimally in polynomial-time by solving an f-factor problem [9], we show that already a heuristic refinement of the "local exchange" heuristic due to Liu and Terzi [11] is able to succeed in most cases. Thus, theoretically and also in our experiments, the "hard part" is to find a good degree-vertex mapping. Roughly speaking, the difficulties are that, according to $\mathcal{D}'$, there is more than one possibility of how many vertices from degree i are increased to degree $j > i$. Even having settled this it is not clear which vertices to choose from block i. See Subsection 3.3 for a detailed description of our algorithm for Phase 2.

3.2 Phase 1: Exact k-Anonymization of Degree Sequences

We start with providing a formal problem description of k-anonymizing a degree sequence $\mathcal{D}$ and describe our dynamic programming algorithm to find such sequences $\mathcal{D}'$. We then describe the criteria that we implemented to improve the lower bound $|\mathcal{D}' - \mathcal{D}|$.

Basic Number Problem. The decision version of the degree sequence anonymization problem reads as follows.

k-DEGREE SEQUENCE ANONYMITY (k-DSA)
Input: A block sequence $\mathcal{B}$ and integers $k, s \in \mathbb{N}$.
Question: Is there a k-anonymous block sequence $\mathcal{B}' \ominus \mathcal{B}$ such that $\|\mathcal{B}' \ominus \mathcal{B}\| = s$?

The requirements on $\mathcal{B}'$ in the above definition ensure that $\mathcal{B}'$ can be obtained by performing exactly s many increases to the degrees in $\mathcal{B}$. Liu and Terzi [11] gave a dynamic programming algorithm that solves k-DSA optimally in $\mathcal{O}(nk)$ time and space. Here, besides using block instead of degree sequences, we added another dimension to the dynamic programming table storing the cost of a solution.

Lemma 1. k-DEGREE SEQUENCE ANONYMITY *can be solved in* $\mathcal{O}(\Delta \cdot k^2 \cdot s)$ *time and* $\mathcal{O}(\Delta \cdot k \cdot s)$ *space.*

There might be multiple minimum solutions for a given k-DSA instance while only one of them is realizable, see Figure 2 for an example. Hence, instead of just computing one minimum-size solution, we iterate through these minimum-size solutions until one solution is realizable or *all* solutions are tested. Observe that there might be exponentially many minimum-size solutions: In the block sequence $\mathcal{B} = \{0, 3, 1, 3, 1, \ldots, 3, 1, 3\}$, for $k = 2$, each subsequence $3, 1, 3$ can be either changed to $2, 2, 3$ or to $3, 0, 4$. We use a data reduction rule to reduce the amount of considered solutions in such instances.

Criteria on the Realizability of k-DSA Solutions. A difficulty in the solutions provided by Phase 1, encountered in our preliminary experiments and as already observed by Lu et al. [12] on a real-world network, is the following: If a solution increases the degree of one vertex v by some amount, say 100, and the overall number of vertices with increased degree is at most 100, then there are not enough neighbors for v to realize the solution. We overcome this difficulty as follows: For a k-DSA-instance $(\mathcal{B}, k)$ and a corresponding solution $\mathcal{B}'$, let S be a k-insertion set for G such that the block sequence of $G + S$ is $\mathcal{B}'$. By definition, the block sequence of the graph induced by the edges S is $\mathcal{B}' \ominus \mathcal{B}$. Hence, it is a necessary condition (for success in Phase 2) that $\mathcal{B}' \ominus \mathcal{B}$ is a *realizable* block sequence, that is, there is a graph with block sequence $\mathcal{B}' \ominus \mathcal{B}$. Tripathi and Vijay [17] have shown that it is enough to check to following *Erdős-Gallai characterization* of realizable degree sequence just once for each block.

Lemma 2 ([7]). *Let* $\mathcal{D} = \{d_1, \ldots, d_n\}$ *be a degree sequence sorted in descending order. Then* $\mathcal{D}$ *is realizable if and only if* $\sum_{i=1}^{n} d_i$ *is even and for each* $1 \leq r \leq n - 1$ *it holds that*

$$\sum_{i=1}^{r} d_i \leq r(r-1) + \sum_{i=r+1}^{n} \min(r, d_i). \tag{1}$$

We call the characterization provided by Lemma 2 the *Erdős-Gallai test*. Unfortunately, there are k-anonymous sequences $\mathcal{D}'$, passing the Erdős-Gallai test, but still or not realizable in the input graph G (see Figure 2 for an example).

We thus designed an advanced version of the Erdős-Gallai test that takes the structure of the input graph into account. To explain the basic idea behind, we first discuss how Inequality (1) in Lemma 2 can be interpreted: Let V^r be the set of vertices corresponding to the first r degrees. The left-hand side sums over the degrees of all vertices in V^r. This amount has to be at most as large as the number of edges (counting each twice) that can be "obtained" by making V^r a clique ($r(r-1)$) and the maximum number of edges to the vertices in $V \setminus V^r$ (a degree-d_i vertex has at most $\min\{d_i, r\}$ neighbors in V^r). The reason why the Erdős-Gallai test might not be sufficient to determine whether a sequence can be realized in G is that it ignores the fact that some vertices in V^r might be already adjacent in G and it also ignores the edges between vertices in V^r and $V \setminus V^r$. Hence, the basic idea of our *advanced Erdős-Gallai test* is, whenever some of the vertices corresponding to the degrees can be uniquely determined, to subtract the corresponding number of edges as they cannot contribute to the right-hand side of Inequality (1).

While the difference between using just the Erdős-Gallai test and the advanced Erdős-Gallai test resulted in rather small differences for the lower bound (at most 10 edges), this small difference was important for some of our instances to succeed in Phase 2 and to optimally solve the instance. We believe that further improving the advanced Erdős-Gallai test is the best way to improve the rate of success in Phase 2.

Complete Strategy for Phase 1. With the above described restriction for realizable k-anonymous degree sequences, we finally arrive at the following problem for Phase 1, stated in the optimization form we solve:

REALIZABLE k-DEGREE SEQUENCE ANONYMITY (k-RDSA)
Input: A degree sequence $\mathcal{B}$ and an integer $k \in \mathbb{N}$.
Task: Compute all k-anonymous degree sequences $\mathcal{B}'$ such that $\mathcal{B}' \ominus \mathcal{B}$, $\|\mathcal{B}' \ominus \mathcal{B}\|$ is minimum, and $\mathcal{B}' \ominus \mathcal{B}$ is realizable.

Our strategy to solve k-RDSA is to iterate (for increasing solution size) through the solutions of k-DSA and run for each of them the advanced Erdős-Gallai test. Thus, we step-wisely increase the respective lower bound $\mathcal{B}' - \mathcal{B}$ until we arrive at some $\mathcal{B}'$ passing the test. Then, for each solution of this size we test in Phase 2 whether it is realizable (if so, then we found an optimal solution). If the realization in Phase 2 fails, then, for each such block sequence $\mathcal{B}'$, we compute how many degrees have to be "wasted" in order to get a realizable sequence. Wasting means to greedily increase some degrees in $\mathcal{B}'$ (while preserving k-anonymity) until the resulting degree sequence is realizable in the input graph. The cost $\mathcal{B}' - \mathcal{B}$ plus the amount of degrees needed to waste in order to realize $\mathcal{B}'$ is stored as an upper-bound. A minimum upper-bound computed in this way is the result of our heuristic.

Due to the power law degree distribution in social networks, the degree of most of the vertices is close to the average degree, thus one typically finds in such instances two large blocks B_i and B_{i+1} containing many thousands of vertices. Hence, "wasting" edges is easy to achieve by increasing degrees from B_i by one to B_{i+1} (this is optimal with respect to the Erdős-Gallai characterization). For the case that two such blocks cannot be found, as a fallback we also implemented a straightforward dynamic programming to find all possibilities to waste edges to obtain a realizable sequence.

Remark. We do not know whether the decision version of k-RDSA (find only one such solution $\mathcal{B}'$) is polynomial-time solvable and resolving this question remains as challenge for future research.

3.3   Phase 2: Realizing a k-Anonymous Degree Sequence

Let (G, k) be an instance of DEGREE ANONYMIZATION and let $\mathcal{B}$ be the block sequence of G. In Phase 1 a k-anonymization $\mathcal{B}'$ of $\mathcal{B}$ is computed such that $\mathcal{B}' \ominus \mathcal{B}$. In Phase 2, given G and $\mathcal{B}'$, the task is to decide whether there is a set S of edge insertions for G such that the block sequence of $G + S$ is equal to $\mathcal{B}'$. We call this the DEGREE REALIZATION problem and first prove that it is NP-hard.

Theorem 1. DEGREE REALIZATION *is NP-hard even on cubic planar graphs.*

We next present our heuristics for solving DEGREE REALIZATION. First, we find a degree-vertex mapping, that is, for $\mathcal{D}' = d'_1, \ldots, d'_n$ being the degree sequence corresponding to $\mathcal{B}'$, we assign each value d'_i to a vertex v in G such that $d'_i \geq \deg_G(v)$ and set $\mathrm{d}(v)$, the demand of v, to $d'_i - \deg_G(v)$. Second, we try to find, mainly by the local exchange heuristic, an edge insertion set S such that in $G + S$ the amount of incident new edges for each vertex v is equal to its demand $d(v)$. The details in the proof of Theorem 1 indeed show that already finding a realizable degree-vertex mapping is NP-hard. This coincides with our experiments, as there the "hard part" is to find a good degree-vertex mapping and the local exchange heuristic is quite successful in realizing it (if possible). Indeed, we prove that "large" solutions can be always realized by it. As a first step for this, we prove that any demand function can be assumed to require to increase the vertex degrees at most up to $2\Delta^2$.

Lemma 3. *Any minimum-size k-insertion set for an instance of* Degree Anonymization *yields a graph with maximum degree at most $2\Delta^2$.*

Theorem 2. *A demand function d is always realizable by the local exchange heuristic in a maximum degree-Δ graph $G = (V, E)$ if $\sum_{v \in V} d(v) \geq 20\Delta^4 + 4\Delta^2$.*

4   Experimental Results

Implementation Setup. All our experiments are performed on an Intel Xeon E5-1620 3.6GHz machine with 64GB memory under the Debian GNU/Linux 6.0 operating system. The program is implemented in Java and runs under the OpenJDK runtime environment in version 1.7.0_25. The time limit for one instance is set to one hour per k-value and we tested for $k = 2, 3, 4, 5, 7, 10, 15, 20, 30, 50, 100, 150, 200$. After reaching the time limit, the program is aborted and the upper and lower bounds computed so far by the dynamic program for Phase 1 are returned. The source code is freely available.[1]

Real-World Instances. We considered the five social networks from the co-author citation category in the 10th DIMACS challenge [5].

We compared the results of our upper bounds against an implementation of the clustering-heuristic provided by Lu et al. [12] and against the lower bounds given by the dynamic program. Our algorithm could solve 26% of the instances to optimality within one hour. Interestingly, our exact approach worked best with the coPapersCiteseer graph from the 10th DIMACS challenge although this graph was the largest one considered (in terms of $n + m$). For all tested values of k except $k = 2$, we could optimally k-anonymize this graph and for $k = 2$ our upper bound heuristic is just two edges away from our lower bound. The coAuthorsDBLP graph is a good representative for the results on the DIMACS-graphs, see Table 1: A few instances could be solved optimally and for the remaining ones our heuristic provides a fairly good upper bound. One can also see that the running times of our algorithms increase (in general) exponentially in k. This behavior captures the fact that our dynamic program for Phase 1 iterates over all minimal solutions and for increasing k the number of these solutions increases dramatically. Our heuristic also suffers from the following effect: Whereas the maximum running time of the clustering-heuristic heuristic was one minute, our heuristic could solve 74% of the instances within one minute and did not finish

---
[1] http://fpt.akt.tu-berlin.de/kAnon/

Table 1. Experimental results on real-world instances. We use the following abbreviations: CH for clustering-heuristic of Lu et al. [12], OH for our upper bound heuristic, OPT for optimal value for the DEGREE ANONYMIZATION problem, and DP for dynamic program for the k-RDSA problem. If the time entry for DP is empty, then we could not solve the k-RDSA instance within one-hour and the DP bounds display the lower and upper bounds computed so far. If OPT is empty, then either the k-RDSA solutions could not be realized or the k-RDSA instance could not be solved within one hour.

graph	k	solution size			DP bounds		time (in seconds)		
		CH	OH	OPT	lower	upper	CH	OH	DP
coAuthorsDBLP	2	97	62		61	61	1.47	0.08	0.043
($n \approx 2.9 \cdot 10^5$,	5	531	321	317	317	317	1.41	0.29	26.774
$m \approx 9.7 \cdot 10^5$,	10	1,372	893		869	869	1.03	0.48	1.58
$\Delta = 336$)	100	21,267	15,050		10,577	11,981	1.13	885.79	
coPapersCiteseer	2	203	80		78	78	9.9	0.1	0.394
($n \approx 4.3 \cdot 10^5$,	5	998	327	327	327	327	10.32	0.19	0.166
$m \approx 1.6 \cdot 10^7$,	10	2,533	960	960	960	960	8.83	0.74	0.718
$\Delta = 1188$)	100	51,456	22,030	22,007	22,007	22,007	5.97	263.95	264.553
coPapersDBLP	2	1,890	1,747		950	1,733	11.28	2.13	
($n \approx 5.4 \cdot 10^5$,	5	9,085	8,219		4,414	8,121	10.66	28.83	
$m \approx 1.5 \cdot 10^7$,	10	19,631	17,571		9,557	17,328	9.95	149.56	
$\Delta = 3299$)	100	258,230			128,143	233,508	22.16		

within the one-hour time limit for 12% of the tested instances. However, the solutions produced by our upper bound heuristic are always smaller than the solutions provided by the clustering-heuristic, on average the clustering-heuristic results are 72% larger than the results of our heuristic.

Random Instances. We generated random graphs according to the model by Barabási–Albert [1] using the implementation provided by the AGAPE project [2] with the JUNG library[2]. Starting with $m_0 = 3$ and $m_0 = 5$ vertices these networks evolve in $t \in \{400, 800, 1200, \ldots, 34000\}$ steps. In each step a vertex is added and made adjacent to m_0 existing vertices where vertices with higher degree have a higher probability of being selected as neighbor of the new vertex. In total, we created 170 random instances.

Our experiments reveal that the synthetic instances are particular hard. For example, even for $k = 2$ and $k = 3$ we could only solve 14% of the instances optimal although our dynamic program produces solutions for Phase 1 in 96% of the instances. For higher values of k the results are even worse (for example zero exactly solved instances for $k = 10$). This indicates that the current lower bound provided by Phase 1 needs further improvements. However, the upper bound provided by our heuristic are not far away: On average the upper bound is 3.6% larger than the lower bound and the maximum is 15%. Further enhancing the advanced Erdős-Gallai test seem to be the most promising step towards closing this gap between lower and upper bound. Comparing our heuristic with the clustering-heuristic reveal similar results as for real-world instances. Our heuristic always beats the clustering-heuristic in terms of solution size, see Figure 3

---

[2] http://jung.sourceforge.net/

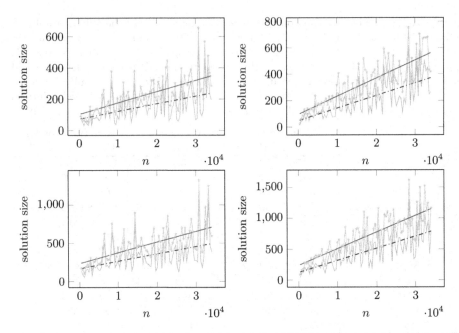

Fig. 3. Comparison of our heuristic (always the light blue line without marks) with the clustering-heuristic (always the light red line with little star as marks) on random data with different parameters: Top row is for $k = 2$, bottom row for $k = 3$; the left column is for $m_0 = 3$, and the right column for $m_0 = 5$. The linear, solid dark red line and dash-dotted blue line are linear regressions of the corresponding data plot. One can see that our heuristic produces always smaller solutions.

for $k = 2$ and $k = 3$. We remark that for larger values of k the running time of the heuristic increases dramatically: For $k = 30$ our algorithm provides upper bounds for 96% of the instances, whereas for $k = 150$ this value drops to 18%.

5   Conclusion

We have demonstrated that our algorithm framework is suitable to solve DEGREE ANONYMIZATION on real-world social networks. The key ingredients for this is an improved dynamic programming for the task to k-anonymize degree sequences together with certain lower bound techniques, namely the advanced Erdős-Gallai test. We have also demonstrated that the local exchange heuristic due to Liu and Terzi [11] is a powerful algorithm for realizing k-anonymous sequences and provided some theoretical justification for this effect.

The most promising approach to speedup our algorithm and to overcome its limitations on the considered random data, is to improve the lower bounds provided by the advanced Erdős-Gallai test. Towards this, and also to improve the respective running times, one should try to answer the question whether one can find in polynomial-time a minimum k-anonymization $\mathcal{D}'$ of a given degree sequence $\mathcal{D}$ such that $\mathcal{D}' - \mathcal{D}$ is realizable.

References

[1] Barabási, A., Albert, R.: Emergence of scaling in random networks. Science 286(5439), 509–512 (1999)

[2] Berthomé, P., Lalande, J.-F., Levorato, V.: Implementation of exponential and parametrized algorithms in the AGAPE project. CoRR, abs/1201.5985 (2012)

[3] Casas-Roma, J., Herrera-Joancomartí, J., Torra, V.: An algorithm for k-degree anonymity on large networks. In: Proc. ASONAM 2013, pp. 671–675. ACM Press (2013)

[4] Chester, S., Gaertner, J., Stege, U., Venkatesh, S.: Anonymizing subsets of social networks with degree constrained subgraphs. In: Proc. ASONAM 2012, pp. 418–422. IEEE Computer Society (2012)

[5] DIMACS 2012. Graph partitioning and graph clustering. 10th DIMACS challenge (2012), http://www.cc.gatech.edu/dimacs10/ (accessed April 2012)

[6] Easley, D., Kleinberg, J.: Networks, Crowds, and Markets. Cambridge University Press (2010)

[7] Erdős, P., Gallai, T.: Graphs with prescribed degrees of vertices. Math. Lapok 11, 264–274 (1960) (in Hungarian)

[8] Fung, B.C.M., Wang, K., Chen, R., Yu, P.S.: Privacy-preserving data publishing: A survey of recent developments. ACM Computing Surveys 42(4), 14:1–14:53 (2010)

[9] Hartung, S., Nichterlein, A., Niedermeier, R., Suchý, O.: A refined complexity analysis of degree anonymization in graphs. In: Fomin, F.V., Freivalds, R., Kwiatkowska, M., Peleg, D. (eds.) ICALP 2013, Part II. LNCS, vol. 7966, pp. 594–606. Springer, Heidelberg (2013)

[10] Hartung, S., Hoffmann, C., Nichterlein, A.: Improved upper and lower bound heuristics for degree anonymization in social networks. CoRR, abs/1402.6239 (2014)

[11] Liu, K., Terzi, E.: Towards identity anonymization on graphs. In: Proc. SIGMOD 2008, pp. 93–106. ACM (2008)

[12] Lu, X., Song, Y., Bressan, S.: Fast identity anonymization on graphs. In: Liddle, S.W., Schewe, K.-D., Tjoa, A.M., Zhou, X. (eds.) DEXA 2012, Part I. LNCS, vol. 7446, pp. 281–295. Springer, Heidelberg (2012)

[13] Samarati, P.: Protecting respondents identities in microdata release. IEEE Transactions on Knowledge and Data Engineering 13(6), 1010–1027 (2001)

[14] Samarati, P., Sweeney, L.: Generalizing data to provide anonymity when disclosing information. In: Proc. PODS 1998, pp. 188–188. ACM (1998)

[15] Sweeney, L.: k-anonymity: A model for protecting privacy. International Journal of Uncertainty, Fuzziness and Knowledge-Based Systems 10(5), 557–570 (2002)

[16] Thompson, B., Yao, D.: The union-split algorithm and cluster-based anonymization of social networks. In: Proc. 4th ASIACCS 2009, pp. 218–227. ACM (2009)

[17] Tripathi, A., Vijay, S.: A note on a theorem of Erdös & Gallai. Discrete Math. 265(1-3), 417–420 (2003)

[18] Zhou, B., Pei, J.: The k-anonymity and l-diversity approaches for privacy preservation in social networks against neighborhood attacks. Knowledge and Information Systems 28(1), 47–77 (2011)

[19] Zhou, B., Pei, J., Luk, W.: A brief survey on anonymization techniques for privacy preserving publishing of social network data. ACM SIGKDD Explorations Newsletter 10(2), 12–22 (2008)

Search Space Reduction through Commitments in Pathwidth Computation: An Experimental Study

Yasuaki Kobayashi[1], Keita Komuro[2], and Hisao Tamaki[2]

[1] Gakushuin University, Toshima-ku, Japan 171-8588
yasuaki.kobayashi@gakushuin.ac.jp
[2] Meiji University, Kawasaki, Japan 214-8571
{kouki-metal,tamaki}@cs.meiji.ac.jp

Abstract. In designing an XP algorithm for pathwidth of digraphs, Tamaki introduced the notion of commitments and used them to reduce the search space with naively $O(n!)$ states to one with $n^{O(k)}$ states, where n is the number of vertices and k is the pathwidth of the given digraph. The goal of the current work is to evaluate the potential of commitments in heuristic algorithms for the pathwidth of *undirected graphs* that are aimed to work well in practice even for graphs with large pathwidth. We classify commitments by a simple parameter called depth. Through experiments performed on TreewidthLIB instances, we show that depth-1 commitments are extremely effective in reducing the search space and lead to a practical algorithm capable of computing the pathwidth of many instances for which the exact pathwidth was not previously known. On the other hand, we find that the additional search space reduction enabled by depth-d commitments with $2 \leq d \leq 10$ is limited and that there is little hope for effective heuristics based on commitments with such depth.

1 Introduction

Pathwidth [16] and treewidth [17] are among the central notions in the graph minor theory developed by Robertson and Seymour and have numerous applications in algorithm design. In particular, many NP-hard graph problems are fixed parameter tractable [11] when parameterized by pathwidth or treewidth: they have algorithms with running time $f(w)n^{O(1)}$, where n is the instance size, w is the pathwidth or treewidth of the instance graph, and f is a typically exponential function of w. Such algorithms are often practical when w is small and therefore computing these width parameters (and constructing associate graph decompositions) is of great practical importance. Theoretically, the problems of computing the treewidth and the pathwidth are both NP-hard [1,12], although they are fixed parameter tractable admitting algorithms with running time linear in the graph size [5,4].

Unfortunately, the running time of the fixed parameter algorithm given by [5,4] has huge dependence on the width parameter and generally considered impractical. From the practical point of view, however, "finding a tree-decomposition of

J. Gudmundsson and J. Katajainen (Eds.): SEA 2014, LNCS 8504, pp. 388–399, 2014.

small width is far from hopeless" [8] and there has been a considerable amount of effort on turning this hope into reality [15,2,8]. For example, van den Broek and Bodlaender provide a benchmark suite TreewidthLIB [3] and lists known upper and lower bounds on the treewidth of most of the graph instances therein (see [2] for a method of computation used to derive such bounds). According to the description of the library [3], the instances there are collected with the criterion that finding tree-decompositions of small width is useful for solving some problem of real interest on those instances.

Compared with this situation for treewidth, the research effort on practically computing the pathwidth seems to be much more scarce. In particular, computing the pathwidth of TreewidthLIB instances would seem a valid goal of experimental research but no such report can be found in the literature.

The work reported in this paper significantly improves this situation. Our experimental results show that "finding a path-decomposition of small width is also far from hopeless" in practical situations. More specifically, our improved algorithms for pathwidth perform well on TreewidthLIB instances. Their performances are comparable to those of treewidth algorithms reported in the TreewidthLIB site. On some instances, they are even able to improve the best known treewidth upper bound by finding a path-decomposition, a special case of a tree-decomposition, with a smaller width.

Our basic algorithm is a backtrack search for a vertex sequence corresponding to an optimal path-decomposition, employing the standard memoization technique (see [10], for example). This basic algorithm may be viewed as a practical implementation of the standard vertex ordering approach for pathwidth [6] (see also [7] for a similar approach for treewidth): the dynamic programming algorithm of [6] stores in the table the solutions of all possible subproblems, while the corresponding table in our algorithm stores solutions of only those subproblems that are encountered in the backtrack search. To reduce the search space of this basic algorithm, we use the notion of *commitments* introduced by Tamaki [19]. He used commitments to obtain a theoretical result: an algorithm for the pathwidth of digraphs that runs in $n^{O(k)}$ time, where n is the number of vertices and k is the pathwidth of the given graph. This notion is also used in [14] to derive an $O(1.89^n)$ time algorithm for the pathwidth of directed and undirected graphs, which improves on the $O(1.9657^n)$ time algorithm of Suchan and Villanger [18] for the pathwidth of undirected graphs. The goal of this paper is to evaluate the potential of commitments in heuristic algorithms that are aimed at performing better in practice than guaranteed by the theoretical bounds.

A commitment occurs between a pair of search states S and T, where T is a descendant of S in the search tree. Under a certain condition, we discard all descendants of S but T, knowing that S leads to a successful computation if and only if T does (see Section 3 for more details of commitments). We define the *depth* of the commitment to be the depth of T in the search tree minus the depth of S. Some variants of depth-1 commitments can be found in theoretical work on pathwidth [18,14]. Our experiments on TreewidthLIB show that depth-1 commitments are extremely effective in reducing the search space. Depth-1 commitments also have the advantage of being "cheap" in that they can be detected with very small computational effort. Indeed, the memoized backtrack search with depth-1 commitments performs well on small to medium-sized TreewidthLIB instances: out of the total of 162 instances therein with 300 or fewer vertices, our algorithm is successful in computing the exact pathwidth of 145 instances. This number compares favorably with 65, which is the number of instances, out of the same set, on which the exact treewidth is known. For fair comparisons, we need to note that those exact bounds for treewidth are rather old (obtained in 2007 or earlier). Despite continued efforts for improvements (see [9], for example), however, no new exact bounds have been reported in TreewidthLIB or elsewhere, to the best of the authors' knowledge.

Depth-d commitments for larger d are more costly. Naively, to find if there is a depth-d commitment from state S, we need an exhaustive search of depth d from S. In most practical situations, this cost is much greater than the gain we obtain from reducing the search space. There may be, however, some lower-cost heuristics that can be used to detect deep commitments not always but often enough to be useful. Our experiments on depth-d commitments for $d \geq 2$ give a ground for evaluating such heuristic potential of commitments. In the experiments, we assume an oracle that, given a search state S, detects a commitment from S within specified depth if one exists. If the reduction of the search space is found significant in these experiments, then it would be worthwhile to look for heuristic implementations of these oracles. Our experiments on TreewidthLIB instances show, however, that the space reduction effect of depth-d commitments for $2 \leq d \leq 10$ is rather limited and suggest that there is little hope of improvements by such heuristics over the algorithm with depth-1 commitments.

A byproduct of our experiments is a finding that the gap between the pathwidth and the treewidth may be dramatically smaller on practical instances than theoretically possible. See Subsection 4.5 for details.

The rest of this paper is organized as follows. Section 2 gives definitions and notation used in this paper. Section 3 describes the general principle of commitments together with necessary definitions. Section 4 describes our experiments. We conclude the paper with Section 5.

2   Preliminaries

Let G be an undirected graph with vertex set $V(G)$ and edge set $E(G)$. For $v \in V(G)$, the set of neighbors of v is denoted by $N(v)$ and the number of them is denoted by $d(v)$. We extend this notation to sets: for $X \subseteq V(G)$, $N(X) = \bigcup_{v \in X} N(v) \setminus X$ and $d(X) = |N(X)|$.

A *path decomposition* of G is a sequence of subsets $(X_1, X_2, \ldots, X_t)$ of $V(G)$ satisfying the following conditions:

1. $\bigcup_{1 \leq i \leq t} X_i = V(G)$,
2. for each $\{u, v\} \in E(G)$, there is an index i such that $\{u, v\} \subseteq X_i$, and
3. for each $v \in V(G)$, the set of indices i such that $v \in X_i$ forms a single interval.

The *width* of a path decomposition $(X_1, X_2, \ldots, X_t)$ is $\max_{1 \le i \le t} |X_i| - 1$. The *pathwidth* $\mathrm{pw}(G)$ of G is the smallest k such that G has a path decomposition of width k. In this paper, we use an alternative characterization of pathwidth, known as the *vertex separation number*, which we define below.

Let σ be a sequence of vertices. We assume all the sequences of vertices in this paper are without repetitions, i.e., all the elements in σ are distinct from each other. We denote by $V(\sigma)$ the set of vertices in σ and the length $|V(\sigma)|$ of σ by $|\sigma|$. When $V(\sigma) = V(G)$, we call σ a *permutation* of $V(G)$. Suppose σ and η are sequences with $V(\sigma) \cap V(\eta) = \emptyset$ and τ is the result of concatenating η after σ. Then, σ is a *prefix* of τ (a *proper prefix* if η is nonempty) and τ is an *extension* of σ (a *proper extension* if η is nonempty). For a non-negative integer k, σ is *k-feasible* if $d(V(\sigma')) \le k$ for each prefix σ' of σ and is *strongly k-feasible* if there is a k-feasible extension of σ that is a permutation of $V(G)$. These notions are extended for sets: a set $S \subseteq V(G)$ is (strongly) k-feasible if there is some (strongly) k-feasible sequence σ such that $V(\sigma) = S$.

The *vertex separation number* of G is the minimum integer k such that $V(G)$ is k-feasible. It is known that the pathwidth of G equals the vertex separation number of G [13]. Our algorithm works on the vertex separation number, rather than directly on the pathwidth. We also note that our algorithm outputs a k-feasible sequence for $k = \mathrm{pw}(G)$, from which it is straightforward to construct a path decomposition of G of width k.

3   Commitments

Fix G and k. Our algorithm looks for k-feasible permutations working on the search tree whose nodes are k-feasible sets. To prune the search tree, we use the following notion of *commitments* introduced in [19].

Let σ be a k-feasible sequence. Following [19], we say that an extension τ of σ is a *k-committable extension* of σ if τ is a proper extension of σ, τ is k-feasible, and $d(X) \ge d(V(\tau))$ for every X with $V(\sigma) \subseteq X \subseteq V(\tau)$. We also use the set version of this definition: T is a k-committable extension of S if there is a k-feasible sequence σ and a k-committable extension τ of σ such that $V(\sigma) = S$ and $V(\tau) = T$.

Lemma 1 ([19]). *Suppose sequence σ is strongly k-feasible and τ is a k-committable extension of σ. Then τ is also strongly k-feasible.*

Corollary 1. *Suppose $S \subset V(G)$ is strongly k-feasible and T is a k-committable extension of S. Then T is also strongly k-feasible.*

Thus, if a search-tree node S has a k-committable extension T then we may commit to T: all the descendants of S but T and its descendants may be ignored without losing the completeness of the search.

Although the use of commitments is a powerful pruning strategy leading to theoretical results in [19,14], our preliminary experiments showed that the use of commitments in their full generality does not result in practically efficient algorithms. The reason of this is the huge overhead of finding k-committable extensions. A naive method of finding a k-committable extension of a given search node S is to do an exhaustive search through the descendants of S and the cost of this auxiliary search overweighs the gain in the reduction of the main search space.

However, in practical algorithms, it is not necessary to use a method that finds a k-committable extension whenever one exists. It is possible that a less costly heuristic method is effective if it succeeds in finding k-committable extensions often enough.

The goal of the current paper is not to evaluate the effectiveness of various heuristics in finding committable extensions but to evaluate the *potential* of the heuristic approach itself. For this goal, we assume oracles that, given a k-feasible set S, return either S itself or a k-committable extension of S. In this manner, we decouple the effect of search space reduction enabled by commitments from the cost of finding commitments. Only after we confirm significant reduction in the search space size, we may pursue efficient, but probably partial, implementations of those oracles. The details of the oracles we use in our experiments are described in the next section.

4   Experiments

4.1   Algorithm

The pseudocode listed below describes our recursive search procedure. For a fixed pair of graph G and positive integer k, it decides if the input vertex set S is strongly k-feasible or not. This procedure implements a standard backtrack search with memoization: it uses a table, called a *failure table*, which stores sets that are found not-strongly k-feasible in order to avoid duplicated search from such sets. It also uses an oracle f, an additional parameter given to the algorithm, for finding k-committable extensions: f can be an arbitrary function such that, for each k-feasible vertex set S, $f(S)$ is either a k-committable extension of S or is equal to S itself. We denote by f^* the limit of this function: for each S, $f^*(S) = f^m(S)$ where m is such that $f^m(S) = f^{m+1}(S)$.

The input for the initial call of this procedure is the empty set: the empty set is strongly k-feasible if and only if the vertex separation number of G is k or smaller.

4.2   Oracles

In the experiments reported here, we use the following oracles. We say that a k-committable extension T of S is of *depth d* where $d = |T| - |S|$. For each non-negative integer d, we define function f_d as follows: $f_d(S)$ is a k-committable extension of the smallest depth d' with $1 \leq d' \leq d$ if one exists; $f_d(S) = S$ otherwise. For convenience, we are allowing d to be 0: f_0 is an "empty oracle" that, given S, always returns S itself.

Algorithm 1. Decides whether S is strongly k-feasible or not

```
1: procedure STRONGLY-FEASIBLE(S)
2:     T ← f*(S).
3:     if T is in the failure table then
4:         return false
5:     end if
6:     if T = V(G) then
7:         return true
8:     end if
9:     for all v ∈ V \ T such that d(T ∪ {v}) ≤ k do
10:         if STRONGLY-FEASIBLE(T ∪ {v}, k) then
11:             return true
12:         end if
13:     end for
14:     Store T in the failure table.
15:     return false
16: end procedure
```

In our experiments, the oracle f_d is implemented by an exhaustive search that costs $O(n^d)$ time. Recall that the purpose of the experiments is not to evaluate the overall efficiency of the search incorporating these oracles but to measure the search space reduction enabled by free uses of those oracles.

4.3   Search Space Reduction

The first part of our experiments measures the size of the search space in terms of the number of vertex sets stored in the failure table. Note that the number of successful vertex sets is at most the number of vertices of the graph and is negligibly smaller than that of unsuccessful ones in typical situations. We use the oracles f_0, f_1, f_5 and f_{10} in these experiments.

We have performed this experiment on 108 instances of TreewidthLIB. Remaining roughly 100 instances, for which running the algorithm with the empty oracle f_0 is already prohibitively time- or space-consuming, are excluded.

The parameter k given to the algorithm is exactly the pathwidth minus one. This is usually the last and the most time consuming step in the pathwidth computation: we know that the instance is $k+1$-feasible from the previous steps and have to confirm the infeasibility for the current k.

Table 1 shows the size of the search space generated by our algorithm with oracles f_0, f_1, f_5, and f_{10}, in terms of the number of sets stored in the failure table, for some sample instances. Columns ltw and utw are the lower and upper bounds on the treewidth. The numbers for other instances show similar tendencies. Table 2 summarizes the effect of search space reduction on all of the tested instances. In each row of this table labeled by oracle f, instances are classified by the search space size with oracle f relative to the search space size with the empty oracle. The parenthesized breakdown (a, b) means that a is the number of instances with 100 or fewer vertices and b is the number of larger instances.

As seen from the second table, for half of all the instances, the reduced search space by depth-1 commitments has size within 5% of the original search space without commitments and, for about 85% of all the instances, within 20% of the original. For 90% of the instances with more than 100 vertices, the reduced search space has size within 5% of the original. We can also see that using deeper commitments does not change this situation significantly.

Table 1. Search space size of sample instances

instance	$|V(G)|$	$|E(G)|$	pw(G)	ltw	utw	f_0	f_1	f_5	f_{10}
1bx7	41	195	11	11	11	898	63	59	56
1g6x	52	405	19	19	19	1572	148	130	129
1bbz	57	543	25	25	25	5700	546	516	516
1a8o	64	536	25	23	25	33522	1350	1155	1149
queen9_9	81	1056	58	50	58	372842	256364	253980	253980
1aba	85	886	28	28	29	73212	1561	1456	1429
1d4t	102	1145	34	32	35	625556	8170	7159	7090
1f9m	109	1349	43	38	45	13299246	68317	59588	59062
ch150.tsp	150	432	13	8	15	663258	24236	22213	21814
u159.tsp	159	431	12	8	12	7801396	20231	17413	16434
kroA200.tsp	200	586	13	9	14	1371893	33807	32129	31870
tsp225.tsp	225	622	13	11	15	12079238	85824	80728	78654
diabetes	413	819	6	4	4	125888	95224	84718	84571

Table 2. Relative search space size: summary of 108 instances

Oracle used	0-5%	5-10%	10-20%	20-40%	40 - 70%	70 - 100%
f_0	0	0	0	0	0	108 (88,20)
f_1	54 (36,18)	27 (26,1)	12 (12,0)	5 (5,0)	4 (4,0)	6 (5,1)
f_5	57 (38,19)	30 (30,0)	9 (9,0)	3 (3,0)	4 (3,1)	6 (5,0)
f_{10}	57 (38,19)	30 (30,0)	9 (9,0)	3 (3,0)	4 (3,1)	6 (5,0)

There are instances for which the reduction is small. For some of them, such as queen9_9 listed in Table 1, we know the reason: they are full of large cliques. For example, queen9_9, which consists of 81 vertices, contains 20 cliques with 9 vertices.

Table 3 shows some statistics on the outcome of the oracle calls made in the execution of the algorithm with oracle f_{10} for the same sample instances. For each instance, the column "#calls" shows the total number of vertex sets S for which the oracle is invoked; column 0 shows the number of such sets for which the oracle returned the given vertex set S itself (which equals the number of sets S for which f^* is invoked), and column d, $1 \leq d \leq 10$, shows the number of such sets for which the oracle returned a k-committable extension of depth d. Note that the behavior of the backtrack search is such that call $f(S)$ for the same set S can be performed many times. The figures in the table are the number of calls after this multiplicity is removed.

The figures in the table consistently show that the opportunities for commitments become more scarce as the depth of commitments increases.

Table 3. Outcomes of oracle calls

instance	#calls	0	1	2	3	4	5	6	7	8	9	10
1bx7	434	230	198	2	0	1	1	2	0	0	0	0
1g6x	1164	674	473	11	1	4	0	1	0	0	0	0
1bbz	4956	3238	1688	19	8	3	0	0	0	0	0	0
1a8o	15068	8611	6265	116	52	8	10	2	3	1	0	0
queen9_9	1454528	1337350	114794	1968	116	236	64	0	0	0	0	0
1aba	15551	9305	6115	45	20	33	7	9	5	6	4	2
1d4t	102199	60376	40801	724	125	64	42	15	17	14	13	8
1f9m	1096583	642067	445541	5269	1721	1104	372	198	144	69	57	41
ch150.tsp	211702	126760	82858	773	615	123	216	120	71	87	41	38
u159.tsp	234363	106988	124579	969	911	168	115	209	167	81	144	32
kroA200.tsp	290108	180055	108257	364	800	254	150	71	71	36	32	18
tsp225.tsp	836232	466287	363612	798	1528	1233	988	511	800	75	144	256
diabetes	212280	200408	10526	1072	232	35	4	0	0	1	2	0

Table 4 shows some statistics on the pruning effect of commitments during the execution of the algorithm with oracle f_{10} for the sample instances. For each vertex set S for which the oracle is invoked and returned a k-committable extension T, $|T| > |S|$, we count the number of descendants of S with cardinality $|T|$ in the search tree when oracle f_1 instead of f_{10} is used. This number, which we denote by p_S, can be seen as indicating the pruning effect of the commitment from S to T. The column d, $1 \le d \le 10$, in the table shows the average of p_S, over all S with $|f_{10}(S)| - |S| = d$.

The pruning effect of deep commitments as shown by the figures in the table is not as large as one might expect. A possible interpretation is that depth-1 commitments are already effective in reducing the branching factor of the search tree and leave small room for further pruning by deeper commitments. Together with Table 3 which shows the small opportunities of deeper commitments, it explains the small effect of additional search space reduction achieved by deeper commitments over depth-1 commitments.

4.4   Performance of the Algorithm with Depth-1 Commitments

The second part of our experiments focuses more on the actual performance of our algorithm with depth-1 commitments, in contrast to the first part which is concerned only with the size of the search space. Here, we are concerned with the actual running time, memory usage, and whether the algorithm is capable of solving each instance in a reasonable amount of time.

Table 4. Pruning effect of commitments

instance	1	2	3	4	5	6	7	8	9	10
1bx7	1	3.5	-	1	2	2.5	-	-	-	-
1g6x	1	2.64	1	3.25	-	1	-	-	-	-
1bbz	1	4.37	3	4	-	-	-	-	-	-
1a8o	1	7.05	6.54	5.13	4	1	1.33	1	-	-
queen9_9	1	1.78	1.10	1.05	1	-	-	-	-	-
1aba	1	5.69	6.85	7.48	5.71	2.56	2	2	2.25	1
1d4t	1	12.28	8.9	8.88	7.62	5.07	5.47	5.5	5.31	3.25
1f9m	1	17.05	17.89	18.74	13.34	11.53	7.97	7.77	5.84	4.95
ch150.tsp	1	10.69	7.9	6.13	6.82	6.85	4.66	5.11	5.98	3.13
u159.tsp	1	15.67	16.66	11.29	12.06	12.73	13.86	10.62	9.1	5.38
kroA200.tsp	1	7.89	6.94	5.87	4.54	4.62	3.76	4.36	3.13	1.89
tsp225.tsp	1	12.57	14.33	10.92	10.47	8.16	7.72	5.64	6.94	7.77
diabetes	1	6.47	2.1	11.23	1.5	-	-	13	1	-

This part of the experiment is run on a machine with Intel Xeon E5606 (2.14GHz × 4) processor, 5.8GB RAM, and Ubuntu Linux. We do not use multi-thread for the execution of our algorithms. The heap space allocated for the Java VM is 1GB for all instances.

For each instance, we first compute an upper bound on the pathwidth by a simple greedy heuristic, set initial value of k to this upper bound, and repeat our algorithm for k-feasibility, decreasing k one by one until we find the instance not k-feasible: the final value of k is the pathwidth minus one. If this process is not completed in 30 minutes, we stop the execution and report the current upper bound on the pathwidth.

In the description of the experimental results below, when we say upper or lower bounds on the treewidth, they mean those bounds listed in TreewidthLIB. We also say that the exact treewidth is known for some instance, if the listed upper and lower bounds for the instance match.

We have run our algorithms on basically all instances in TreewidthLIB, for which the upper and lower bounds on the treewidth are listed. We have excluded, however, those instances that are the result of preprocessing, which simplifies the graph without changing the treewidth (but possibly changing the pathwidth). We have also excluded 3 weighted instances. We have selected 207 instances from TreewidthLIB by these criteria.

Table 5 summarizes the number of instances for which the pathwidth computation was successful, in the sense the exact pathwidth was obtained in 30 minutes, with oracle f_0 or f_1. Column 'total' shows the total number of tested instances in the specified range. The basic algorithm with the empty oracle is already effective for small instances. For instances with over 100 vertices, however, the algorithm with the depth-1 oracle clearly outperforms the basic algorithm.

For comparison, Table 5 also lists the number of instances for which the exact treewidth is known. Note that this comparison is not meant for exactly

Table 5. The number of instances for which the exact pathwidth computation is successful

$|V(G)|$	total	f_0	f_1	treewidth known
1 – 50	17	17	17	17
51 – 100	81	70	80	32
101 – 200	56	19	42	15
201 – 300	7	1	5	3
301 –	46	1	2	24

measuring the relative difficulty of computing the exact pathwidth and the exact treewidth: the amount of time spent for the bounds listed in TreewidthLIB is typically smaller while the amount of time spent for unsuccessful efforts for improving the bounds is not known.

Table 6 lists more computational details of some selected instances. Columns ltw and utw are the lower and upper bounds on the treewidth; ipw is the pathwidth achieved by the initial greedy solution; pw_d, for $d = 0, 1$ is the upper bound on the pathwidth obtained by the iteration using Algorithm 1 with oracle f_d, which is the exact pathwidth unless the computation is aborted; t_d, for $d = 0, 1$ is the time, in seconds, consumed by oracle f_d. TLE means that the computation is aborted with the time limit of 30 minutes and MLE means that the computation is aborted because the heap space is exhausted.

From the results for larger instances in this list, we can see that the search space reduction effect of depth-1 commitments observed in the first part of the experiments indeed leads to dramatic improvements on the running time.

The largest two instances in this list are in contrast to each other: even though the exact pathwidth is not obtained for either of the instances, the upper bound obtained for u2319.tsp still improves the upper bound on the treewidth, while for BN_26 the computed upper bound on the pathwidth is far above the upper bound on the treewidth.

4.5   Comparisons between the Pathwidth and the Treewidth

Table 7 compares the pathwidth and the treewidth of the tested instances. For each range of the number of vertices, column 'total' shows the number of instances in the range for which the exact treewidth is known and moreover the exact pathwidth is obtained by our computation; other columns show the breakdown of this total number according to the difference between the pathwidth and the treewidth. For the columns with $pw = tw + 1$ and $pw = tw + 2$, the numbers in the parentheses show the treewidth of the individual instances counted in that column. Among all the tested instances, there are 16 instances with treewidth smaller than 10, 10 of which have 50 or fewer vertices. It is surprising that the two width parameters are identical for most of the instances in this category.

Table 6. Computational details for some selected instances

G	$\|V(G)\|$	$\|E(G)\|$	ltw	utw	ipw	pw_0	pw_1	t_0	t_1
1g6x	52	405	19	19	23	19	19	0.13	0.12
1bbz	57	543	25	25	26	25	25	0.13	0.21
1a8o	64	536	23	25	27	25	25	0.24	0.46
1cc8	70	813	27	32	33	32	32	0.20	0.89
queen9_9	81	1056	50	58	61	58	58	2.5	94
1aba	85	886	28	29	30	28	28	0.56	0.96
1c5e	95	1148	33	36	38	34	34	1.3	2.4
1d4t	102	1145	32	35	41	34	34	4.4	4.3
1f9m	109	1349	38	45	47	43	43	251	60
bier127.tsp	127	368	8	15	22	15	15	39	4.4
ch150.tsp	150	432	8	15	17	13	13	7.5	1.5
u159.tsp	159	431	8	12	19	12	12	111	1.5
kroA200.tsp	200	586	9	14	25	13	13	25	2.3
tsp225.tsp	225	622	11	15	21	13	13	685	5.1
diabetes	413	819	4	4	31	6	6	5.6	2.6
celar06	100	350	11	11	18	11	11	TLE	5.1
graph01	100	358	21	24	38	23	23	TLE	1272
1bkb	131	1485	26	30	31	29	29	MLE	1.4
anna	138	493	12	12	24	15	14	TLE	TLE
a280.tsp	280	788	12	14	19	14	14	TLE	28
fpsol2.i.1	496	11654	66	66	79	75	67	TLE	323
u2319.tsp	2319	6869	41	56	70	50	47	TLE	TLE
BN_26	3025	14075	9	9	1005	103	103	TLE	TLE

Table 7. Classifying the instances based on the comparison between the pathwidth and the treewidth

$\|V(G)\|$	total	pw = tw	pw = tw + 1	pw = tw + 2	pw $\geq$ tw + 3
1 – 50	17	12	5(3, 3, 4, 9, 19)	0	0
51 – 100	32	31	1(9)	0	0
101 – 200	10	8	2(6, 9)	0	0
201 – 300	3	3	0	0	0
300 –	2	0	1(66)	1(4)	0

5 Conclusion

The results of our experiments show that depth-1 commitments are quite effective in reducing the search space and lead to dramatic improvements of the performance of the vertex ordering approach for pathwidth. On the other hand, the results on deeper commitments are negative. Even assuming the oracles that detect those commitments without any cost, the improvement over the depth-1 commitments would be slim. This suggests that looking for heuristics for detecting deeper commitments is probably not worth the effort. This conclusion, however, is not final, as experiments are done only for commitments of depth up to 10. There still remains a small possibility that commitments of much larger depth are useful in heuristic algorithms.

References

1. Arnborg, S., Corneil, D., Proskurowski, A.: Complexity of finding embeddings in a k-tree. SIAM Journal on Matrix Analysis and Applications 8(2), 277–284 (1987)
2. Bachoore, E.H., Bodlaender, H.L.: A branch and bound algorithm for exact, upper, and lower bounds on treewidth. In: Cheng, S.-W., Poon, C.K. (eds.) AAIM 2006. LNCS, vol. 4041, pp. 255–266. Springer, Heidelberg (2006)
3. van den Broek, J., Bodlaender, H.L.: TreewidthLIB, http://www.cs.uu.nl/research/projects/treewidthlib/ (accessed February 9 2014)
4. Bodlaender, H.L., Kloks, T.: Efficient and constructive algorithms for the pathwidth and treewidth of graphs. Journal of Algorithms 21, 358–402 (1996)
5. Bodlaender, H.L.: A linear-time algorithm for finding tree-decompositions of small treewidth. SIAM Journal on Computing 25(6), 1305–1317 (1996)
6. Bodlaender, H.L., Fomin, F.V., Koster, A.M.C.A., Kratsch, D., Thilikos, D.M.: A note on exact algorithms for vertex ordering problems on graphs. Theory of Computing Systems 50(3), 420–432 (2012)
7. Bodlaender, H.L., Fomin, F.V., Koster, A.M.C.A., Kratsch, D., Thilikos, D.M.: On exact algorithms for treewidth. ACM Transactions on Algorithms 9(1), 12 (2012)
8. Bodlaender, H.L., Grigoriev, A., Koster, A.M.C.A.: Treewidth lower bounds with brambles. Algorithmica 51(1), 81–98 (2008)
9. Bodlaender, H.L., Koster, A.M.C.A.: Treewidth Computations II. Lower Bounds. Information and Computtation 209(7), 1103–1119 (2011)
10. Cormen, T.H., Leiserson, C.E., Rivest, R.L., Stein, C.: Introduction to algorithms. The MIT Press, Boston (2001)
11. Downey, R.G., Fellows, M.R.: Parameterized complexity. Springer, Berlin (1998)
12. Kashiwabara, T., Fujisawa, T.: NP-completeness of the problem of finding a minimum-clique-number interval graph containing a given graph as a subgraph. In: Proceedings of International Symposium on Circuits and Systems, pp. 657–660 (1979)
13. Kinnersley, G.N.: The vertex separation number of a graph equals its path-width. Information Processing Letters 42(6), 345–350 (1992)
14. Kitsunai, K., Kobayashi, Y., Komuro, K., Tamaki, H., Tano, T.: Computing directed pathwidth in $O(1.89^n)$ time. In: Thilikos, D.M., Woeginger, G.J. (eds.) IPEC 2012. LNCS, vol. 7535, pp. 182–193. Springer, Heidelberg (2012)
15. Koster, A.M.C.A., Bodlaender, H.L., van Hoesel, S.P.: Treewidth: computational experiments. Electronic Notes in Discrete Mathematics 8, 54–57 (2001)
16. Robertson, N., Seymour, P.D.: Graph minors. I. Excluding a forest. Journal of Combinatorial Theory, Series B 35(1), 39–61 (1983)
17. Robertson, N., Seymour, P.D.: Graph minors. II. Algorithmic aspects of tree-width. Journal of Algorithms 7(3), 309–322 (1984)
18. Suchan, K., Villanger, Y.: Computing pathwidth faster than 2^n. In: Chen, J., Fomin, F.V. (eds.) IWPEC 2009. LNCS, vol. 5917, pp. 324–335. Springer, Heidelberg (2009)
19. Tamaki, H.: A polynomial time algorithm for bounded directed pathwidth. In: Kolman, P., Kratochvíl, J. (eds.) WG 2011. LNCS, vol. 6986, pp. 331–342. Springer, Heidelberg (2011)

Efficient Representation
for Online Suffix Tree Construction

N. Jesper Larsson, Kasper Fuglsang, and Kenneth Karlsson

IT University of Copenhagen, Denmark
{jesl,kfug,kkar}@itu.dk

Abstract. Suffix tree construction algorithms based on *suffix links* are popular because they are simple to implement, can operate *online* in linear time, and because the suffix links are often convenient for pattern matching. We present an approach using *edge-oriented* suffix links, which reduces the number of branch lookup operations (known to be a bottleneck in construction time) with some additional techniques to reduce construction cost. We discuss various effects of our approach and compare it to previous techniques. An experimental evaluation shows that we are able to reduce construction time to around half that of the original algorithm, and about two thirds that of previously known branch-reduced construction.

1 Introduction

The *suffix tree* is arguably the most important data structure in string processing, with a wide variety of applications [1,10,14], and with a number of available construction algorithms [3,5,9,17,22,23], each with its benefits. Improvements in its efficiency of construction and representation continues to be a lively area of research, despite the fact that from a classical asymptotic time complexity perspective, optimal solutions have been known for decades. Pushing the edge of efficiency is critical for indexing large inputs, and make large amounts of experiments feasible, e.g., in genetics, where lengths of available genomes increase. Much work has been dedicated to reducing the memory footprint with representations that are compact [13] or compressed (see Cánovas and Navarro [3] for a practical view, with references to theoretical work), and to alternatives requiring less space, such as suffix arrays [15]. Other work adresses the growing performance-gap between cache and main memory, frequently using algorithms originally designed for secondary storage [4,6,20,21].

While memory-reduction is important, it typically requires elaborate operations to access individual fields, with time overhead that can be deterring for some applications. Furthermore, compaction by a reduced number of pointers per node is ineffective in applications that use those pointers for pattern matching. Our work ties in with the more direct approach to improving performance of the conventional primary storage suffix tree representation, taken by Senft and Dvořák [19]. Classical representations required in Ukkonen's algorithm [22]

J. Gudmundsson and J. Katajainen (Eds.): SEA 2014, LNCS 8504, pp. 400–411, 2014.

and the closely related predecessor of McCreight [17] remain important in application areas such as genetics, data compression and data mining, since they allow online construction as well as provide *suffix links*, a feature useful not only in construction, but also for string matching tasks [10, 12]. In these algorithms, a critically time-consuming operation is *branch*: identifying the correct outgoing edge of a given node for a given character [19]. This work introduces and evaluates several representation techniques to help reduce both the number of branch operations and the cost of each such operation, focusing on running time, and taking an overall liberal view on space usage.

Our experimental evaluation of runtime, memory locality, and the counts for critical operations, shows that a well chosen combination of our presented techniques consistently produce a significant advantage over the original Ukkonen scheme as well as the branch-reduction technique of Senft and Dvořák.

2 Suffix Trees and Ukkonen's Algorithm

We denote the *suffix tree* (illustrated in fig. 1) over a string $T = t_0 \cdots t_{N-1}$ of length $|T| = N$ by $\mathcal{ST}$. Each edge in $\mathcal{ST}$, directed downwards from the root, is labeled with a substring of T, represented in constant space by reference to position and length in T. We define a *point* on an $\mathcal{ST}$ edge as the position between two characters of its label, or – when the point coincides with a node – after the whole edge label. Each point in the tree corresponds to precisely one nonempty substring $t_i \cdots t_j$, $0 \le i \le j < N$, obtained by reading edge labels on the path from the root to that point. A consequence is that the first character of an edge label uniquely identifies it among the outgoing edges of a node. The point corresponding to an arbitrary pattern can be located (or found non-existent) by scanning characters left to right, matching edge labels from the root down. For convenience, we add an auxiliary node $\perp$ above the root (following Ukkonen), with a single edge to the root. We denote this edge $\vdash$ and label it with the empty string, which is denoted by ϵ. (Although $\perp$ is the topmost node of the augmented tree, we consistently refer to the root of the unaugmented tree as the root of $\mathcal{ST}$.) Each leaf corresponds to some suffix $t_i \cdots t_{N-1}$, $0 \le i < N$. Hence, the label endpoint of a leaf edge can be defined implicitly, rather than updated during construction. Note, however, that any suffix that is not a unique substring of T corresponds to a point higher up in the tree. (We do not, as is otherwise common, require that t_{N-1} is a unique character, since this clashes with online construction.)

Except for $\vdash$, all edges are labeled with nonempty strings, and the tree represents exactly the substrings of T in the minimum number of nodes. This implies that each node is either $\perp$, the root, a leaf, or a non-root node with at least two outgoing edges. Since the number of leaves is at most N (one for each suffix), the total number of nodes cannot exceed $2N + 1$ (with equality for $N = 1$).

We generalize the definition to $\mathcal{ST}_i$ over string $T_i = t_0 \cdots t_{i-1}$, where $\mathcal{ST}_N = \mathcal{ST}$. An *online* construction algorithm constructs $\mathcal{ST}$ in N updates, where update i reshapes $\mathcal{ST}_{i-1}$ into $\mathcal{ST}_i$, without looking ahead any further than t_{i-1}.

We describe suffix tree construction based on Ukkonen's algorithm [22]. Please refer to Ukkonen's original, closer to an actual implementation, for details such as correctness arguments.

Define the *active point* before update $i > 1$ as the point corresponding to the longest suffix of T_{i-1} that is not a unique substring of T_{i-1}. Thanks to the implicit label endpoint of leaf edges, this is the point of the longest string where update i might alter the tree. The active point is moved once or more in update i, to reach the corresponding start position for update $i+1$. (This diverges slightly from Ukkonen's use, where the active point is only defined as the start point of the update.) Since any leaf corresponds to a suffix, the label end position of any point coinciding with a leaf in $\mathcal{ST}_i$ is $i - 1$. The tree is augmented with *suffix links*, pointing upwards in the tree: Let v be a non-leaf node that coincides with the string aA for some character a and string A. Then the suffix link of v points to the node coinciding with the point of A. The suffix link of the root leads to $\perp$, which has no suffix link. Before the first update, the tree is initialized to $\mathcal{ST}_0$ consisting only of $\perp$ and the root, joined by $\vdash$, and the active point is set to the endpoint of $\vdash$ (i.e. the root). Update i then procedes as follows:

1. If the active point coincides with $\perp$, move it down one step to the root, and finish the update.
2. Otherwise, attempt to move the active point one step down, by scanning over character t_i. If the active point is at the end of an edge, this requires a *branch* operation, where we choose among the outgoing edges of the node. Otherwise, simply try matching the character following the point with t_i. If the move down succeeds, i.e., t_i is present just below the active point, the update is finished. Otherwise, keep the current active point for now, and continue with the next step.
3. Unless the active point is at the end of an edge, split the edge at the active point and introduce a new node. If there is a saved node v_p (from step 5b), let v_p's suffix link point to the new node. The active point now coincides with a node, which we denote v.
4. Create a new leaf w and make it a child of v. Set the start pointer of the label on the edge from v to w to i (the end pointer of leaf labels being implicit).
5. If the active point corresponds to the root, move it to $\perp$. Otherwise, we should move the active point to correspond to the string A, where aA is the string corresponding to v for some character a. There are two cases:
   (a) Unless v was just created, it has a suffix link, which we can simply follow to directly arrive at a node that coincides with the point we seek.
   (b) Otherwise, i.e. if v's suffix link is not yet set, let u be the parent of v, and follow the suffix link of u to u'. Then locate the edge below u' containing the point that corresponds to A. Set this as the active point. Moving down from u' requires one or more branch operations, a process referred to as *rescanning* (see fig. 2). If the active point now coincides with a node v', set the suffix link of v to point to v'. Otherwise, save v as v_p to have its suffix link set to the node created next.
6. Continue from step 1.

3   Reduced Branching Schemes

Senft and Dvořák [19] observe that the *branch* operation, searching for the right outgoing edge of a node, typically dominates execution time in Ukkonen's algorithm. Reducing the cost of branch can significantly improve construction efficiency. Two paths are possible: attacking the cost of the branch operation itself, through the data structures that support it, which we consider in section 4, and reducing the *number* of branch operations in step 5b of the update algorithm.

We refer to Ukkonen's original method of maintaining and using suffix links as *node-oriented top-down* (NOTD). Section 3.1 discusses the *bottom-up* approach (NOBU) of Senft and Dvořák, and sections 3.2–3.3 present our novel approach of *edge-oriented* suffix links, in two variants *top-down* (EOTD) and *variable* (EOV).

3.1   Node-Oriented Bottom-Up

A branch operation comprises the rather expensive task of locating, given a node v and character c, v's outgoing edge whose edge label begins with c, if one exists. By contrast, following an edge in the opposite direction can be made much cheaper, through a parent pointer. Senft and Dvořák [19] suggests the following simple modification to suffix tree representation and construction:

- Maintain parents of nodes, and suffix links for leaves as well as non-leaves.
- In step 5b of update, follow the suffix link of v to v' rather than that of its parent u to u', and locate the point corresponding to A moving up, *climbing* from v' rather than rescanning from u' (see fig. 2).

Senft and Dvořák experimentally demonstrate a runtime improvement across a range of typical inputs. A drawback is that worst case time complexity is not linear: a class of inputs with time complexity $\Omega(N^{1.5})$ is easily constructed, and it is unknown whether actual worst case complexity is even higher. To circumvent degenerate cases, Senft and Dvořák suggest a hybrid scheme where climbing stops after c steps, for constant c, falling back to rescan. (As an alternative, we suggest bounding the number of edges to climb to by using rescan iff the remaining edge label length below the active point exceeds constant c'.) Some of the space overhead can be avoided in a representation using clever leaf numbering.

3.2   Edge-Oriented Top-Down

We consider an alternative branch-saving strategy, slightly modifying suffix links.

For each split edge, the NOTD update algorithm follows a suffix link from u to u', and immediately obtains the outgoing edge e' of u' whose edge label starts with the same character as the edge just visited. We can avoid this *first* branch operation in rescan (which constitutes a large part of rescan work) , by having e' available from e directly, without taking the detour via u and u'.

Define the string that *marks* an edge as the shortest string represented by the edge (corresponding to the point after one character in its label). For edge e, let

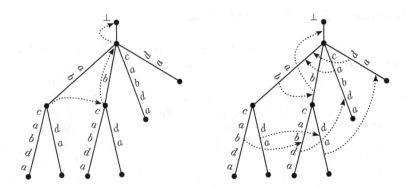

Fig. 1. Suffix tree over the string *abcabda*, with dotted lines showing node-oriented suffix links for internal nodes only, as in Ukkonen's original scheme (left), and edge-oriented suffix links (right)

aA, for character a and string A, be the shortest string represented e such that A marks some other edge e'. (The same as saying that aA marks e, except when e is an outgoing edge of the root and $|A| = 1$, in which case a marks e.) Let the *edge oriented suffix link* of e point to e' (illustrated i fig. 1).

Modifying the update algorithm for this variant of suffix links, we obtain an *edge-oriented top-down* (EOTD) variant. The update algorithm is analogous to the original, except that edge suffix links are set and followed rather than node suffix links, and the first branch operation of each rescan avoided as a result. The following points deserve special attention:

- When an edge is split, the top part should remain the destination of incoming suffix links, i.e., the new edge is the bottom part.
- After splitting one or more edges in an update, finding the correct destination for the suffix link of the last new edge (the bottom part of the last edge split) requires a *sibling lookup* branch operation, not necessary in NOTD.
- Following a suffix link from the endpoint of an edge occasionally requires one or more extra rescan operation, in relation to following the node-oriented suffix link of the endpoint.

The first point raises some implementation issues. Efficient representations (see e.g. Kurtz's [13]) do not implement nodes and edges as separate records in memory. Instead, they use a single record for a node and its incoming edge. Not only does this reduce the memory overhead, it cuts down the number of pointers followed on traversing a path roughly by half. The effect of our splitting rule is that while the top part of the split edge should retain incoming suffix links, the new record, tied to the bottom part should inherit the children. We solve this by adding a level of indirection, allowing all children to be moved in a single assignment. In some settings (e.g., if parent pointers are needed), this induces a one pointer per node overhead, but it also has two potential efficiency benefits. First, new node/edge pairs become siblings, which makes for a natural

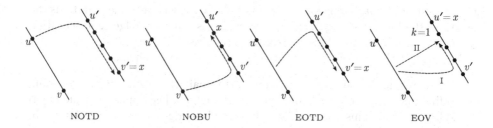

Fig. 2. Examples of moving the active point across a suffix link in four schemes, where in each case x is a node that coincides with the active point after the move. For EOV, we see two cases, before (I) and after (II) the destination of the suffix link is moved

memory-locality among siblings (cf. *child inlining* in section 4). Second, the original bottom node stays where it was in the branching data structure, saving one replace operation. These properties are important for the efficiency of the EOTD representation.

The latter two points go against the reduction of branch operations that motivated edge-oriented suffix links, but does not cancel it out. (Cf. table 1.)

These assertions are supported by experimental data in section 5. Furthermore, worst case time complexity is maintained by the modifications, which we state as the following:

EOTD retains the $O(N)$ total construction time of NOTD. To see this, note first that the modification to edge-oriented suffix links clearly adds at most constant-time operation to each operation, except possibly with regards to the extra rescan operations after following a suffix link from the endpoint of an edge. But Ukkonen's proof of total $O(N)$ rescan time still applies: Consider the string $t_j \cdots t_i$, whose end corresponds to the active point, and whose beginning is the beginning of the currently scanned edge. Each downward move in rescanning deletes a nonempty string from the left of this string, and characters are only added to the right as i is incremented, once for each online suffix tree update. Hence the number of downward moves are bounded by N, the total number of characters added.

3.3   Edge-Oriented Variable

Let e be an edge from node u to v, and let v' and u' be nodes such that node suffix links would point from u to u' and from v to v'. If the path from u' to v' is longer than one edge, the EOTD suffix link from e would point to the first one, an outgoing edge of u'. Another edge-oriented approach, more closely resembling NOBU, would be to let e's suffix link to point to the *last* edge on the path, the incoming edge of v', and use climb rather than rescan for locating the right edge after following a suffix link. But this approach does not promise any performance gain over NOBU.

An approach worth investigating, however, is to allow some freedom in where to on the path between u' and v' to point e's suffix link. We refer to the path from

u' to v' as the *destination path* of e's suffix link. Given that an edge maintains the length of the longest string it represents (which is a normal edge representation in any case), we can use climb or rescan as required.

We suggest the following *edge-oriented variable* (EOV) scheme:

- When an edge is split, let the bottom part remain the destination of incoming suffix links, i.e., let the top part be the new edge. (The opposite of the EOTD splitting rule.) This sets a suffix link to the last edge on its destination path, possibly requiring climb operations after the link is followed.
- When a suffix link is followed and c edges on its destination path climbed, if $c > k$ for a constant k, move the suffix link $c - k$ edges up.

Intuitively, this approach looks promising, in that it avoids spending time on adjusting suffix links that are never used, while eliminating the $\Omega(N^{1.5})$ degeneration case demonstrated for NOBU [19]. Any node further than k edges away from the top of the destination path is followed only once per suffix link, and hence the same destination path can only be climbed multiple times when multiple suffix links point to the same path, and each corresponds to a separate occurrence of the string corresponding to the climbed edge labels. We conjecture that the amortized number of climbs per edge is thus $O(1)$. However, our experimental evaluation indicates that the typical savings gained by the EOV approach are relatively small, and are surpassed by careful application of EOTD.

4　Branching Data Structure

Branching efficiency depends on the input alphabet size. Ukkonen proves $O(N)$ time complexity only under the assumption that characters are drawn from an alphabet Σ where $|\Sigma|$ is $O(1)$. If $|\Sigma|$ is not constant, *expected* linear time can be achieved by hashing, as suggested in McCreight's seminal work [17], and more recent dictionary data structures [2,11] can be applied for bounds very close to deterministic linear time. Recursive suffix tree construction, originally presented by Farach [5] achieves the same asymptotic time bound as character sorting, but does not support online construction.

We limit our treatment to simple schemes based on linked lists or hashing since, to our knowledge, asymptotically stronger results have not been shown to yield a practical improvement. Kurtz [13] observed in 1999 that linked lists appear faster for practical inputs when $|\Sigma| \leq 100$ and $N \leq 150\,000$. For a lower bound estimate of the alphabet size breaking point, we tested suffix tree construction on random strings of different size alphabets. We used a hash table of size $3N$ with linear probing for collision resolution, which resulted in an average of less than two hash table probes per insert or lookup across all files. The results, shown in table 3, indicate that hashing can outperform linked lists for alphabet sizes at least as low as 16, and our experiments did indeed show hashing to be advantageous for the *protein* file, with this size of alphabet. However, for many practical input types that produce a much lower average suffix tree node out-degree, the breaking point would be at a much larger $|\Sigma|$.

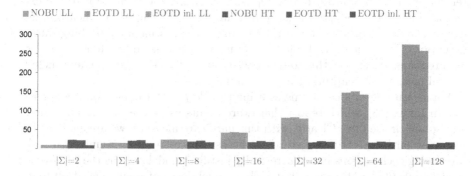

Fig. 3. Comparing linked list (LL) to hash table implementations (HT) for random files different size alphabets. Each file is 50 million characters long, and the vertical axis shows runtime in seconds.

Child Inlining. An internal node has, by definition, at least two children. Naturally occurring data is typically repetitive, causing some children to be accessed more frequently than others. (This is the basis of the PPM compression method, which has a direct connection to the suffix tree [14].) By a simple probabilistic argument, children with a high traversal probability also have a high probability of being encountered first. Hence, we obtain a possible efficiency gain by storing the first two children of each node, those that cause the node to be created, as *inline* fields of the node record together with their first character, instead of including them in the overall child retrieval data structure. The effect should be particularly strong for EOTD, which, as noted in section 3.2, eliminates the *replace child* operation that otherwise occurs when an edge is split, and the record of the original child hence remains a child forever. Furthermore, if nodes are laid out in memory in creation order, EOTD's consecutive creation of the first two children can produce an additional caching advantage. Note that inline space use is compensated by space savings in the non-inlined child-retrieval data structure.

When linked lists are used for branching, we can achieve an effect similar to inlining by always inserting new children at the back of the list. This change has no significant cost, since an addition is made only after an unsuccessful list scan.

5 Performance Evaluation on Input Corpora

Our target is to keep the number of branch operations low, and their cost low through lookup data structures with low overhead and good cache utilization. The overall goal is reducing construction time. Hence, we evaluate these factors.

5.1 Models, Measures, and Data

Practical runtime measurement is, on the one hand, clearly directly relevant for evaluating algorithm behavior. On the other hand, there is a risk of exaggerated effects dependent on specific hardware characteristics, resulting in limited relevance for future hardware development. Hence, we are reluctant to use execution time as the sole performance measure. Another important measure, less dependent on conditions at the time of measuring, is memory probing during execution. Given the central role of main memory access and caching in modern architectures, we expect this to be directly relevant to the runtime, and include several measures to capture it in our evaluation.

We measure level 3 cache misses using the *Perf* performance counter subsystem in Linux [18], which reports hardware events using the performance monitoring unit of the CPU. Clearly, with this hardware measure, we are again at the mercy of hardware characteristics, not necessarily relevant on a universal scale. Measuring cache misses in a theoretically justified model such as the *ideal-cache model* [8] would be attractive, but such a model does not easily lend itself to experiments. Attempts of measuring emulated cache performance using a virtual machine (*Valgrind*) produced spurious results, and the overhead made large-scale experiments infeasible. Instead, we concocted two simple cache models to evaluate the locality of memory access: one minimal cache of only ten 64 byte cache lines with a *least recently used* replacement scheme (intended as a baseline for the level one cache of any reasonable CPU), and one with a larger amount of cache lines with a simplistic direct mapping without usage estimation (providing a baseline expected to be at least matched by any practical hardware).

We measure runtimes of Java implementations kept as similar as possible in regards to other aspects than the techniques tested, with the 1.6.0_27 Open JDK runtime, a common contemporary software environment. With current *hotspot* code generation, we achieve performance on par with compiled languages by freeing critical code sections of language constructs that allocate objects (which would trigger garbage collection) or produce unnecessary pointer dereference. We repeat critical sections ten times per test run, to even out fluctuation in time and caching. Experiment hardware was a Xeon E3-1230 v2 3.3GHz quadcore with 32 kB per core for each of data and instructions level 1 cache, 256 kB level 2 cache per core, 8 MB shared level 3 cache, and 16 GB 1600 MHz DDR3 memory. Note that this configuration influences only runtime and physical cache (table 2 and the first two bars in each group of fig. 4); other measures are system independent.

We evaluate over a variety of data in common use for testing string processing performance, from the *Pizza & Chili* [7] and *lightweight* [16] corpora. In order to evaluate a degenerate case for NOBU, we also include an adversary input constructed for the pattern $T = ab^{m^2}abab^2ab^3 \cdots ab^m a$ (with $m = 4082$ for a 25 million character file), which has $\Omega(N^{1.5})$ performance in this scheme [19].

5.2   Results

Fig. 4 shows performance across seven implementations and five performance measures (explained in section 5.1), which we deem to be relevant for comparison. It summarizes the runtimes (also in table 2) and memory access measures by taking averages across all files except *adversary*, with equal weight per file. The bars are scaled to show percentages of the measures for the basic NOTD implementation, which is used as the benchmark. The order of the implementations

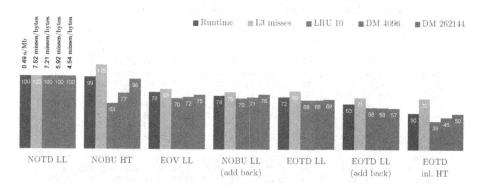

Fig. 4. Main comparison diagram for performance measures across the file set, excluding *adversary*. Branch data structures are either linked lists (LL, where new entries are added at the front unless *add back* is stated), or a single hash table.

Table 1. Operation counts. *rs*: rescan branch operations, *sl*: extra sibling lookup (see section 3.2), *move down*: branch operations outside of rescan. Files from the *Pizza and Chili Corpus* ([A]) and *Lightweight Corpus* ([B]). File categories are DNA ([1]), XML ([2]), source code ([3]), text ([4]), MIDI ([5]), proteins ([6]), database ([7]), and NOBU adversary ([8]).

File	size· 10^{-6}	NOTD rs	EOTD rs+sl	EOV rs	EOV climb	NOBU climb	move down branch ops
chr22[B1]	34.55	29 569 178	18 927 812	318 499	33 064 133	33 669 019	35 053 371
dna[A1]	104.86	87 681 116	58 585 203	236 634	100 172 270	100 736 053	111 372 537
dblp[A2]	104.86	54 757 925	14 743 305	32 980	55 418 573	55 594 399	73 784 654
rctail96[B2]	114.71	74 993 651	20 777 946	86 190	71 863 312	72 128 546	70 211 436
jdk13c[B3]	69.73	50 659 938	6 678 647	54 174	49 300 828	49 413 385	28 044 490
sources[A3]	104.86	80 270 528	30 753 392	191 755	75 419 031	75 953 764	70 537 447
w3c2[B3]	104.20	80 056 887	12 933 438	57 108	75 773 161	75 904 742	41 111 077
english[A4]	104.86	86 528 338	43 803 204	109 151	78 451 578	78 998 269	85 577 767
etext[B4]	105.28	73 446 539	40 782 335	106 811	73 482 182	74 097 636	99 131 563
howto[B4]	39.42	28 590 381	13 523 660	89 650	27 703 460	27 944 722	32 676 237
rfc[B4]	116.42	88 716 588	32 739 584	452 280	83 618 480	84 486 767	77 334 572
pitches[A5]	55.83	47 744 716	21 303 582	279 505	42 615 067	43 081 419	46 777 918
proteins[A6]	104.86	74 912 821	39 662 469	31 075	70 942 644	71 016 405	111 979 688
sprot34[B7]	109.62	70 190 029	20 737 274	45 034	69 274 605	69 425 197	78 927 702
adversary[A8]	25.00	41 662 928	16 323	8 313 003	41 654 774	68 033 898 010	12 249

when ranked by performance is fairly consistent across the different measures, with some deviation in particular for the hardware cache measure and smaller-cache models. The hardware cache measurement comes out as a relatively poor predictor of performance; by the numbers reported by Perf, the hardware cache even appears to be outperformed by our simplistic theoretical cache model.

We detect only a minor improvement of EOTD LL implementations in relation to NOBU LL, while inline EOTD HT provides a more significant improvement. Note, however that for NOBU, the HT implementation is much worse than the LL implementation, while the reverse is true for EOTD. This can be attributed to the different hash table use and the particular significance of inlining, noted in

Table 2. Running times in seconds for the same files as table 1

File	NOTD	NOTD HT	NOBU	NOBU back	NOBU HT	EOV LL	EOV HT	EOTD LL	EOTD HT	EOTD back	EOTD inl. HT
chr22	11.43	16.73	8.66	9.00	13.72	9.08	14.26	8.96	14.40	8.80	8.91
dblp	29.31	35.56	22.41	21.90	30.60	23.60	32.15	20.35	26.55	17.67	16.91
dna	40.37	60.76	30.65	32.12	51.66	32.70	53.77	31.60	53.37	30.97	32.89
english	64.26	50.99	45.65	46.36	42.11	47.47	43.34	42.70	42.77	36.64	26.21
etext	64.96	50.15	47.68	46.44	42.43	50.37	44.30	45.56	43.38	39.06	27.67
howto	21.74	15.43	16.12	15.09	12.64	16.48	12.92	15.33	12.50	12.56	7.61
jdk13c	7.97	23.24	6.39	6.53	19.29	6.97	20.24	5.72	14.46	5.27	6.76
pitches	46.65	21.34	34.66	28.98	18.40	35.38	19.07	34.08	17.26	26.34	10.55
proteins	104.49	49.60	74.30	74.46	41.95	76.49	44.27	75.73	46.18	70.55	31.97
rctail96	35.67	44.76	26.72	26.54	37.44	27.61	38.35	24.59	31.32	21.18	18.35
rfc	52.58	50.81	38.78	37.22	42.96	40.55	44.14	37.18	39.99	29.45	21.82
sources	44.21	44.23	32.71	30.12	37.24	34.27	38.63	31.49	34.28	24.70	17.76
sprot34	50.19	42.65	38.40	37.92	37.37	39.82	38.50	36.71	33.66	33.24	21.24
w3c2	18.98	39.84	14.38	15.19	33.13	14.89	33.73	12.91	24.98	11.41	10.47
adversary	1.30	7.89	267.50	266.16	296.42	1.64	8.07	1.40	5.10	1.39	1.34

section 4. The fact that EOTD HT without inlining (not in the diagram) is not clearly better than NOBU HT stands to confirm this. Although table 2 shows that EOTD LL beats its HT counterpart for files producing a low average out-degree in $\mathcal{ST}$ (because of a small alphabet and/or high repetitiveness), the robustness of hashing (cf. fig 3) has the greater impact on average. We have included results to show the impact of the *add to back* heuristic in EOTD LL, which also produced a slight improvement for NOBU (not shown in diagram), as expected.

The operation counts shown in table 1 generally confirm our expectations. (Branch counts include moves down from ⊥ to the root, in order to match Senft and Dvořák's corresponding counts [19].) EOV yields a large rescan reduction, even for the adversary file, which makes it an attractive alternative to NOBU when branching is very expensive. We found the exact choice of the k parameter of EOV not to be overly delicate. All values shown were obtained with $k = 5$.

6   Conclusion

It is possible to significantly improve online suffix tree construction time through modifications that target reducing branch operations and cache utilization, while maintaining linear worst-case time complexity. In many applications, our representation variants should be directly applicable for runtime reduction. Interesting topics remaining to explore are how our techniques for, e.g., suffix link orientation, fit into the compromise game of time versus space in succinct representations such as compressed suffix trees, and comparison to off-line construction.

References

1. Apostolico, A.: The myriad virtues of subword trees. In: Apostolico, A., Galil, Z. (eds.) Combinatorial Algorithms on Words. NATO ASI Series, vol. F 12, pp. 85–96. Springer (1985)

2. Arbitman, Y., Naor, M., Segev, G.: Backyard cuckoo hashing: Constant worst-case operations with a succinct representation. In: Proc. 51st Ann. IEEE Symp. Foundations of Comput. Sci., pp. 787–796 (2010)
3. Cánovas, R., Navarro, G.: Practical compressed suffix trees. In: Festa, P. (ed.) SEA 2010. LNCS, vol. 6049, pp. 94–105. Springer, Heidelberg (2010)
4. Clark, D.R., Munro, J.I.: Efficient suffix trees on secondary storage. In: Proc. Seventh Ann. ACM–SISM Symp. Discrete Algorithms, pp. 383–391 (1996)
5. Farach, M.: Optimal suffix tree construction with large alphabets. In: Proc. 38th Ann. IEEE Symp. Foundations of Comput. Sci., pp. 137–143 (October 1997)
6. Ferragina, P.: Suffix tree construction in hierarchical memory. In: Encyclopedia of Algorithms, pp. 922–925. Springer (2008)
7. Ferragina, P., Navarro, G.: Pizza & chili corpus (2005),
   `http://pizzachili.dcc.uchile.cl/`
8. Frigo, M., Leiserson, C., Prokop, H., Ramachandran, S.: Cache-oblivious algorithms. In: Proc. 40th Ann. IEEE Symp. Foundations of Comput. Sci., pp. 285–297 (1999)
9. Giegerich, R., Kurtz, S., Stoye, J.: Efficient implementation of lazy suffix trees. Software – Practice and Experience 33(11), 1035–1049 (2001)
10. Gusfield, D.: Algorithms on Strings, Trees, and Sequences. Cambridge University Press (1997)
11. Hagerup, T., Miltersen, P.B., Pagh, R.: Deterministic dictionaries. Journal of Algorithms 41(1), 69–85 (2001)
12. Kiełbasa, S.M., Wan, R., Sato, K., Horton, P., Frith, M.C.: Adaptive seeds tame genomic sequence comparison. Genome Research 21(3), 487–493 (2011)
13. Kurtz, S.: Reducing the space requirement of suffix trees. Software – Practice and Experience 29(13), 1149–1171 (1999)
14. Larsson, N.J.: Extended application of suffix trees to data compression. In: Proc. IEEE Data Compression Conf., pp. 190–199 (March-April 1996)
15. Manber, U., Myers, G.: Suffix arrays: A new method for on-line string searches. J. Comput. 22(5), 935–948 (1993)
16. Manzini, G., Ferragina, P.: Lightweight corpus (2004),
   `http://people.unipmn.it/manzini/lightweight/corpus/`
17. McCreight, E.M.: A space-economical suffix tree construction algorithm. J. ACM 23(2), 262–272 (1976)
18. Perf: Linux profiling with performance counters, `https://perf.wiki.kernel.org/`
19. Senft, M., Dvořák, T.: On-line suffix tree construction with reduced branching. Journal of Discrete Algorithms 12(0), 48–60 (2012)
20. Tian, Y., Tata, S., Hankins, R.A., Patel, J.M.: Practical methods for constructing suffix trees. The VLDB Journal 14(3), 281–289 (2005)
21. Tsirogiannis, D., Koudas, N.: Suffix tree construction algorithms on modern hardware. In: Proc. 13th International Conference on Extending Database Technology, pp. 263–274 (2010)
22. Ukkonen, E.: On-line construction of suffix trees. Algorithmica 14(3), 249–260 (1995)
23. Weiner, P.: Linear pattern matching algorithms. In: Proc. 14th Ann. IEEE Symp. Switching and Automata Theory, pp. 1–11 (1973)

LCP Array Construction in External Memory[*]

Juha Kärkkäinen and Dominik Kempa

Department of Computer Science, University of Helsinki, and
Helsinki Institute for Information Technology HIIT, Helsinki, Finland
{firstname.lastname}@cs.helsinki.fi

Abstract. One of the most important data structures for string processing, the suffix array, needs to be augmented with the longest-common-
prefix (LCP) array in numerous applications. We describe the first
external memory algorithm for constructing the LCP array given the
suffix array as input. The only previous way to compute the LCP array
for data that is bigger than the RAM is to use a suffix array construction algorithm with complex modifications to produce the LCP array as
a by-product. Compared to the best prior method, our algorithm needs
much less disk space (by more than a factor of three) and is significantly
faster. Furthermore, our algorithm can be combined with any suffix array
construction algorithm including a better one developed in the future.

1 Introduction

The suffix array [16,8], a lexicographically sorted array of the suffixes of a text,
is the most important data structure in modern string processing. It is the basis
of powerful text indexes such as enhanced suffix arrays [1] and many compressed
full-text indexes [18]. Modern text books spend dozens of pages in describing
applications of suffix arrays, see e.g. [20]. In many of those applications, the suffix array needs to be augmented with the longest-common-prefix (LCP) array,
which stores the lengths of the longest common prefixes between lexicographically adjacent suffixes (see e.g. [1,20]).

The construction of these data structures is a bottleneck in many of the applications. There are numerous suffix array construction algorithms (SACAs) [21]
including linear time internal memory SACAs [13,19] as well as external memory SACAs with the optimal I/O complexity [4,3]. There are also simple, linear
time internal memory LCP array construction algorithms (LACAs) [14,11] that
take the suffix array and the text as input, but external memory LCP array
construction remains a problem. In this paper, we describe the first external
memory LACA.

Related Work. The first LACA by Kasai et al. [14] is simple and runs in linear
time but needs a lot a space (the text plus $3n$ integers). Several later algorithms

---

[*] This research is partially supported by the Academy of Finland through grant 118653
(ALGODAN).

J. Gudmundsson and J. Katajainen (Eds.): SEA 2014, LNCS 8504, pp. 412–423, 2014.

aimed at reducing the space [15,17,22,11,7,2]. Some of the algorithms can even be made semi-external, i.e., they keep most of the data structures on disk but need to have at least the full text in RAM [22,11].

When the text size exceeds the RAM size, the only prior option for constructing the LCP array is to use an external memory SACA modified to compute the LCP array too during the construction [12,3] — we call these SLACAs — but there are several drawbacks to this approach. First, they add a substantial amount of complication to already complicated algorithms, and this complication is repeated for each SLACA, whereas a proper LACA can be combined with any SACA without a modification. Second, while theoretically the complexity of the algorithms does not change, adding the LCP computation increases the running time significantly in practice. Third, the SLACAs need a lot of disk space, so much in fact that the disk space is likely to be the biggest problem in scaling the algorithms for bigger data. For a text of length n, the best SLACA implementation, eSAIS [3], needs $54n$ bytes of disk space when computing the suffix and LCP arrays compared to only $28n$ bytes of disk space when computing only the suffix array.

Our Contribution. The new LACA, called LCPscan, is the first external memory LACA that is independent of any suffix array construction algorithm. The algorithm combines elements from several internal memory LACAs such as the original LACA [14], the Φ algorithm [11] and the irreducible LCP algorithm [11], adds some new twists such as a new method for identifying irreducible LCP values, and implements everything using external memory scanning and sorting. The main new idea, however, is to divide the text into blocks that are small enough to fit in RAM and then scan the rest of the text once for each block. A similar approach has been recently applied to computing the Burrows–Wheeler transform [6], the Lempel–Ziv factorization [10], and the suffix array [9]. This approach leads to a quadratic complexity in theory: $\mathcal{O}\left(\frac{n^2}{M\log_\sigma n} + n\log_{\frac{M}{B}}\frac{n}{B}\right)$ time and $\mathcal{O}\left(\frac{n^2}{MB(\log_\sigma n)^2} + \frac{n}{B}\log_{\frac{M}{B}}\frac{n}{B}\right)$ I/Os. However, in practice the size of the text would have to be more than about 100 times the size of the RAM, before the quadratic part of the computation would start to dominate. Up to that point, LCPscan is the fastest way to construct the LCP array in external memory as shown by our experiments. Furthermore, LCPscan needs just $16n$ bytes of disk space, which is less than a third of the disk usage of the best previous method [3].

2   Preliminaries

Strings. Throughout we consider a string $X = X[0..n) = X[0]X[1]\ldots X[n-1]$ of $|X| = n$ symbols drawn from the alphabet $[0..\sigma)$. Here and elsewhere we use $[i..j)$ as a shorthand for $[i..j-1]$. For $i \in [0..n]$, we write $X[i..n)$ to denote the *suffix* of X of length $n - i$, that is $X[i..n) = X[i]X[i+1]\ldots X[n-1]$. We will often refer to suffix $X[i..n)$ simply as "suffix i". Similarly, we write $X[0..i)$ to denote the *prefix* of X of length i. $X[i..j)$ is the *substring* $X[i]X[i+1]\ldots X[j-1]$ of X that starts at position i and ends at position $j - 1$.

Suffix Array. The *suffix array* [16] SA of X is an array SA[0..n] which contains a permutation of the integers [0..n] such that X[SA[0]..n) < X[SA[1]..n) < ⋯ < X[SA[n]..n). In other words, SA[j] = i iff X[i..n) is the (j + 1)^{th} suffix of X in ascending lexicographical order. The *inverse suffix array* ISA is the inverse permutation of SA, that is ISA[i] = j iff SA[j] = i. Conceptually, ISA[i] tells us the position of suffix i in SA. Another representation of the permutation is the Φ *array* [11] Φ[0..n) defined by Φ[SA[j]] = SA[j − 1] for j ∈ [1..n]. In other words, the suffix Φ[i] is the immediate lexicographical predecessor of the suffix i.

LCP Array. Let lcp(i, j) denote the length of the longest-common-prefix (LCP) of suffix i and suffix j. For example, in the string X = ccccatcat, lcp(0, 3) = 2 = |cc|, and lcp(4, 7) = 3 = |cat|. The *longest-common-prefix array* [14], LCP[1..n], is defined such that LCP[i] = lcp(SA[i], SA[i − 1]) for i ∈ [1..n]. The *permuted LCP array* [11] PLCP[0..n) is the LCP array permuted from the lexicographical order into the text order, i.e., PLCP[SA[j]] = LCP[j] for j ∈ [1..n]. Then PLCP[i] = lcp(i, Φ[i]) for all i ∈ [0..n). The following result is the basis of all efficient LACAs.

Lemma 1. *Let* i, j ∈ [0..n). *If* i ≤ j, *then* i + PLCP[i] ≤ j + PLCP[j]. *Symmetrically, if* Φ[i] ≤ Φ[j], *then* Φ[i] + PLCP[i] ≤ Φ[j] + PLCP[j].

Proof. As shown in [14,11], i+PLCP[i] ≤ (i+1)+PLCP[i+1] for all i ∈ [0..n−2], an iterative application of which results the first part of the claim. The second part follows by symmetry.   □

3   Basic Algorithm

In this section, we describe the basic LCPscan algorithm for computing the LCP array of a string X given X and the suffix array SA of X. In the next two sections, we describe further optimizations and analyze the theoretical and practical properties of the algorithm.

The basic approach of the algorithm is similar to the original linear time LACA by Kasai et al. [14] and to the Φ algorithm introduced in [11]. The main steps in the computation are:

1. Compute ISA and Φ from SA.
2. Compute PLCP from X and Φ.
3. Compute LCP from ISA and PLCP.

The first and the third step are easy to implement in external memory using sorting. In step 1, we scan the suffix array creating a triple (i, SA[i], SA[i − 1]) for each i ∈ [1..n]. When the triples are sorted by the middle component, the sequence of first components forms ISA[0..n) and the sequence of third components forms Φ[0..n). In the third step, we similarly sort the pairs (ISA[i], PLCP[i]), i ∈ [0..n), by the first component obtaining LCP[1..n] as the sequence of the second components.

In the middle step, we partition the text into $\mathcal{O}(n/m)$ blocks of size at most m and process them one at a time. The block size m is chosen so that one block of text fits in RAM together with a constant number of disk buffers. For each block $X[s..e]$, $0 \leq s \leq e < n$, we want to compute $PLCP[s..e]$ from $\Phi[s..e]$ and X. Only the block $X[s..e]$ is kept in RAM and the rest of X is scanned once. During the computation, we maintain triples (i, j, ℓ), where $i \in [s..e)$, $j = \Phi[i]$ and $\ell \in [0..\text{lcp}(i, j)]$. Each triple starts with $\ell = 0$ and ends with $\ell = \text{lcp}(i, j)$ allowing us to set $PLCP[i] = \ell$. The main idea is to process the triples in the order of the second component so that we can complete all the lcp computations during a single scan of X.

A small complication in the computation is dealing with the block boundaries. The triple (i, j, ℓ) is first created when processing the block that contains i but its computation may be finished when processing a different block, the one that contains $i + \text{lcp}(i, j)$. We keep a set R_y to hold triples that cross a block boundary y. Thus, the processing of a block $X[s..e]$ has R_s as an extra input and R_e as an extra output. Furthermore, the main output is not necessarily $PLCP[s..e]$ but $PLCP[s'..e')$ for some $s' \leq s$ and $e' \leq e$. The full algorithm for processing a block is given in Fig. 1.

ProcessBlock$(s, e, X, \Phi[s..e], R_s)$
1: $Q \leftarrow R_s \cup \{(i, \Phi[i], 0) : i \in [s..e)\}$
2: sort Q by the second component
3: $j_{\text{prev}} \leftarrow s$, $\ell_{\text{prev}} \leftarrow 0$
4: $L \leftarrow R_e \leftarrow \emptyset$
5: **for** $(i, j, \ell) \in Q$ **do**
6:       $\ell \leftarrow \max(\ell, \ell_{\text{prev}} + j_{\text{prev}} - j)$
7:       **while** $i + \ell < e$ **and** $j + \ell < n$ **and** $X[i + \ell] = X[j + \ell]$ **do** $\ell \leftarrow \ell + 1$
8:       **if** $i + \ell \geq e$ **and** $j + \ell < n$ **then** $R_e \leftarrow R_e \cup \{(i, j, \ell)\}$
9:       **else** $L \leftarrow L \cup \{(i, \ell)\}$
10:      $j_{\text{prev}} \leftarrow j$, $\ell_{\text{prev}} \leftarrow \ell$
11: sort L by the first component
12: $e' \leftarrow \min(\{e\} \cup \{i : (i, j, \ell) \in R_e\})$
13: $s' \leftarrow \min(\{e'\} \cup \{i : (i, \ell) \in L\})$
14: $PLCP[s'..e') \leftarrow \{\ell : (i, \ell) \in L\}$
15: **return** $PLCP[s'..e')$, R_e

Fig. 1. Process a text block $X[s..e]$

A subtle point in the algorithm is line 6, where we may set $\ell = \ell_{\text{prev}} + j_{\text{prev}} - j$. First, this is safe because $j_{\text{prev}} + \ell_{\text{prev}} \leq j + \text{lcp}(i, j)$ by Lemma 1. Second, this ensures that $j + \ell$ never decreases during the algorithm and thus the accesses to $X[j + \ell]$ are purely sequential. The other access to the text, $X[i + \ell]$, can be non-sequential, but we always have that $i + \ell \in [s..e)$. Everything else in the algorithm can be done by scanning and sorting and can thus be implemented efficiently in external memory.

4   Irreducible LCP Values

For highly repetitive texts, the longest common prefixes can be very long and cross several block boundaries, and conversely, a single block boundary may be crossed by LCPs starting from several blocks. Thus the sets R_s and R_e in the algorithm can grow big and the sum of their sizes over the whole algorithm could be as large as $\Theta(n^2/m)$. To prevent this, we will employ the irreducible LCP technique introduced in [11].

An LCP value PLCP[i] is said to be *reducible* if $X[i-1] = X[\Phi[i]-1]$. Otherwise, in particular when $i = 0$ or $\Phi[i] = 0$, PLCP[i] is *irreducible*. The following key properties of (ir)reducible LCP values were proved in [11].

Lemma 2 ([11]). *If* PLCP[i] *is reducible, then* PLCP[i] = PLCP[$i-1$] − 1.

Lemma 3 ([11]). *The sum of all irreducible lcp values is* $\leq 2n \log n$.

We will modify the procedure in Fig. 1 to compute only the irreducible LCP values in the main loop. The first lemma above shows that the reducible values are easy to obtain afterwards. The second lemma above ensures that the total number of boundary crossings of irreducible LCP values is $\mathcal{O}(n + (n \log n)/m)$, which is $\mathcal{O}(n)$ under the reasonable assumption that $m = \Omega(\log n)$.

In the procedure ProcessBlock, we need to discard a triple (i, j, ℓ) if PLCP[i] is reducible, i.e., if $X[i-1] = X[\Phi[i]-1]$, which could be done in the main loop when the text scan reaches $X[j-1]$. However, we can do it already earlier using the following alternative characterization of the irreducible LCP values.

Lemma 4. PLCP[i] *is a reducible value iff* $i > 0$ *and* $\Phi[i-1] = \Phi[i] - 1$ *and* PLCP[$i-1$] > 0.

Proof. If $i > 0$ and $\Phi[i-1] = \Phi[i] - 1$ and PLCP[$i-1$] = lcp($i-1, \Phi[i-1]$) > 0, then $X[i-1] = X[\Phi[i-1]] = X[\Phi[i]-1]$ and thus PLCP[i] is reducible. PLCP[i] is irreducible if $i = 0$ (by definition) or if $\Phi[i-1] = \Phi[i]-1$ and PLCP[$i-1$] $= 0$ (since then $X[i-1] \neq X[\Phi[i-1]] = X[\Phi[i]-1]$). The only thing left to prove is that if PLCP[i] is reducible then $\Phi[i-1] = \Phi[i]-1$.

Assume by contradiction that PLCP[i] is reducible but $\Phi[i-1] \neq \Phi[i]-1$. Since $X[i-1] = X[\Phi[i]-1]$ and $X[\Phi[i]..n) < X[i..n)$, we have that $X[\Phi[i]-1..n) < X[i-1..n)$. Since the suffix $\Phi[i-1]$ is the immediate lexicographical predecessor of the suffix $i-1$, we must have have $X[\Phi[i]-1..n) < X[\Phi[i-1]..n) < X[i-1..n)$. But this implies that $X[\Phi[i-1]] = X[i-1]$ and that $X[\Phi[i]..n) < X[\Phi[i-1]+1..n) < X[i..n)$, which contradicts the suffix $\Phi[i]$ being the immediate lexicographical predecessor of the suffix i.   □

Note that PLCP[$i-1$] $= 0$ implies that the suffix $i-1$ is the lexicographically smallest suffix starting with the character $X[i-1]$. Thus there are at most σ such positions and they can be easily computed from the text and the suffix array. We scan (if σ is small) or sort (if σ is large) the text to compute the character frequencies, which then identify the positions of the relevant suffixes in SA. The other reducible positions can be recognized on line 1 of ProcessBlock while scanning Φ and only the irreducible positions are added into Q. The only other modification to Algorithm ProcessBlock is on line 14. Since L contains only irreducible LCP values, the missing values in PLCP are filled using Lemma 2.

We are now ready to analyze the complexity of the algorithm in the standard external memory model (see [23]) with RAM size M and disk block size B, both measured in units of $\Theta(\log n)$-bit words. We assume that $M = \Omega(\log n)$, $M = \mathcal{O}(n)$ and $\sigma = \mathcal{O}(n)$.

Theorem 1. *Given a text of length n over an alphabet of size σ and its suffix array, the associated LCP array can be computed with the algorithm described above in*

$$\mathcal{O}\left(\frac{n^2}{M \log_\sigma n} + n \log_{\frac{M}{B}} \frac{n}{B} \right) \ time$$

and

$$\mathcal{O}\left(\frac{n^2}{MB(\log_\sigma n)^2} + \frac{n}{B} \log_{\frac{M}{B}} \frac{n}{B} \right) \ I/Os.$$

Proof. The text is divided into $\mathcal{O}(n/m)$ blocks, where m is chosen so that m characters fit in RAM. The size of the RAM is $M \log n$ bits and we can fit $m = \Theta(M \log_\sigma n)$ characters into that space. For each block, we scan the text once with each scan requiring $\mathcal{O}(n)$ time and $\mathcal{O}(n/(B \log_\sigma n))$ I/Os. This gives the first terms in the complexities.

Everything else in the algorithm involves scanning and sorting (tuples of) integers and the total number of integers involved in these scans and sorts is $\mathcal{O}(n)$. This gives the second terms in the complexities.   □

5   Practical Improvements

In this section, we describe some practical improvements and analyze the practical properties of LCPscan. In the analysis we assume that the implementation uses one byte for each character and five bytes for each integer.

As championed by the STXXL library [5], pipelining is an important technique for external memory algorithms. That is, instead of writing the output of one stage to disk and then reading it from the disk in the next stage, we execute the two stages simultaneously so that the input of the first stage is fed directly to the next stage. For example, when in the first step of the algorithm we obtain Φ as the third components of the sorted triples, we do not write the Φ values directly to disk. Instead, we identify the reducible positions and discard the corresponding entries of Φ already at this stage. For each irreducible position i, we form the pair $(i, \Phi[i])$ and feed these pairs to a sorter. The sorter uses the block number $\lfloor i/m \rfloor$ as the primary key and the value $\Phi[i]$ as the secondary key. This accomplishes most of the work on lines 1 and 2 in procedure ProcessBlock. The beginning of the actual ProcessBlock procedure will then just read the sorted $(i, \Phi[i])$ pairs, add the third components (0) and merge with R_s to obtain the (i, j, ℓ) triples, which are then immediately processed in the loop on lines 5–10.

Using pipelining throughout and assuming that n elements in the algorithm can be sorted in one pass of multiway mergesort,[1] the total I/O volume of LCPscan is $71n + 40r + \lceil n/m \rceil n$ bytes, where r is the number of irreducible LCP values.

The peak disk usage of the algorithm occurs during the stage described above, where we read the sorted $(i, \mathsf{SA}[i], \mathsf{SA}[i-1])$ triples and output ISA and the irreducible $(i, \varPhi[i])$ pairs. All that data requires $20n + 10r$ bytes of disk space, which is $30n$ bytes in the worst case. We also have the suffix array and the text on disk occupying a further $6n$ bytes for a total of $36n$ bytes.

To reduce the disk usage, we divide the text into four superblocks of sizes $0.31n$, $0.27n$, $0.23n$ and $0.19n$, and run the algorithm separately for each superblock. That is, in step 1, we scan SA forming the triple $(i, \mathsf{SA}[i], \mathsf{SA}[i-1])$ only when $\mathsf{SA}[i]$ belongs to the current superblock. The output of the superblock computation is a subsequence of LCP containing only the entries $\mathsf{LCP}[i]$ such that $\mathsf{SA}[i]$ belongs to the current superblock. Once all superblocks have been processed, the LCP subsequences are merged using SA to determine the merging order. In the rest of the algorithm, processing a superblock instead of the full text changes little as each superblock is a contiguous segment in the main data structures $\varPhi$, ISA and PLCP. With the superblock division, the peak disk usage is reduced to $16n$ bytes. For example, when processing the third superblock, we need $6n$ bytes for the text and the suffix array, $0.23 \times 30n = 6.9n$ bytes for the $(i, \mathsf{SA}[i], \mathsf{SA}[i-1])$ triples, $(i, \varPhi[i])$ pairs and ISA for the superblock, and $(0.31 + 0.27) \times 5n = 2.9n$ bytes for the LCP subsequences for the first two superblocks, which sums up to $15.8n$ bytes. The full $16n$ bytes is needed when merging the LCP subsequences into the final LCP array.

With the division into superblocks, the extra scans of the suffix array and the merging of the LCP subsequences add $30n$ bytes to the I/O volume for a total of $101n + 40r + \lceil n/m \rceil n$ bytes. We are willing to accept the slightly increased running time because the lack of disk space is likely to be a more serious limitation than time.

6   Experimental Results

We have implemented LCPscan as described above using the STXXL library [5] for external memory sorting. As there are no previous external memory LACAs, we compare it to eSAIS [3], the fastest external memory SLACA in previous studies. eSAIS can compute either SA and LCP arrays or only SA, and it is the difference between the two modes that we compare LCPscan against. This reveals if the combination of eSAIS as SACA plus LCPscan is better than eSAIS as SLACA. Combining LCPscan with another SACA such as the recent SAscan [9] might be an even better combination, but we want to focus on the LCP computation alone. The C++ implementations are available at http://www.cs.helsinki.fi/group/pads/.

---

[1] Considering the amount of RAM on modern computers, one pass of merging is a reasonable assumption in practice.

Table 1. Statistics of data used in the experiments. In addition to basic parameters, we show the percentage of irreducible LCP values among all LCP values (expression $100r/n$, where r denotes the number of irreducible LCPs) and the average length of the irreducible LCP value (Σ_r/r, where Σ_r is the sum of all irreducible LCPs).

Name	$n/2^{30}$	σ	$100r/n$	Σ_r/r
countries	64	210	0.14	1294.0
wiki	120	213	17.9	27.1
dna	64	6	20.7	22.0
debruijn	64	2	99.2	34.0

Data Set. For the experiments we selected a range of testfiles varying in the number and length of irreducible LCP values:

- countries: a concatenation of all versions (edit history) of Wikipedia articles about 78 largest countries in the XML format[2]. It contains a small number of large irreducible LCP values,
- wiki: latest English, German and French Wikipedia dumps[3] in the XML format concatenated and truncated to 120GiB. It represents a natural text,
- dna: a collection of DNA reads from a human genome[4] filtered from symbols other than $\{A, C, G, T, N\}$ and newline. The irreducible LCP values are very short but relatively frequent ($\sim 30\%$),
- debruijn: a binary De Bruijn sequence of order k is an artificial sequence of length $2^k + k - 1$ than contains all possible binary k-length substrings. It contains $\Theta(n)$ irreducible LCPs of maximal possible total length $\Theta(n \log n)$ [11]. It represents the worst case for LCPscan.

Table 1 gives detailed statistics about the data.

Experiments Setup. We performed experiments on 2 different machines referred to as Platform S (small) and Platform L (large).

Platform S was equipped with a 3.16GHz Intel Core 2 Duo CPU with 6MiB L2 cache and two 320GiB hard drives with a total of 480GiB of usable disk space. For experiments, we artificially restricted the RAM size to 2GiB using the Linux boot option mem, and the algorithms were allowed to use at most 1.5GiB.

Platform L was equipped with 1.9GHz Intel Xeon E5-2420 CPU with 15MiB L2 cache and 7.2TiB of disk space striped with RAID0 across 4 local disks of size 1.8TiB. For experiments we restricted the RAM size to 4GiB, and the algorithms were allowed to use at most 3.5GiB.

On both platforms we used a disk block of size 1MiB. The OS was Linux (Ubuntu 12.04, 64bit). All programs were compiled using g++ version 4.6.4 with -O3 -DNDEBUG options. In all experiments we used only a single thread of execution. All reported runtimes are wallclock (real) times.

---

[2] http://www.mediawiki.org/wiki/Parameters_to_Special:Export
[3] http://dumps.wikimedia.org/
[4] http://www.1000genomes.org/

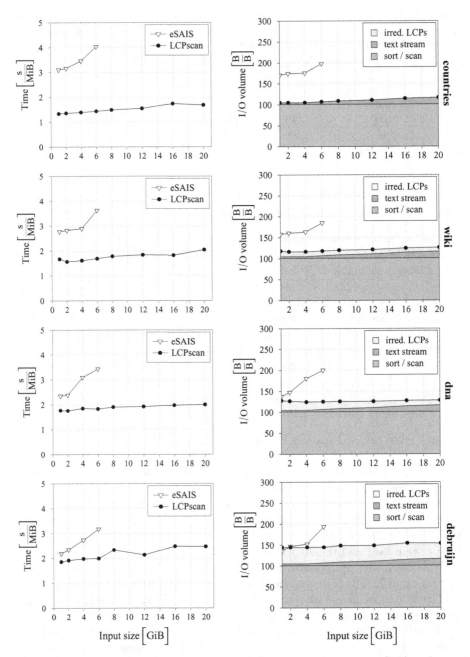

Fig. 2. Experimental results on platform S: comparison of the runtime (left) and normalized I/O volume (right) of LCPscan and eSAIS. For eSAIS, the values are the difference between constructing both SA and LCP arrays and constructing SA only. For the LCPscan I/O volume we show a detailed breakdown into: (bottom) I/Os that involve sorting/scanning of $\Theta(n)$ elements, (middle) I/Os resulting from text scans of which there are $\Theta(n/m)$ and (top) I/Os that only involve processing irreducible values.

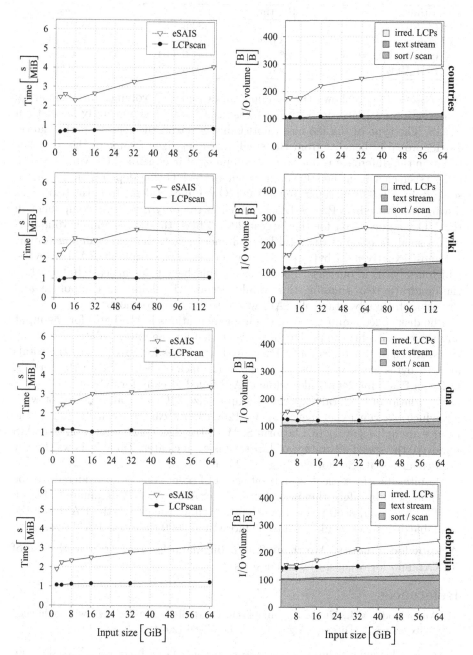

Fig. 3. Experimental results on platform L (analogous to Fig. 2)

Table 2. Summary of experiments on the 120GiB wiki test file. Disk space usage includes input and output.

Algorithm	Runtime	I/O volume	Disk space usage
eSAIS (SA only)	5.0 days	30.5 TiB	3.4 TiB
LCPscan	1.6 days	17.0 TiB	2.3 TiB
eSAIS (SA) + LCPscan	6.6 days	47.5 TiB	3.4 TiB
eSAIS (SA+LCP)	9.9 days	60.3 TiB	7.2 TiB

Discussion. Fig. 2 shows the running times and I/O volumes for LCP array construction on Platform S. In most cases, LCPscan is significantly faster than eSAIS. The type of the file has a noticeable effect on the relative performance. eSAIS probably benefits from the small alphabet size of some of the files. For LCPscan the difference is due to r, the number of irreducible LCP values, which can be seen clearly in the I/O volumes, where the I/O that depends on r is shown separately. Also shown separately is the I/O volume resulting from the text scans during the procedure ProcessBlock, which is the only quadratic component in the I/O volume. Even for the largest 20GiB files, the text scanning volume is only a small fraction of the total I/O volume. In fact, the text scanning volume is almost exactly n/m bytes per byte, where m is the available RAM in bytes, which is 1.5GiB in our case. The file size would have to be at least $100m$ before the quadratic text scanning time would become the dominant component and the asymptotic advantage of eSAIS would really start to show.

The disk space requirement of the implementations is $11n$ bytes for the input and the output plus the peak disk space needed for the intermediate data structures, which is $10n$ bytes for LCPscan and $54n$ bytes for eSAIS; thus the totals are $21n$ bytes and $65n$ bytes.[5] Because of the lack of disk space on Platform S, LCPscan failed for 24GiB files while eSAIS failed already for 8GiB files.

For Platform L we ran experiments up to size 64GiB for all files except wiki, for which we run experiments up to 120GiB. As can be seen in Fig. 3, the results are quite similar to Platform S. With the disk space limitation of eSAIS removed, the speed advantage of LCPscan is even clearer. Table 2 shows the key performance statistics for the largest 120GiB file.

Despite its theoretical disadvantage, LCPscan can be considered to be the more scalable algorithm in practice. For eSAIS to clearly dominate LCPscan, the file size would need to be more than 100 times the size of the RAM and the available disk space would need to be more than 6000 times the RAM size.

Acknowledgements. We thank Timo Bingmann for guiding us in the use of STXXL and for other useful discussions.

References

1. Abouelhoda, M.I., Kurtz, S., Ohlebusch, E.: Replacing suffix trees with enhanced suffix arrays. J. Discrete Algorithms 2(1), 53–86 (2004)

---

[5] This is a bit more than in the analysis earlier and in [3] because STXXL does not release disk space it has once allocated. In both algorithms, the peak intermediate disk usage occurs at a time when no disk space is needed for the output yet, but STXXL keeps occupying that space even when the output is written to disk.

2. Beller, T., Gog, S., Ohlebusch, E., Schnattinger, T.: Computing the longest common prefix array based on the Burrows-Wheeler transform. J. Discrete Algorithms 18, 22–31 (2013)
3. Bingmann, T., Fischer, J., Osipov, V.: Inducing suffix and lcp arrays in external memory. In: Proc. ALENEX 2013, pp. 88–102. SIAM (2013)
4. Dementiev, R., Kärkkäinen, J., Mehnert, J., Sanders, P.: Better external memory suffix array construction. ACM J. Experimental Algorithmics 12 (2008)
5. Dementiev, R., Kettner, L., Sanders, P.: STXXL: standard template library for XXL data sets. Softw., Pract. Exper. 38(6), 589–637 (2008)
6. Ferragina, P., Gagie, T., Manzini, G.: Lightweight data indexing and compression in external memory. Algorithmica 63(3), 707–730 (2012)
7. Gog, S., Ohlebusch, E.: Fast and lightweight lcp-array construction algorithms. In: Proc. ALENEX 2011, pp. 25–34. SIAM (2011)
8. Gonnet, G.H., Baeza-Yates, R.A., Snider, T.: New indices for text: Pat trees and Pat arrays. In: Frakes, W.B., Baeza-Yates, R. (eds.) Information Retrieval: Data Structures & Algorithms, pp. 66–82. Prentice–Hall (1992)
9. Kärkkäinen, J., Kempa, D.: Engineering a lightweight external memory suffix array construction algorithm. In: Proc. ICABD 2014, pp. 53–60 (2014)
10. Kärkkäinen, J., Kempa, D., Puglisi, S.J.: Lempel-Ziv parsing in external memory. In: Proc. DCC 2014, pp. 153–162. IEEE CS (2014)
11. Kärkkäinen, J., Manzini, G., Puglisi, S.J.: Permuted longest-common-prefix array. In: Kucherov, G., Ukkonen, E. (eds.) CPM 2009 Lille. LNCS, vol. 5577, pp. 181–192. Springer, Heidelberg (2009)
12. Kärkkäinen, J., Sanders, P.: Simple linear work suffix array construction. In: Baeten, J.C.M., Lenstra, J.K., Parrow, J., Woeginger, G.J. (eds.) ICALP 2003. LNCS, vol. 2719, pp. 943–955. Springer, Heidelberg (2003)
13. Kärkkäinen, J., Sanders, P., Burkhardt, S.: Linear work suffix array construction. J. ACM 53(6), 918–936 (2006)
14. Kasai, T., Lee, G., Arimura, H., Arikawa, S., Park, K.: Linear-time longest-common-prefix computation in suffix arrays and its applications. In: Amir, A., Landau, G.M. (eds.) CPM 2001. LNCS, vol. 2089, pp. 181–192. Springer, Heidelberg (2001)
15. Mäkinen, V.: Compact suffix array — a space efficient full-text index. Fundamenta Informaticae 56(1-2), 191–210 (2003)
16. Manber, U., Myers, G.W.: Suffix arrays: a new method for on-line string searches. SIAM J. Comp. 22(5), 935–948 (1993)
17. Manzini, G.: Two space saving tricks for linear time lcp array computation. In: Hagerup, T., Katajainen, J. (eds.) SWAT 2004. LNCS, vol. 3111, pp. 372–383. Springer, Heidelberg (2004)
18. Navarro, G., Mäkinen, V.: Compressed full-text indexes. ACM Computing Surveys 39(1), article 2 (2007)
19. Nong, G., Zhang, S., Chan, W.H.: Two efficient algorithms for linear time suffix array construction. IEEE Trans. Computers 60(10), 1471–1484 (2011)
20. Ohlebusch, E.: Bioinformatics Algorithms: Sequence Analysis, Genome Rearrangements, and Phylogenetic Reconstruction. Oldenbusch Verlag (2013)
21. Puglisi, S.J., Smyth, W.F., Turpin, A.: A taxonomy of suffix array construction algorithms. ACM Computing Surveys 39(2), 1–31 (2007)
22. Puglisi, S.J., Turpin, A.: Space-time tradeoffs for Longest-Common-Prefix array computation. In: Hong, S.-H., Nagamochi, H., Fukunaga, T. (eds.) ISAAC 2008. LNCS, vol. 5369, pp. 124–135. Springer, Heidelberg (2008)
23. Vitter, J.S.: Algorithms and data structures for external memory. Foundations and Trends in Theoretical Computer Science 2(4), 305–474 (2006)

Faster Compressed Suffix Trees
for Repetitive Text Collections*

Gonzalo Navarro[1] and Alberto Ordóñez[2]

[1] Dept. of Computer Science, Univ. of Chile, Chile
gnavarro@dcc.uchile.cl
[2] Lab. de Bases de Datos, Univ. da Coruña, Spain
alberto.ordonez@udc.es

Abstract. Recent compressed suffix trees targeted to highly repetitive text collections reach excellent compression performance, but operation times in the order of milliseconds. We design a new suffix tree representation for this scenario that still achieves very low space usage, only slightly larger than the best previous one, but supports the operations within microseconds. This puts the data structure in the same performance level of compressed suffix trees designed for standard text collections, which on repetitive collections use many times more space than our new structure.

1 Introduction

Suffix trees [33] are a favorite data structure in stringology, with a large number of applications in bioinformatics [3,15,27], thanks to their versatility. Their main problem is their space usage, which can be as much as 20 bytes per text character. On DNA text, where each character can be represented in 2 bits, the suffix tree takes 80 times the text size! On the other hand, most suffix tree algorithms traverse it across arbitrary access paths, and thus secondary memory representations are not efficient. This restricts the applicability of suffix trees to small text collections only, for example, a machine with 1GB of RAM can handle the suffix tree of collections of up to 50 million bases.

Sadakane [30] was the first in introducing a compressed suffix tree (CST) representation, which requires slightly more than 2 bytes per text character, a giant improvement over the basic representation. A recent, well engineered implementation, has been developed by Gog [14]. Fischer et al. [12,11] developed a new CST using even less space, between 1 and 1.5 bytes per character, as shown in practical implementations by Ohlebusch et al. [28] and Cánovas and Navarro [7,1]. Their main idea was to avoid the explicit representation of the tree topology. Their operation times, as a consequence, are slower than Sadakane's, but still within microseconds. Russo et al. [29] introduced an even smaller CST, using about half a byte per character, yet raising operation times to milliseconds.

All these CSTs use space proportional to the empirical entropy of the text collection [23] and perform well on standard text collections (although in most DNA

---

* Funded in part by Fondecyt Grant 1-140796, Chile, Xunta de Galicia (co-funded with FEDER) ref. GRC2013/053, and by MICINN (PGE and FEDER) grants TIN2009-14560-C03-02, TIN2010-21246-C02-01, andAP2010-6038 (FPU Program).

J. Gudmundsson and J. Katajainen (Eds.): SEA 2014, LNCS 8504, pp. 424–435, 2014.
© Springer International Publishing Switzerland 2014

collections the entropy is also close to 2 bits per character, i.e., they are not much compressible with statistical compressors). The fastest growing DNA collections, however, are formed by the sequenced genomes of hundreds or thousands of individuals of the same species. This makes those collections highly repetitive, since for example two human genomes share more than 99.9% of their sequences. Statistical compression does not take proper advantage of repetitiveness [16], but other techniques like grammar or Lempel-Ziv compression do.

There have been some indexes aimed at performing pattern matching on repetitive collections based on those techniques [17,16,8,10,13]. However, they do not provide the versatile suffix tree functionality, and they do not seem to yield a way to obtain it. Instead, the so-called run-length compressed suffix array [20] (run-length CSA or RLCSA), although based in principle on weaker compression techniques, yields a data structure that is useful to achieve CSTs for repetitive collections (because CST implementations always build on a CSA).

Based on the RLCSA, Abeliuk and Navarro [2,1] introduced the first CST for repetitive collections. The space on the repetitive biological collections tested is around 1-2 *bits* per character (bpc), well below the spaces achieved with the CSTs for general text collections. Their operation time was, however, in the order of milliseconds. Their structure is based on the ideas of Fischer et al. [12].

In this paper we introduce a new CST called GCT, for "grammar-compressed topology", that achieves times in the order of microseconds, close to the times of those CSTs using 8 to 12 bpc on general text collections described above. However, GCT uses much less space on repetitive collections, around 2–3 bpc. This is slightly larger than the previous structure for repetitive collections [2,1] but one to two orders of magnitude faster than it. On an extremely repetitive collection, GCT uses even less space than that previous structure, near 0.5 bpc.

To achieve this result, we build on Sadakane's CST [30], but use grammar compression on the tree structure, instead of representing it plainly with parentheses. More precisely, we use string grammar compression on the sequence of parentheses that represents the suffix tree topology (an idea briefly sketched by Bille et al. [5] for arbitrary trees). A repetitive text collection turns out to have a suffix tree with repetitive topology, and having the tree represented in this form allows us to speed up many operations that are very slow to simulate without the explicit topology [12].

2   Basic Concepts

2.1   Succinct Tree Representations

We describe the tree representation of Sadakane and Navarro [31]. The tree topology is represented using a sequence of parentheses. We traverse the suffix tree in preorder, writing an opening parenthesis when we first arrive at a node, and a closing one when we leave it. Thus a tree of t nodes is represented with $2t$ parentheses, as a sequence $P[1, 2t]$. Each node is identified with the offset of its opening parenthesis in P. We define the *excess* of a position, $E(i)$, as the number of opening minus closing parentheses in $P[1, i]$. Note that $E(i)$ is the depth of node i. Many tree navigation operations can be carried out with two operations related to the excess: $fwd(i, d)$ is the smallest $j > i$ such that $E(j) = E(i) - d$, and $bwd(i, d)$ is the largest $j < i$ such that $E(j) = E(i) - d$. For example the parenthesis closing the one that opens at position i is at $fwd(i, 1)$, so the next sibling of node i is $j = fwd(i, 1) + 1$ if $P[j] ='$ (', else i is the last child of its parent. Analogously, the previous sibling is $bwd(i - 1, 0) + 1$ if $P[i - 1] =')'$. The parent of node i is $bwd(i, 2) + 1$ and the h-th level ancestor is $bwd(i, h + 1) + 1$. Other operatons, like the lowest common ancestor between two nodes, $LCA(i, j)$ requires operation RMQ on the virtual array of depths: $RMQ(i, j)$ is the position of a minimum in $E(i \ldots j)$ and $LCA(i, j)$ is the parent of node $RMQ(i, j) + 1$. To convert between nodes and preorder values we need operation $preorder(i)$, which is the number of opening parentheses in $P[1, i]$, and $node(j)$, which is the position in P where the jth opening parenthesis appears. Many other operations are available with these primitives [31].

To implement those operations, the sequence $P[1, 2t]$ is cut into blocks of $b \log t$ parentheses (we use base 2 logarithms by default) and for each block k we store $m[k]$, the minimum excess within the block, and $e[k]$, the total excess within the block (we also need $p[k]$, the number of opening parentheses in the block, but this is implicit as $p[k] = (e[k] + b)/2$). The blocks are the leaves of a perfect binary tree of higher-level blocks, for which we also store $m[k]$ and $e[k]$. Then, operation $fwd(i, d)$ is solved in $O(b + \log t)$ time by first scanning the block k of the node using precomputed tables, then (if the answer was not found within block k) climbing up the balanced tree to search for the lowest ancestor of block k containing the desired excess difference d' to the right of k (i.e., where k descends from its left child and its right child k' holds $-m[k'] \geq d'$, being d' the value of d plus the excess from $i + 1$ to the end of block k), then going down to the leftmost leaf node k'' that descends from k' and such that $-m[k''] \geq d''$, where d'' is again d adjusted to the beginning of k'', and finally scanning block k'' to find the exact answer position. Operations bwd, RMQ, $preorder$ and $node$ are solved analogously, see the article [31] for more details. By using, for example, $b = \Theta(\log t)$, one obtains $O(\log t)$ time for the operations and $2t + o(t)$ bits to store the the parentheses plus the balanced tree of $m[]$ and $e[]$ values. (They [31] obtain constant times, but the practical implementation [4] reaches logarithmic times.)

2.2 Compressed Suffix Trees

Let $T[1, n]$ be a text (or the concatenation of the texts in a collection) over alphabet $[1, \sigma]$. The character at position i of T is denoted $T[i]$, whereas $T[i, j]$ denotes $T[i]T[i + 1] \ldots T[j]$, a substring of T. A *suffix* of T is a substring of the form $T[i, n]$ and a *prefix* of T is of the form $T[1, i]$. The *suffix trie* of T is the digital tree formed by inserting all the suffixes of T, so that any substring of T labels the path from the root to a node of the suffix trie, and any suffix of T, in particular, labels the path from the root to a leaf of the suffix trie. We consider that the labels in the suffix trie are on the edges, and each leaf corresponding to suffix $T[i, n]$ is labeled i. The *suffix tree* of T [33] is formed by compressing the unary paths of the suffix trie into a unique edge labeled with the concatenation of the labels of the compressed edges, that is, with a string. The first characters of the labels of the edges that lead to the children of any node are distinct, and we assume they are sorted by increasing value left to right. The *suffix array* [21] of T is an array $A[1,n]$ of values in $[1,n]$, formed by collecting the leaf labels of the suffix tree in left-to-right order. Alternatively, $A[1,n]$ can be seen as the array of all the suffixes of T sorted in lexicographic order.

Sadakane [30] showed that a functional compressed suffix tree (CST) could be represented with three components: (1) a compressed suffix array (CSA), (2) a compressed longest common prefix (LCP) array, and (3) a compressed representation of the topology of the suffix tree (thus, elements like the string labels could be deduced from these components without representing them).

There are many CSAs in the literature [25]. The basic functionality they offer is (a) given a pattern $p[1,m]$, find the suffix array interval $A[sp,ep]$ of the suffixes of T that start with p (therefore $A[sp], A[sp+1], \ldots, A[ep]$ is the list of occurrences of p in T), (b) given a suffix array position i, return $A[i]$, (c) given a text position j, return $A^{-1}[j]$, that is, the position in A that points to the suffix $T[j,n]$, and (d) given $[l,r]$, obtain the text substring $T[l,r]$. Most CSAs achieve times of the form $O(m)$ to $O(m\log n)$ for operation (a), $O(\text{polylog}\, n)$ for (b) and (c), and at most $O((l-r)\log\sigma + \text{polylog}\, n)$ for (d). They require space $O(n\log\sigma)$ *bits* (as opposed to $O(n\log n)$ of classical suffix arrays), and in most cases close to the empirical entropy of T [23] (a measure of compressibility with statistical compressors). Note that, within this space, CSAs can reproduce any substring of T, so T does not need to be stored separately. Mäkinen et al. [20] introduced the *run-length CSA*, or *RLCSA*, which compresses better when T is *repetitive* (i.e., it can be represented as the concatenation of a few different substrings). Statistical compressors do not take proper advantage of repetitiveness [16].

The *longest common prefix (LCP)* array, $LCP[1,n]$, stores in $LCP[i]$ the length of the longest common prefix between the suffixes $T[A[i],n]$ and $T[A[i-1],n]$ (with $LCP[1]=0$). Sadakane [30] showed how to represent LCP using just $2n$ bits, by representing instead $PLCP[1,n]$, where $PLCP[j] = LCP[A^{-1}[j]]$ (or $LCP[i] = PLCP[A[i]]$), that is, $PLCP$ is LCP represented in text order, not in suffix array order. The key property is that $PLCP[j+1] \geq PLCP[j]-1$, which allows $PLCP$ be represented using a bitvector $H[1,2n]$, at the price of having to compute $A[i]$ in order to compute $LCP[i]$. Fischer et al. [12] proved that H was in addition compressible when the text was statistically compressible, but Cánovas and Navarro [7,1] found out that the compressibility was not significant on standard texts. Instead, Abeliuk and Navarro [2,1] showed that the technique proposed to compress H [12] made a significant difference on repetitive texts.

Finally, Sadakane [30] represented the tree topology using succinct trees, taking $2n$ to $4n$ bits since the suffix tree has $t = n$ to $2n$ nodes. A study of such succinct tree representations [4] shows that the one described in Section 2.1 is well suited for the operations required on a suffix tree.

Fischer et al. [12,11] showed that one can operate without explicitly representing the tree topology, because each suffix tree node corresponds to a distinct suffix array interval. One can operate directly on those intervals, and all the tree operations can be simulated with three primitives on the intervals: $RMQ(i,j)$ finds the (leftmost) position of the smallest value in $LCP[i,j]$, and $PSV/NSV(i)$ finds the position in LCP preceding/following i with a value smaller than $LCP[i]$. Cánovas and Navarro [8,1] implemented this theoretical proposal, speeding up the operations RMQ and PSV/NSV by building the balanced tree described in Section 2.1 on top of the LCP array (instead of array E) and using ideas similar to those used to navigate trees [31] (albeit the application is quite different). Ohlebusch et al. [28] presented an alternative implementation that is more efficient when sufficient space is available.

Abeliuk and Navarro [2,1] proposed the first CST for repetitive text collections. They build on the representation of Fischer et al., using the RLCSA and the compressed version of H to represent LCP. The only obstacle was the balanced tree used to speed up RMQ and PSV/NSV operations, which was not compressible. They instead used the fact that the differential LCP array ($LCP[i] - LCP[i-1]$) is grammar-compressible as much as the differential suffix array is, and that it compresses particularly well on repetitive text collections. They applied RePair compression [18] to the differential LCP array and used the tree grammar (which is compressed, by definition) instead of an incompressible balanced tree, storing the needed information in the nodes of the grammar tree. As a result, they obtain very low space usage on repetitive texts (from 0.6 to 4 bits per character, depending on the repetitiveness of the real-life collections used). A drawback is that the operations require milliseconds, instead of the microseconds required by most CSTs designed for standard text collections [1].

2.3 Grammar Compression of Strings and Trees

Grammar compression of a string S is the task of finding a (context-free) grammar G that generates (only) S. RePair [18] is a compression algorithm that finds such a grammar in linear time and space. It finds the most frequent pair ab of characters in S, creates a rule $X \rightarrow ab$, replaces all ab in S by X, and iterates until the most frequent pair appears only once (in subsequent iterations, a and/or b maybe nonterminals). The final product of RePair is a set R of rules of the general form $X \rightarrow YZ$ and a sequence C of terminals and nonterminals corresponding to the final reduced version of S after all the replacements.

Grammar compression can also be applied to trees, by using grammars that generate trees instead of strings [9]. The simplest grammar is one that replaces full trees, so the associated grammar compression seeks for the minimal DAG (directed acyclic graph) equivalent to the tree. More powerful variants allow nonterminals with variables, with which grammar compression can replace connected subgraphs of the tree [22,19]. In general, supporting even the most basic traversal operations on those compressed trees is not trivial, even in the simplest DAG compression. Bille et al. [5] sketch a simple idea that retains all the full power of navigational operations of succinct trees (see Section 2.1). They basically propose to grammar-compress the string of parentheses $P[1, 2t]$ that represents the tree, attaching $m[]$ and $e[]$ (and the other) values to the nonterminals in order to support efficient navigation. They prove this compression is a powerful as the simple DAG tree compression, provided some small fixes are applied to the grammar.

Note that this theoretical idea is what was implemented in practice by Abeliuk and Navarro [2,1], as described in Section 2.2, for solving queries on the LCP array: using the RePair grammar tree instead of a balanced tree for storing $m[]$ and $e[]$ information. In this paper we implement the idea on the excess array of an actual tree — the suffix tree of the text. Unlike Bille et al., we do not alter the grammar given by RePair, but use it directly.

3 A New CST for Repetitive Text Collections

We introduce a new CST tailored to repetitive texts, building on Sadakane's original proposal [30]. We use the RLCSA as the suffix array, and the compressed representation of H [12,2] for the LCP array. Unlike the previous CST of Abeliuk and Navarro, we do represent the suffix tree topology, to avoid paying the high price in time of omitting it. As anticipated, this tree topology will be grammar-compressed to exploit repetitiveness. As a result, our CST will use slightly more space than that of Abeliuk and Navarro, but it will be orders of magnitude faster. We call it GCT, for "grammar-compressed topology".

Let $R[1,r]$ be the rules (including void rules for the terminals '(' and ')') and $C[1,c]$ the final sequence resulting from applying RePair compression to the parentheses sequence $P[1,2t]$. We use a version of RePair that yields balanced grammars (i.e., of height $O(\log t)$) in most cases.[1]

The r rules will be stored using $r \log r + O(r)$ bits (as opposed to the $2r \log r$ bits needed by a plain storage) by simplifying a technique described by Tabei et al. [32]. The grammar is seen as a DAG where the nodes are the nonterminals and terminals, and each rule $X \to YZ$ induces arrows from X to Y and to Z. Now all the arrows from nodes to their left children, seen backward, form a tree T_L, and those to their right children, seen backward, form a tree T_R. We represent T_L and T_R using a succinct representation [31,4] in $O(r)$ bits, and $r \log r + O(r)$ bits are used to map preorders from T_L to T_R and vice-versa, using a permutation representation [24] that computes the mapping in constant time and its reverse in time $O(\log r)$. We use preorders in T_L as nonterminal identifiers. Therefore, to find Y and Z given X, we compute the T_L node with preorder X, find its parent, and then Y is its preorder number in T_L. To find Z we map the node of X to its T_R node, find its parent in T_R, map back that parent to T_L, and then Z is its preorder value. The total time is $O(\log r)$.

In addition, we will store for each nonterminal k the values $m[k]$ and $e[k]$, as well as $s[k]$ (the size of the string generated by nonterminal k), $l[k]$ (the number of leaf nodes — i.e., substrings '()' — in the string), $pl[k]$ and $pr[k]$ (the first and last parentheses of the string). To induce small numbers, $e[k]$ will be stored as a difference with $m[k]$, and all the numbers will be stored with DACs, a technique to represent small numbers while supporting direct access [6]. We use the variant that uses optimal space. To further save space, only some nonterminals k will

---

[1] From www.dcc.uchile.cl/gnavarro/software

store this information, guaranteeing that, if a terminal k is not sampled, we can obtain its information by combining that of its descendants, such that we will not have to recursively expand more than y nonterminals, for a parameter y [26].

Sequence C is sampled every $z = \Theta(\log t)$ positions. For each sample, we store (i) the cumulative length of the expansion of C up to that position, in array C_s, (ii) the cumulative excess of the corresponding string, in array C_e, (iii) the minimum excess within the corresponding string, and (iv) the number of leaves in the corresponding string, in array C_l. This adds $O(c)$ bits to the $c \log r$ bits used to store sequence C.

The total space is $(r+c) \log r + O(r+c)$ bits, asymptotically equal to just the plain grammar-compressed representation of the compressed sequence $P[1, 2t]$. Now we describe how to solve operation $fwd(i, d)$, where i is a position in P.

1. We binary search C_s for position i, to find the largest sampled position $j \leftarrow C_s[u] < i$ in C; the excess up to that position is $e \leftarrow C_e[u]$.
2. We sequentially traverse the terminals and nonterminals $k \leftarrow C[zu + 1 \ldots]$, updating $j \leftarrow j + s[k]$ and $e \leftarrow e + e[k]$, where we remind that $s[k]$ and $e[k]$ are the total length and excess, respectively, of the string represented by the nonterminal k. We stop at the position $C[v]$ where j would exceed i.
3. Now we navigate the expansion of the nonterminal $X = C[v]$. Let $X \to YZ$. If $j + s[Y] < i$, then we add $j \leftarrow j + s[Y]$ and $e \leftarrow e + e[Y]$, and continue recursively with Z; otherwise we continue recursively with Y. When we finally reach a terminal node, we know the excess e of node i and start looking for a negative difference d of excess to the right of position $j = i$.
4. We traverse back (returning from recursion) the path in the grammar followed in point 3. If we went towards the right child, we just return. If, instead, we went towards Y in a rule $X \to YZ$, then we check whether $-m[Z] \geq d$. If so, the answer is within Z, otherwise we add $j \leftarrow j + s[Z]$ and $d \leftarrow d + e[Z]$, and return to the parent in the recursion.
5. If in the previous point we have established that the answer is within a nonterminal Z, we traverse the expansion of $Z \to VW$. If $-m[V] < d$, then we add $j \leftarrow j + s[V]$ and $d \leftarrow d + e[V]$, and continue recursively with W; else we continue recursively with V. When we reach a leaf, the answer is j.
6. If in point 4 we return from the recursion up to the root symbol $C[v]$ without yet finding the desired excess difference d, we scan the nonterminals $C[v+a]$ for $a = 1, 2, \ldots$, increasing $j \leftarrow j + s[C[v+a]]$ and $d \leftarrow d + e[C[v+a]]$ until finding an a such that $-m[C[v+a]] \geq d$. At this point we look for the final answer within the symbol $C[v+a]$ just as in point 5.
7. If we reach the next sampled position, $v + a = z(u + 1)$ without yet finding the answer, we sequentially traverse the samples $u' = u + 1 \ldots$, updating $j \leftarrow C_s[u']$ and $d \leftarrow d + C_e[u'] - C_e[u' - 1]$, until reaching u' such that $-C_m[u'] \geq d$. Then we traverse C sequentially from $C[zu' + 1]$ as in point 6.

To avoid the sequential scan in point 7, we can build a balanced tree with cumulative C_s, C_e and C_m values on top of the sequence of samples, incurring a negligible extra space of $O(c)$ bits, but this makes imperceptible difference in practice. In complexity terms, the operation requires $O(\log t)$ steps because the grammar is balanced, and the most expensive part of a step is computing component Z in the expansions $X \to YZ$. Thus the overall time complexity is $O(\log^2 t)$. This could be reduced to $O(\log t)$ time by using a plain representation of the grammar rules, but in practice our representation is already very fast and reducing space is more interesting.

In practice, we regularly sample the universe $[1, n]$ instead of the extent $[1, c]$ of C, so as to avoid the binary search in point 1. Although this introduces an overhead of $O(n)$ extra bits in theory (to achieve guaranteed logarithmic time), it works better in practice in both space and time.

The operation $bwd(i, d)$ is analogous. For $LCA(i, j)$ we traverse all the $O(\log t)$ grammar nodes between positions i and j and locate the point where the minimum excess occurs; we omit the details for lack of space. Operations $preorder(i)$ and $node(j)$ require a similar traversal considering the values $s[k]$ and the (virtual) values $p[k]$. Finally, the storage of $l[k]$, $pl[k]$ and $pr[k]$ is necessary to convert suffix tree leaves into suffix array positions, and back; once again the details are omitted but the reader can work them out easily once the concept behind $fwd(i, d)$ is understood.

4   Experimental Results and Discussion

To facilitate comparisons with previous work [1], we use the same computer and datasets. The computer is an Intel Core2 Duo at 3.16GHz, with 8GB of RAM and 6MB cache, running Linux version 2.6.24-24. We use various DNA collections from the Repetitive Corpus of *Pizza&Chili*.[2] Collections Influenza (148MB) and Para (410MB) are the most repetitive ones, whereas Escherichia (108MB) is less repetitive. These are collections of genomes of various bacteria of those species. In order to show how results improve with higher repetitiveness, and although it is not a biological collection, we have also included Einstein (89MB), corresponding to the Wikipedia versions of the article about Einstein in German. We use their same mechanism for choosing random suffix tree nodes, and averaged over 10,000 runs to obtain each data point.

For our CST (called "GCT" in the plots), we used various combinations of parameters y and z, obtained a cloud of points, and chose the dominant ones. We used the balanced version of RePair, which consistently gave us better results. We used the RLCSA with parameters $blockSize = 32$ and $sample = 128$. The previous CST for repetitive collections [2,1] is called "NPR-Repet" in the plots, and is obtained with the best reported point among their use of balanced and unbalanced RePair. We include the smallest CST for general collections [29] (called "FCST" in the plots), which is basically insensitive to the repetitiveness and achieves times within milliseconds, close to those of NPR-Repet. Since our CST achieves times within microseconds, it is worthy to compare its performance with the smallest CST that reaches that time range [8,1]; we choose their variant FMN-RRR and call it "NPR". We do not include Sadakane's CST because even a good implementation [14] uses too much space for our interests. For the same

---

[2] http://pizzachili.dcc.uchile.cl/repcorpus

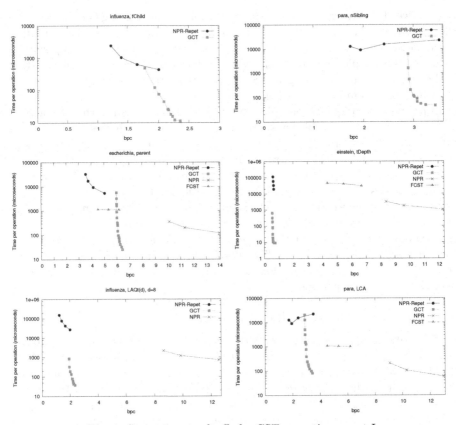

Fig. 1. Space-time tradeoffs for CST operations, part I

reason we do not use the faster and larger variants of NPR, as they represent LCP values directly and these become very large on repetitive collections (≈ 27 bpc only the LCPs!). For lack of space, we show each operation on one collection. See Fig. 1 and 2, with times in logscale. Not all the previous CSTs implement all the operations, so they may not appear in some plots.

The first figure is sufficient to discuss the space. On the repetitive DNA collections, our GCT achieves 2–3 bpc, compared to 1–2 bpc reached by NPR-Repet (in exchange, our times are up to 2 orders of magnitude faster, as discussed soon). On the less repetitive DNA sequence, GCT uses more than 6 bpc, whereas NPR-Repet uses 4–5 bpc. For this collection, NPR-Repet is already reached by FCST, the smallest CST for general texts, which uses about 4.5 bpc and is faster than NPR-Repet (but way slower than GCT). Interestingly, on the most repetitive collection, GCT achieves even less space than NPR-Repet, 0.5 bpc. Finally, NPR always uses more than 8 bpc, well beyond the others.

The plots of Fig. 1 measure simple tree traversal operations: going to the first child (*fChild*), to the next sibling (*nSibling*), to the parent (*parent*), computing the tree depth of the node (*tDepth*), a level ancestor of the node (*LAQt*), and the lowest common ancestor of two nodes (*LCA*). Our GCT excells on those opera-

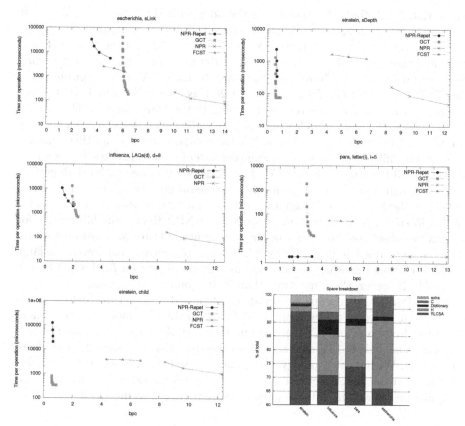

Fig. 2. Space-time tradeoffs for CST operations, part II, and space breakdown

tions because it represents the tree topology explicitly. In the simplest operations (the first four), GCT runs in 10–50 microseconds (μsec), whereas NPR-Repet takes 300 to over 10,000 μsec (i.e., 10 milliseconds, msec). The FCST uses around 1-50 msec and NPR takes 0.1–5 msec. Operation $LAQt$ is implemented directly on CGT, but in others requires a linear search for the proper ancestor node. It takes around 50 μsec on GCT, 1–3 msec on NPR, and around 50 msec on NPR-Repet. Operation LCA is relatively complex, and it takes around 100 μsec on our GCT and NPR, and 1–3 msec on NPR-Repet and FCST. Thus, our GCT is 1–2 orders of magnitude faster than comparable CSTs on those operations.

Fig. 2 includes operations that are exclusive of suffix trees, and access the other CSA components. The suffix link operation ($sLink$) requires, in our case, to map nodes to suffix array leaves, compute function Ψ on the RLCSA [20], map back to suffix tree nodes, and compute an LCA. Our GCT and NPR take near 200 μsec to complete this operation, whereas NPR-Repet and FCST use 2–5 msec, an order of magnitude slower. Operation $sDepth$ computes the string depth of a node, and is crucial for other suffix tree operations. It requires mapping nodes to the CSA and accessing the LCP data (i.e., bitvector H). It takes around 70 μsec on the GCT, 100–200 μsec on NPR, and 300–1000 μsec on NPR-Repet and FCST.

Operation *LAQs* finds the ancestor of the node with the given string depth. It requires a binary search on *sDepth* on the GCT, but is computed directly on NPR and NPR-repet [1]. It takes 500 μsec on GCT, 50–200 μsec on NPR, and around 2 msec on NPR-Repet. Operation *letter* gives the ith letter of the string represented by a node. It requires mapping to the CSA and computing Ψ^{i-1} on the RCLSA. Our GCT solves it in 10–20 μsec, while NPR and NPR-Repet require 2 μsec (they do not require mapping to the CSA), and FCST takes 50 μsec. Finally, the most complex operation is *child*, which descends to a child by an edge labeled with a given letter. It must compute *sDepth* and then traverse linearly the children of the node, computing *letter* for each. It takes 300–500 μsec on GST, 1–3 msec on NPR and FCST, and around 20 msec on NPR-Repet.

As a conclusion, GCT outperforms the general-purpose CSTs on repetitive collections by 1–2 orders of magnitude in time in most operations, and by a factor of 2–4 in space. It uses some more space than NPR-Repet, the alternative for repetitive collections, but it is 2 orders of magnitude faster for most operations. The times obtained on larger CSTs [30,14,1] are, of course, much lower: the large NPR [1] reaches 1 μsec in most operations (except 10 μsec on *LCA* and 100 μsec on *child*). However, these use more than 25 bpc on our collections.

Fig. 2 finishes with a space breakdown of the GCT structure (note it starts at 60%). In all cases, the sum of the RLCSA and the H components (i.e., those inherited from previous CSTs) account for 85–95% of the space, so the tree topology adds only 5–15% of space, which is responsible for speedups of orders of magnitude. Within the topology, the grammar itself (C+Dictionary) dominates the space on the least repetitive collection, whereas the extra data we insert for speeding up operations gains importance as repetitiveness increases.

References

1. Abeliuk, A., Cánovas, R., Navarro, G.: Practical compressed suffix trees. Algorithms 6(2), 319–351 (2013)
2. Abeliuk, A., Navarro, G.: Compressed suffix trees for repetitive texts. In: Calderón-Benavides, L., González-Caro, C., Chávez, E., Ziviani, N. (eds.) SPIRE 2012. LNCS, vol. 7608, pp. 30–41. Springer, Heidelberg (2012)
3. Apostolico, A.: The myriad virtues of subword trees. Combinatorial Algorithms on Words. NATO ISI Series, pp. 85–96. Springer (1985)
4. Arroyuelo, D., Cánovas, R., Navarro, G., Sadakane, K.: Succinct trees in practice. In: Proc. ALENEX, pp. 84–97 (2010)
5. Bille, P., Landau, G., Raman, R., Sadakane, K., Rao, S.S., Weimann, O.: Random access to grammar-compressed strings. In: Proc. SODA, pp. 373–389 (2011)
6. Brisaboa, N., Ladra, S., Navarro, G.: DACs: Bringing direct access to variable-length codes. Inf. Proc. Manag. 49(1), 392–404 (2013)
7. Cánovas, R., Navarro, G.: Practical compressed suffix trees. In: Festa, P. (ed.) SEA 2010. LNCS, vol. 6049, pp. 94–105. Springer, Heidelberg (2010)
8. Claude, F., Navarro, G.: Improved grammar-based compressed indexes. In: Calderón-Benavides, L., González-Caro, C., Chávez, E., Ziviani, N. (eds.) SPIRE 2012. LNCS, vol. 7608, pp. 180–192. Springer, Heidelberg (2012)
9. Comon, H., Dauchet, M., Gilleron, R., Löding, C., Jacquemard, F., Lugiez, D., Tison, S., Tommasi, M.: Tree Automata Techniques and Applications. INRIA (2007)

10. Do, H.-H., Jansson, J., Sadakane, K., Sung, W.-K.: Fast relative Lempel-Ziv self-index for similar sequences. In: Snoeyink, J., Lu, P., Su, K., Wang, L. (eds.) FAW-AAIM 2012. LNCS, vol. 7285, pp. 291–302. Springer, Heidelberg (2012)
11. Fischer, J.: Wee LCP. Inf. Proc. Lett. 110, 317–320 (2010)
12. Fischer, J., Mäkinen, V., Navarro, G.: Faster entropy-bounded compressed suffix trees. Theor. Comp. Sci. 410(51), 5354–5364 (2009)
13. Gagie, T., Gawrychowski, P., Kärkkäinen, J., Nekrich, Y., Puglisi, S.J.: A faster grammar-based self-index. In: Dediu, A.-H., Martín-Vide, C. (eds.) LATA 2012. LNCS, vol. 7183, pp. 240–251. Springer, Heidelberg (2012)
14. Gog, S.: Compressed Suffix Trees: Design, Construction, and Applications. PhD thesis, Univ. of Ulm, Germany (2011)
15. Gusfield, D.: Algorithms on Strings, Trees and Sequences: Computer Science and Computational Biology. Cambridge University Press (1997)
16. Kreft, S., Navarro, G.: On compressing and indexing repetitive sequences. Theor. Comp. Sci. 483, 115–133 (2013)
17. Kuruppu, S., Puglisi, S.J., Zobel, J.: Optimized relative Lempel-Ziv compression of genomes. In: Proc. ACSC, CRPIT, vol. 113, pp. 91–98 (2011)
18. Larsson, J., Moffat, A.: Off-line dictionary-based compression. Proc. of the IEEE 88(11), 1722–1732 (2000)
19. Lohrey, M., Maneth, S., Mennicke, R.: Tree structure compression with repair. In: Proc. DCC, pp. 353–362 (2011)
20. Mäkinen, V., Navarro, G., Sirén, J., Välimäki, N.: Storage and retrieval of highly repetitive sequence collections. J. Comp. Biol. 17(3), 281–308 (2010)
21. Manber, U., Myers, E.: Suffix arrays: a new method for on-line string searches. In: SIAM J. Comp., pp. 935–948 (1993)
22. Maneth, S., Busatto, G.: Tree transducers and tree compressions. In: Walukiewicz, I. (ed.) FOSSACS 2004. LNCS, vol. 2987, pp. 363–377. Springer, Heidelberg (2004)
23. Manzini, G.: An analysis of the Burrows-Wheeler transform. J. ACM 48(3), 407–430 (2001)
24. Munro, J., Raman, R., Raman, V., Srinivasa Rao, S.: Succinct representations of permutations. In: Baeten, J.C.M., Lenstra, J.K., Parrow, J., Woeginger, G.J. (eds.) ICALP 2003. LNCS, vol. 2719, pp. 345–356. Springer, Heidelberg (2003)
25. Navarro, G., Mäkinen, V.: Compressed full-text indexes. ACM Comp. Surv. 39(1), article 2 (2007)
26. Navarro, G., Puglisi, S., Valenzuela, D.: Practical compressed document retrieval. In: Pardalos, P.M., Rebennack, S. (eds.) SEA 2011. LNCS, vol. 6630, pp. 193–205. Springer, Heidelberg (2011)
27. Ohlebusch, E.: Bioinformatics Algorithms: Sequence Analysis, Genome Rearrangements, and Phylogenetic Reconstruction. Oldenbusch Verlag (2013)
28. Ohlebusch, E., Fischer, J., Gog, S.: CST++. In: Chavez, E., Lonardi, S. (eds.) SPIRE 2010. LNCS, vol. 6393, pp. 322–333. Springer, Heidelberg (2010)
29. Russo, L., Navarro, G., Oliveira, A.: Fully-compressed suffix trees. ACM Trans. Alg. 7(4), article 53 (2011)
30. Sadakane, K.: Compressed suffix trees with full functionality. Theor. Comp. Sys. 41(4), 589–607 (2007)
31. Sadakane, K., Navarro, G.: Fully-functional succinct trees. In: Proc. SODA, pp. 134–149 (2010)
32. Tabei, Y., Takabatake, Y., Sakamoto, H.: A succinct grammar compression. In: Fischer, J., Sanders, P. (eds.) CPM 2013. LNCS, vol. 7922, pp. 235–246. Springer, Heidelberg (2013)
33. Weiner, P.: Linear pattern matching algorithms. In: IEEE Symp. Swit. and Aut. Theo., pp. 1–11 (1973)

Improved and Extended Locating Functionality on Compressed Suffix Arrays*

Simon Gog[1] and Gonzalo Navarro[2]

[1] Department of Computing and Information Systems,
The University of Melbourne, Australia
`simon.gog@unimelb.edu.au`
[2] Department of Computer Science, University of Chile, Chile
`gnavarro@dcc.uchile.cl`

Abstract. Compressed Suffix Arrays (CSAs) offer the same functionality as classical suffix arrays (SAs), and more, within space close to that of the compressed text, and in addition they can reproduce any text fragment. Furthermore, their pattern search times are comparable to those of SAs. This combination has made CSAs extremely successful substitutes for SAs on space-demanding applications. Their weakest point is that they are orders of magnitude slower when reporting the precise positions of pattern occurrences. SAs have other well-known shortcomings, inherited by CSAs, such as retrieving those positions in arbitrary order. In this paper we present new techniques that, on one hand, improve the current space/time tradeoffs for locating pattern occurrences on CSAs, and on the other, efficiently support extended pattern locating functionalities, such as reporting occurrences in text order or limiting the occurrences to within a text window. Our experimental results display considerable savings with respect to the baseline techniques.

1 Introduction

Suffix arrays [12,21] are text indexing data structures that support various pattern matching functionalities. Built on a text $T[1, n]$ over an alphabet $[1, \sigma]$, the most basic functionality provided by a suffix array (SA) is to *count* the number of times a given pattern $P[1, m]$ appears in T. This can be done in $O(m \log n)$ and even $O(m + \log n)$ time [21]. Once counted, SAs can report each of the *occ* positions of P in T in $O(1)$ time. A suffix array uses $O(n \log n)$ *bits* of space and can be built in $O(n)$ time [19,18,17].

The space usage of suffix arrays, albeit "linear" in classical terms, is asymptotically larger than the $n \lg \sigma$ bits needed to represent T itself. Since the year 2000, two families of compressed suffix arrays (CSAs) emerged [25]. One family, simply called CSAs [14,15,30,31,13], built on the compressibility of a so-called Ψ function (see details in the next section), and simulated the basic SA procedure for pattern searching, achieving the same $O(m \log n)$ counting time of basic SAs. A second family, called FM-indexes [5,6,7,1], built on the Burrows-Wheeler transform [3] of T and

---

* Funded by Fondecyt grant 1-140796, Chile and ARC grant DP110101743, Australia.

J. Gudmundsson and J. Katajainen (Eds.): SEA 2014, LNCS 8504, pp. 436–447, 2014.

on a new concept called backward-search, which allowed $O(m \log \sigma)$ and even $O(m)$ time for counting occurrences. The counting times of all CSAs are comparable to those of SAs in practical terms as well [4]. Their space usage can be made asymptotically equal to that of the *compressed* text under the k-th order empirical entropy model, and in all cases it is below $n \lg \sigma + o(n \lg \sigma)$ bits. Within this space, CSAs support even stronger functionalities than SAs. In particular, they can reconstruct any text segment $T[l, r]$, as well as to compute "inverse" suffix array entries (again, details in the next section), efficiently. Reproducing any text segment allows CSAs to *replace* T, further reducing space.

The weakest part of CSAs in general is that they are much slower than SAs at retrieving the *occ* positions where P occurs in T. SAs require basically *occ* contiguous memory accesses. Instead, both CSA families use a sampling parameter s that induces an extra space of $O((n/s) \log n)$ bits (and therefore s is typically chosen to be $\Omega(\log n)$); then Ψ-based CSAs require $O(s)$ time per reported position and FM-indexes require $O(s \log \sigma)$. In practical terms, all CSAs are orders of magnitude slower than SAs when reporting occurrence positions [4], even when the distribution of the queries is known [8]. Text extraction complexities for windows $T[l, r]$ are also affected by s, but to a lesser degree, $O(s + r - l)$.

Although widely acknowledged as a powerful and flexible tool for text searching activities, the SA has some drawbacks that can be problematic in certain applications. The simplest one is that it retrieves the occurrence positions of P in an order that is not left-to-right in the text. This complicates displaying the occurrences in order (unless one obtains and sorts them all), as for example when displaying the occurrences progressively in a document viewer. A related one is that there is no efficient way to retrieve only the occurrences of P that are within a window of T unless one uses $\Omega(n \log n)$ bits of space [20,2,26,16]. This is useful, for example, to display occurrences only within some documents of a collection (T being the concatenation of the documents), only recent news in a collection of news documents, etc.

In this paper we present new techniques that speed up the basic pattern locating functionalities of CSAs, and also efficiently support extended functionalities. Our experimental results show that the new solutions outperform the baseline solutions by a wide margin, in some cases.

1. We unify the samplings for pattern locating and for displaying text substrings into a single data structure, by using the fact that they are essentially the inverse permutations of each other. This yields improved space/time trade-offs for locating pattern positions and displaying text substrings, especially in memory-reduced scenarios where large values of s must be used.
2. The *occ* positions of P have variable locating cost on a CSA. We use a data structure that takes $2n + o(n)$ additional bits to report the occurrences of P from cheapest to most expensive, thereby making reporting considerably faster when only some occurrences must be displayed (as in search engine interfaces, or when one displays results progressively and can show a few and process the rest in the background). Our experiments show that, when reporting less than around 15% of the occurrences, this technique is faster

438   S. Gog and G. Navarro

than reporting random occurrences, even when the baseline uses those extra $2n + o(n)$ bits to reduce s. A simple alternative that turns out to be very competitive is just to report first the occurrences that are sampled in the CSA, and thus can be reported at basically no cost.

3. Variants of the previous idea have been used for document listing [24] and for reporting positions in text order [26]. We study this latter application in practice. While for showing *all* the occurrences in order it is better to extract and partially sort them, one might need to show only the first occurrences, or might have to show the occurrences progressively. Our implementation becomes faster than the baseline when we report a fraction below 25% of the occurrences. The improvement increases linearly, reaching for example three times faster when reporting 5% of the occurrences. Again, we let the baseline spend those $2n + o(n)$ extra bits on a denser sampling.

4. Finally, we extend this second idea to report the text positions that are within a given text window. While the result is not competitive for windows located at random positions of T, our method is faster than the baseline of filtering the text positions by brute force when the window is within the first 15% of T. This is particularly useful in versioned collections or news archives, when the latest versions/dates are those most frequently queried.

The improved sampling we proposed is available in this branch of SDSL: https://github.com/simongog/sdsl-lite/tree/better_sampling.

2   Compressed Suffix Arrays

Let $T[0, n-1]$ be a text over alphabet $[0, \sigma - 1]$. Then a substring $T[i, n-1]$ is called a *suffix* of T, and is identified with position i. A *suffix array* $SA[0, n-1]$ is a permutation of $[0, n-1]$ containing the positions of the n suffixes of T in increasing lexicographic order (thus the suffix array uses at least $n \lg n$ bits). Since the positions of the occurrences of $P[0, m-1]$ in T are precisely the suffixes of T that start with P, and those form a lexicographic range, *counting* the number of occurrences of P in T is done via two binary searches using SA and T, within $O(m \log n)$ time. Once we find that $SA[sp, ep]$ contain all the occurrences of P in T, their number is $occ = ep - sp + 1$ and their positions are $SA[sp], SA[sp+1], \ldots, SA[ep]$. With some further structures adding up to $O(n \log n)$ bits, suffix arrays can do the counting in $O(m + \log n)$ time [21]. This can be reduced to $O(m)$ by resorting to suffix trees [34], which still use $O(n \log n)$ bits but too much space in practice.

Our interest in this paper is precisely using less space while retaining the SA functionality. A *compressed suffix array* (CSA) is a data structure that emulates the SA while using $O(n \log \sigma)$ bits of space, and usually less on compressible texts [25]. One family of CSAs [14,15,30,31,13] builds on the so-called Ψ function: $\Psi(i) = SA^{-1}[SA[i]+1]$, where SA^{-1} is the inverse permutation of the suffix array (given a text position j, $SA^{-1}[j]$ tells where in the suffix array is the pointer to the suffix $T[j, n-1]$). Thus, if $SA[i] = j$, $\Psi(i)$ tells where is $j + 1$ mentioned in SA, $SA[\Psi(i)] = SA[i] + 1 = j + 1$. It turns out that array Ψ is compressible up to the k-th order empirical entropy of T [22]. With small additional data structures, Ψ-based CSAs find the range $[sp, ep]$ for $P[0, m - 1]$ in $O(m \log n)$ time.

A second family, FM-indexes [5,6,7,1], build on the Burrows-Wheeler transform [3] of T, denoted T^{bwt}, which is a reversible permutation of the symbols in T that turns out to be easier to compress. With light extra structures on top of T^{bwt}, one can implement a function called $LF(i) = SA^{-1}[SA[i] - 1]$, the inverse of Ψ, in time at most $O(\log \sigma)$. An extension to the LF function is used to implement a so-called backward-search, which allows finding the interval $[sp, ep]$ corresponding to a pattern $P[0, m - 1]$ in $O(m \log \sigma)$ and even $O(m)$ time [1].

Once the range $SA[sp, ep]$ is found (and hence the counting problem is solved), locating the occurrences of P requires finding out the values of $SA[k]$ for $k \in [sp, ep]$, which are not directly stored in CSAs. All the practical CSAs use essentially the same solution for locating [25]. Text T is sampled at regular intervals of length s, and we store those sampled text positions in a *sampled suffix array* $SA_s[0, n/s]$, in suffix array order. More precisely, we mark in a bitmap $B[0, n-1]$ the positions $SA^{-1}[s \cdot j]$, for all j, with a 1, and the rest are 0s. Now we traverse B left to right, and append the value $SA[i]/s$ to SA_s for each i such that $B[i] = 1$. Array SA_s requires $(n/s) \lg(n/s) + O(n/s)$ bits of space, and B can be implemented in compressed form using $(n/s) \lg s + O(n/s) + o(n)$ bits [28,27], for a total of $(n/s) \lg n + O(n/s) + o(n)$ bits.

To compute $SA[i]$ at search time, we proceed as follows on a Ψ-based CSA. If $B[i] = 1$, then the position is sampled and we know its value is in SA_s, precisely at position $rank_1(B, i)$, which counts the number of 1s in $B[1, i]$ (this function is implemented in constant time in the compressed representation of B [28]). Otherwise, we test $B[\Psi(i)]$, $B[\Psi^2(i)]$, and so on until we find $B[\Psi^k(i)] = B[i'] = 1$. Then we find the corresponding value at SA_s; the final answer is $SA[i] = SA_s[rank_1(B, i')] \cdot s - k$. The procedure is analogous on an FM-index, using function LF, which traverses T backwards instead of forwards. The sampling guarantees that we need to perform at most s steps before obtaining $SA[i]$.

To display $T[l, r]$ we use the same sampling positions $s \cdot j$, and store a *sampled inverse suffix array* $SA_s^{-1}[1, n/s]$ with the suffix array positions that point to the sampled text positions, in text order. More precisely, we store $SA_s^{-1}[j] = SA^{-1}[j \cdot s]$ for all j. This requires other $(n/s) \lg s + O(n/s)$ bits of space. Then, in order to display $T[l, r]$ with a Ψ-based CSA, we start displaying slightly earlier, at text position $\lfloor l/s \rfloor \cdot s$, which is pointed from position $i = SA_s^{-1}[\lfloor l/s \rfloor]$ in SA. The first letter of a suffix $SA[i]$ is easily obtained on all CSAs if i is known. Therefore, displaying is done by listing the first letter of suffixes pointed from $SA[i], SA[\Psi(i)], SA[\Psi^2(i)], \ldots$ until covering the window $T[l, r]$. The process is analogous on FM-indexes. In total, we need at most $s + r - l$ steps.

This mechanism is useful as well to compute any $SA^{-1}[j]$ value. If j is a multiple of s then the answer is at $SA_s^{-1}[j/s]$. Otherwise, on a Ψ-based CSA, we start at $i = SA_s^{-1}[\lfloor j/s \rfloor]$ and the answer is $\Psi^k(i)$, for $k = j - \lfloor j/s \rfloor \cdot s$ (analogously

i	SA	SA^{-1}	B	T
0	13	6	0	\$
1	12	7	1	a\$
2	4	9	0	atenatsea\$
3	8	5	0	atsea\$
4	11	2	0	ea\$
5	3	12	1	eatenatsea\$
6	0	8	1	eeleatenatsea\$
7	1	10	0	eleatenatsea\$
8	6	3	1	enatsea\$
9	2	13	0	leatenatsea\$
10	7	11	0	natsea\$
11	10	4	0	sea\$
12	5	1	0	tenatsea\$
13	9	0	1	tsea\$

$$SA_s = \quad 4 \quad 1 \quad 0 \quad 2 \quad 3$$

$$SA_s^{-1} = \quad 6 \quad 5 \quad 8 \quad 13 \quad 1$$

$$SA_s^{-1*} = \quad 2 \quad 1 \quad 3 \quad 4 \quad 0$$

$$SA^{-1}[i \cdot s] = SA_s^{-1}[i]$$
$$= select(B, SA_s^{-1*}[i])$$

Fig. 1. Example of a suffix array, its inverse, and the sample arrays SA_s, the conceptional and formerly used SA^{-1}, and SA^{-1*} for $s = 3$

on an FM-index), taking up to s steps. Computing $SA^{-1}[j]$ is useful in many scenarios, such as compressed suffix trees [32,10] and document retrieval [33].

3 A Combined Structure for Locating and Displaying

In order to have a performance related to s in locating and displaying text, the basic scheme uses $2(n/s) \lg n + O(n/s) + o(n)$ bits. In this section we show that this can be reduced to $(1 + \epsilon)(n/s) \lg n + O(n/s) + o(n)$ bits, retaining the same locating cost and increasing the display cost to just $1/\epsilon + s + r - l$ steps.

The key is to realize that SA_s and SA_s^{-1} are essentially inverse permutations of each other. Assume we store, instead of the value $i = SA_s^{-1}[j]$, the smaller value $i' = SA_s^{-1*}[j] = rank_1(B, i)$. Since $B[i] = 1$, we can retrieve i from i' with the operation $i = select_1(B, i')$, which finds the i'th 1 in B and is implemented in constant time in the compressed representation of B [28]. Now, at the cost of computing $select_1$ once when displaying a text range, we can store SA_s^{-1*} in $(n/s) \lg(n/s) + O(n/s)$ bits. What is more important, however, is that SA_s and SA_s^{-1*} arrays are two permutations on $[0, n/s]$, and are inverses of each other. Fig. 1 shows an example.

Lemma 1. *Permutations SA_s and SA_s^{-1*} are inverses of each other.*

Proof. $SA_s[SA_s^{-1*}[j]] = SA_s[rank_1(B, SA_s^{-1}[j])] = SA[SA_s^{-1}[j]]/s = SA[SA^{-1}[j \cdot s]]/s = (j \cdot s)/s = j$.

Munro et al. [23] showed how to store a permutation $\pi[0, n'-1]$ in $(1+\epsilon)n' \lg n' + O(n')$ bits, so that any $\pi(i)$ can be computed in constant time and any $\pi^{-1}(j)$ in time $O(1/\epsilon)$. Basically, they add a data structure using $\epsilon n' \lg n' + O(n')$ bits on top of a plain array storing π. By applying this technique to SA_s, we retain the same fast access time to it, while obtaining $O(1/\epsilon)$ time access to SA_s^{-1*}

without the need to represent it directly. This yields the promised result (more precisely, the space is $(1 + \epsilon)(n/s) \lg(n/s) + (n/s) \lg s + O(n/s) + o(n)$ bits). We choose to retain faster access to SA_s because it is more frequently used, and for displaying the extra $O(1/\epsilon)$ time cost is blurred by the remaining $O(s + r - l)$ time. One is free, of course, to choose the opposite.

Our experiments will show that this technique is practical and yields a significant improvement in the space-time tradeoff of CSAs, especially when the space is scarce and relatively large s values must be used.

4   A Structure for Prioritized Location of Occurrences

Assume we want to locate only some, say t, occurrences of P in $SA[sp, ep]$, as for example in many interfaces that show a few results. In a Ψ-based CSA, the number of steps needed to compute $SA[k]$ is its distance to the next multiple of s, that is, $D[k] = 0$ if $SA[k]$ is a multiple of s and $D[k] = s - (SA[k] \bmod s)$ otherwise. In an FM-index, the cost is $D[k] = SA[k] \bmod s$. We would like to use this information to choose low-cost entries $k \in [sp, ep]$ instead of arbitrary ones, as current CSAs do. The problem, of course, is that we do not yet know the value of $SA[k]$ before computing it! (some CSAs actually store the value $SA[k] \bmod s$ [29], but they are not space-competitive with the best current CSAs).

However, we do not need that much information. It is sufficient to know, given a range $D[x, y]$, *which is the minimum value* in that range. This is called a *range minimum query* (RMQ): $\text{RMQ}(x, y)$ is the position of a minimum value in $D[x, y]$. The best current (and optimal) result for RMQs [9] preprocesses D in linear time, producing a data structure that uses just $2n + o(n)$ bits. Then the structure, *without accessing D*, can answer $\text{RMQ}(x, y)$ queries in $O(1)$ time.

The Basic Method. We use the RMQ data structure to progressively obtain the values of $SA[sp, ep]$ from cheapest to most expensive, as follows [24]. We compute $k = \text{RMQ}(sp, ep)$, which is the cheapest value in the range, retrieve and report $SA[k]$ as our first value, and compute $D[k]$ from it. The tuple $\langle sp, ep, k, D[k] \rangle$ is inserted as the first element in a min-priority queue that sorts the tuples by $D[k]$ values. Now we iteratively extract the first (cheapest) element from the queue, let it be the tuple $\langle lp, rp, k, v \rangle$, compute $k_l = \text{RMQ}(lp, k-1)$ and $k_r = \text{RMQ}(k+1, rp)$, then retrieve and report $SA[k_l]$ and $SA[k_r]$, and insert tuples $\langle lp, k-1, k_l, D[k] \rangle$ and $\langle k+1, rp, k_r, D[k] \rangle$ in the priority queue (unless they are empty intervals). We stop when we have extracted the desired number of answers or when the queue becomes empty. We carry out $O(t)$ steps to report t occurrences [24].

A Stronger Solution Using B. Recall bitmap B that marks the sampled positions. The places where $D[k] = 0$ are precisely those where $B[k] = 1$. We can use this to slightly reduce the space of the data structure. First, using operations $rank_1$ and $select_1$ on B, we spot all those $k \in [sp, ep]$ where $B[k] = 1$. Only then we start reporting the next cheapest occurrences using the RMQ data structure as above. This structure, however, is built only on the entries of array

D', which contains all $D[k] \neq 0$. Using $rank_0$ operations on B (which counts 0s, $rank_0(B,i) = i - rank_1(B,i)$), we map positions in $D[lp, rp]$ to $D'[lp', rp']$. Mapping back can be solved by using a $select_0$ structure on B, but we opt for an alternative that is faster in practice and spends little extra memory: we create a sorted list of pairs $\langle k, rank_0(B,k) \rangle$ for the already spotted k with $B[k] = 1$, and binary search it for mapping the positions back.

Refining Priorities. The process can be further optimized by refining the ordering of the priority queue. Our method sorts the intervals $[sp, ep]$ only according to the minimum possible value $u\ (= D[k])$. Assuming that the values in $D[sp, ep]$ are distributed uniformly at random in $[u, s)$ we can calculate the value of the expected minimum $\eta(lp, rp, u) = u + \sum_{v=u+1}^{s-1} \left(\frac{s-v}{s-u} \right)^z$, where $z = rp - lp + 1$ is the range size. This can be used as a refined priority value.

Experiments. In the experimental section we will explore the performance of four solution variants: The 'standard' method, which extracts the first t entries in $SA[sp, ep]$; a variant we call 'select', which enhances the baseline by using $rank$ and $select$ to first report all $SA[k]$ with $B[k] = 1$; and the described RMQ approach on D', with the priority queue ordering according to the minimum value $D[k]$ ('RMQ') or the expected minimum in the intervals ('RMQ+est.min.').

Locating Occurrences in Text Position Order. By giving distinct semantics to the D array, we can use the same RMQ-based mechanism to prioritize the extraction of the occurrences in different ways. An immediate application, already proposed in the literature (but not implemented) [26], is to report the occurrences in text position order, that is, using $D[k] = SA[k]$. In the experimental section we show that our implementation of this mechanism is faster than obtaining all the $SA[sp, ep]$ values and sorting them, even when a significant fraction of the occurrences is to be reported.

5   Range-Restricted Location of Occurrences

We now extend the mechanism of the previous section to address, partially, the more complex problem of retrieving the occurrences of $SA[sp, ep]$ that are within a text window $T[l, r]$. Again, we focus on retrieving some of those "valid" occurrences, not all of them. We cannot guarantee a worst-case complexity (as it would not be possible in succinct space [16]), but expect that in practice we perform faster than the baseline of scanning the values left to right and reporting those that are within the range, $l \leq SA[k] \leq r$, until reporting the desired number of occurrences.

If, as in the end of Section 4, we obtain the occurrences in increasing text position order, we will eventually report the leftmost occurrence within $T[l, r]$, and since then we will report all valid occurrences. As soon as we report the first occurrence position larger than r, we can stop. Although introducing ordering in the process, this mechanism is unlikely to be very fast, because it must traverse all the positions to the left of l before reaching any valid occurrence.

We propose the following heuristic modification, in order to arrive faster to the valid occurrences. We again store tuples $\langle lp, rp, k, SA[k] \rangle$, where k gives the minimum position in $SA[lp, rp]$. But now we use a max-priority queue sorted according to $SA[k]$, that is, it will retrieve first the *largest* minima of the enqueued ranges. After inserting the first tuple as before, we iteratively extract tuples $\langle lp, rp, k, SA[k] \rangle$. If $SA[k] > r$, then the extracted range can be discarded and we continue. If $SA[k] < l$, then we split the interval into two as before and reinsert both halves in the queue (position $SA[k]$ is not reported). Finally, if $l \le SA[k] \le r$, we run the algorithm of the end of Section 4 on the interval $SA[lp, rp]$, which will give all valid positions to report. This process on $SA[lp, rp]$ finishes when we extract the first value larger than r, at which point this segment is discarded and we continue the process with the main priority queue.

Note that, although this heuristic is weaker in letting us know when we can stop, it is likely to reach valid values to report sooner than using the algorithm of Section 4. In the experimental section we will show that, although our technique is slower than the baseline for general intervals (e.g., near the middle of the text), it is faster when the desired interval is close to the beginning (or the end, as desired). This biased range-restricted searching is useful, for example, in versioned systems, where the latest versions are those most frequently queried.

6 Experimental Results

All experiments were run on a server equipped with 144 GB of RAM and two Intel Xeon E5640 processors each with a 12 MB L3 cache. We used the PIZZA&CHILI corpus[1], which contains texts from various application domains.

Our implementations are based on structures of version 2.0.2 of SDSL[2]. The CSAs of SDSL can be parameterized with the described traditional sampling method, which uses a bitmap B to mark the sampled suffixes. It has recently been shown [11] that this sampling strategy, when B is represented as sd_vector [27], gives better time/space tradeoffs than a strategy that does not use B but samples every $SA[i]$ with $i \equiv 0 \mod s$. In our first experiment, we compare the traditional sampling using SA_s, SA_s^{-1} and B to the new solution that replaces SA_s^{-1} by just $\epsilon \cdot n/s$ samples plus a bitmap B' of length n/s to mark those samples in SA_s. We opted for $\epsilon = 1/8$, so that every SA^{-1} value can be retrieved in at most 8 steps, and B' is represented as an uncompressed bitmap (bit_vector). The underlaying CSA is a very compact FM-index (csa_wt) parameterized with a Huffman-shaped wavelet tree (wt_huff) and a compressed bitmap (rrr_vector). We choose this FM-index deliberately, since the ratio of space assigned to samples is especially high. With the new method we save so much space that we can also afford to represent B as bit_vector. Fig. 2 shows the time/space tradeoffs for accessing SA and SA^{-1} on one text (the results were similar on the others).

The points representing the same s value lie further left in the new solution, since we saved space. Replacing B with an uncompressed bitmap slightly reduces

[1] Available under http://pizzachili.dcc.uchile.cl/

[2] Publicly available under https://github.com/simongog/sdsl-lite

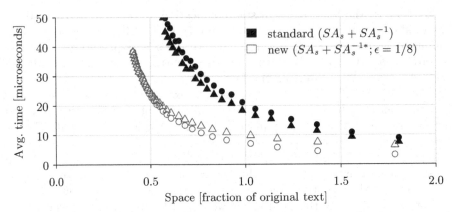

Fig. 2. Time/space tradeoffs to extract a SA (∘) respectively SA^{-1} value (△) from indexes over input `english.2108MB`. Sampling density s was varied between 1 and 32.

access time, in addition. We only report this basic experiment, but we note that these better tradeoffs directly transfer to applications that need simultaneous access to SA and SA^{-1}, like the child operation in compressed suffix trees.

In the second experiment, we measure the time to extract $t = 50$ arbitrary $SA[k]$ values from a range $[sp, ep]$. We use the same FM-index of the previous paragraph. We create one index with $s = 6$ and one using an RMQ structure on D' (`rmq_succinct_sct`). Setting $s' = 10$ for the latter index results in a size of 2,261 MB, slightly smaller than the $s = 6$ index (2,303 MB).

Fig. 3 (top) shows time and the average distance of the retrieved values to their nearest sample. Solution 'standard' spends the expected $(s'-1)/2 = 2.5\,LF$ steps per SA value, independently of the range size. Method 'select' first reports all sampled values in the range, hence the average distance linearly decreases and is close to zero at $s \times 50 = 300$. The RMQ based indexes spend the expected $(s-1)/2 = 4.5$ steps for $z = 50$. Using the RMQ information helps to decrease time and distance faster than linearly. The version using minimum estimation performs fewer LF steps for ranges in $[150, 220]$, but the cost of the RMQs in this case is too high compared to the saved LF steps.

In scenarios where LF is more expensive, 'RMQ+est.min.' can also outperform 'select' in runtime. The cost of one LF step depends logarithmically on the alphabet size σ, while the RMQ cost stays the same. Thus, using a text over a larger alphabet yields a stronger correlation between the distance and runtime, as shown in Fig. 3 (bottom), where we repeat the experiment using an FM-index (`csa_wt` parameterized with `wt_int`) on a word parsing `enwiki.4646MB` of the English Wikipedia ($\sigma = 3,903,703$). The RMQ supported index takes 3362 MB for $s' = 10$ and we get 3393 MB for $s = 6$.

Using almost the same index on `english.2108MB` (RMQ is built on SA this time, using $s' = 32$ and $s = 10$, obtaining sizes 1,554 MB and 1,590 MB), we now evaluate how long it takes to report the $t = 10$ smallest $SA[k]$ values in a range $SA[sp, ep]$. The standard version sequentially extracts all values in $SA[sp, ep]$, while keeping a max-priority queue of size k with the minima.

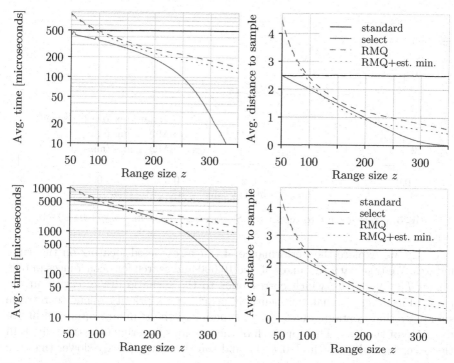

Fig. 3. Left: Time to report 50 values in the range $SA[sp, sp+z-1]$. Right: Distance of a reported $SA[k]$ value to its nearest sample. Input: `english.2108MB` (top) and `enwiki.4646MB` (bottom).

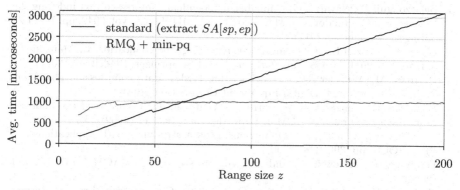

Fig. 4. Average time to report the ten smallest $SA[k]$ values in $SA[sp, sp+z-1]$

The RMQ based method uses a min-priority queue that is populated with ranges and corresponding minimum values. Fig. 4 contains the results. For range size $z = 10$, the standard method is about 3 times faster, since we decode 10 values in both methods and the sampling of the standard method is 3.2 times denser than that of the RMQ supported index. The RMQ index extracts $2t - 1$ values in the worst case, when there are $t - 1$ left in the priority queue. Therefore it is not surprising that the crossing point lies at about $60 \approx 3.2 \times (2t - 1)$.

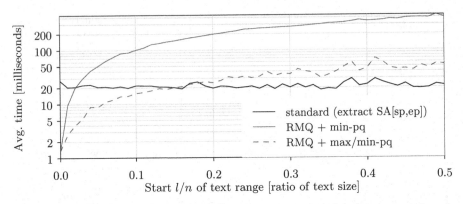

Fig. 5. Average time to report ten $SA[k] \in [l, r]$ for k in $[sp, sp+10000-1]$

Lastly, we explore the performance of range-restricted locating on the same indexes. We take pattern ranges of size 10,000 and search for occurrences in text ranges $T[l, l + 0.01n]$, which corresponds to the scenario drawn earlier in the paper. Fig. 5 shows that our heuristic using a max-priority queue to retrieve subranges that contain values $\geq l$, is superior to the standard approach in the first 15% of the text. The approach of the previous experiment extracts first all the occurrences located in $T[0, l-1)$, and thus becomes quickly slower than the standard approach.

References

1. Belazzougui, D., Navarro, G.: Alphabet-independent compressed text indexing. In: Demetrescu, C., Halldórsson, M.M. (eds.) ESA 2011. LNCS, vol. 6942, pp. 748–759. Springer, Heidelberg (2011)
2. Bille, P., Gørtz, I.L.: Substring range reporting. In: Giancarlo, R., Manzini, G. (eds.) CPM 2011. LNCS, vol. 6661, pp. 299–308. Springer, Heidelberg (2011)
3. Burrows, M., Wheeler, D.: A block sorting lossless data compression algorithm. Technical Report 124, Digital Equipment Corporation (1994)
4. Ferragina, P., González, R., Navarro, G., Venturini, R.: Compressed text indexes: From theory to practice. ACM J. Exp. Alg. 13, article 12 (2009)
5. Ferragina, P., Manzini, G.: Opportunistic data structures with applications. In: Proc. FOCS, pp. 390–398 (2000)
6. Ferragina, P., Manzini, G.: Indexing compressed texts. J. ACM 52(4), 552–581 (2005)
7. Ferragina, P., Manzini, G., Mäkinen, V., Navarro, G.: Compressed representations of sequences and full-text indexes. ACM Trans. Alg. 3(2), article 20 (2007)
8. Ferragina, P., Sirén, J., Venturini, R.: Distribution-aware compressed full-text indexes. Algorithmica 67(4), 529–546 (2013)
9. Fischer, J., Heun, V.: Space-efficient preprocessing schemes for range minimum queries on static arrays. SIAM J. Comp. 40(2), 465–492 (2011)
10. Fischer, J., Mäkinen, V., Navarro, G.: Faster entropy-bounded compressed suffix trees. Theor. Comp. Sci. 410(51), 5354–5364 (2009)
11. Gog, S., Petri, M.: Optimized succinct data structures for massive data. In: Soft. Prac. & Exp. (2013) (to appear), http://dx.doi.org/10.1002/spe.2198

12. Gonnet, G., Baeza-Yates, R., Snider, T.: New indices for text: Pat trees and Pat arrays. In: Information Retrieval: Data Structures and Algorithms, ch. 3, pp. 66–82. Prentice-Hall (1992)
13. Grossi, R., Gupta, A., Vitter, J.S.: High-order entropy-compressed text indexes. In: Proc. SODA, pp. 636–645 (2003)
14. Grossi, R., Vitter, J.: Compressed suffix arrays and suffix trees with applications to text indexing and string matching. In: Proc. STOC, pp. 397–406 (2000)
15. Grossi, R., Vitter, J.: Compressed suffix arrays and suffix trees with applications to text indexing and string matching. SIAM J. Comp. 35(2), 378–407 (2006)
16. Hon, W.-K., Shah, R., Thankachan, S., Vitter, J.: On position restricted substring searching in succinct space. J. Discr. Alg. 17, 109–114 (2012)
17. Kärkkäinen, J., Sanders, P., Burkhardt, S.: Linear work suffix array construction. J. ACM 53(6), 918–936 (2006)
18. Kim, D., Sim, J., Park, H., Park, K.: Constructing suffix arrays in linear time. J. Discr. Alg. 3(2-4), 126–142 (2005)
19. Ko, P., Aluru, S.: Space efficient linear time construction of suffix arrays. J. Discr. Alg. 3(2-4), 143–156 (2005)
20. Mäkinen, V., Navarro, G.: Rank and select revisited and extended. Theor. Comp. Sci. 387(3), 332–347 (2007)
21. Manber, U., Myers, G.: Suffix arrays: a new method for on-line string searches. SIAM J. Comp. 22(5), 935–948 (1993)
22. Manzini, G.: An analysis of the Burrows-Wheeler transform. J. ACM 48(3), 407–430 (2001)
23. Munro, J., Raman, R., Raman, V., Rao, S.: Succinct representations of permutations. In: Baeten, J.C.M., Lenstra, J.K., Parrow, J., Woeginger, G.J. (eds.) ICALP 2003. LNCS, vol. 2719, pp. 345–356. Springer, Heidelberg (2003)
24. Muthukrishnan, S.: Efficient algorithms for document retrieval problems. In: Proc. SODA, pp. 657–666 (2002)
25. Navarro, G., Mäkinen, V.: Compressed full-text indexes. ACM Comp. Surv. 39(1), article 2 (2007)
26. Nekrich, Y., Navarro, G.: Sorted range reporting. In: Fomin, F.V., Kaski, P. (eds.) SWAT 2012. LNCS, vol. 7357, pp. 271–282. Springer, Heidelberg (2012)
27. Okanohara, D., Sadakane, K.: Practical entropy-compressed rank/select dictionary. In: Proceedings of the Workshop on Algorithm Engineering and Experiments. SIAM (2007)
28. Raman, R., Raman, V., Rao, S.S.: Succinct indexable dictionaries with applications to encoding k-ary trees, prefix sums and multisets. ACM Trans. Alg. 3(4), article 43 (2007)
29. Rao, S.: Time-space trade-offs for compressed suffix arrays. Inf. Proc. Lett. 82(6), 307–311 (2002)
30. Sadakane, K.: Compressed text databases with efficient query algorithms based on the compressed suffix array. In: Lee, D.T., Teng, S.-H. (eds.) ISAAC 2000. LNCS, vol. 1969, pp. 410–421. Springer, Heidelberg (2000)
31. Sadakane, K.: New text indexing functionalities of the compressed suffix arrays. J. Alg. 48(2), 294–313 (2003)
32. Sadakane, K.: Compressed suffix trees with full functionality. Theor. Comp. Sys. 41(4), 589–607 (2007)
33. Sadakane, K.: Succinct data structures for flexible text retrieval systems. J. Disc. Alg. 5(1), 12–22 (2007)
34. Weiner, P.: Linear pattern matching algorithm. In: Proc. 14th IEEE Symposium on Switching and Automata Theory, pp. 1–11 (1973)

Author Index